Space Stations

Springer-Verlag Berlin Heidelberg GmbH

Ernst Messerschmid · Reinhold Bertrand

Space Stations

Systems and Utilization

Springer

Prof. Dr. Ernst Messerschmid
Dr. Reinhold Bertrand
Universität Stuttgart
Institut für Raumfahrtsysteme
Pfaffenwaldring 31
D-73760 Stuttgart
Germany

Translation:
Tanja Freyer
Tangastraße 46
D-81827 München

DOI 10.1007/978-3-662-03974-8

Catalogin-in-Publication Data applied for

Messerschmid, Ernst:
Space Stations : systems and utilization / Ernst Messerschmid ; Reinhold Bertrand. - Berlin ; Heidelberg ; New York ; Barcelona ; Hongkong ; London ; Mailand ; Paris ; Singapur ; Tokio : Springer, 1999

Originally published by Springer-Verlag Berlin Heidelberg New York in 1999.
MyCopy version of the original edition 1999

Coverdesign: Struwe & Partner, Heidelberg
Typesetting: Computer to plate by authors

62/3012 -5 4 3 2 - Printed on acid-free paper
www.springer.com/mycopy

DaimlerChrysler Aerospace

Space Infrastructure

Columbus, the European contribution to the International Space Station

Zarya and Unity, the first and second elements of the International Space Station, were launched on November 20, 1998, and December 4, 1998, respectively

Preface

This book on *space stations* is a result of a series of lectures held over the past several years at the University of Stuttgart in Germany. After the first NASA/ESA Spacelab mission SL1 in 1983 and the first German D1 Spacelab mission in 1985, the interest in human spaceflight of both the scientific community and the public at large reached another peak in the post-Apollo era. Participation in an international space station project became a possible prospect of European space policy. A complementary step logical in this development was to offer seminars and lectures on the subject of space stations and spaceflight application at universities, as well. In the 1987/88 winter term, we (E. Messerschmid and F. Pohlemann who was followed by R. Bertrand in 1992) offered a lecture series on "Space Stations" for the first time. Since the 1990/91 winter term, this lecture series has not only been held at the University of Stuttgart, but has also been delivered in English at the *École Nationale Supérieure de L'Aéronautique et de l'Espace* (ENSAE) in Toulouse, France, and in 1995/96 at the *International Space University* (ISU) in Strasbourg, France as part of the newly introduced course for the *Master of Space Studies* (MSS) degree.

To complement these lectures, the Space Systems Institute (IRS) of the University of Stuttgart offers students at the universities mentioned above a *Space Station Design Workshop* (SSDW) covering various aspects of a space station. The students can design their own space stations using software tools developed at the IRS. Experts are at their disposal to help them combine the requirements of "virtual customers" with previously defined boundary conditions, to support optimization of their conceptual space station design, and eventually to "let it fly" via computer-based simulation tools.

The knowledge and experience compiled together with F. Pohlemann through many years of both lecturing and the development of the Space Station Design Workshop went into our first book (*Raumstationen – Systeme und Nutzung*) in 1996. This has now been revised, expanded and translated into English.

All the different aspects of a space station – from design to assembly to eventual operation, i.e. practical implementation in numerous disciplines – could not have been investigated and cohesively assembled for lecture, had it not been for the experts in space research and industry who have supported us right from the beginning by giving lectures on individual aspects. We owe them a great debt of gratitude; their lectures and support were an invaluable asset to the following chapters: *History and Current Development* (B. Burkhalter, Auburn University), *Orbital Environment* (H. Hamacher, DLR Köln-Porz), *Power and Thermal Control Systems* (C. Audy, DLR Stuttgart; J. Krüger, IRS/DASA Bremen), *Utilization* (B. Feuerbacher, DLR Köln-Porz; D. Isakeit and H. Walter, ESA), *Microgravity* (T. Rösgen, IRS/ETH Zurich), *Human Factors* (P. Granseuer, ESA), *Logistics,*

Communications and Operation (G. Hirzinger, DLR; F. Huber, IRS; H. Lenski, DASA; K. Meinzer, University of Marburg; U. Schöttle, IRS) and *The International Space Station* (D. Andresen, K. Knott and G. Seibert, ESA).

We would also like to express our gratitude to our colleagues A. Hinüber, J. Krüger and J. Osburg from the IRS who have helped us through their hard work and dedication to organize and improve the corresponding lecture series, this book, and the Space Station Design Workshop, itself. We would like to thank T. Freyer for her tireless efforts in translating the text (with language style work and proofreading by J. Amos Osburg and further support on Chapter 5 by I. Hrbud) as well as D. Cibis and A. Zocholl for their support in editing text and illustrations. We are grateful to our reviewers, in particular, G. Biddis, and we thank them for the valuable remarks submitted. Dr. D. Merkle of Springer Publishing accompanied us excellently throughout the entire process, all the way from the preliminary phases to the final publication of this book.

We gratefully acknowledge the substantial support for the making of this book given by the European Space Agency (ESA), DaimlerChrysler Aerospace AG, Bremen, and M. Bedorf, Bonn.

The Europeans decided in October 1995 to participate in the International Space Station after the end of a period marked by numerous conflicts and by generally fierce competition between East and West; that same competitiveness also made its mark in the field of space technology. Now the time has come to work together constructively to build a mutually rewarding and truly international space station. This is the apex of a long period of development which was predicted by many spaceflight pioneers. For the first time in history, it will combine interdisciplinary and international cooperation in an extremely interesting environment which is both multicultural and extraterrestrial. We are convinced that this space station will be followed by many others, and we hope that it will help to bring about a lasting era of peaceful cooperation in space. The intention of this book is to help its readers to become more familiar with "Space Stations – Systems and Utilization" and many other aspects of space systems engineering and applications.

Stuttgart, February 1999

Ernst Messerschmid
Reinhold Bertrand

Table of Contents

1 Introduction

The idea of a space station, i.e. a permanently habitable orbital structure, has existed since the very early ideas of spaceflight itself were conceived. As early as 1903 the "father of cosmonautics", Konstantin Tsiolkovsky from Russia, dealt with inhabited stations in space close to Earth, with an autonomous power supply and bioregenerative life support systems. While he summarized his ideas in the book "Raketa v mezhplanetnoe prostranstvo" ("The Rocket into Interplanetary Space"), two other great space flight pioneers, Robert Goddard and Hermann Oberth, had already developed similar ideas as a result of their own work. In his book "Die Rakete zu den Planetenräumen" ("The Rocket into Interplanetary Space") published in 1923, Hermann Oberth was the first to mention the expression "Raumstation" (German for "space station"), and in the third edition of his book, published only shortly after the first two, he described the space station as the engineering project of the future and already mentioned nearly all the areas of application which are of interest today. Space stations challenged engineers, scientists and journalists to deal with concepts, sometimes rather unusual ones, for the realization of differing areas of application. It was not until 1952 that the subject attracted more public attention thanks to Wernher von Braun's study "Across the Space Frontier", in which he described a large, wheel-shaped space station.

The historical development of space activities can be outlined very clearly. The real beginning of space flight was the launch of the satellite "Sputnik" in 1957 and only four years later the first manned vehicle was launched into orbit, and after a further eight years, the first lunar landing took place. These events were followed by the first missions of space probes to other planets, with some of these even landing on the planets' surfaces. The first space station "Salyut 1" was launched in 1971 by the former Soviet Union and two years later, the USA placed its space station "Skylab" in orbit and this relied mainly on existing Apollo hardware. The first commercial communication satellites followed; for the first time, satellites were repaired on orbit and the first series of experiments was run as part of the newly developed discipline of "microgravity research". The latter was particularly supported by "Spacelab" as the European contribution to the US Space Shuttle program. All of these factors were important steps on the way to exploring the space environment, experiencing work and research in space and gaining an idea of how the space close to Earth could be used effectively. Under the leadership of the USA, these preparatory steps lead to plans for the (initially US with Western partners only) Space Station Freedom and, after the end of the Cold War with the integration of Russia into the project, plans for a truly global International Space Station.

Orbital stations have always stood their ground as far as plans for a future in space are concerned – from the first fantastic visions of space flight pioneers, through the idea's temporary stagnation during the race to the Moon, until today's large-scale program for a space station. Leading nations involved in spaceflight have become aware of a space station being an inevitable milestone of the long-term, well-founded evolution of research and development of and for humankind.

How does a space station differ from other space systems placed in orbit, such as satellites and platforms? To help answer this question we can introduce four main characteristics:

A space station is

- an orbiting system,
- large and usually to be assembled on orbit,
- intended to serve long-duration multi-user missions, and
- a crewed system.

Being an *orbiting system*, a space station must be robust enough to withstand the stresses of launch and still perform its functions in space. It must be equipped with a remote control system, a communication system for ground contact, a propulsion system, an attitude and orbit control system (AOCS), etc.

Given that a *large* space station will usually exceed the payload capacity of a single space transport vehicle, it will be necessary to design several space station modules and assemble them directly on orbit. Extensive dimensions also pose dynamic problems, which are typical of such large structures: a problem which does not occur when dealing with relatively compact capsules or satellites.

Compared to a transportation system or satellite, a space station is usually intended to serve many and different users for *long-duration missions*. That means, its subsystems and components are not only to operate continuously over a certain period of time: they must also be repaired or exchanged easily and quickly in case of malfunctions or nominal degradation. Moreover, space stations depend on the resupply of goods and sustaining operational and structural expenses that do not occur when dealing with satellite systems.

Finally, a space station is *inhabited by a crew*, either permanently or temporarily (otherwise it would be called a space platform). This is probably the most demanding feature for the design of the station. The crew needs a pressurized environment and a life support system, which in turn determines the amount of logistics supplies. The crew must also be provided with corresponding safety measures, which often require additional constructions such as shields to protect the crew against radiation and meteoroids, and/or additional procedures. In the case of an emergency, rapid return-to-Earth transportation is needed, e.g. by special crew rescue vehicles.

What makes the construction of a space station such a challenge for engineers? A space station of a considerable size is one of the most complex technical systems known today. Its interdisciplinary design and construction require a knowledge covering most of the disciplines in science and technology, e.g. mechanics, statics, thermodynamics, process engineering, electrical engineering, telecommunications, computer science, medical science, psychology, and systems engineering – only to mention the most important ones.

Due to this manifold range of disciplines, the effective cooperation and co-ordination of experts is inevitable. This is exactly the reason why the subject presents such a challenge for engineers, economists and political leaders, as well as for all those who plan and construct a space station, ensure its operation and, as a consequence, have to justify all decisions taken in the course of the project's life cycle. Considering the size and complexity of the corresponding tasks, one can easily imagine that the management of a space station sometimes develops a certain dynamics of its own.

When being confronted with the engineering problem "space station" for the first time, it is a good idea to categorize the overall problem. At the top level, two different views are clear: overall system design and subsystem design. Table 1.1 presents these two views and identifies the relevant chapters of this book where an elaboration can be found.

Table 1.1. Two Different Views of the Engineering Problem "Space Station"

Overall System Design	**Subsystem Design**	**Chapter**
Crew (Safety, Ergonomics, and Habitability)	Environmental Control and Life Support System	4, 11
Energy Balance	Power and Thermal Control System	3, 5
Attitude and Orbit Control Strategy	Attitude and Orbit Control System	6
Utilization Aspects	Payload Systems	7, 8
Layout and Mass Distribution	Mechanisms	9, 10
Systems Integration	Structures	9
Logistics and Maintenance	EVA Systems, Robotics	12, 13
Command, Control, and Communication Architecture	Flight Operation and Ground Support System, Communication and Data Management System	12, 13

When designing a subsystem and its components, technology that is already known and available has to be combined within the given framework of specifications.

At system level, assessments are necessary to find out whether a space station consisting of different subsystems is able to perform all the tasks of the mission objectives. This will not be the case in the early design stages, and changes in mission requirements and thus new specifications for the subsystems will have to be defined.

During this iterative process of designing a space station, it is impossible to independently deal with overall system aspects and subsystem design as they are not completely isolated from one another. Both sides depend on each other and only their successful interaction will yield a useful concept and product.

One example for such an interaction is the choice of an appropriate propulsion system. Before choosing a technology for its realization, several questions must be answered, such as the following:

- What are the propulsive requirements (determined by mass distribution, aerodynamic drag and further parameters)?
- Are there any additional restrictive conditions such as maximum allowable acceleration (microgravity), or the exclusion of certain propellants due to safety precautions or compatibility with the environment, that have to be observed?

Being aware of such conditions, the subsystem engineers will be in a position to design an appropriate propulsion system. They may arrive at a point where there are two technologies as viable options. In this case, the subsystem engineers have to consult the system level engineers. The system level engineers may have to decide whether, for example, to choose an electrical propulsion system (due to its smaller propellant requirement), or an H_2/O_2 engine (which could receive part of its fuels from the life support system). In the end, all such aspects have to be weighed against one another until either a solution is found and accepted by all sides involved, or, when some new framework conditions (e.g. cost reduction) arise, the whole process begins anew.

Why do we need a space station at all? As already mentioned earlier in this introduction, the idea of a space station evolved in connection with the theoretical option of crewed spaceflight. Hermann Oberth, for instance, thought as early as 1925 about a space station to be used as a platform for Earth observation, or as a communications platform or as a mirror to illuminate the Earth's surface. Space stations as possible means intended to solve specific (i.e. technical and scientific) problems did not receive attention until the real beginning of the space age. Basically, four areas of application can be characterized for a space station today:

- A permanently available, multidisciplinary research facility in a low Earth orbit for basic and applied research
- A test facility for new technologies in the space environment
- A platform for the observation of the Earth environment, the solar system and the universe
- A starting point and traffic node for further space exploration and use, i.e. as a place of assembly, maintenance and resupply of space vehicles

Of course, the prioritization of these areas of application has not always been viewed equally throughout history. As a consequence, in most of the cases it is possible to derive underlying application scenarios from the different space station designs of the past decades or vice versa. Other reasons usually offered for building a space station – potential scientific payoff, high-tech employment, educational motivation, foreign policy benefits – have not been compelling, but together they have created enough support for the programs to survive. Though seldom clearly articulated and perhaps not fully understood, the essential foundation is the belief that sustained human activity, both in Earth orbit and beyond, has significant tangible payoffs.

In order to improve the understanding of these historical contexts and their significant influences on future concepts, Chapter 2 introduces a chronology of concepts, plans and projects for space stations and their applications: from the very first ideas of living in space up to the very detailed project of the International Space Station. Since the space station's architecture and applications always depend on the station's orbit and its physical conditions, the space environment close to Earth (i.e. at an altitude of some 100 km) is described in Chapter 3.

Chapters 4 to 6 introduce the main subsystems of the space station. Unlike other spaceflight systems, space stations are mainly characterized by the permanent presence of a crew, as already mentioned above. Therefore, it seems logical to describe the different systems in order of their importance: "Environmental Control and Life Support System" (Chapter 4); "Power and Thermal Control System" (Chapter 5); "Attitude and Orbit Control System" (Chapter 6). Chapters 7 and 8 titled "Utilization" and "Microgravity" cover the different areas of applications and their characteristics, especially for research in weightlessness. Only by considering these application areas can the required mission objectives be deduced from the space station's design or vice versa. This is one of the main objectives of this book: enabling the advanced reader to design his or her "own" space station according to a defined framework of requirements, e.g. mission profile, crew size, possible space transportation vehicles, logistic scenarios, etc. This iterative design process is described in Chapter 9 "System Design" and shows that the amount of supply goods from Earth can assume immense proportions. In order to minimize supply, future space stations will make use of synergistic effects resulting from different subsystems linked to one another. This process, as described in Chapter 10 "Synergisms", will help to achieve nearly closed regenerative process cycles.

Chapters 11 "Human Factors" and 12 "Logistics, Communications and Operation" address components and subjects which are certainly important, but which do not dominate the space station design, such as human factors, operation and maintenance of space stations, current space transportation vehicles (to handle logistics), communications and data systems, automation and maintenance. In conclusion, Chapter 13 describes "The International Space Station" with its utilization peculiarities, layout and general aspects of access and operation. At the end of the book, a detailed bibliography and an index are included.

2 History and Current Development

Long before scientists and engineers in our century began to develop visions about space stations and their applications, authors from the end of the last century had already laid down their ideas in short stories and novels. This period will be shortly addressed in Section 2.1 "Visions, Concepts and Early Designs of Space Stations (1865–1957)". At the dawn of the real "space age", i.e. in 1957 when the first artificial satellite Sputnik was launched, development in the field of space stations progressed in two different ways: the space programs of the USA on one hand and the space programs of the former Soviet Union on the other. Their efforts climaxed in the so-called "space race", when the achievements of both nations followed each other in rapid succession and so each aiming to be the first nation on the moon, resulting in considerable progress on both sides: The USA conducted a significant number of studies and developed concepts leading up to the Apollo vehicles and, as a follow-up program, the space station Skylab. In the former Soviet Union, a series of Salyut space stations was developed and successfully operated. These achievements are covered in Sections 2.2 and 2.3. Important scientific and technological expertise was acquired in Europe during the 1980's which can be attributed to the development and operation of Spacelab (which was launched by a Space Shuttle). The Spacelab program will be briefly introduced in Section 2.4 and continued in Chapter 7 where its application is characterized. It was not until the end of the Cold War and the subsequent dissolution of the Soviet Union that from 1990–1995 the idea of a joint project had evolved: building and operating the large International Space Station under the leadership of the USA, with important contributions from Russia as well as Europe, Japan and Canada. With the Russian space station Mir, the USA and its later partners of the International Space Station (ISS) learned to cooperate in the fields of science, technology and the joint use of a space station. This period of political change as well as the assembly and use of the International Space Station will be addressed in Section 2.5. The chapter will be concluded with a comparison of several space stations in Section 2.6.

2.1 Visions, Concepts and Early Designs of Space Stations (1865–1957)

Early science fiction authors were inspired by the idea of crewed Earth satellites and rockets. In 1869, the American clergyman and author Edward Everett Hale (1822–1909) of Boston was the first to write a novel published in the "Atlantic Monthly" describing a white painted "brick moon space station" serving as navigational aid for seafarers. This is how the story goes: During an earthquake, this spherical "satellite" built from fireclay begins to roll, eventually reaching the launch platform made from flywheels already in full operation. It is prematurely launched into an Earth orbit with the construction workers and their families on board. But the tragic accident takes a favorable turn as the "crew" makes its home aboard the satellite, and, while orbiting the Earth, this "seed" of extraterrestrial civilization, transmits messages to its home planet Earth.

Another imaginative author worth mentioning is the famous Jules Verne (1828–1905) from Nantes, France. In his books "De la Terre à la Lune" ("From the Earth to the Moon") and "Autour de la Lune" ("Round the Moon"), he already foresaw many details of the lunar landing the way it was to take place 100 years later [Verne, Walter 92]. He not only described concentrated and preservable nutrition, constant air purification and oxygen resupply in detail, but he also thought of observation windows, a library, tools and animals being aboard (Fig. 2.1). It is well known that many space pioneers of later times were inspired by his novels.

The third author to be mentioned as representative of the past century is the German teacher Kurd Laßwitz. In his book "Auf zwei Planeten" ("Two Planets") he presented a rather poetic and utopian view of a space station without taking its technical aspects into account.

The Russian mathematics teacher Konstantin Eduardovich Tsiolkovsky (1857–1935) was the first of three great space flight pioneers our century has seen. The father of the theoretical basis of space travel had already made plans for a "satellite rocket" around 1903, but it was not until 1911 that he thought of using it to transport human beings. In 1933, he published his work "Album of Space Travels" in which

Fig. 2.1. Jules Verne's Vision of a Crewed Space Vehicle: Looking Back on Earth and Victorian Comfort on the Way to the Moon [Verne]

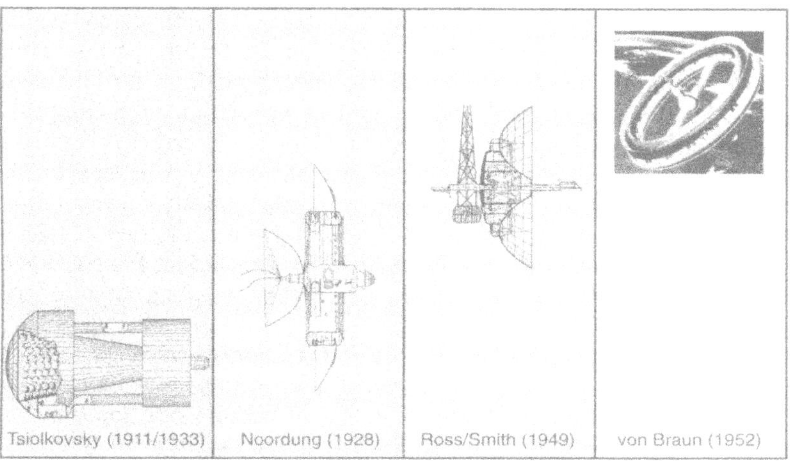

Fig. 2.2. Ideas of Space Stations from 1911–1952

he presented a concept of building a large habitation module in orbit (Fig. 2.2). Special attention should be paid to his theories of simulating gravity by rotating the station on its longitudinal axis and creating a park inside the station with vegetable beds and trees as an important part of a bioregenerative life support system. [Ordway 92, Puttkamer, Walter 92]

Around 1920, the second great space pioneer, the American Robert Goddard (1882–1945), described how our civilization could flee the dying solar system aboard a nuclear powered "ark" and suggested the use of extraterrestrial resources to manufacture the vehicle and its propellants. It is interesting to note certain parallels in Tsiolkowsky's and Goddard's work: on one hand, they both made valuable and detailed contributions to the scientific and technological development of space flight, but on the other hand many of their ideas turned out to be rather utopian. This is also true for the third space flight pioneer, the German Hermann Oberth (1894–1989). Yet, compared to Tsiolkovsky and Goddard, he was more interested in scientific principles and the applications of space stations in lower Earth orbits. Oberth suggested an (orbital) altitude of approximately 1000 km and pointed out the possibilities of an orbiting station for astronomical and Earth observation tasks. Additionally, he thought of a large solar mirror of 100 m in diameter for concentrating sunlight and solar heat and reflecting it back to the Earth. He also discussed the importance of a space station for several purposes: observation tasks in case of military conflicts, support of rescue operations, telegraphic or meteorological application, or even a space station as an "extraterrestrial refueling station" for interplanetary flights.

Early reflections on space stations were based on the assumption that when in orbit, humans would need "artificial gravity" to survive. Such simulated gravity obviously could be achieved by rotating a cylindrical or toroidal space station body. In this context, the works of two officers of the Austrian Imperial Army, Baron Guido von Pirquet and Hermann Noordung (pen name of Hermann Potocnik), published in 1928, are worth mentioning: In the journal "Die Rakete", Baron Guido von Pirquet presented three different concepts of a space station: the first for Earth observation in a 750 km-orbit, the second serving as a launching platform for interplanetary space vehicles in a 5000 km-orbit and the third a space station on an elliptical orbit intersecting the orbits of the first two stations. Hermann Noordung prepared a detailed study of a space station, consisting of three elements: a radial construction as habitation module, a power supply system (with solar collectors, evaporation and condenser tubes) and an observatory (cf. Fig. 2.2). He was the first to calculate a geostationary orbit in which the station was to revolve around the Earth and perform tasks like predicting the weather, military observation, warning ships of icebergs and mapping the Earth.

From 1930 until the end of World War II, rocket scientists and engineers concentrated mainly on developing missiles instead of space station ideas. Engineers of the large military development facility at the village of Peenemünde, located in northeastern Germany on the Baltic Sea, were the first to find a solution for problems concerning propulsion and navigation of a large rocket. The long-range ballistic missile A4 (meaning "Aggregat 4"), designated by the Propaganda Ministry as V-2 (meaning "Vengeance Weapon 2") was demonstrated for the first time in 1942 on the occasion of the rocket's initial flight. At the conclusion of World War II, the engineers from Peenemünde "moved" into Soviet and American projects of research and development (e.g. the US military operation called "Project Paperclip"). With the help of those engineers, substantial progress was made in rocket technology simultaneously in both the East and West. In addition to its military character, rocket technology came to civilian application, manifesting itself most clearly in the realization of space station concepts.

On the basis of the knowledge achieved by the middle of the century, Ross and Smith from Great Britain came to the conclusion that a massive space station could not be placed into orbit as a whole. Instead they advised to divide it into several separate discrete elements to be assembled in orbit. Inspired by Noordung's and Arthur C. Clarke's publications on the subject of geostationary orbits, H.E. Ross in January, 1949 wrote an article published in the Journal of the British Interplanetary Society describing the advantages of a space station revolving in such an orbit. For operation he suggested a 24-person crew. What is amusing is that this crew did not only include scientists and engineers, but, marked by the British-imperialist view at that time, two cooks and four orderlies as well!

In 1952, Wernher von Braun contributed greatly to the concept and creation of space stations by designing his famous wheel-shaped model, composed of several modules (Fig. 2.2 and Fig. 2.3). Von Braun's space station was to revolve in a polar orbit around the Earth at an altitude of 1600 km, accompanied by a free-flying astronomical telescope. Resupply was provided from the Earth by a winged reusable space transportation vehicle. The wheel-shaped structure of the space station was partially inflatable, had an outer diameter of about 85 m, and was triple-decked. Although Wernher von Braun considered his space station only a stage of development on his way to long-term exploration plans for Mars, he related a wide variety of fundamental technical ideas and principles which are still valid today. For example: the

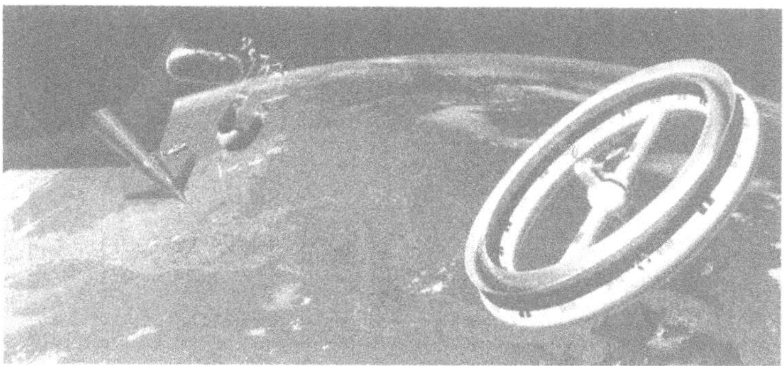

Fig. 2.3. The Concept of Wernher von Braun (1952) [Pioneering 86]

use of a micrometeoroid shield and progressive methods of "Concurrent Engineering" as well as management of technically demanding projects.

However, the space age saw its true beginning with the launch of the first artificial satellite Sputnik on October 4, 1957. This event enabled the visions and speculations of space flight pioneers and engineers to evolve into realistic space station programs, that were government-funded in the East and the West.

2.2 US Space Station Studies (1957–1985) and Skylab

After the ground for other ideas on space stations had been laid by the pioneer concepts presented in the previous section, more concrete technical ideas became public shortly before 1960. But still, these new concepts and the previous ones had one thing in common: they were far from being realistic in terms of feasibility of transportation, assembly or long-term logistics. But a solution would not be long in coming.

Around 1960, NASA's long-term plans considered the translunar flight and a crewed space station as parallel objectives to be realized around 1970. The Mercury program (at that time shortly before its initial flight) with its one-person-capsule was to serve as the basis of their development. This path was to start and finish in the 1970's with either a permanently crewed platform or with a planetary landing (Fig. 2.4).

According to these plans, the National Aeronautics and Space Administration NASA (founded October 1, 1958) undertook studies for a crewed research laboratory, parallel to their crewed spaceflight program (later on the Apollo program) [Bekey 85].

During the 1960's, all studies regarding space station concepts conducted by NASA field centers and associated enterprises were based on the assumption that large launch vehicles were available, i.e. Saturn V and its projected successor. This assumption manifests itself in their structural concept consisting of at least one element of the size of a Saturn V third stage as well as branched expansion modules.

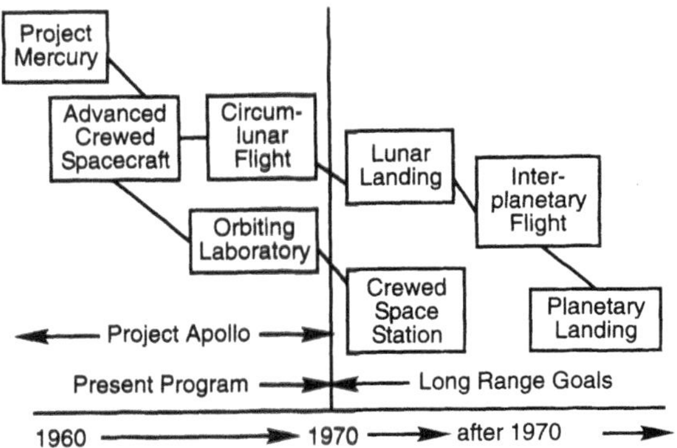

Fig. 2.4. NASA Crewed Space Flight Program Plans around 1960

An early, widely publicized concept had its origins in a survey of the newspaper "Daily Mail" in London. The survey was to address aerospace companies to solicit ideas and plans for a wooden mock-up of a space station for display at the London Home Show in 1959 on the subject "A Home in Space".

A concept of the Douglas Aircraft Company was selected with a crewed orbital observatory, consisting of the upper stage of a two-stage launch system. The idea was that, while on its way to a low Earth orbit, the launcher's second stage carried hydrogen and oxygen for the purpose of propulsion. Once the launcher reached its orbit, the four astronauts in a reentry capsule mounted on top of the launcher would convert the spent second stage into a space station. During this time, ideas ranging from live-in-tank concepts to crew restraint systems, equipment items and sleeping bunks (as they were later applied on Skylab) were born, as well as the perpetually revived discussion to use the numerous external tanks of the Space Shuttle to build large space stations.

On the occasion of a space station symposium in 1960, several enterprises presented their ideas including:

- Lockheed (modular design, launch by Saturn V)
- North American Aviation (rigid but self-deploying structure)
- Others (inflatable, partially launched by smaller rockets, partially driven by nuclear reactors)

At that time, it still was not clear whether priority would be given to a lunar landing or the construction of a space station.

This discussion was terminated in May 1961 with the famous words of the then US President John F. Kennedy: "I believe we should go to the Moon". NASA planners accepted the challenge choosing the lunar-orbit rendezvous technique as approach. Finally, the dream of a lunar landing came true in 1969. As a consequence, an Earth-orbiting space station as intermediate station for the lunar expedition "Apollo" was no longer necessary and this plan was abandoned in the 1960's. Remaining studies anticipated that a space station would be the logical post-Apollo program.

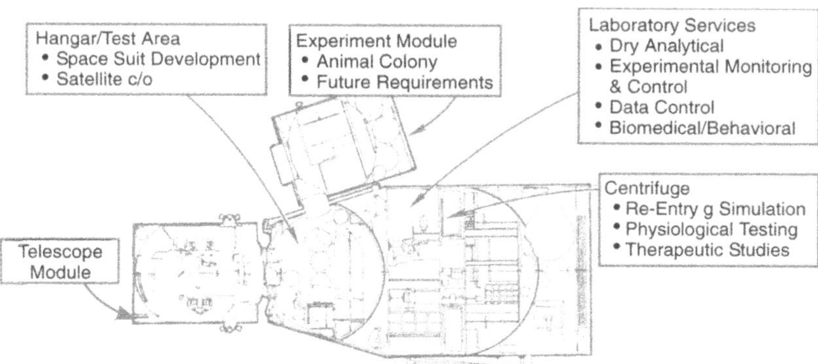

Fig. 2.5. Manned Orbiting Research Laboratory [Woodcock 86]

A joint study undertaken by NASA-Langley and contractors was further detailed by Douglas Aircraft from 1963–66 and lead up to a concept called Manned Orbiting Research Laboratory (MORL). The MORL was to have a crew of nine and revolve in an orbit at 300 km altitude and at 50° inclination. MORL was designed for biomedical experiments and was to serve as a telescope platform (Fig. 2.5). In the course of the studies, the MORL steadily became larger and more complex since no one had limited the logistics and consequently the costs either. This error was repeatedly committed in the course of the next two years. The Olympus studies undertaken by NASA-MSC Houston completed in 1962 on the basis of a two-stage Saturn V rocket suffered the same fate and ended up with a crew of 24 [Logsdon 85].

Douglas Aircraft was awarded a contract by the US Air Force to develop a similar concept MOL which had already taken into consideration the philosophy of the Apollo Applications Program conducted by NASA in 1965: components designed for the lunar landing should also be used for subsequent missions. It was not until then that the top levels of NASA management began to take up plans for a space station again and supported further internal studies during the years 1966–68, undertaken by groups headed by C. Donlan and E.Z. Gray. Again and again these groups had to deal with several conflicts between the different requirements of a space station such as:

- "Gravity simulated by rotation" vs. "Research in microgravity"
- "Astronomy (i.e. inertial flight mode)" vs. "Earth observation (i.e. gravity gradient stabilization)"
- "LEO servicing" vs. "Logistics for interplanetary flights"

Eventually, the "minimum station concept" resulting from this discussion was based on intermittent operation with changing flight modes, crewed (8–12 astronauts) or uncrewed, 320 km orbital altitude, 50° inclination and a 12 years design life.

Between 1969–70, NASA issued a statement of work for a space station "Definition" study (Phase B, cf. Chapter 9 "System Design"). Of the three firms who had submitted proposals, NASA awarded contracts to North American Rockwell and McDonnell Douglas. The space station was to be a cylinder, 10 m in diameter and 16 m in length, and it was supposed to be launched in 1977. Its design life was to be 10 years, and it was to be equipped with a photovoltaic or solardynamic isotope/Brayton cycle system. The station was to be resupplied every 45 days, and rotation of the US crew was to take place every 90 days. Costs, at that time, were estimated at $ 8–15 billion (US). But even before NASA could issue a contract for the development of the station, the Saturn V program was "stopped" due to US congressional budget decisions. The only launch vehicle available now was the Space Shuttle. As a consequence, cylindrical space station modules of 10 m in diameter had to be reduced to the dimensions of the Shuttle's cargo bay, i.e. 4.5 m in diameter and 18 m maximum in length. The same restraints applied to other subsystems and components. Yet, even in this new situation, the contractors continued their study activities. However, in summer 1970 it had become obvious that after the Apollo mission there was to be only one large program, and this would be the Space Shuttle. Despite this development, studies were furthered until 1972 including the idea of a Shuttle based Sortie Lab. On the basis of this idea and in pursuance of it, the development of Spacelab was finally realized by the ESA, supported and operated by the NASA.

With the end of the Saturn V program and the decision to develop, from 1972 onwards, the Space Transportation System (STS, also called Space Shuttle) and to use it as the main transportation system, NASA altered its concepts of a space station. The changes now involved a modular design of the station so that any single component could be transported by the Space Shuttle. Rotating systems with large dimensions, assembled in space over thousands and thousands of crew-hours were, due to steadily increasing flight practice, replaced by smaller, zero gravity concepts for modules and stations which were designed rather with concern for their application than for human habitation.

During the second phase of US space flight development (cf. Fig. 2.6), sometime during the 1970's, concepts of large space stations existed more or less as marginal projects whereas the Space Shuttle program received full attention by NASA. The project of a space station remained interesting though, not with least thanks to the orbiting laboratory Skylab, which emerged from the Apollo Applications Program and was more or less assembled using parts from the Apollo Program. With a total of three crews, however, it turned out to be outstandingly successful [Skylab, Woodcock 86].

In the beginning, two alternatives were taken into consideration for the construction of Skylab: First, the so-called "Wet Workshop", launched with a Saturn IB. The Astronauts could enter the third stage after residual propellants had been purged and convert the spent stage into a space laboratory. Second, the "Dry Workshop" in which the third stage was to be outfitted on the ground before being launched by a Saturn V. During the first lunar landing in July 1969, the second alternative was selected and was to be brought into orbit with the last launch of a Saturn V rocket.

Figure 2.6 shows the chronology of American space station studies on their thorny way to the first American space station project Skylab. Two decades had passed since the studies of Wernher von Braun by Skylab's initial launch. And another two decades had yet to pass until the project to design a second space station with major US involvement was settled: the International Space Station.

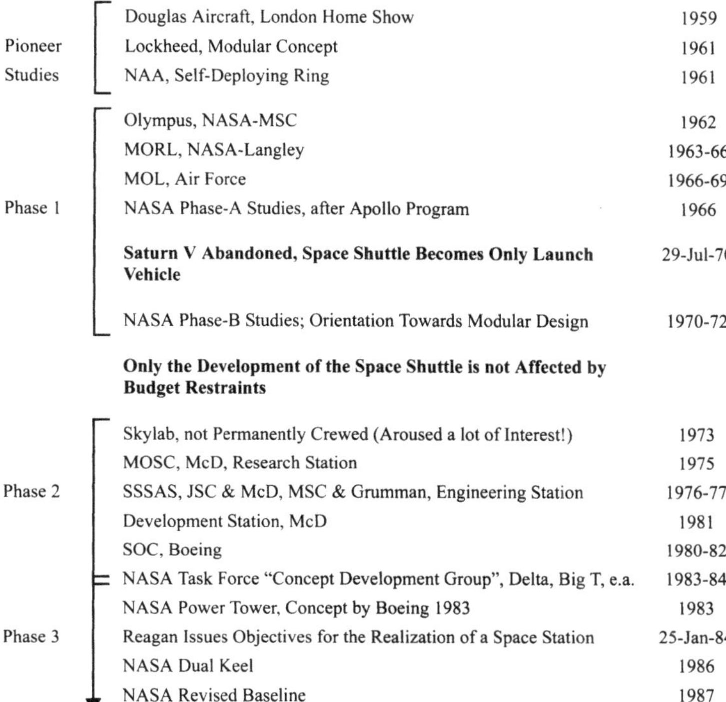

Fig. 2.6. Chronology of US Space Station Studies until 1990

Skylab was launched on May 14, 1973 into an orbit at 432 km altitude and 50° inclination. It was composed of a converted Saturn V third stage, the so-called "Orbital Workshop", a telescope mount, an airlock (from Gemini components) and a docking adapter (Fig. 2.7), respectively. Three crewed missions (28, 59 and 84 days in duration) were flown between May 25, 1973 and February 8, 1974, the three-person crews being transported by Apollo Command Modules. The first mission turned out to be an emergency repair service: During the launch of Skylab, the meteoroid shield failed due to aerodynamic forces and in turn damaged a solar panel and part of the thermal insulation. Additionally, the second solar panel could not be fully deployed. The first crew, however, succeeded in providing satisfactory operation of Skylab by carrying out several EVA's and installing an umbrella-like sunshade.

In terms of size and success, Skylab was a remarkable station, although it was not designed as a permanent system with provisions for resupply and therefore many did not accept it as a true space station. For example, the potable water for all three missions was stored in tanks aboard Skylab. Furthermore, Skylab had no propulsion system for reboost maneuvers. Due to the fact that there was no vehicle available (i.e. the Space Shuttle) for reboost maneuvers, Skylab could not stay in orbit. After having orbited the Earth for six years (with only about 171 days of true operation)

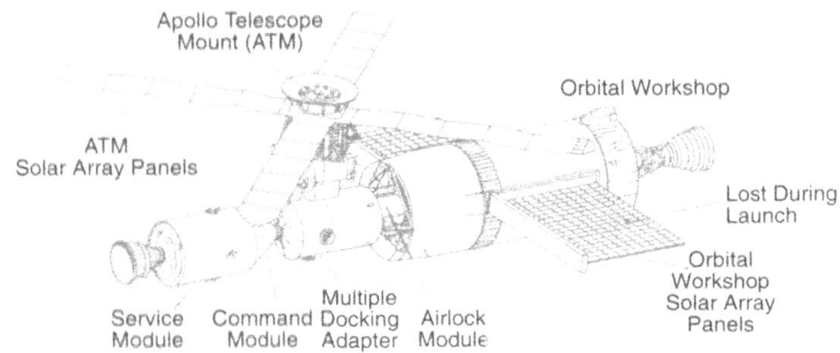

Fig. 2.7. Skylab

the space laboratory fell below its minimum orbital altitude and reentered the Earth's atmosphere over Australia on July 11, 1979 [Ruppe 80].

The configuration of the Skylab subsystems resulted from an interesting mixture of influences:

- The desire to keep development costs low by using already existing equipment
- A strong motivation to make the Skylab missions as productive as possible from the scientific point of view, and, as a consequence
- The necessity to develop intelligent operation concepts on the basis of already existing technology

Nevertheless, the use of devices that had not seen prior operational use in spaceflight was encouraged, e.g. large momentum wheels (Control Momentum Gyros, CMG) or molecular sieves for the removal of CO_2 (even at an atmospheric pressure of 35 kPa).

Even more clearly than the Apollo missions, Skylab missions showed that humans are able to work and react to unforeseen events in space just as effective as they can on Earth on the condition that the appropriate tools and instruments are provided. Moreover, the Skylab program proved that a space laboratory is in the position to meet the majority of engineering and scientific requirements. All three Skylab missions concentrated on solar and Earth observation, life sciences, astrophysics, research in the field of physical sciences (including fluids) and newly developed materials [Skylab, Puttkamer].

Space station concepts in 1975 and later basically concentrated on two ideas: first, the introduction of a modular design for the realization of the transportation of components by the Space Shuttle; second, the use of a space station as an assembly base for extensive orbital structures, e.g. large antenna platforms, solar energy satellites, etc. The Manned Orbital Systems Concept (MOSC) was to use elements from the Skylab, Spacelab and Space Shuttle programs ("insofar as practical") in order to meet the requirements of space station development as well as to investigate the increasing effect of using limited terrestrial resources. It was at that time that the question of where limits of development were to be set, arose. Nevertheless, the ideas of Gerard O'Neil (colonization of space) and Peter Glaser (Solar Power Satellites SPS) received considerable public interest at that time. The Space Station

System Analysis Study (SSSAS) initiated in 1976, continued the studies. In the meantime, concepts were being examined dealing with prolonging Space Shuttle flights. This was to be achieved by launching additional, permanently orbiting solar power modules, e.g. the Power Extension Package (PEP) and the Orbital Support Module (OSM) and, from 1980 onwards, orbital platforms like the Science and Space Application platform (SSAP) by MSFC/McD, the Advanced Science and Application Platform, and the "System Z Concept". Their electrical power ranged from 25 kW to 35 kW, and they were partially equipped with solar-electrical propulsion. Europe, represented by the European Space Agency (ESA), simultaneously developed the Spacelab and the two platforms, SPAS and EURECA (to be launched by the Space Shuttle, cf. Chapter 8), under the leadership of the German company ERNO Raumfahrttechnik GmbH, Bremen (now DaimlerChrysler Aerospace). This era of research in the field of primarily modular space station concepts was continued in the USA through the Evolutionary Space Station Concept (ESSC by MSFC/McD) and the Space Operation Center (SOC by JSC/Boeing) from 1981–82.

The studies undertaken by Boeing from 1980–82 on the Space Operations Center were based on the use of a space station as a transfer point, like Oberth had suggested years ago. A space station could e.g. support missions to higher and interplanetary trajectories and act as a servicing base for satellites and upper stages. The idea on servicing-in-orbit turned out to be a favorable development, since it underlined the efficient use of space technology, which in turn positively influenced the discussion on the extremely high costs of transporting payloads by the Space Shuttle [Boeing 82].

The SOC (see Fig. 2.8) was to orbit at an altitude of 370 km. As to its size and elements, it already resembled the concept of the Space Station Freedom (SSF). Although the pressurized modules (similar to those of Spacelab), the Interconnecting Elements (ICE), the Logistics Module (module cluster) and the manipulator arm had already been part of earlier studies, the SOC was the first program to combine them into a cluster. This philosophy also served as a basis for SSF.

After this transitional period during the 1970's, the third phase of US space flight development was initiated. From 1983–84 NASA, fully concentrating on the Space Shuttle as the only transportation system, examined and evaluated a number of space station designs (Fig. 2.9) by introducing the Concept Development Group (CDG). The main result of the CDG's work, besides more concrete concepts, was the idea that the arrangement of the pressurized modules can be isolated from a station's overall structure. From the concepts suggested by the CDG like "Planar", "Big T", "Delta" and "Power Tower" designs, all included a similar cluster of four pressurized modules. These are mounted on a large main truss structure which comprises the station's subsystems and onto which the radiators and solar collectors are mounted. The only differences among these four concepts were to be found in the rigidity of their truss structures and in their mode of attitude control and power supply.

On January 25, 1984, Ronald Reagan issued objectives for the realization of a space station. From the above mentioned CDG concepts, the "Power Tower" (Fig. 2.9) was selected on the basis of a proposal by Boeing in 1983 to act as reference concept for the space station. For this project, Reagan invited allied and other

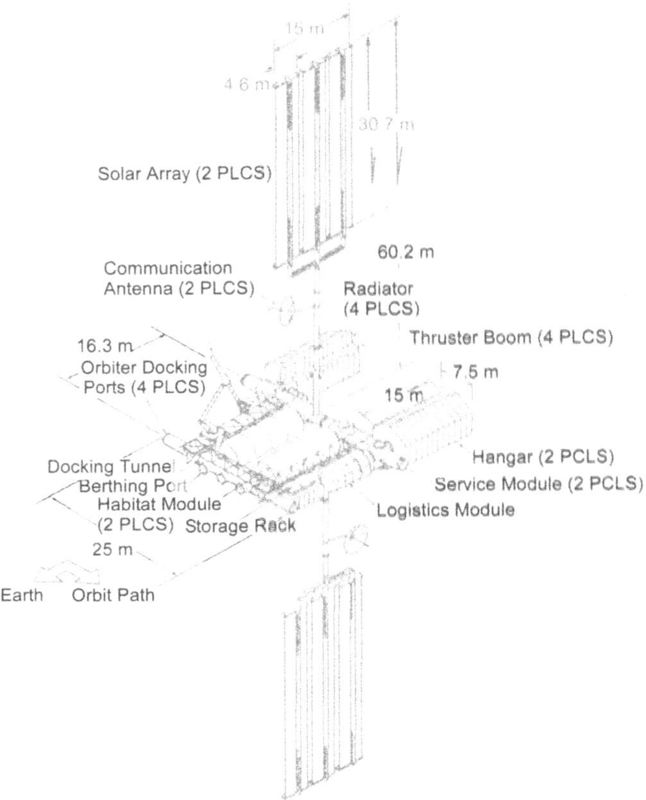

Fig. 2.8. Space Operations Center [Woodcock 86]

friendly nations to participate in the station's design, development, operation and use, all for the promotion of "peace, prosperity and freedom" [Mark 87].

In 1986, the station's configuration was altered with the integration of elements from the newly gained international partners: Canada, Japan and the ESA, the latter acting as one single partner. The new concept was called "Dual Keel" (Fig. 2.10), the name being derived from its two main truss structures, one perpendicular to the other. Elements crucial for operation such as power supply, pressurized modules and subsystems, were mounted on the horizontal truss whereas the vertical truss offered attachment points for astronomical and Earth observation payloads. The "Dual Keel" concept was to include sufficient possibilities for future orbital extension stages, e.g. a place for the Orbit Maneuvering Vehicle (OMV), hangar, and satellite servicing facilities.

In 1987, the complete realization of the Dual-Keel concept fell victim to budget restraints and was continued as "Revised Baseline". As shown in Fig. 2.10, this "Revised Baseline" had only one truss and the number of attachment points was reduced and placed on the remaining horizontal truss. If at this point we look at the effects

CDG Planar IOC

Features
- Gravity Gradient or Inertial Flight Mode
- Articulated Solar Array & Radiator (min. Size w/ Techn. Transparency)
- Separate Activity Zones
- Module-to-Module Connections

Big "TEE" IOC

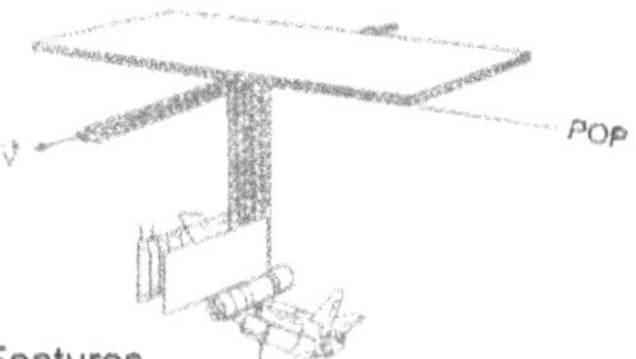

Features
- Gravity Gradient Flight Mode
- Minimum Drag
- Articulated Solar Array, Fixed Radiators
- Generous Activity Zones
- Truss-Mounted Pressurized Modules

Delta IOC

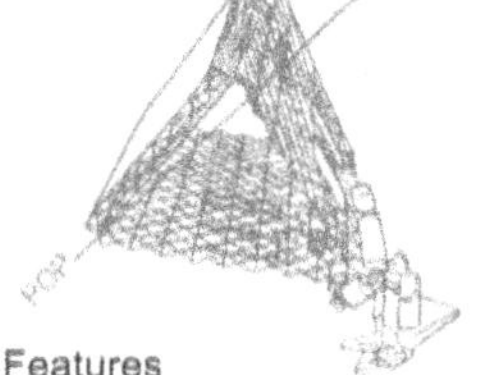

Features
- Inertial Mode about P.O.P. Axis
- Fixed Solar Array & Radiator SA Size with 10% Penalty
- Very Stiff Structure
- "Natural" Service Hangar
- Truss-Mounted Pressurized Modules

Power Tower IOC

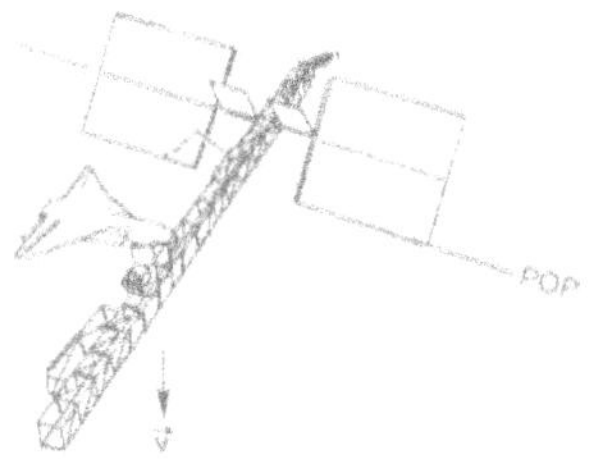

Features
- Gravity Gradient Flight Mode
- Articulated Solar Array & Radiators (min. Size w/ Techn. Transparency)
- Access to Whole Base via Single Mobile Crane
- Widely Separated Activity Zones
- Good Observations Sites
- Truss-Mounted Pressurized Modules

Fig. 2.9. Overall Structure Design (NASA CDG) (for Acronyms See Glossary)

of the gravity gradient (cf. Chapter 6), it is remarkable that the flight orientation of the Dual Keel could be maintained with the "Revised Baseline", although the stabilizing effect of the second truss was missing. Although there had already been irritations among some of the space station partners because NASA had very often changed the space station's design without having consulted its partners first, the USA, Japan, Canada and the ESA signed an Intergovernmental Agreement (IGA) on September 29, 1988.

Due to US fiscal policy, the program constantly had to fight for its budget, especially when presidential elections were about to take place as it was the case e.g. in 1988. As a consequence of continuous economical measures, the space station dubbed "Space Station Freedom" in the meantime was reduced once more in 1989 and its implementation schedule prolonged.

The subsequent time witnessed a steady stream of financial curtailments by the US Congress, NASA internal revisions, and a reduction of the whole concept. The number of resupply flights by the Space Shuttle and the crewtime for EVAs necessary to assemble the station were especially subject to constant criticism. To make things worse, the pressure on NASA had increased after the two incidents in 1990

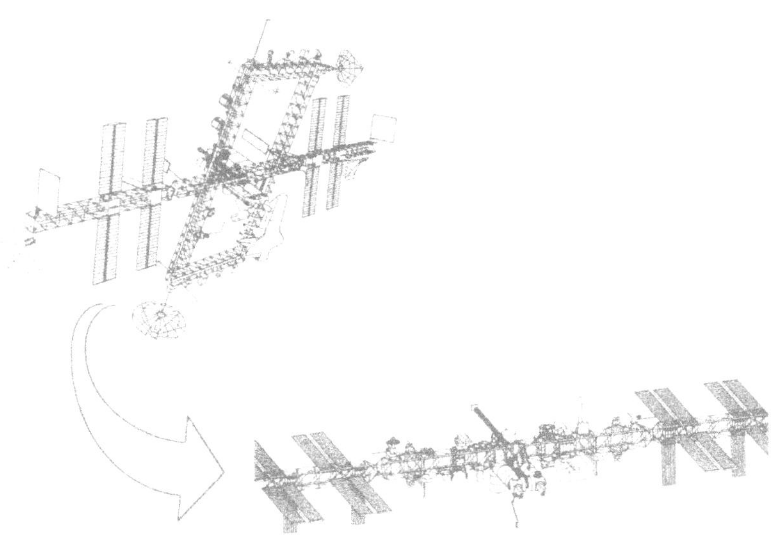

Fig. 2.10. Dual Keel and Revised Baseline

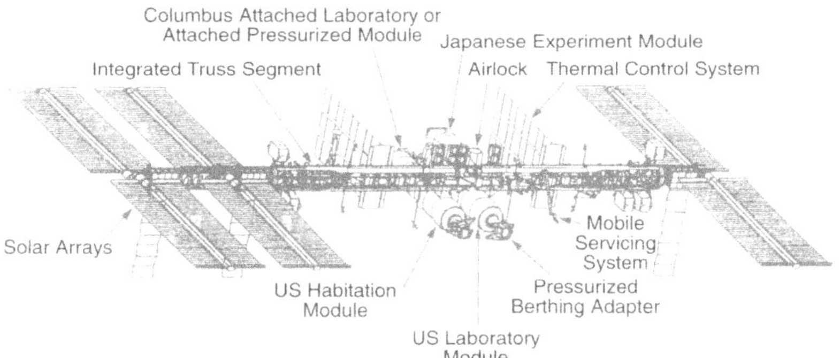

Fig. 2.11. Space Station Freedom (Plan Dates from 1989)

(Hubble Space Telescope and technical problems with Space Shuttle). The legislature regularly threatened to cancel the whole program, the last time in 1991. The resulting programs were marked by these considerable curtailments. Compared to the "Revised Baseline", they were designed to be extendible interim solutions for the Space Station Freedom. They included markedly smaller systems with the SSF module cluster (Fig. 2.11 and Fig. 2.12) as the core element, but their successful implementation remained uncertain from a political point of view [SSF 88].

The development of the originally planned *Space Station Freedom* (SSF, Fig. 2.11) between 1986–93 was scheduled for the international partners as follows:

The *USA*, from the mid-1990's onwards, should have launched the modules with the participation of the international partners into an orbit at an altitude of 352–463 km at 28.5° inclination. In 1989, the completion of assembly was planned for 1999 and the station was to have a design life of 30 years. The station's orbital altitude was to be coarsely adapted to solar activity so that the station would always fly in an environment of constant atmospheric density. The station's orbital decay caused by atmospheric drag was to be compensated after each visit of the Space Shuttle (i.e. every 90 days) by means of a cyclic reboost maneuver.

For SSF, a photovoltaic electrical power system with an overall performance of 75 kW (30 kW alone for payloads) was planned. To facilitate the distribution of electrical power, a DC system was to be included. The thermal control system was initially planned as a distributed system with a thermal bus (on two temperature levels with a two-phase-loop each) which was to conduct the excess heat to two large radiators mounted on the truss. The propulsion system for attitude control was to be constructed on the basis of hydrazine thrusters: they were to serve as backup for the control moment gyros and perform their desaturation. Originally, the entire propulsion system was to be based on H_2/O_2 thrusters and resistojets in order to profit from the reuse of electrolyzed water and waste gases from the life support system. This plan, however, was abandoned because planners considered costs and risks of development too high.

The station's truss included mounting points for so-called attached payloads, i.e. remotely operated systems such as telescopes. They were to be provided with electrical power up to 12 kW each. The arrangement of the four pressurized modules (two US, one ESA, one NASDA, see Fig. 2.12) was installed on the mid-truss and had an overall mass of about 136 tons. Due to their similar dimensions, every single module could be transported inside the cargo bay of a Space Shuttle. They were linked to each other by smaller elements, the so-called "resource nodes". When arranging the modules, special attention was paid to safety. The "race-track" arrangement enabled the crew to enter each module from two different sides and offered two different escape routes in case of emergencies, but only in the US part.

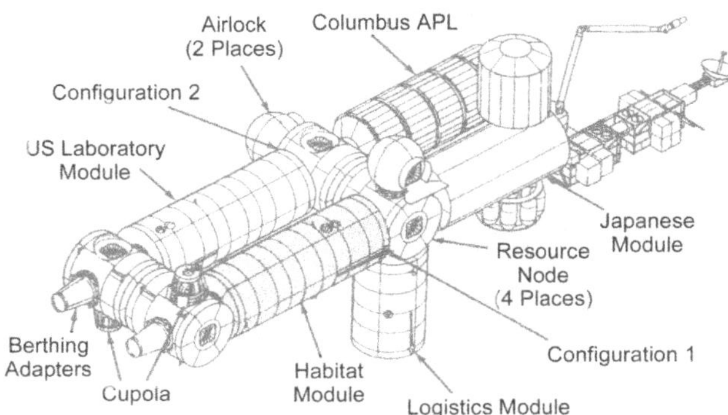

Fig. 2.12. SSF Pressurized Module Configuration

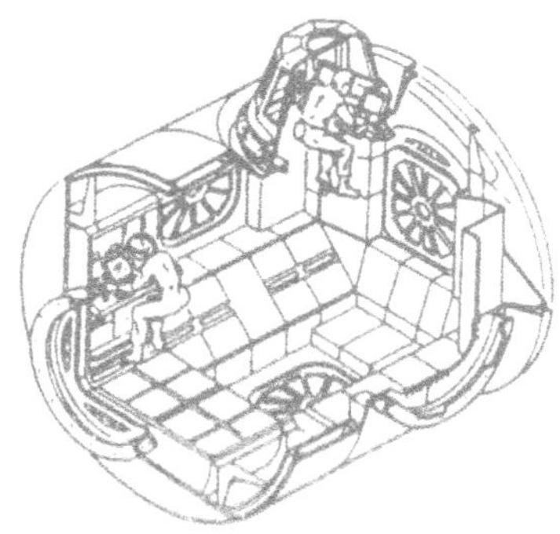

Fig. 2.13. SSF Resource Node

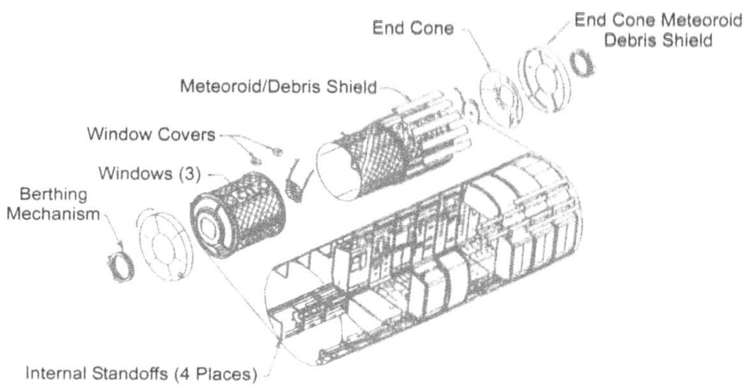

Fig. 2.14. Laboratory Module Structures

Each resource node included six berthing adapters and additional space for subsystems (Fig. 2.13). On two of the adapters, cupolas were to be installed for observation of the station's environment and air locks were mounted on two other adapters.

The planned US laboratory module had a total volume of 166 m^3 (inside diameter 4.22 m, length 12.8 m) and included a primary structure (able to carry mechanical loads, see Fig. 2.14) and a surrounding meteoroid shield. Although thin, this second outer skin provides for tiny particles which, although penetrating the shield, evaporate before hitting the primary structure. Inside, the secondary structure includes the feed lines as well as the stand-offs for the racks containing subsystems and payloads.

The Laboratory Module was to accommodate 44 double racks (one double rack is 1.05 m wide, 1.89 m high and 0.8 m deep): 17 racks for payloads only, 14 for general laboratory systems (vacuum, workbenches, etc.), and 13 for housekeeping. The housekeeping racks belonged to the "Common Subsystems" (ECLS; Communication, Power Distribution, etc.) spread over both US modules.

The Habitation Module was identical in construction and included 25 double racks for crew systems and 19 racks for housekeeping. The Habitation Module was

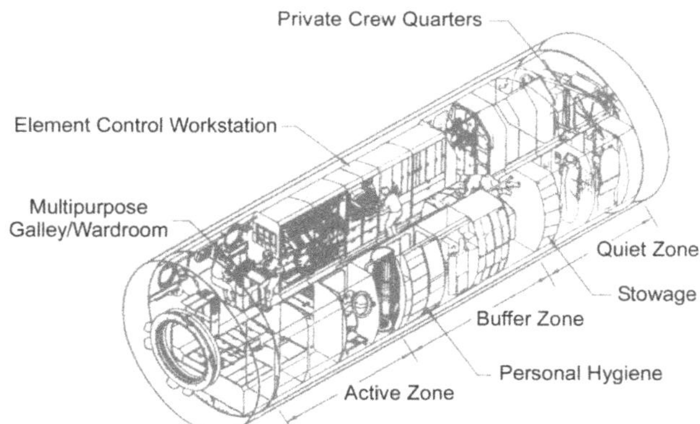

Fig. 2.15. Habitation Module Layout

designed for a total of eight crew members (Fig. 2.15). In the beginning, it provided a galley/wardroom, hygiene compartment, health care & exercise system and crew quarters only for a crew of four before it was extended to house a crew of eight after the installation of laundry/dishwasher and freezer.

Canada. The Canadian contribution to SSF was to be the Mobile Servicing Center (MSC). It comprised a mobile platform mounted on the truss, the Space Station Remote Manipulator System and the Special Purpose Dextrous Manipulator (a small additional manipulator arm capable of performing tiny, precise movements). The MSC was to be operated from control panels inside the cupolas or from an EVA workstation directly on the platform's surface. The "Flight Telerobotic Servicer" could have been used as additional robotic system as well. This two-armed device had been developed by NASA for assembly of the station and could either be mounted on the Space Shuttle's manipulator or on the station. Later on, however, its construction was indefinitely postponed.

Japan. The Japanese Space Agency NASDA planned to contribute the "Japanese Experiment Module" (JEM). It was divided into a pressurized part (smaller than the US or European modules) and the so-called "Exposed Facility" where experiments were to be performed under direct space conditions (Fig. 2.16). The Exposed Facility had its own small manipulator arm and equipment could be transferred into the pressurized part through an airlock. Moreover, NASDA wanted to include an own logistics module (JELM), compatible with the Shuttle and able to be carried by the Japanese launcher H-II.

In the pressurized part, 10 double racks with experiments could be accommodated. From a total of 15 kW, 6 kW were needed for housekeeping. Plans dating from 1989 intended that the JEM should have been launched in February 1998.

Europe. According to the plans of the ESA Council meeting at Ministerial Level in The Hague in 1987, ESA wanted to contribute several programs to the SSF: Colum-

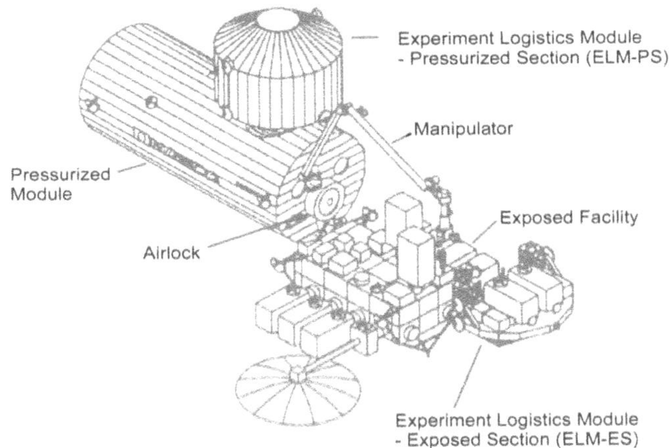

Fig. 2.16. Japanese Experiment Module

bus, Hermes and Ariane 5. Columbus was originally comprised of a laboratory docked at the station (Attached Pressurized Module, APM; ESA's name was Columbus Attached Laboratory, CAL); a smaller, free-flying laboratory (Columbus Free Flying Laboratory, CFFL); and a polar platform. The last component was later on excluded from the program since it was not directly related to SSF.

As matters stood in 1989, the Columbus Free Flying Laboratory (see Fig. 2.17) was to be docked at SSF in July 1998. It was laid out as a laboratory for all disciplines related to microgravity. From a total of 20 kW for the whole station, about 10 kW were to be transferred to payloads and systems. With an outer diameter of 4.46 m and a length of 12.8 m, CAL had the same dimensions as the US Laboratory Module.

Contrary to the CAL, the Columbus Free Flying Laboratory was to be launched independently of the Space Station. One to two years after the launch of CAL, it was to be brought into an orbit by an Ariane 5 and co-orbit with SSF. During normal op-

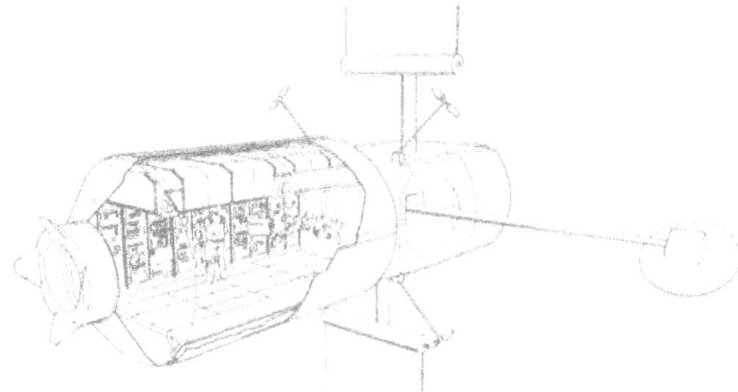

Fig. 2.17. Columbus Free Flying Laboratory

eration, it should have been sun-oriented and approached the Space Station due to natural orbital evolution every 180 days. The CFFL was composed of the pressurized module (about half the size of the CAL) and a resource module. The latter included all subsystems such as solar generators, propulsion, antennae, etc. As the laboratory was to satisfy stringent μg requirements, it was planned to be uncrewed and remotely operated (manipulation of samples was to be effected by robots). The European space plane Hermes was to visit it every six months for a duration of about 9 days and the CFFL was to dock at the SSF every five years for servicing and refueling. Before launch, the vehicle would have had a weight of 18.4 tons. In orbit, experiments would have been added until a total of 3 tons of payloads and 2.7 tons of propellants had been reached. CFFL's total length was 11.95 m, its outer diameter 4.46 m and the span of the fully deployed solar panels measured 59 m. The projected minimum power was about 10 kW, of which 5 kW were needed for the payloads.

All plans for the development of CFFL and the space vehicle Hermes were canceled after the ESA Ministerial Conference in Granada in 1992. It is quite remarkable that the Japanese contribution to SSF survived all SSF-modifications completely unaltered and was even integrated into the International Space Station Program (cf. Sect. 2.5.4) without being subject to any modification. The American modules have experienced only slight changes, and the European Columbus Attached Laboratory (CAL), due to ESA's constant curtailment of expenses, was reduced to a length of 6.7 m and renamed Columbus Orbital Facility (COF).

2.3 Russian Space Stations: Salyut (1971–1991) and Mir (Development until 1994)

The history of Soviet and now Russian space stations can be characterized by systems that have been built and operated with increasing success and duration as time went on. In 1971, while, in the USA, the lunar landing program Apollo still was in full swing, the Soviet Union launched and tested its first space station Salyut 1. And yet, this successful mission came to a tragic end when all three cosmonauts lost their lives due to sudden decompression of the Soyuz capsule during reentry. After a brief pause the Soviet Union resumed crewed space flight creating several redesigns of Salyut 1. The last one, Salyut 7, was in orbit for nine years, from 1982 to 1991 [Bekey 85, Harland 97].

The development of Soviet space stations excluding total failures and military projects named Almaz (Salyut 2 could not be docked and was abandoned; an unnamed station was lost during launch in 1973) were rather successful and can be seen by studying the Salyut family, presented in Fig. 2.18 and Fig. 2.19. Within a period of 24 years (since the launch of Salyut 1), astronauts spent a total of 30 years together aboard the space stations Salyut and Mir. Table 2.1 gives a chronology of Soviet space station projects.

The Salyut family included a cylindrical, pressurized main module (all of them more than 13 m long and between 3.6 and 4.2 m in diameter; see Fig. 2.18) where the electrical power system and all other subsystems were integrated. The station

Table 2.1. Chronology of Soviet Space Station Projects

First Generation of Soviet Space Stations (1971–77)			**Crews**	**Progress Flights (Transfer Mass)**
Salyut 1	1971	First Space Station	1 (Soyuz 10, 11)	
without name	1972	Failure During Launch		
Salyut 2	1973	First Almaz -Station		
Cosmos 557	1973	Failure		
Salyut 3	1974–75	Second Almaz-Station	1 (Soyuz 14)	
Salyut 4	1974–75	Three Solar Panels	2 (Soyuz 17, 18)	Soyuz 20 (0)
Salyut 5	1976–77	Including Solar Cells Last Almaz Station	2 (Soyuz 21, 24)	
Second Generation of Soviet Space Stations (1977–87)				
Salyut 6	1977-82	Second Docking Possibility	16 (Soyuz 26-32, 35-40, T-2 to T-4)	12 (20 tons)
Salyut 7	1982-87	Deorbit in 1991	10 (Soyuz T-5 to T-7, T-9 to T-15)	13 (25 tons)
Third Generation of Soviet/Russian Space Stations (1986–99)				
Mir-Station	20-Feb-86	Core Module		2 (4 tons) 1986
Kvant	31-Mar-87	Extension Module		13 (26 tons) 1987+88
Kvant 2	26-Nov-89	Astrophysics Module		4 (9 tons) 1989
Kristall	31-May-90	Materials Production		4 (9.5 tons) 1990
Spektr	20-May-95	Atmosphere & Astronomy Research	2+2x Atlantis	cf. Details in Table 2.6.
Priroda	23-Apr-96	Earth Remote Sensing Module	2+2x Atlantis	cf. Details in Table 2.6.

had a berthing adapter where the Soyuz capsule docked when cosmonauts were aboard the station. This basic concept was expanded with Salyut 4. It was equipped with three photovoltaic panels instead of two. They could be rotated to face the Sun whilst permitting the station flexibility in orientation, thus increasing the power up to 4 kW. Salyut 4 tested the new Delta semi-automatic navigation system with a set of Sun sensors and new computers as well as new attitude and thermal control systems. The military station Salyut 5 was structurally identical with Salyut 3 but had no requirement for replenishment, cargo delivery or crew exchange. The main instrument aboard Salyut 5 was an improved telescopic camera for military reconnaissance.

With the civilian and military stations flying alternatively and so combining subsystems of two different design bureaus, Korolev and Chelomei, it was generally expected that the next station would have several docking ports, could be replenished, and would involve resident and visiting crews. It was time for the orbital station program to switch from basic engineering development to routine operations.

Salyut 6 introduced the second generation of Russian space stations. It included an additional aft docking port for the new Progress freighters. With these freighters, crews could stay aboard the station for a longer period of time; visiting crews were able to join them, and supplies could be provided. Progress freighters are similar to Soyuz capsules, but automated and uncrewed. In their pressurized part, supplies for the interior of the station can be stored. Supervised by a ground station, the station's tanks may be refilled by pumping propellants out of the freighter's tanks. Moreover, the Progress module may be used for propulsion maneuvers and the disposal of waste, since it burns up when reentering the Earth's atmosphere.

Thanks to these additional logistics possibilities, mainly scientifically trained cosmonauts from other Soviet-bloc countries were able to visit the station during the so-called Intercosmos Program. The first "foreign" space traveler was Vladimir Remek of Czechoslovakia. He was followed by an astronaut from Poland and the first German astronaut, Sigmund Jähn from the then GDR. Each of them worked aboard the station for seven days in 1978 [Harland 97, Hooper 90]. Salyut 6 was home to a total of 16 crews, including cosmonauts from Czechoslovakia, Poland, GDR, Bulgaria (although in this case, the capsule was not able to dock with the station), Vietnam, Cuba, Mongolia, Romania and, of course, the Soviet Union. From these 16 crews, six experienced long-duration stays of up to 185 days. Whenever crews visited the station, they traded their Soyuz capsule for the one already docked at the station since the capsules' operational life was relatively short at that time. Later on, their period of operation was gradually extended to 60, and, in the end, to 90 days. The Soyuz capsules in operation at present, achieve an in-orbit lifetime of up to 180 days, and in view of their use for ISS, this will again be extended up to over one year.

Such a space complex consisting of Soyuz/Salyut 6/Progress already weighed about 33 tons at that stage of development. A Progress capsule could carry a maximum of two tons of material, including nutrition, clothing, water, air, propellants, equipment, spare parts, experiment units, etc. With the improved Progress-M capsule (initially operated in 1989), the payload capacity was raised to 2.5 tons. A total of 12 Progress capsules provided supplies for Salyut 6 consisting of 20 tons of additional equipment and expendables. In Fig. 2.18, the evolution of Salyut space stations is shown up to Salyut 6.

The last space station in this family, Salyut 7 (see Fig. 2.19) was launched in 1982. On this occasion, the Kosmos-1443 modules were finally put to use. During unoccupied modes, they were berthed at the aft docking adapter and had additional space for scientific experiments as well as a return vehicle laid out for payloads of 450 kg. Furthermore, the Kosmos modules were equipped with an additional propulsion system for the entire station. This was truly necessary for such a complex, since with a length of 13 m and a mass of 20 tons, these modules nearly had the dimensions of the Salyut 7 core module.

Salyut 7 failed during an uncrewed phase of operation in 1985: due to a sensor defect, the station's batteries had not been recharged. In the course of a daring mission, a Soyuz capsule was able to dock, and the two cosmonauts found the failure in the station's main module, which in the meantime had frozen. Finally, they were able to bring the station back to normal operation. In 1987, Salyut 7 was again the scene of an interesting maneuver: the first crew of the space station Mir flew to

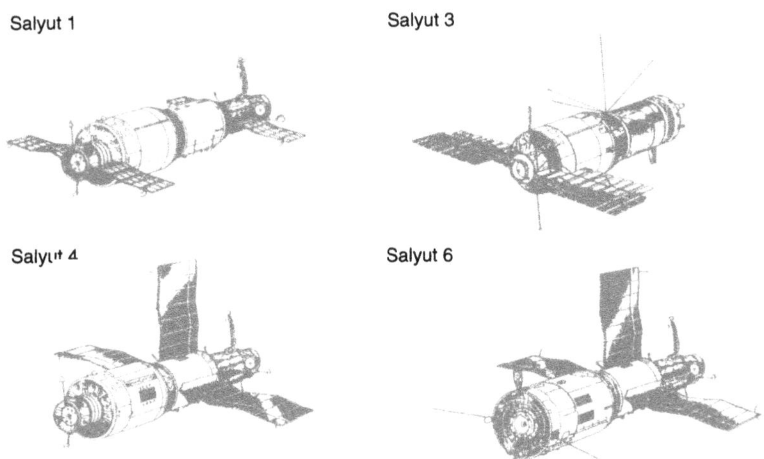

Fig. 2.18. The Evolution of Salyut Space Stations

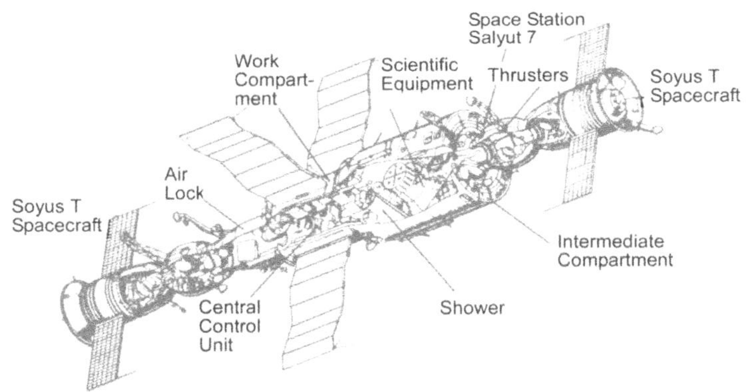

Fig. 2.19. Salyut 7 (1982–1987)

Salyut 7, dismounted all devices of use and returned to Mir where these devices were again put to use. Despite ground control efforts, the technical condition of Salyut 7 (which had been shut down in the meantime) deteriorated in 1990 and the station could no longer be kept in orbit: in February 1991 it reentered Earth's atmosphere over Argentina. Salyut 7 had been home to 10 crews, including six long-duration crews (with stays up to 237 days) and cosmonauts from France and India. 13 Progress capsules had delivered more than 25 tons of equipment and propellants. Moreover, two logistics vehicles (Kosmos-1443 and Kosmos-1686) came to use in order to acquire knowledge for future stations with the experimental modules [Hooper 90, Harland 97].

In February 1986, the third generation of Russian space stations was introduced with the launch of the space station Mir. Except for two short breaks (from July 1986 to

February 1987 and from April to September 1989) this station has been permanently crewed. In December 1989 and July 1990, two large orbital stages were added. Mir is the result of consequently developing the Salyut family. Its core module weighed 21 tons when launched and is about 13 m long. The station flies at an inclination of 51.6° and at orbital altitudes of between 350 and 400 km. Mir's initial flight mode was Earth-oriented, but later on, its orientation has been adapted for berthing maneuvers or specific mission tasks.

Contrary to earlier stations, Mir has at its front a sixfold berthing adapter (Fig. 2.20), on which (besides the Soyuz capsules), four additional modules can be installed. The first true expansion, however, took place at the station's aft docking port, where in 1987 the "Kvant"-astrophysics laboratory was mounted. The Kvant module measures 5.8 m long, weighs 11 tons and has a second berthing adapter, so the number of docking possibilities for Progress freighters can remain unchanged.

A second expansion followed in December 1989 with the module "Kvant 2". It has the same dimensions as Mir's core module (length 13.73 m, diameter 4.35 m, weight 19.6 tons), but includes a larger airlock, accommodations for the Russian "Manned Maneuvering Unit" (MMU), a shower, an electrolytic O_2-system, more storage area for supplies, additional tanks, new scientific facilities including a stabilized platform and further gyro systems for the station's attitude control.

The Kristall module, again the same in size and dimensions, was added in July 1990 and carries e.g. facilities for experiments in material sciences, externally retractable solar panels as well as the so-called "Docking Module" (DM). This module was developed in 1975 for the "Apollo-Soyuz Test Project" (ASTP) and includes two special, identical docking ports (Androgynous Peripheral Docking System, APDS) designed to receive vehicles having a mass up to 100 tons. Originally, it was planned that the Russian space orbiter Buran would have docked at Mir using the APDS. The Kristall module was based on a logistics vehicle designed to carry cosmonauts to the Almaz military stations. Kristall and other modules (delivered later on) docking at Mir's radial ports initially docked at the front port. Afterwards, their position was changed: a manipulator mechanism carried by each module pivots the module into place at the proper radial port. In order to balance mass, a symmetric arrangement was provided. By 1990, the T-shaped Mir station (Fig. 2.20) measured 33 m long, 28 m high and had an overall weight of 70 tons. Its initial gravity gradient stabilized flight mode was gradually replaced by the inertial flight mode, which allowed the solar collectors to remain aligned towards the Sun.

Between 1990 and 1991, further alterations were carried out. For example, some of the solar wings were installed in different places in order to prepare the station for further expansion. Assembly was planned for completion with the launch of two additional large modules ("Priroda" and "Spektr", with optical devices and payloads for remote sensing), but it was not until 1995 and 1996 that these modules were launched.

In the medium term planning, the Soviet Union aimed at a slow but steady increase in Mir's use: its operation is to reach the end of the 1990's and it even could have been the destination of the first crewed Buran missions. In the long run, however, the idea of larger, permanently orbiting space stations occurred, especially with a view to the preparation for a Mars mission. At the end of the 1980's, plans emerged in the West, covering a Mir-2 station at 65° inclination. These plans were

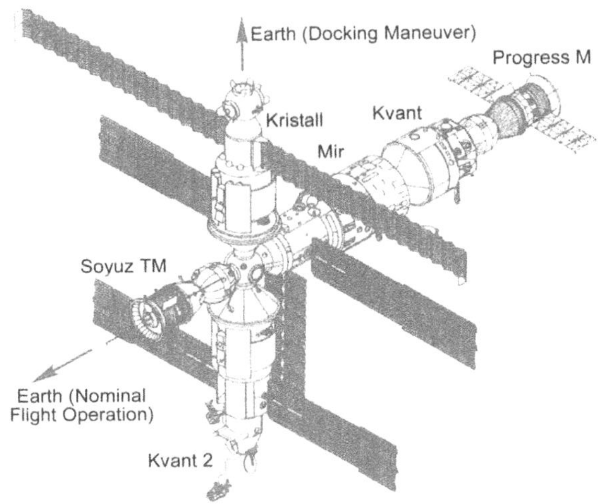

Fig. 2.20. Mir Configuration starting from July 1990

an obvious compromise: they combined increasing possibilities for Earth observation, thanks to increasing degrees of latitude on one hand, but increasing space radiation in these areas, injurious to humans and equipment alike, on the other.

In 1989, despite all continuity and technologically rather conservative actions, Soviet space flight also had to face financial problems. The Soviet Union's political and social restructuring had already given rise to public discussion on the aim and objective of crewed space flight. After the dissolution of the Soviet Union, this discussion became even more intense among those states of the Commonwealth of Independent States (CIS) who continued space flight. The discussion resulted in an agreement on the necessity of two points: international cooperation with other space flight centers and the re-orientation of Mir's operation towards more economic objectives.

2.4 The European Space Laboratory Spacelab and the US Module Spacehab

2.4.1 The European Spacelab Program

Significant advances in space flight technology have been achieved by the development and use of two special and very important modules: the European space laboratory Spacelab and the US Spacehab modules. Both covered one third of all Space Shuttle flights and both prepared the ground for further progress: e.g. the utilization concepts of the International Space Station and the realistic assessment of scientific and technical disciplines' potential yield. Some important parts and components of Spacelab and Spacehab payload systems were chosen to be put into operation aboard ISS with some modifications.

It was not until the beginning of the 1970's that, with Spacelab, Europe decided to take part in human space flight. Europe's aim was to develop a space laboratory for a crew of at least six (including the orbiter's pilot and commander) who could conduct scientific missions of up to nine days duration. In the beginning, NASA had clearly stated that important parts of the Space Shuttle were not to be developed in Europe. In the course of negotiations, the design of the space laboratory Sortie Can (which could be "inserted" into the Space Shuttle) by the Europeans was agreed upon, and eventually, it was concluded that a space laboratory (mounted in the Space Shuttle's cargo bay) was to be delivered by Europe: the Spacelab [Hoffmann 84, Spacelab 83, Shapland 84].

Europe, at that time represented by the European Space Research Organisation ESRO – predecessor to today's ESA, developed and built Spacelab, whereas the USA fully concentrated on the development and operation of the Space Shuttle as a transportation system. Contrary to Skylab, which had mostly been integrated from existing hardware, Spacelab was a new construction offering a much wider range of applications. In the course of its numerous missions, Spacelab turned out to be the most important large payload and boasts to have been the most often flown cargo element [Spacelab 93] (cf. Table 2.4).

The Spacelab System

The space laboratory Spacelab is a modular construction and its basis consists of two kinds of elements: modules and pallets.

Modules (cf. Table 2.2) are composed of either one segment (short module) or two segments (long module) and they are fixed in the Space Shuttle's cargo bay [Shapland 84, Spacelab 83]. This modular design allows the individual parts to be assembled in many different ways to form different flight configurations. As a consequence, maximum flexibility and possibilities of operation can be offered, and integration, as well as testing of racks and experiments, can be carried out separately from the Space Shuttle.

Table 2.2. Characteristics of Spacelab Elements

Characteristic Parameters:		
Diameter	[m]	4.06
Module Length (One Segment)	[m]	2.70
Module Volume (One/Two Segment)	[m^3]	8 / 22
Pallet Length (One Segment, Max. Five)	[m]	2.90
Maximum Mass	[kg]	11340
Payload Mass	[kg]	4500–9300
Design Life	[Years]	10
Operation Cycles	[–]	50
Diameter	[m]	4.06

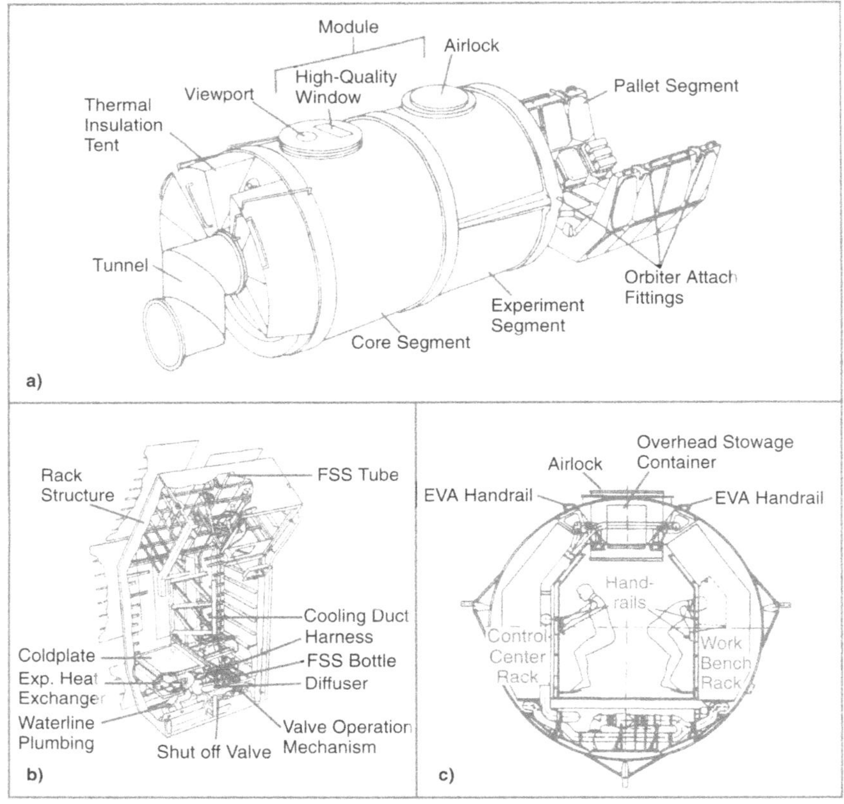

Fig. 2.21. a) Spacelab Including Long Module and a Pallet
b) Experiment Rack
c) Cross-Section of Work Compartment

Spacelab's short module consists of a core segment carrying the subsystems and experiment facilities needed by the crew. By adding the experiment segment, a long module is achieved and space for experiments is more than doubled (cf. Fig. 2.21). In order to economize on space, a limited number of experiments can be performed directly inside the Space Shuttle.

From 1974–84, parts for two long modules were developed and produced by a consortium of European enterprises under the leadership of ERNO (today DaimlerChrysler Aerospace in Bremen, northern Germany). The first module flew with the first Spacelab mission (STS-9, from November 28 to December 8, 1983 with six astronauts, among them was Ulf Merbold from Germany as the first non-US astronaut on a Space Shuttle flight). The second Spacelab flight module flew for the first time with the first German Spacelab mission D1 (STS-61A, from October 30 to November 11, 1985, aboard were the two German astronauts Rein-

Table 2.3. Available Resources for Different Configurations

Resources Available		Spacelab Configuration				
		Short Module 3 Pallets	Long Module	5 Pallets (3+2)	3 Pallets (1+1+1)	Long Module 1 Pallet
• Payload Mass (incl. 100% MDE*)	[kg]	5500	4500	8500	9300	4900
• Inside Volume	[m³]	7.6	22.2	—	—	22.2
• Mounting Area for Pallets	[m²]	51.0	—	85.0	51.0	17.0
Electrical Power						
• No MDE* (Max. Permanent Power)	[W]	3600	3900	5800	5800	3500
• All MDE* (Max. Permanent Power)	[W]	2600	2600	4900	5000	2100
• Peak	[W]	6500	6500	9200	9200	6100
Data Management						
• Down Link	[Mbps]	← up to 50 →				
• Digital Data Storage	[Mbps]	← up to 32 →				
• Telecommand	[Kbps]	← from 70 to 2 →				

*MDE: Mission Dependent Equipment

hard Furrer and Ernst Messerschmid, and from the Netherlands Wubbo Ockels). Each Spacelab module is designed to withstand about 50 flights.

The second group of Spacelab elements are the so-called "*pallets*". They are also laid out in a modular design, have a U-shaped cross section and can be composed of up to five segments. Due to its modular design, Spacelab is a multipurpose module. The three basic possibilities are module-only, module-and-pallet, and pallet-only configuration (row arrangement). A comparison of the available resources in the different configurations is given in Table 2.3.

The module's structure is made from a special aluminum alloy. One side is defined as "floor", so the crew has a reference point to distinguish the "floor" from the "ceiling".

Inside the laboratory as well as, for EVAs, on its outer skin and on the pallets, handholds and handrails are mounted, supporting the astronauts' work in microgravity. The racks (cf. Fig. 2.21) on both sides of the laboratory are designed in such a way that experiments and units of standard size (19 inches) fit into them. Racks exist in two forms: single racks (56 cm wide) and double racks (112 cm wide). Both structures are 76 cm deep and can carry payloads with a total mass of 290 or 580 kg, respectively. Additionally, they are equipped with cooling loops, water loops, power supply lines and data lines. A long module contains a total of twelve racks which serve various purposes: e.g., one double rack housing the Spacelab subsystem's electronics and another rack serving as a crew workbench.

The core segment has two windows made from special glass. Furthermore, a valve for evacuation of waste experimental gases is provided. An airlock can be in-

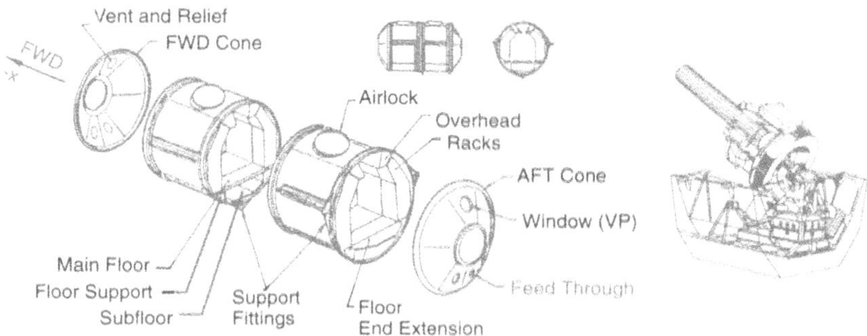

Fig. 2.22. Module Structure and Instrument Pointing System (IPS)

stalled on the upper side of the experiment segment to enable the exposure of experiments directly to space and retrieve them.

Pallets with their U-shaped cross section are able to carry up to 3000 kg of payloads, directly exposed to environmental space conditions. The experiments mounted on the pallets receive their power from the orbiter and are also connected by data links. In the course of the Shuttle program, the pallets gained importance: depending on the mission, they can either carry several small experiments or one large device, e.g. telescopes for astronomical tasks. Supported by the Instrument Pointing System (IPS), such telescopes can be pointed and maintained on target with a very high degree of accuracy (cf. Fig. 2.22).

The rather massive pallet was exchanged for the lighter Unique Support Structure (USS) during many of the Spacelab missions which had a long module (as was the case with the German D1 and D2 missions).

Crew access from the orbiter to the laboratory's pressurized module is provided via a tunnel, one meter in diameter. The tunnel offers the same ambient conditions as the Space Shuttle and the space laboratory. By adding a second tube, the tunnel's length, and at the same time Spacelab's position in the Space Shuttle's cargo bay, may be adapted to the existing mission requirements. This is important because during landing, the Space Shuttle's center of mass must be adjusted with utmost accuracy. The Table 2.4 offers an overview of all Spacelab missions.

Subsystems

Environmental Control Subsystem (ECS). The ECS may be divided into two different areas, one covering the different functions of the life support system (ECLS) and the other dealing with thermal control.

Temperatures in the laboratory can vary between 18–27°C, relative humidity is kept between 30%–70%. Cabin air velocity inside the pressurized module is kept within a range of 5–12 m/min. For more detailed parameters, cf. Table 2.5.

These environmental conditions are maintained by the atmosphere storage and control system as well as the so-called Atmosphere Revitalization System (ARS).

Table 2.4. Spacelab Missions

STS-	Mission	Date	AR	AP	SR	EO	PP	MS	LS	Others	Total
2	OSTA-1	Nov 81				6			1		7
3	OSS-1	Mar 82		1	2		2		1	3	9
7	OSTA-2	Jun 83						6			6
9	**SL-1**	Nov 83	4+(3)	3	(3)	2	5	36	16	1	70
12	OAST-1	Aug 84								3	3
13	OSTA-3	Oct 84				3					3
17	**SL-3**	Apr 85	1	2				5	2		10
19	SL-2	Jul 85	(1)	3	3+(1)		3	1	2		13
22	**SL-D1**	Oct 85						54	26	2	82
23	EASE	Nov 85								2	2
24	MSL-2	Jan 86						3			3
35	ASTRO-1	Dec 90		4							4
40	**SLS-1**	Jun 91							20		20
42	**IML-1**	Jan 92						8	28		36
45	ATLAS-1	Mar 92	6+(4)	1	(4)		3				14
50	**USML-1**	Jun 92						13	3		16
46	TSS-1	Jul 92					13			TSS-1	13
47	**SL-J**	Sep 92						23	18		41
52	USMP-1	Oct 92						3			3
56	ATLAS-2	Apr 93	3+(4)		(4)						7
55	**SL-D2**	Apr 93	1	1		3		38	41	4	88
58	**SLS-2**	Oct 93							13		>13
62	USMP-2	Mar 94						4			>4
59	SRL-1	Apr 94				2					2
65	**IML-2**	Jul 94						24	53	2	79
71	**SL-M**	Jul 95						X	X		>30
69	**USML-2**	Oct 95						X	X		>30
75	USMP-3	Feb 96					6	6		TSS-1R	>30
78	**LMS**	Jun 96						24	16	3	43
83	**MSL-1**	Apr 97	premature landing					19		14	33
94	**MSL-1R**	Jul 97	reflight of MSL-1					19		14	33
87	**USMP-4**	Nov 97						X	X		
90	**Neurolab**	Apr 98							26		>26

AR: Atmospheric Research; AP: Astrophysics; SR: Solar Research; EO: Earth Observation;
PP: Plasma Physics; MS: Materials Science; LS: Life Science;
Bold STS-number: Spacelab (SL) Mission including Long Module

Two fans operating in alternating mode are mounted under the main floor on a so-called "subfloor". The air, heated and polluted by crew activities and equipment, enters the subfloor through openings in the floor, with particle contaminants being retained by special filters. Metabolic CO_2 is adsorbed by LiOH-canisters that contain activated charcoal beds to remove the remaining contaminants (cf. Sect. 4.2.2 or Fig. 4.5, respectively).

A heat exchanger cools down the air below the dew point and the resulting condensate is separated. The atmosphere storage and control system consists of a high

Table 2.5. ECS Design Parameters

Average Radiation Temperature	[°C]	≤ 30
Maximum Contact Temperature	[°C]	≤ 45
Atmospheric Losses due to Leakage	[kg/Day]	1.35
CO_2 Control	[bar] [bar]	≤ 0.0067 Nominal ≤ 0.0101 Peak
Air Filtration	[m] [m]	$280 \cdot 10^{-6}$ Nominal Filter $300 \cdot 10^{-6}$ Absolute
Pressurization of Airlocks	[m^3]	0.87 per Day

pressure nitrogen tank (mounted on the outer skin of the module) and of corresponding controls for different pressure levels. The oxygen required for air revitalization, however, is fed through a line from the orbiter to the laboratory. Cabin pressure is automatically controlled within a range of 1.013 ± 0.013 bar, with the oxygen partial pressure varying between 0.220 ± 0.017 bar.

The heat resulting from the subsystem and experiment avionics has to be disposed of in order to keep the rack's temperature within the permissible range. This is performed by the so-called "avionics air loop" which can provide up to 12 racks with cooling air. The subsystem racks are directly cooled by the water loop. The water loop conducts the heat to the payload heat exchanger of the orbiter. Due to this loop, an overall performance of up to 5.8 kW may be permanently dissipated (cf. Fig. 4.5).

Another component of thermal control is the cooling loop which contains halogenated hydrocarbons; it serves as the cooling loop for the pallet-mounted experiments and subsystems. This additional loop is necessary, since water would freeze immediately when exposed to the extreme thermal conditions, in which all lines leading to the pallets are installed.

Large radiators at the inside of the orbiter's cargo bay doors emit the heat transported by all three cooling loops. Spacelab's permanent total heat dissipation capacity amounts to 8.5 kW. In order to protect the laboratory against the extreme thermal conditions of space, Spacelab is provided with a multilayer isolation coated with gold.

Electrical Power Supply. The Space Shuttle's three or four fuel cells cause oxygen and hydrogen to react to form electrical power (cf. Fig. 4.8) for Spacelab as well as for the Shuttle systems. During nominal operation, two fuel cells are needed for the orbiter and the remaining two provide power for the laboratory. Each of the fuel cells has a continuous performance of 7 kW at a voltage of 27 V DC. If necessary, their performance can be increased so that they are able to produce a 12 kW peak for 15 minutes every three hours. The 400 Hz converters aboard Spacelab convert part of the electrical power into AC with effective voltages of 115 V and 200 V.

Electrical power distribution for the various experiments (pallet or rack mounted) is effected by computer-based power regulation systems. The Experiment Power Distribution Boxes (EPDB) distributes the power by means of a cable network.

An additional 750 W unit supplies the laboratory's interior illumination, caution and warning signals and other components with low power requirements. In case of emergency, a 400 W supply is guaranteed to ensure operation of all vital subsystems and important experiments. During launch and landing of the Space Shuttle, the maximum power for Spacelab is limited to 1 kW, which is sufficient, since most of the devices are shut down during these phases of the mission.

Additional fuel cells (which can be installed in the Space Shuttles Columbia and Endeavour) increased the typical mission durations of Spacelab (with a long module) from one week initially, up to more than two weeks.

Data Management. Experiments performed during a mission, as well as the different monitoring and control systems aboard the laboratory, produce a huge amount of data which has to be transmitted either in real time or stored aboard until its evaluation once the mission is completed.

Real-time data transfer is effected at NASA by two different systems: the Space Tracking and Data Network (STDN), and the Tracking and Data Relay Satellite System (TDRSS). The STDN allows relatively low data transfer rates with a 192 kbit/s down-link and a 72 kbit/s up-link. With TDRSS, much higher data transfer rates can be achieved with up to a 50 million bit/s down-link. TDRSS comprises two geostationary satellites and a ground station in White Sands, New Mexico, USA. The ground station acts as a link between all NASA centers involved and Spacelab aboard the orbiter. As a result, the data link can be maintained for 85% of mission time. During that portion of the orbit where the radio link (e.g. visual contact) with at least one of the two satellites is interrupted, the obtained flight data is stored aboard the laboratory. The uplink data received from TDRSS or STDN is transferred to the Space Shuttle's communications system where the Command and Data Management System (CDMS) monitors and controls the data flow (cf. Fig. 2.23).

The Shuttle/Spacelab System. The US Space Shuttle was launched for the first time on April 12, 1981. Three further test flights followed, before, for the first time in the history of space flight, a reusable space transportation system was regularly available for crewed missions.

The first four space-qualified Space Shuttles (Columbia, Challenger, Discovery and Atlantis) were complemented by Endeavour after Challenger (STS-51L, on January 28, 1986) was lost. From November 1981 to April 1998, Spacelab or Spacelab elements were flown in 33 missions, 17 of which included the long pressurized Spacelab module covering 375 days of flight. NASA sources account for 764 distinct scientific investigations, resulting in more than 1000 publications in refereed journals and many thousands of Master and Ph.D. theses. Table 2.4 summarizes flown missions as well as the scientific disciplines involved. One of the Space Shuttles with a pressurized module, STS-71 with Atlantis, docked June 29, 1995 at the Mir station for five days and transferred crew members, consumables and experiment hardware. Table 2.10 (Sect. 2.6) presents the most important parameter of the Shuttle-Spacelab system.

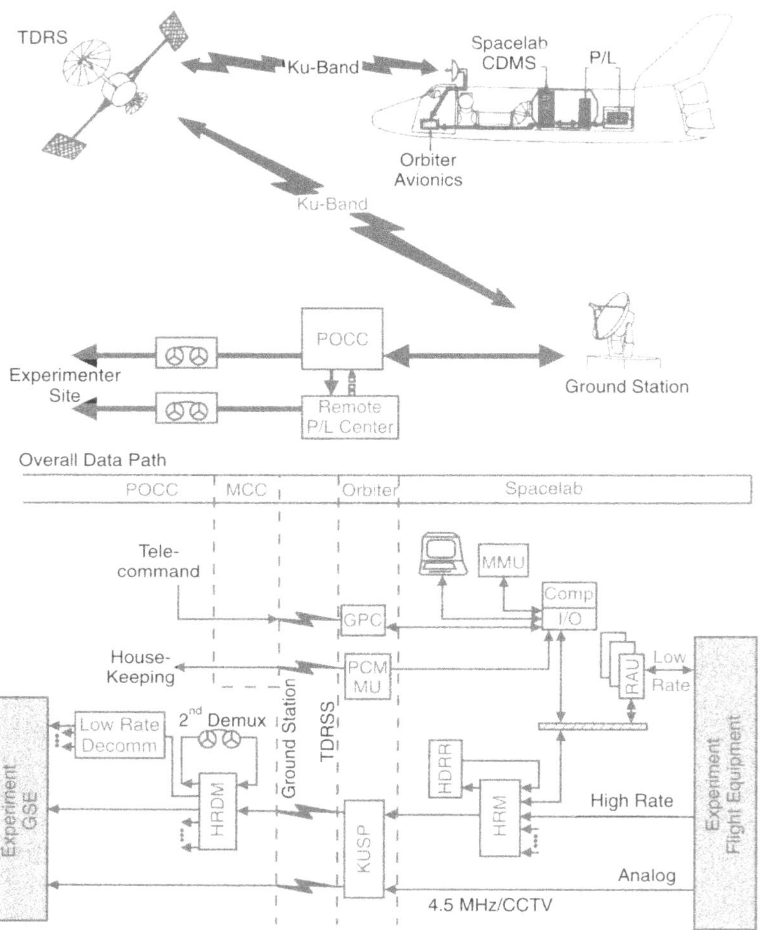

Fig. 2.23. Data Transmission and Telecommunication of the Space Shuttle – Spacelab System

2.4.2 The US Spacehab Module

In order to expand the storage and experimental possibilities of Space Shuttle's pressurized mid-deck into the cargo bay, the American enterprise Spacehab, Inc. developed the so-called Spacehab module in 1986. Since then, three Spacehab modules have been commercially produced. The Spacehab module can be connected to the rear part of the mid-deck by a modified Spacelab connecting tunnel. It has a total volume of 31 m^3 and 104 m^3 including mid-deck plus connecting tunnel. This remains by far under the volume of Spacelab (166 m^3, cf. Table 2.10).

In June 1993, Spacehab flew for the first time. Until 1998, it was launched eight times and it will continue to be used: Spacehab was to support Mir logistics and of-

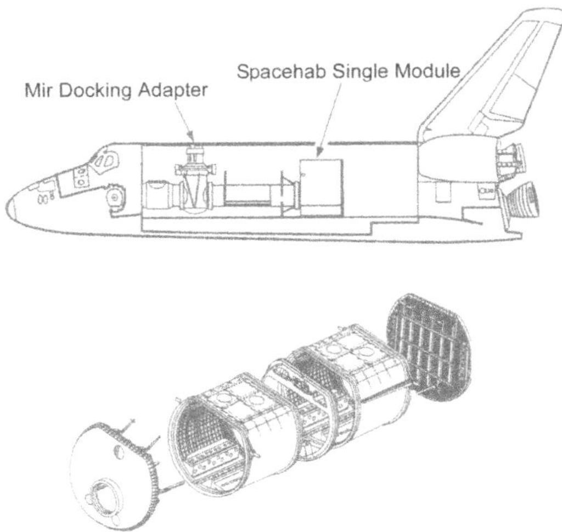

Fig. 2.24. Space Shuttle including Spacehab Module and Spacehab Double Module

fer a parallel possibility to conduct experiments (but limited, when compared to Spacelab), see Table 2.6, Sect. 2.5. Figure 2.24 shows the Spacehab module, firmly mounted inside the Space Shuttle during flight and the Spacehab double module which had its initial flight in September 1996.

Spacehab can carry mid-deck lockers, experiments, payload elements or racks, an optical window and externally mounted payloads, varying in size and number. Similar to Spacelab, it can be internally and externally supplied with electrical power, cooling and CDMS.

Concerning Space Shuttle logistics flights to Mir, NASA gradually replaced Spacelab missions by Spacehab missions, mainly due to shorter turn-around cycle times. For that reason, a fourth module was produced and connected to an already existing one so that they formed an correspondingly equipped Double Module. Five Spacehab flights were carried out to the Mir station between March 1996 (STS-76, with Single Module) and January 1998 (STS-79, 81, 84, and 89, with Double Module, see also Table 2.6). Another great advantage is the fact that Spacehab can still be loaded with containers even when the Space Shuttle is already standing on the launch platform. However, with regard to resources, volume and flexibility for experiments, Spacelab was superior to Spacehab.

2.5 From Mir to the International Space Station ISS (1994–2004)

After the Soviet Union broke up into the CIS and other states during the years from 1989–91, the USA and Russia agreed in 1992 on a treaty embedded in a more broad-based political agreement. This treaty comprised a wide range of points such as exploration for oil in Siberia, non-proliferation of Russian liquid rocket propulsion

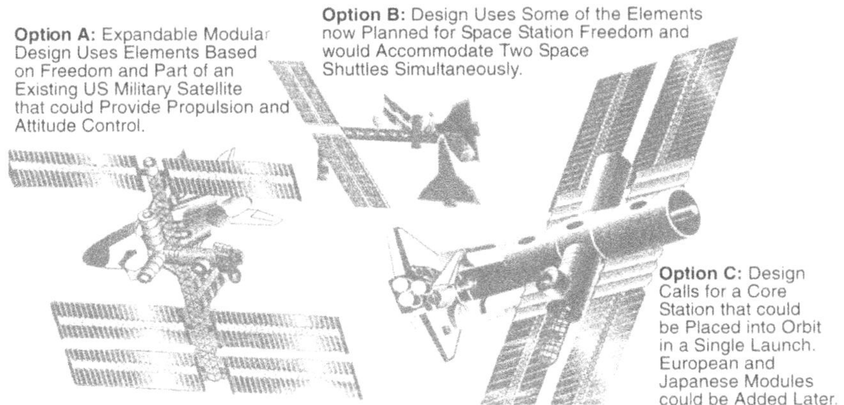

Fig. 2.25. Design Options for the International Space Station

systems to India, Iran etc. (Russian compliance with the Missile Technology Control Regime), access of Russian rockets to the commercial launcher market, cooperation in the field of nuclear technology, etc. Both states agreed to cooperate in the field of assembly and operation of space stations. As a result of this agreement, US President Bill Clinton demanded in 1993 once more that NASA modify existing space station concepts and reduce costs. Furthermore, Russia should be included among the partners for the Space Station Freedom. In order to generate more than one innovative proposal, three NASA centers were simultaneously commissioned to develop three different station concepts. Figure 2.25 shows all three of the options for a space station [Space News 93] which had become public in 1993 and which still resembled previous project studies. The White House opted for option A (A as in Alpha), thus the station's name: Space Station Alpha.

Only under political pressure did NASA allow more Russian modules and finally, after Russia had guaranteed to provide further hardware initially designed for the station Mir-2, plans for the International Space Station (ISS) as it is known today came into being. With this new form, 75% of the hardware technology from the Freedom program could be directly transferred to use. The orbit inclination was to be adapted: initially, it was at 28.5° corresponding to the Cape Kennedy launch site in Florida, USA, but it was changed to 51.6°, which is an inclination more favorable for Russia and identical to Mir's inclination. The core module and the first component to be launched is the Russian basic module FGB (= Functional Cargo Block, the acronym being derived from the Russian term). The FGB includes propulsion systems, attitude and orbit control, docking ports and will ensure further fundamental tasks for operation and research. All this is much more than any of the elements developed for SSF would have been able to perform. Program management was modified as well. Only one contractor was left (Boeing in cooperation with NASA-JSC in Houston, Texas) from all industrial contractors working with the three NASA centers involved, and program costs were reduced to $ 2.1 billion (US) per year. In 1998, the NASA Advisory Council's Cost Assessment and Validation Task Force, led by Mr. Jay Chabrow, "examined the nuts and bolts" of the project and concluded that the space station will now cost about 24 billion $, almost 7 billion $ over NASA's 1993 estimate. Completion of the space station, according to the task force's report, is likely to be delayed until 2005 or 2006 [Space Times 98].

The ISS program agreed upon by all space station partners is divided into three phases: *Phase 1 (1994–1998)*. Space Shuttle missions improve the utilization of Mir by transporting mixed US-Russian teams composed of astronauts, engineers and flight controllers as well as by providing and testing new modules and payloads. The acquired experiences and the common hardware will be put to use in *Phase 2 (1998–2000)* and *3 (2000–2004)*. Those phases start simultaneously with the beginning of the station's assembly and thus of the Early Utilization Phase. In 2004, i.e. 12 years after implementation of the US-Russian treaty, the station is scheduled to be completely assembled and shall be operated for at least ten additional years.

2.5.1 Phase 1 (1994–1998): Further Expansion and Operation of Mir

The Russian Space Station Mir was continuously assembled over a period of ten years, introduced by the launch of its Core Module in February 1986. At present, Mir's operation is supported by the Russian government, organized by the Russian Space Agency RKA and carried out by the RKK (Rocket Space Corporation).

By the end of 1996, 23 crewed Soyuz vehicles have docked at the station and 20 crews stayed for more than one month aboard Mir, among them was the current holder of the world record in mission time in space, Dr. Valeri Polyakov, who reached Mir in January 1994 aboard a Soyuz-TM 18 and left it with a Soyuz-TM 20 in March 1995. Including his earlier space flights, he spent a total of 438 days in space! Further scientific astronauts were sent by Afghanistan, Bulgaria, Germany (Klaus Flade and Reinhold Ewald), the ESA (Dr. Ulf Merbold on EuroMir 94 and Thomas Reiter on EuroMir 95 – with his stay of 179 days, the latter boasts the longest stay among non-US or non-Russian astronauts), France (four astronauts), Great Britain, Japan, Kazakhstan, Austria and Syria as well as the USA (six astronauts). 18 Progress capsules (36 tons), 34 Progress-M capsules (85 tons) and four Space Shuttles (16 tons) until end of 1996 have delivered a total of about 137 tons of equipment and consumables to Mir. Table 2.6 gives an overview of the missions to Mir from 1986 to 1999 [Gugerell 97]. A "Raduga" capsule transported by a Progress-M can return about 150 kg experiment samples to Earth and additionally, each crewed return flight was able to carry up to 20 kg of test material. Since the first Space Shuttle/Mir docking in 1995 (STS-71), it was possible to return not only astronauts but also sufficient amounts of equipment (station components such as solar panels and test material) to Earth (in total nine Space Shuttle dockings). It is surely true that, according to the definitions established in Chapter 1, Mir is the first space station to completely fulfill the criteria that make a real space station.

Assembly was completed with the launch of the modules Spektr, Priroda and the Russian "Docking Module" (DM, Androgynous Peripheral Docking System APDS) in 1995 and 1996, with the configuration dating from April 1996 onwards (cf.

Table 2.6. Russian and/or US Missions to Mir from 1986–1999

Launch Date	Flight/Orbiter	Stay aboard Mir [Days]	Crew	Remarks
20-Feb-86	Mir Core Module	Permanent		
13-Mar-86	Soyuz T15	125	L. Kizim V. Solovyow	
21-May-86	Soyuz TM-1	Test	uncrewed	1986: 2 Progress Flights
06-Feb-87	Soyuz TM-2	170	Y. Romanenko	
31-Mar-87	Kvant	Permanent	A. Laveikin	
21-Jul-87	Soyuz TM-3	158	A. Viktorenko A. Alexandrov M. Faris (SYR)	
21-Dec-87	Soyuz TM-4	176	V. Titov M. Manarov A. Levchenko	1987: 7 Progress Flights
07-Jun-88	Soyuz TM-5	88	A. Solovyev V. Saviniykh A. Alexandrov	
29-Aug-88	Soyuz TM-6	111	V. Lyakhov V. Polyakov A. Mohmand (AFG)	
26-Nov-88	Soyuz TM-7	149 25	A. Volkov S. Krikalev J. -L. Chretien (F)	1988: 6 Progress Flights
06-Sep-89	Soyuz TM-8	167	A. Viktorenko A. Serebrov	
26-Nov-89	Kvant 2 Module	Permanent		1989: 2 Progress + 2 Progress-M Flights
11-Feb-90	Soyuz TM-9	159	A. Solovyev A. Balandin	
31-May-90	Kristall Module	Permanent		
01-Aug-90	Soyuz TM-10	129	G. Manakov G. Strekalov	
02-Dec-90	Soyuz TM-11	172	V. Afanasyev M. Manarov T. Akiyama (J)	1990: 1 Progress + 3 Progress-M Flights
18-May-91	Soyuz TM-12	143	A. Artsebarsky S. Krikalev H. Sharman (GB)	
02-Oct-91	Soyuz TM-13	172	A. Volkov T. Aubakirov F. Viehböck (A)	1991: 5 Progress-M Flights
17-Mar-92	Soyuz TM-14	144	A. Viktorenko A. Kaleri K.-D. Flade (D)	Mir '92
27-Jul-92	Soyuz TM-15	187	A. Solovyev S. Avdeyev M. Tognini (F)	1992: 5 Progress-M Flights
24-Jan-93	Soyuz TM-16	177	G. Manakov A. Polishchuk	
01-Jul-93	Soyuz TM-17	195	V. Tsibiliyev A. Serebrov J.-P. Haignere (F)	1993: 5 Progress-M
08-Jan-94	Soyuz TM-18	180	V. Afanasyev Y. Usachov V. Polyakov	
01-Jul-94	Soyuz TM-19	124	Y. Malachenko T. Musabayev	

Table 2.6. Russian and/or US Missions to Mir from 1986–1999 (Ctd.)

Launch Date	Flight/Orbiter	Stay aboard Mir [Days]	Crew	Remarks
04-Oct-94	Soyuz TM-20	167 167 31	A. Viktorenko E. Kondakova U. Merbold (D)	EuroMir 94 1994: 5 Progress-M, 30 Experiments: 23 LS, 7 MS and TE
02-Feb-95	STS-63, Discovery		cf. Soyuz TM-20	Spacehab-SM, Rendezvous 10 m
15-Feb-95	Progress M-26			
14-Mar-95	Soyuz TM-21 (Mir-18)	115 115 115	V. Dezhurov G. Strekalov N. Thagard (USA)	Crew Return with STS-71 28 LS and MS
09-Apr-95	Progress M-27			
20-May-95	Spektr	Permanent	Uncrewed	incl. 335 kg Equipment
27-Jun-95	STS-71, Atlantis Mir-Docking #1	75	cf. Soyuz TM-21 N. Budarin A. Solovyev	Including Spacelab, 7 astronauts to Mir 8 returned to Earth, 200 kg Food + Cloths, 170 kg Water, 50 kg O_2
20-Jul-95	Progress M-28			04-Sep-95 Destructive Reentry
03-Sep-95	Soyuz TM-22 (Mir-18)	179	Y. Gidzenko S. Avdeyev T. Reiter (from ESA)	EuroMir 95: 41 Experiments: TE (10), LS (13), EO (5), MS (8), EA (5)
11-Sep-95	Soyuz TM-21 (Mir-19) Return Flight		1	Return of N. Budarin and A. Solovyev
08-Oct-95	Progress M-29	172		for RKA and ESA (80 kg)
12-Nov-95	STS-74, Atlantis Mir-Docking #2	5 of 8	5	Including the 4.7 m Long APDS† (4100 kg) and equipment (968 kg)
18-Dec-95	Progress M-30			for RKA and ESA (62 kg)
21-Feb-96	Soyuz TM-23 (Mir-21)	193	Y. Onufrienko Y. Usachev	
29-Feb-96	Soyuz TM-22 Return Flight		3	
22-Mar-96	STS-76, Atlantis Mir-Docking #3	5 of 188 9+1	6 to Mir incl. S. Lucid 5 to Earth	Including Spacehab-SM‡, 590 kg of Water (=19 Containers) and 862 kg of Equipment; LS, TE
23-Apr-96	Priroda	Permanent	Uncrewed	
05-May-96	Progress M-31			
31-Jul-96	Progress M-32			
17-Aug-96	Soyuz TM-24 (Mir-22)	197 197 16	V. Korzun A. Kaleri C. André-Deshays (F)	 LS, MS, TE
02-Sep-96	Soyuz TM-23 Return Flight		Y. Onufrienko, Y. Usachev, S. Lucid	
16-Sep-96	STS-79, Atlantis Mir-Docking #4	5 of 9+1 128	6 incl. J. Blaha	Spacehab-DM* and 4082 kg of Clothes, Computers, Gyrodynes; LS, MS
19-Nov-96	Progress M-33	115		
12-Jan-97	STS-81, Atlantis Mir-Docking #5	5 of 9+1 (Day 3–8) 132	5 incl. J. Linenger	Spacehab-DM* (3.356 tons of Equip- ment); Mir Payload of 2.268 tons of which 0.635 tons are H_2O; LS, MS
10-Feb-97	Soyuz TM-25 (Mir-23)	184 184 20	V. Tsibiliyev A. Lazutkin R. Ewald (D)	MIR97, German Aerospace Center DLR; Fire Breaks Out on Feb, 24, while Changing an Air Filter, LS, MS, TE
02-Mar-97	Soyuz TM-24 Return Flight		3	
06-Mar-97	Progress M-35 Fails to Dock, is Dumped			O_2 Generator Fails on Mar, 7

Table 2.6. Russian and/or US Missions to Mir from 1986–1999 (Ctd.)

Launch Date	Flight/Orbiter	Stay aboard Mir [Days]	Crew	Remarks
04-Apr-97	Leak in Cooling System on Apr, 4; Shutdown of CO_2 Removal System; Organic Coolant Leak Causes Health Problems for Crew on Apr, 11; O_2 and O_2 Generator-SM†† from Apr, 14 to 20			
6-Apr-97	Progress M-34			
15-May-97	STS-84, Atlantis Mir-Docking #6	5 of 9+1 143	6 incl.M. Foale	Spacehab-DM*, O_2 Generator
25-Jun-97	Progress M-34 Collides with Spektr Module and Solar Panel			
27-Jun-97	Progress M-35			Station's Computer is Disconnected after Batteries Ran Low
03-Jul-97	Shutdown of Stabilizing Gyroscopes			
05-Jul-97	Unidentified Substance Leaks from the Damaged Spektr Module			
17-Jul-97	Loss of Power after Crew Member Accidently Disconnects Computer Cable Sending Mir into Free Drift			
05-Aug-97	Soyuz TM-26 (Mir-24)	5	A. Solovyev P. Vinogradov	210 kg of Replacement Parts
14-Aug-97	Soyuz TM-25 Return Flight	184	2	Main Computer Fails on Aug, 18
25-Sep-97	STS-86, Atlantis Mir-Docking #7	6 of 9+1	6 incl. D. Wolf	aboard: J. L. Chretien (F); Flight Including Logistics Module and 3.5 tons of Water and Replacement Parts
01-Oct-97	Progress M-36			
06-Oct-97	Undocking of Progress M-35 not Possible			
20-Dec-97	Progress M-37			
22-Jan-98	STS-89, Endeavour Mir-Docking #8	5 of 9+1	6	Flight Including Logistics Module, Spacehab-DM
29-Jan-98	Soyuz TM-27 (Mir-25)		T. Musabayev N. Budarin L. Eyharts (F)	
19-Feb-98	Soyuz TM-26 Return Flight		3	
14-Mar-98	Progress M-40			
15-May-98	Progress M-38			
02-Jun-98	STS-91, Discovery Mir-Docking #9	5 of 9+1	5 to Mir 6 to Earth	Flight Including Logistics Module, AMS‡‡ and 6 tons of Equipment, Spacehab-DM
13-Aug-98	Soyuz TM-28 (Mir-26)		G. Padalka S. Avdeyev Y. Baturin	During 9 STS Docking Missions: 12.8 t Equipment + Supplies 6.8 t Water 4.1 t Docking Module 23.7 t Total 7.9 t Returned to Earth 43 American, 38 Russian and 17 Astronauts from other Countries Visited Mir until June 98
25-Aug-98	Soyuz TM-27 Return Flight		3	
15-Oct-98	Progress M-39			
20-Feb-99	Soyuz TM-29 (Mir-27)		V. Afanasyev J.-P. Heignere I. Bella	
02-Mar-99	Soyuz TM-28 Return Flight		2	
10-Mar-99	Progress TM-41			
02-Apr-99	Progress TM-50			
01-Jun-99	Soyuz TM-29 Return Flight		3	
08-Jun-99	Mir Deorbit (Later if Commercial Customers Provide Resources)			

†APDS: Androgynous Peripheral Docking System ‡SM: Single Module *DM: Double Module
‡‡AMS: Alpha Magnetic Spectrometer ††SM: Service Module
EA: External Activity EO: Earth Observation MS: Materials Science
LS: Life Science TE: Technology Experiments

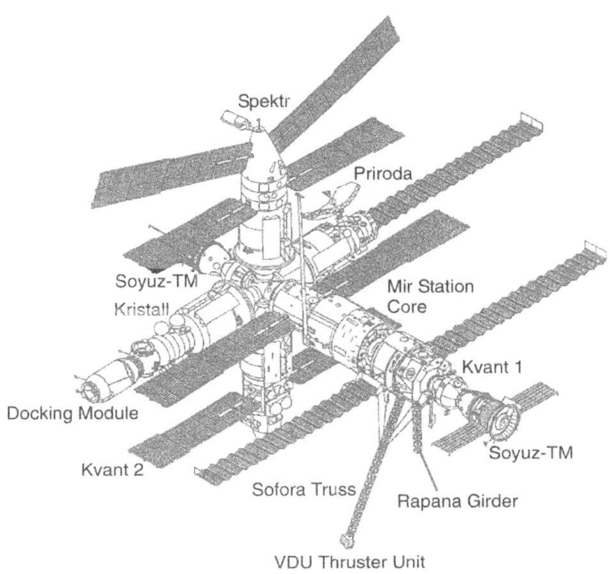

Fig. 2.26. Space Station Mir from late 1996 onwards

Fig. 2.26). With these new modules and the two transport vehicles Soyuz-TM and Progress-M, Mir reached a total mass of 124 tons and a pressurized volume of 398 m^2 (cf. Table 2.7). Fifteen solar panels with a total surface area of 254 m^2 produce electrical power of up to 35 kW.

All missions presented in Table 2.7 were planned and carried out within the framework of Phase 1 of the US-Russian missions to Mir. Mir offers a vast range of possibilities for scientific and technological experiments. At the beginning of its operation, emphasis was mainly on the disciplines of Earth observation and materials science. From 1992–93, the relative portions of the different disciplines were distributed as follows: 40% technology, 24% Earth observation and environmental monitoring, 15% biotechnology, 13% astrophysics, 8% life sciences. The quality of the station and its facilities does not reach Spacelab standards, but due to the long operation periods, important lessons have been learned, which certainly have had a formative influence on the development of the International Space Station. In the course of the US-Russian cooperation of Mir Support and utilization, the six American astronauts had spent 975 days aboard Mir, returned 19000 images of Earth and have carried out many experiments, mainly in the fields of life sciences and station operation.

Chapter 7 presents also important results from research aboard Mir. A reference to a summary of scientific research and facilities can be found in the bibliography [NRC 95]. Since in the past much space has been given to the Research and Development (R&D) of new space flight technologies, some useful examples ought to be mentioned. R&D for space flight technologies does not only include flight systems and components, but also new facilities and utilization possibilities. Many years of

Table 2.7. Space Station Mir after Further Expansion (late 1996) [Harland 97, WWW-OSF]

Module	Mass [tons]	Length [m]	Max. Diameter [m]	Pressurized Volume [m^3]	Solar Arrays/Area [m^2]	Electrical Power [kW]	Function/ Utilization
Mir Core Module	20.9	13.13	4.15	90	3/76	10.1	Habitation, Thermal and Attitude Control, ECLSS, Docking Port, Power Supply
Kvant 1	11.05	5.8	4.15	40	none	—	Astrophysics Equipment, Attitude Control, Some ECLS, Second Docking Port
Kvant 2	18.5	12.4	4.35	61.3	2/53	6.9	EO Equipment, ECLSS, Air Lock, Solar Arrays
Kristall	19.64	11.9	4.35	60.8	2/70	5.5–8.4	Facilities for Materials Science and EO, Docking Node, Solar Arrays
Spektr	19.64	12.0	4.35	61.9	4/35	6.9	Geophysics, EO, US-PL
Priroda	19.7	12.0	4.35	66	none	—	EO and Geophysics
Soyuz-TM	7.1	7.0	2.70	10.3	2/10	1.3	Transport: Max. 3 Persons
Progress-M	7.2	7.0	2.70	7.6	2/10	1.3	Uncrewed Logistics Capsule
Total:	**123.7**			**397.9**	**15/254**	**< 34.9**	

EO: Earth Observation; US-PL: US Payload

developing and testing subsystems continuously improved the space station Mir. This includes the following facilities:

- Elektron and Vika system for electrolytic decomposition of water
- Gyrodyn gyro system for attitude control
- APDS docking system
- Luch satellite data relay system
- Igla and Kurs-rendezvous systems
- Argon 16B, Salyut 5B and EVM computer systems
- Burs, Korona and Tranzit-A communication and data transfer systems
- ASPG-M movable instrument platform
- Orlan-DMA EVA suit
- Ljappa module orientation systems
- Rodnik water supply unit
- YMK astronaut maneuvering system
- VDU propulsion system for roll control
- Strela external manipulator
- Vozdukh unit for removal of carbon dioxide

2.5.2 Phase 2 (1998–2000): Start of Assembly of the International Space Station

Phase 2 of US-Russian cooperation is introduced with the assembly of ISS. With a total of three Russian rockets (2 Proton, 1 Soyuz) and six Space Shuttle orbiters launched from Baikonur (Kazakhstan) and Cape Kennedy (Florida) respectively, the individual modules, important subsystems and, of course, the astronauts will be placed into the ISS orbit (cf. approximate schedule in Table 2.8). The first module was launched by a Proton launcher (Flight 1A/R, November 20, 1998), the Russian autonomous orbital vehicle FGB, christened "Zarya" (mass: 20 tons). It includes attitude and orbit control systems, solar generators and docking ports for additional modules. Basically, it is similar to the Mir modules Kvant 2 and Kristall. Flight 2A (STS-88) was the first Space Shuttle flight to carry a space station module: the Resource Node 1 dubbed "Unity", which stores supply goods, equipment, docking ports for modules, and the ISS truss as well as a docking port for the Space Shuttle. Resource Node 1 will connect the US modules launched later on with the FGB and thus the rest of the Russian modules.

With the second Russian Assembly flight (designated 1R), the Service Module with habitat and work compartments for three crew members will be docked at the aft docking port of the FGB module. The Service Module is based on the Mir Core Module and is therefore flight proven hardware. The automatic Russian Progress-M freighters will periodically dock at the aft docking port of the Service Module to provide food, water, propellants and equipment.

With the third US flight designated Flight 3A (Shuttle mission STS-92), the Endeavour crew will mount the first segment of the station's truss (Z1) onto Node 1. From Flight 2R onwards, i.e. flight six of Phase 2, the International Space Station will be permanently crewed with three astronauts. From now on, a crew will be able to live, work and conduct experiments aboard the station for long periods of time. Whenever the Space Shuttle is absent, a Soyuz capsule (based on the Soyuz-TM capsules from the Mir program) is docked at the station to ensure that the crew can safely return to Earth in case of emergency. The next US flight (Flight 4A) will integrate segment P6 of the truss (it carries a solar generator) into segment Z1. By doing so, for the first time, supply with US solar power aboard ISS will be ensured.

A very important milestone of Phase 2 is the launch of the US Laboratory Module with the Shuttle STS-98 (Flight 5A). This module is outfitted with a total of 5 Experiment Racks, an ECLSS and maintenance and control systems. The Atlantis crew will deliver further equipment with STS-99 transported in the Italian-built Multi-Purpose Logistics Module (MPLM). Last milestone of this phase of assembly is the following Shuttle mission STS-100. The Endeavour crew will mount an airlock onto Resource Node 1 and thus make "routine" EVAs for US and Russian astronauts possible, even when the Shuttle is absent. Additional EVAs will be possible from January 2000 onwards, when the station will be permanently inhabited by a crew of three. Generally, the hatches in the Russian Service Module are used for such EVAs. Thanks to the new airlock, EVAs which are necessary for assembly of ISS will be considerably facilitated. Table 2.8 offers an overview presenting the most important assembly flights of Phases 2 and 3.

Table 2.8. ISS Assembly Sequence (Revision D)

Launch Date	Flight	Delivered Elements	Launch Date	Flight	Delivered Elements
20 Nov 98	1A/R	**FGB (Zarya)**, Launched on Proton	Jun 01	12A.1	P5, MPLM, Radiator OSE
04 Dec 98	2A	**Node 1 (Unity):** 1 SR, PMA1, PMA2, 2 APFRs, on STS-88	Jun 01	13A	S3/4 Truss, PV Array, 4 PAS
Jul 99	1R	**Service Module,** Launched on Proton	Sep 01	10A	**Node 2**, Nitrogen Tank Assembly
May 99	2A.1	Spacehab Double Cargo Module, OTD, on STS-96	Oct 01	1J/A	JEM ELM PS (4 Sys., 3 ISPRs, 1 Stow.), 2 SPP SA w/Truss
Jun 99	3A	Z1 Truss, CMGs, Ku/S-Band Eqpt., PMA3, EVAS, on STS-92	Jan 02	1J	JEM PM (4 JEM Sys. Racks), JEM-RMS
Jul 99	2R	Soyuz, **3 Person Permanent Presence Capability**	Feb 02	9R	Docking & Stowage Module
Aug 99	4A	P6, PV Array / EEATCS Radiators, on STS-97	Feb 02	UF3	**3rd Util. Flight,** ISPRs, 1 SR, 1 RSP, 1 Express Pallet w/PL
Oct 99	5A	**US Lab (Destiny):** (5 Lab Syst. Racks), PDGF, on STS-98	May 02	UF4	**4th Util. Flight,** Truss Attach Site P/L, Express Pallet w/ P/L, ATA, SPDM
Dec 99	6A	US Lab Outfitting (6 Syst., 2 RSRs), 4 RSPs, UHF, SSRMS, on STS-99	Jun 02	2J/A	JEM EF, ELM-ES w/ P/L, 4 PV Battery Sets
Jan 00	7A	Airlock, High Pressure Gas ($2O_2$, $2N_2$), on STS-100	Aug 02	14A	2 SPP SAs, SM MMOD Shields, Cupola, Port Rails
→ **Phase 2 Complete**			Aug 02	8R	**Russian Research Module RM1**
Mar 00	4R	Docking Compartment 1	Sep 02	UF5	**5th Util. Flight,** ISPRs, 1 Stow. Rack, 1 RSP, Express Pallet w/ P/L
Mar 00	7A.1	2 RSRs, 6 RSPs, ISPRs (on MPLM), OTD, APFR	Oct 02	20A	**Node 3** (2 Avionics, 2 ECLSS Racks)
Apr 00	UF1	**1st Util. Flight,** ISPRs, 2 RSR, 2 RSP-2), ORUs, Spares Pallet	Nov 02	10R	Russian Research Mod. RM2
Jun 00	8A	S0 Truss, MT, GPS	Nov 02	17A	5 Syst. Racks, 3 CHeCS Racks, 1 Stow. Rack, ISPRs
Aug 00	UF2	**2nd Util. Flight,** ISPRs, 1 JEM Rack, 3 RSR, 1 RSP-2, MELFI, MBS, PDGF	Feb 03	1E	**COF** (5 ISPRs)
Oct 00	9A	S1 Truss (3 Radiators), TCS, CETA, S-Band	Mar 03	18A	CRV #1, CRV Adapter
Jan 01	9A.1	Science Power Platform w/ 4 SAs and ERA	Jun 03	19A	5 Stow. Racks, 1 RSR, ISPRs, 4 Crew Quarters
Feb 01	11A	P1 Truss (3 Radiators), TCS, CETA, UHF	Jul 03	15A	S6, PV Array (4 Battery Sets)
Apr 01	3R	Universal Docking Module, on Proton	Sep 03	UF6	**6th Util. Flight,** 3 RSRs, 1 RSP, ISPRs, 2 PV Battery Sets
May 01	12A	P3/4, PV Array, 4 PAS	Nov 03	UF7	**7th Util. Flight,** Centrifuge Accommodation Module (CAM), ISPRs
May 01	5R	Docking Compartment 2, on Soyuz	Jan 04	16A	**Hab** (3 Syst. Racks, 2 RSRs, ISPRs)

for Acronyms see Glossary

2.5.3 Phase 3 (2000–2004): Operation of the International Space Station and Further Expansion

Phase 3 is comprised of 18 US, six Russian (two Proton, four Soyuz), and two Japanese, one European, and two US-Japanese assembly flights, all with the Space Shuttle. The third US flight of this phase, however, is not an assembly flight: Shuttle mission STS-104 will be the first of seven Space Station Utilization Flights (UF) in Phase 3. During this period, the Shuttle crew will work for more than two weeks in the US-Lab. The Space Shuttle docked at the station will guarantee safe return to Earth, if necessary. These Utilization Flights shall ensure earliest possible Utilization of the International Space Station.

Phase 3 will see the installation of further international laboratory modules. The first Russian research modules (resembling those of Mir) will be delivered to the station from August 2002 onwards. The Japanese laboratory module JEM and the European laboratory module COF will be delivered during five assembly flights from 2001–2003. Robotic elements within the COF being essentially integrated into the multi-user facilities will support the astronauts in their experiment operations. The JEM carries an additional experiment platform, the so-called "Exposed Facility" which allows direct exposure of experiments and equipment to the space environment. Europe intends to launch its module with the Space Shuttle, whereas Japan wants to make use of its H-II rocket to transport parts of the JEM to the station.

In November 2003, the USA plans to launch the Centrifuge Module with Shuttle mission STS-130. For the first time, this centrifuge will allow investigation of the influences which permanent partial gravity has on life forms, for example to simulate human life on Mars, whose gravitation force is only one third of the Earth's.

The largest single element of ISS, its truss, will continuously be expanded during Phase 3. The 10th and last segment will be mounted during the 15th US assembly flight. The truss with a total length of about 108 m will carry numerous subsystems, which require a direct space environment, e.g. communication antennas, external cameras, external payload accommodations, systems for thermal and attitude control as well as the manipulator arm.

The truss will also include four Sun-tracking pairs of solar double panels. Together with the solar generators of the Russian segments, a total of 110 kW of electrical power for the entire station can be provided. This is nearly double the amount which was planned for the Freedom concept and ten times as much as was available aboard Skylab and Mir.

From 2003 onwards, a second Soyuz capsule will be docked at ISS thus ensuring possibilities for six persons to return, even when no Space Shuttle is docked at the station. This introduces the possibility of having a permanent ISS crew of six astronauts. In the course of 2004, the US Habitation Module will eventually be delivered. It is designed to house four persons, providing a safe operating and living environment which supports eating, sleeping, personal hygiene, relaxation and also offers possibilities, e.g. for conferences and meetings. The arrival of the US Habitation Module with the 20th US Assembly flight also marks the end of Phase 3. In early 2004, the International Space Station will be completely assembled constituting a unique capacity for research in space in the next century (see Fig. 2.27).

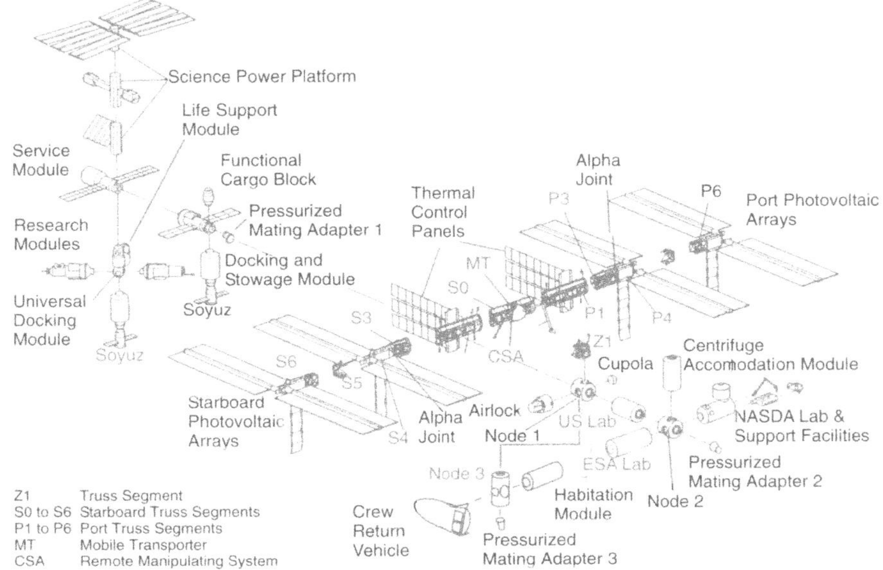

Fig. 2.27. Expanded View of the Completely Assembled International Space Station ISS

2.5.4 General Description of the International Space Station

The International Space Station will have various functions, such as

- multidisciplinary scientific research facility for fundamental and applied research in Low Earth Orbit (LEO),
- test bed for new technologies in space environment,
- platform for remote sensing of targets on Earth and in the sky, and
- stepping stone to further research and exploitation of space.

In order to be prepared for such ambitious long-term objectives, an international crew will permanently live and work in several pressurized modules. Apart from the pressurized modules for experiments and habitation, the space station includes accommodations for external scientific and technological payloads which will be directly exposed to space environment. The mechanical connection of the segments is assured by the truss, nodes and adapters. Electrical power and heat rejection are provided by large solar panels and radiators. All units together ensure the safe flight around the Earth and the successful operation of the station over a period of at least 10 years.

The overall configuration of ISS after its assembly is presented in Fig. 2.28. The truss carries four large double-winged solar generators arranged in pairs. Between them are several booms with radiators (rejecting the heat produced as by-product in the station) and the main group of pressurized modules.

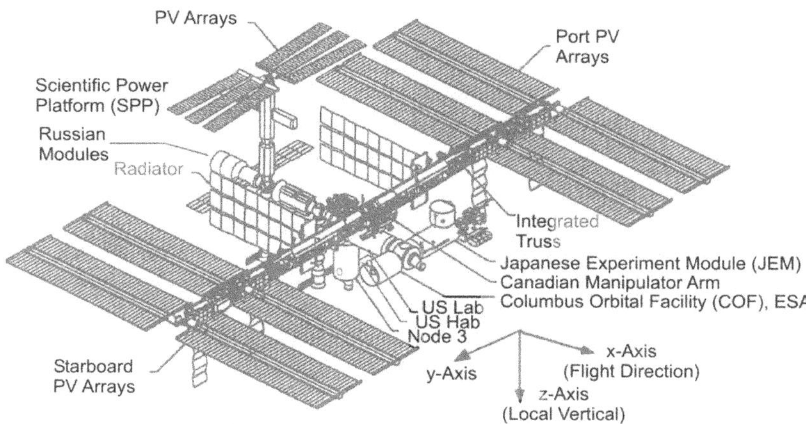

Fig. 2.28. View of the International Space Station

In the center of the Russian modules, there is a second smaller truss: the Science Power Platform (SPP). It carries further solar panels and all units necessary for the station's attitude control. The Russian elements of the station are often referred to as the "Russian Orbital Segment" (ROS) whereas the US elements are sometimes called "United States On-Orbit Segment" (USOS).

The most important data and characteristics of the International Space Station are summarized in Table 2.9 and Table 2.10.

The Pressurized Modules of ISS. The International Space Station includes the following six pressurized modules for scientific and technological research:

- The US Laboratory Module "US Lab"
- The Japanese laboratory module "JEM" (Japanese Experiment Module)
- The European laboratory module "COF" (Columbus Orbital Facility)
- Two Russian Research Modules (designated RM1, RM2)

The US Laboratory Module is situated directly under the truss in transverse direction. The European and Japanese laboratory modules run parallel to the truss and are situated in front of the US Lab. The Russian research modules are situated one level below the US Lab.

In addition to the research modules, the station also comprises habitation modules as well as modules for storage of material and subsystems.

The Station's External Robots. A large robotic arm called "Space Station Remote Manipulator System" (SSRMS), contributed by Canada, is able to move along the truss between those points where the solar panels intersect with the truss. Initially, it will support the station's assembly. Later on, it will not only assume crucial tasks of external maintenance, but also carry out transport and supply of externally mounted experimental facilities and equipment.

Table 2.9. Main Parameters of the International Space Station

Dimensions	108.4 m x 74.1 m
Mass	415 000 kg
Electrical Power	110 kW of which 47–50 kW are for Payloads
Total Pressurized Volume	1140 m^3
External Payload Area	> 50 m^2
Pressurized Accommodations	6 Laboratories (1 US, 3 Russian, 1 Japanese, 1 European) plus US Centrifuge and Habitation Modules
External Payload Accommodations	4 USA, 1 Japan, Various from Russia; Total: > 50 m^2
Crew Availability	3 (1998-2002), 6 (from 2002 Onwards)
Nominal Orbit	Circular at Varying Altitudes (335-460 km)
Flight Attitude (Ideal)	One Axis in Velocity Direction, one in Nadir Direction, one Perpendicular to Orbit Plane
Attitude Deviations	5.0 deg/Axis 3.5 deg/Axis/Orbit
Microgravity Requirements	from 10^{-6} g to 0.1 Hz (10^{-5} x f) g at $0.1 \leq f \leq 100$ Hz
Undisturbed Microgravity	30 Days
Data Transmission	Via TDRSS Downlink 43 Mbps, Uplink 72 kbps, Option: Commercial Satellite Systems
Thermal Control	Water Loop Cooling in the Modules
Venting and Vacuum for Experiments	< 0.13 Pa Available to Pressurized Modules
Mission Duration	10 Years of Permanent Operation after 3–4 Years Limited Operation During Assembly Phase
Late Access/Early Retrieval of Experiments	Including Power Supply During Transport in MPLM*

* for Acronyms see Glossary

A second robot, the "European Robotic Arm" (ERA) which Europe will deliver to Russia, is to be installed on the Russian Science Power Platform (SPP). It will perform all tasks in the field of assembly, maintenance and supply related to the Russian segments of the International Space Station. Just like the Canadian robotic arm, ERA can move on a mobile platform. It is even able to leave the platform and move to another by using the symmetrical grips at its ends either as a "hand" or as a "foot".

A third robot is situated on the Japanese Experiment Module and will be used for maintenance and supply of experiments placed on its platform.

Orbital Parameters of the International Space Station. ISS will fly between at least 335 km (during assembly) and 460 km (during routine operation) above the Earth's surface. This corresponds to a velocity of about 29000 km/h and an orbital period of about 90 minutes.

The inclination of the station's orbit is such that the station will cross the Earth's equator at an angle of 51.6°. The station's groundtrack, i.e. the projection of its orbital trajectory onto the Earth's surface, is a sinusoidal track symmetrical with the Equator with its extreme points located at 51.6° latitude North and South, respectively. As the Earth rotates beneath the station at an angular speed of 360° per day (i.e. 22.5° in 90 minutes), the groundtrack shifts 22.5° to the West during one flight

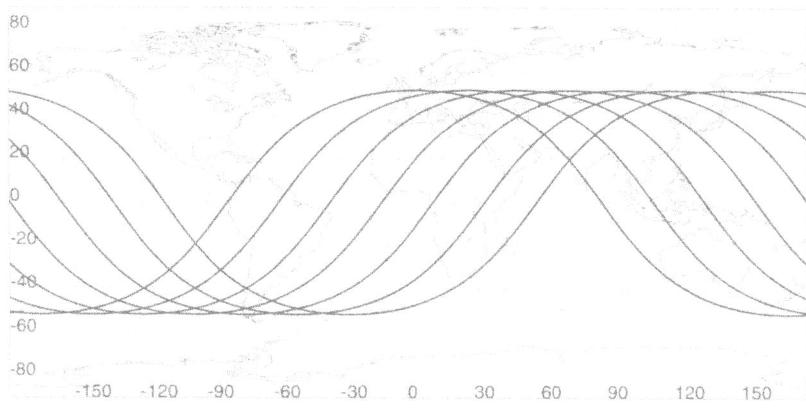

Fig. 2.29. Typical Groundtrack of ISS over 12 Hours

around the Earth. Figure 2.29 illustrates a typical groundtrack of ISS during eight successive orbits. The orbital parameters of ISS allow observation of 85% of the Earth's surface where 95% of the Earth's population lives.

Orbital Altitude. The altitude of ISS is determined by two factors: logistics and safety. On one hand, it will be kept as low as possible to maximize the payload capacities of the transport vehicles which will travel to the station to deliver supplies. On the other hand, it has to be as high as necessary in order to avoid the residual atmospheric drag of the Earth's atmosphere causing the re-entry of the station (cf. Chapter 6, Fig. 6.11). The Russian Soyuz-TM capsules carrying maximum payload, for example, can only reach an altitude of 425 km.

Flight Attitude. The station's flight attitude will be as follows: The European and the Japanese modules are at the front and pointing in the direction in which the station flies. Both modules as well as the central truss, are oriented perpendicular to the flight direction, similar to an airplane's wings.

The plane formed by the truss, the front node, the European and Japanese modules, the US Lab and both Russian elements remains, more or less, is oriented parallel to the Earth's surface. The truss of the Russian SPP is oriented in zenith/nadir direction. Such a flight attitude is called "Local Vertical Local Horizontal" or simply "LVLH".

The joint forces of atmospheric drag and angular acceleration by gravity gradient cause a deviation from the ideal LVLH attitude. This deviation can be limited to 5° per axis by the station's attitude control system. The inertia of the overall configuration ensures that velocities of attitude changes for each of the axes remain under 0.02°/s.

With regard to keeping the level of microgravity aboard the station as stable as possible, the LVLH attitude is preferred since it minimizes the perturbations caused by the gravity gradient in one axis.

For Earth observation, the LVLH attitude is also advantageous. However, for most of the station's applications, deviations of a few degrees from the nominal LVLH attitude are not acceptable, so suitable devices for instrument pointing on locally mounted platforms have to ensure a motion compensation.

For observation sciences, the LVLH attitude, i.e. looking up or down, offers the opportunity to constantly scan the sky and the Earth surface, whereas constantly observing certain points, e.g. the Sun or other stars, requires instrument pointing. Provided this is available, observation durations of up to 30 minutes per orbit can be achieved.

2.6 Space Station Comparison

Table 2.10 offers a comparison of the most important key parameters of the space stations described in the previous sections (for a comparison of scale see Fig. 2.30). The table holds values of ISS after assembly in 2004; for Mir with its configuration from April 1996 onwards; for the Space Shuttle including either the Spacehab or the Spacelab modules; for the Space Station Freedom (SSF) design; and also for Skylab. With the help of these parameters, the capabilities of the different orbiting systems can be compared [NCR 95].

When drawing up Table 2.10, special attention was paid to using objective factors such as quantifiable parameters, or others. Thus, certain parameters were excluded from the comparison, such as

- factors varying over a certain period of time (e.g. level of CO_2 or humidity in the cabin atmosphere),
- misleading factors like lifetime or
- factors not valid for each system (e.g. quality of microgravity).

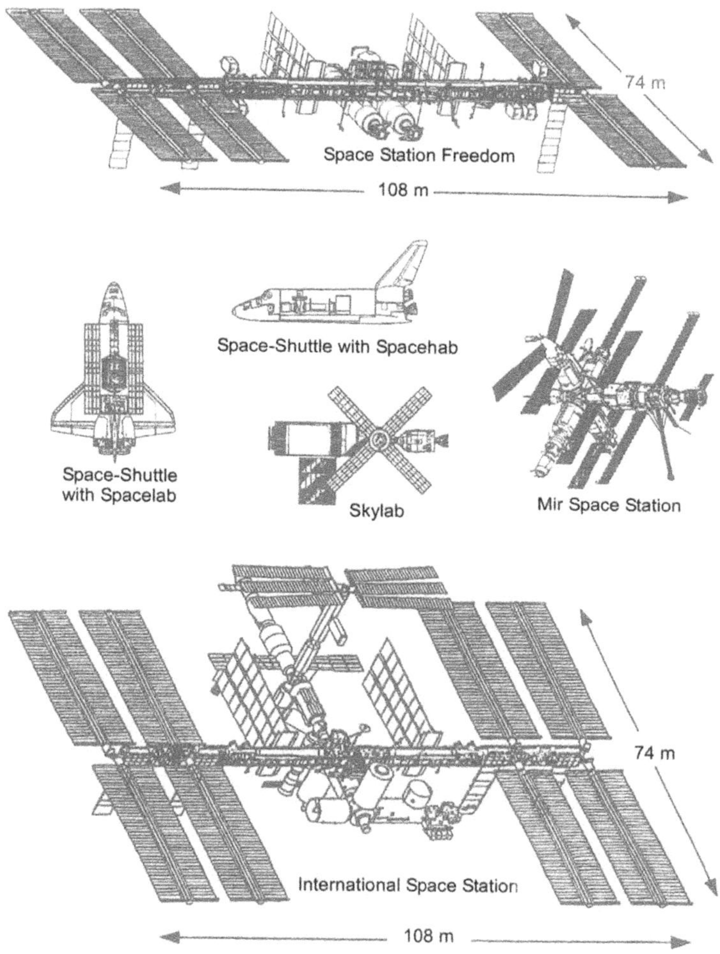

Fig. 2.30. Comparison of Space Stations to Relative Scale [NRC 95]

Table 2.10. Space Station Parameters

Parameter	Mir (1996, with All Modules)	Skylab (1994)	Space-Shuttle with Spacelab Module	Space-Shuttle with Single (Double) Spacehab Modules	Space Station Freedom Assembly 1993 CDR	International Space Station Assembly 1995 IDR
Countries Involved	Russia, USA, ESA Member States	USA	USA, ESA Member States, Japan	USA, Italy, Japan	USA, ESA Member States, Japan, Canada	USA, ESA Member States, Japan, Canada
Availability Date	At least through late 1997	Not Applicable, Deorbited in 1979	—	—	1999 (to be Completed in 2000)	from 1998 (to be Completed in 2002) at least through 2012
Total Pressurized Volume $[m^3]$	410	354	166	104	680	1140
Total Modules	9	2	2	2	8	17
Total Mass $[kg]$	140000	90000	13700	5000	281000	415000
Density $[kg/m^3]$	341.5	254.24	82.53	48.1	413.24	374.1
Length x Width $[m]$	33 x 41	36.1 x 28	6.9 x 4.1	2.8 x 4.1	108 x 74	108 x 74
Associated Launch Vehicle	Soyuz, Proton, Space Shuttle	Saturn V, Saturn IB	Space Shuttle	Space Shuttle	Space Shuttle	Space Shuttle, Soyuz, Proton, Zenit, Ariane 5
Projected Number of Launches to Assemble	6	1	Not Applicable	Not Applicable	27	44
Permanent Crew Capability	Yes	No	No	No	Yes	Yes
Typical Crew Size [Persons]	3	3	up to 7	up to 7	4	6
Crew Duration	4-6 Months typically (Max. 14 Months)	28, 59 and 84 Days	up to 15–20 Days	up to 15–20 Days	3 Months (Standard)	3 Months (Standard)
Total Power $[kW]$	< 35	24/18	7.7	3.15	75	110
User Power $[kW]$	4.5	4/3	3.5 to 7.7	3.15	30	~50
Solar Array Area $[m^2]$	430	216/162	0	0	~1800	~3000
Data Transmission Rate (Down Link) [Mbps]	7	< 1	45	16	50	50
Water Recycling	Yes	No	No	No	Yes	Yes

3 Orbital Environment

Space stations, platforms and vehicles are exposed to various environmental influences. These influences can both directly and indirectly have an impact on the orbit and attitude of the space vehicle, on the properties and condition of the materials used, on the onboard crew, or on experiments and their operation. Table 3.1 shows the existing environmental influences and their possible effects, especially for Low Earth Orbits (LEOs).

Table 3.1. Environmental Influences

Environmental Influences: Types and Origins		Effects on:		
		Flight Altitude and Attitude	**Materials**	**Experiments; Operation; Crew**
Gravitational Field	• Influences by other Planets			○
	• Deviations from the Central-Mass Field and from Inhomogeneities	●		●
	• Further Anomalies			○
Magnetic Field	• Dipole Field	○ (1)		●
	• Deviations from Symmetrical Dipole	○		
	• Influence of Solar Radiation		●	●
Radioactive Radiation	• Particles of Solar Origin		●	● (4)
	• Particles of Cosmic Origin		●	
	• Electromagnetic Radiation			● (4)
Electromagnetic Radiation	• Solar Radiation	●	●	●
	• Solar Radiation Pressure	●		
	• Albedo Radiation			●
	• Thermal Radiation		●	●
	• Radiofrequency Noise			●
Atmosphere	• Density, Fluctuations, Composition	●		○ (2)
	• Atomic Oxygen		●	
Ionosphere	• by Solar Radiation	●	○	●
Solid Matter	• Meteoroids		●	● (3)
	• Space Debris		●	● (3)

Symbols: ● Important Effect; ○ Effect of Secondary Importance; (1) ElectromagneticTethers; (2) μg Quality; (3) Long Duration Exposure; (4) Solar Arrays

We can divide environmental influences into the following two main categories:

- *Natural space environment*, which is characterized by the matter and energy, physical laws and constants existing there (such as the Sun, the Earth, interplanetary matter, neutral atmosphere, charged particles and plasmas, cosmic particles, gravitational fields, radiation fields, and magnetic fields).

- *Induced Environment due to operation of space vehicles and other causes.* It is caused by the presence of space vehicles or the active release of matter and energy (control maneuvers, outgassing, space debris, leakage losses, radiofrequency (RF) noise).

Communications and navigation satellites are mainly operated in the geostationary orbit (GEO), in orbits of mean orbit altitude (MEO), or in the High Eccentricity Orbit (HEO); therefore, we will also see some data for these orbits. The main emphasis in this chapter, however, is put on the space environment for altitudes between 200 km and 600 km (LEO), as they are typical of space stations.

The state of near weightlessness (microgravity or μg) is of paramount importance for the permanent utilization of space stations. The feasibility and success of many experiments depend on the level of residual accelerations, which determine the μg quality. The influence of the atmospheric condition on the μg quality is described in Chapter 8.

3.1 Gravitational Fields

Gravitational fields result from mass and its distribution in stars, planets and other celestial bodies. They trap the neutral atmosphere, constrain its motion, and also control the motion of space vehicles, meteoroids and space debris. In places where electrical forces are present (e.g. in the case of charged particles), gravitational fields have only a slight effect, since electrical forces are much stronger.

3.1.1 Gravitational Field at Large Distances from a Central Body

The gravitational field for spacecrafts in a high-altitude orbit around a celestial body can be described by Newton's law of gravitation due to the field's approximately spherical-symmetrical character. The *sphere of activity* r_A of a planet is that distance from its center up to where it may be called a central body and where the gravity of the Sun may be considered a disturbance. The *point of gravitational equilibrium* r_N indicates the distance at which the gravity of this planet and of the Sun neutralize each other.

Due to the inverse-square relationship of gravity force to distance, mechanical stresses are induced, e.g. in moons, which can result in their destruction within a certain distance called the *Roche limit.* This limit is also supposed to be the cause for the formation of the Saturnian Rings.

Usually, there is no closed-form expression for the movement of more than two bodies in their interacting gravitational fields. An important exception to the rule is the movement of three bodies located in the three corners of an equilateral triangle. In a two-body system, the two possible coordinates of a third body are designated as *stable Lagrange points* L_4 *and* L_5, whereas L_1 to L_3 are unstable Lagrange points. The Lagrange points of the Earth-Moon system are presented in Fig. 3.1. In this system, small celestial bodies, for example, can be kept in a stable orbit, as it is the case with the Trojan asteroids in the Sun-Jupiter system. Lagrange points can serve as storage locations for transfer mission resupply [Beatty 83, Wertz 78, Jursa 85, Smith 82].

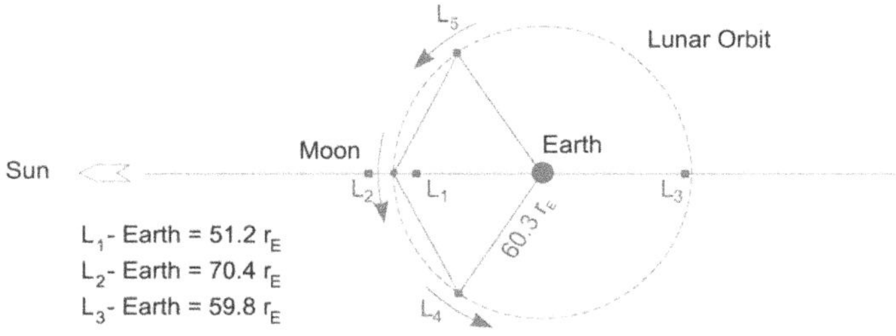

Fig. 3.1. The Lagrange Points in the Earth-Moon System

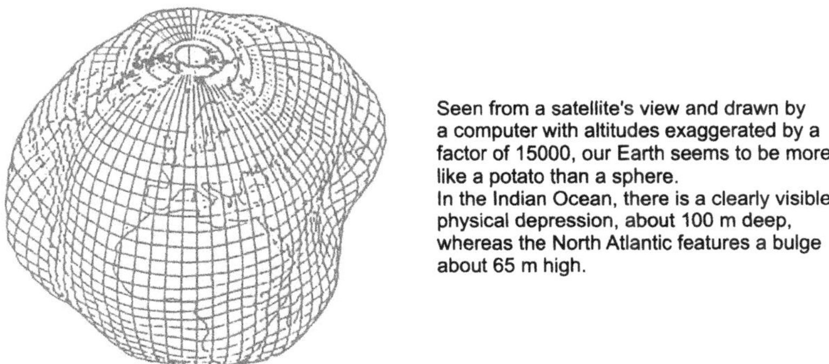

Fig. 3.2. The Earth's Deviation from a Spherical Form (15000 Times Enlarged)

3.1.2 Gravitational Field Near a Central Body

The mass distribution of celestial bodies is mostly neither spherical nor homogeneous (cf. Fig. 3.2). For that reason, amount and direction of the gravity near the celestial bodies may deviate from those results obtained from Newton's law of gravitation.

The effective gravitational potential U can be described with a sufficient degree of accuracy by means of developing zonal (i.e. latitude-dependent) spherical harmonic expansions (cf. Fig. 3.3).

With the help of an exact determination of satellite orbits by navigational aids (GPS, Inertial Navigation) and supporting measurements by radar and laser, complex mathematical models were developed for the Earth describing its shape with an accuracy up to 1 m. The results yield an equatorial radius of R_0 = 6378.140 km and a polar radius of r_P = 6356.755 km.

Development of the Gravitational Potential According to Spherical Harmonic Expansions:

$$U(r, \beta) = -\frac{\mu}{r}\left[1 - \sum_{n=2}^{\infty} J_n\left(\frac{R_0}{r}\right)^n P_n(\sin\beta)\right] \tag{3.1}$$

r : distance from Earh's center — R_0: Earth radius
J_n : Bessel functions — μ : gravitational constant
P_n : Legendre polynomials of degree n
β : gravimetric latitude with respect to the equator

The perturbation terms J_n are also called *zonal harmonics*. As it is the case with most of the celestial bodies, the perturbation term J_2 (the second zonal harmonic) is much larger than those of a higher order (with the order zonal harmonics J_n, $n>2$). The main perturbing influences which the zonal harmonic has on the trajectory of vehicles on Earth orbits are a shift of the ascending node of the orbit and a shift of the line of apsides. The values from J_2 to J_5 obtained for the Earth and the qualitatively sketched deformations resulting from them are as follows:

$J_2 = 1083.9 \cdot 10^{-6}$ $J_3 = -2.4 \cdot 10^{-6}$ $J_4 = -1.3 \cdot 10^{-6}$ $J_5 = -0.2 \cdot 10^{-6}$

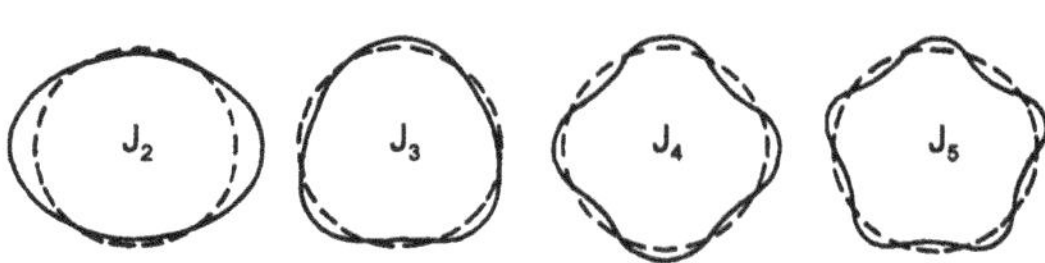

By neglecting the higher harmonic terms (n>2), an approximately homogeneous mass distribution in the form of an *ellipsoid of rotation* can be assumed.
The gravitational vector is defined by the gradient of the potential:

$$\vec{g}(r, \beta) = -\nabla U(r, \beta) \tag{3.2}$$

For the approximation of a central, inverse-quadratic field, the gravitational force for mannss m is $\vec{G} = -[(\mu \cdot m)/r^3] \cdot \vec{r}$. This results in a deviation on an order of magnitude of 0.1% for space station altitudes and, however small this deviation may seem, for the precise determination of the orbit, it is unacceptable.

Fig. 3.3. Development of the Gravitational Potential According to Spherical Harmonic Expansions

3.2 Magnetic Fields

3.2.1 The Earth's Magnetic Field

Our Earth is surrounded by a magnetic field which, according to theories generally approved today, is caused by a dynamo-mechanism inside the Earth, where convection, Coriolis and gravitational forces keep conducting liquid metals in motion. Mobile charges, however, cause a flux and always go together with the occurrence of magnetic fields. At altitudes over about 100 km, the Earth's magnetic field can be described in good approximation with the help of the potential of a dipole field (see Fig. 3.4). In this context, the magnetic momentum M_m can be assumed to have a value of 10^{17} Vsm.

At the same time, the center of the field is offset by about 450 km with respect to the geographic center and the dipole itself is rotated by about 11.3° in reference to

Potential Function of the Centric Dipole Field:

$$\Phi = -M_m \cdot \frac{\sin\vartheta}{4 \cdot \pi \cdot \mu_0 \cdot r^2} \tag{3.3}$$

M_m: Magnetic Dipole Moment
r: Radial Distance from the Dipole Center
ϑ : Angle to the Magnetic Equator (Magnetic Latitude)

By forming the gradient portions, the field assigned to the potential is obtained in the form of the components of the magnetic induction in radial direction (B_r) and in the direction of magnetic north (B_ϑ):

$$B_r = -\mu_0 \cdot \frac{\partial\Phi}{\partial r} = -\frac{M_m \cdot \sin\vartheta}{2 \cdot \pi \cdot r^2} \tag{3.4}$$

$$B_\vartheta = -\mu_0 \cdot \frac{\partial\Phi}{r \cdot \partial\vartheta} = \frac{M_m \cdot \cos\vartheta}{4 \cdot \pi \cdot r^2} \tag{3.5}$$

The total magnetic field intensity amounts to:

$$B = \sqrt{B_r^2 + B_\vartheta^2} = \frac{M_m}{4 \cdot \pi \cdot r^3} \cdot \sqrt{1 + 3 \cdot \sin^2\vartheta}$$

The tilt angle of B (inclination) results from:

$$\tan\Theta_B = \frac{B_r}{B_\vartheta} = -2 \cdot \tan\vartheta \tag{3.6}$$

Fig. 3.4. Potential Function of the Centric Dipole Field

the rotation axis: this is how the different positions of the magnetic and the geographic poles can be explained. Due to the gradient of the magnetic field (caused by the dipole structure), charged particles from interstellar space can be trapped in this field; this is called the effect of the *magnetic bottle* (cf. Fig. 3.5). Due to this effect and also to the eccentricity, there is one region in the South Atlantic with lower electromagnetic field strength and higher radiation, the so-called *South Atlantic Anomaly (SAA),* also known as the Brazilian Anomaly (cf. Fig. 3.6). Mission analyses, especially for crewed space vehicles in LEOs with low inclination, have to take this region into account. In Fig. 3.7, this anomaly is clearly visible because of the minimum in the magnetic induction.

Further deviations from the symmetrical dipole are caused by local concentrations of ferromagnetic minerals and possibly also by irregularities in the system of thermal convection streams inside the Earth. Some of these anomalies as well as the magnetic poles feature a very slow drift; geological examinations of certain minerals indicate a change in polarity in the Earth's magnetic field with periods of 10^5 to 10^6 years.

Measured field strengths deviate by ± 25% from the central dipole field which is idealized and characterized by Eq. 3.4 and Eq. 3.5. Using the model of the eccentric dipole, the deviations are smaller than ± 10%. On the Earth's surface, field strengths in the polar region are about 0.6 Gs and at an altitude of 400 km, they are 0.5 Gs. Near the equator, their value is only about half as large. More exact calculations have to be performed with the help of the model International Geomagnetic Reference Field (IGRF). This model uses fitting parameters (adapted to measured data)

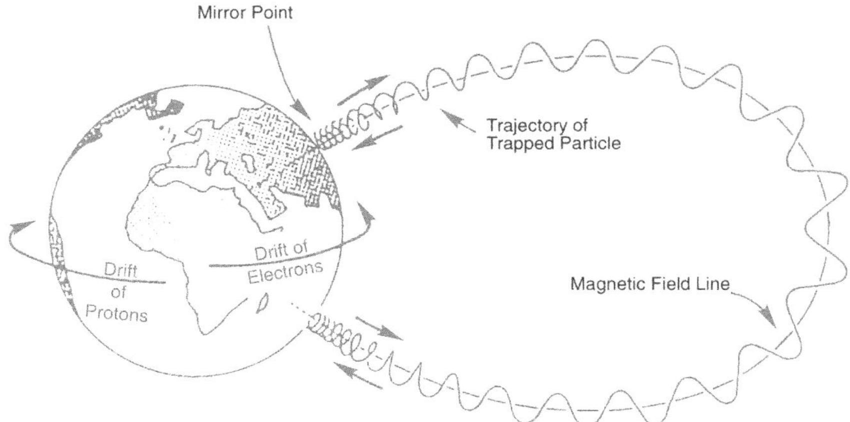

Fig. 3.5. Trajectory of the Particles Trapped in the Earth's Magnetic Field

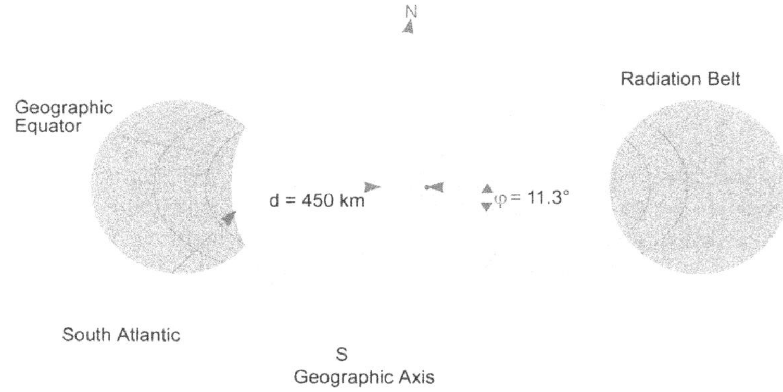

Fig. 3.6. Radiation Belt and South Atlantic Anomaly (exaggerated illustration)

to determine the Gaussian coefficients up to the 10th spherical harmonic [Skrivanek 94]. The influence magnetic storms have on the field strength is smaller than 0.01 Gs.

At high altitudes, i.e. from three to four Earth radii and up, the Earth's magnetic field is not axially symmetrical anymore since the field lines are distorted due to the interaction of the solar wind with the magnetic field. The complex field line structure resulting from this is called the *magnetosphere*, cf. Fig. 3.8.

The solar wind hits the Earth's magnetic field at a velocity of 400 to 500 km/s, transferring a power of about $1.4 \cdot 10^{13}$ W (of which only 2% to 3% is absorbed, though). For that reason, the magnetosphere is somehow similar to a case known from the discipline of gas dynamics: a blunt body in hypersonic flow corresponding to a Mach number of about eight. The magnetopause, which results from the balance

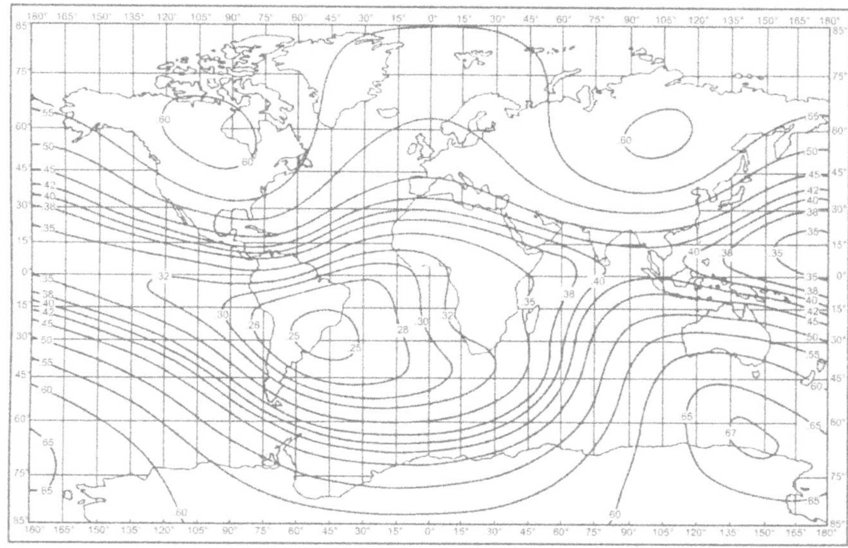

Fig. 3.7. Lines of Constant Induction B in Gs at the Earth's Surface [10^4 Gs = 1 T = 1 Vs/m^2]

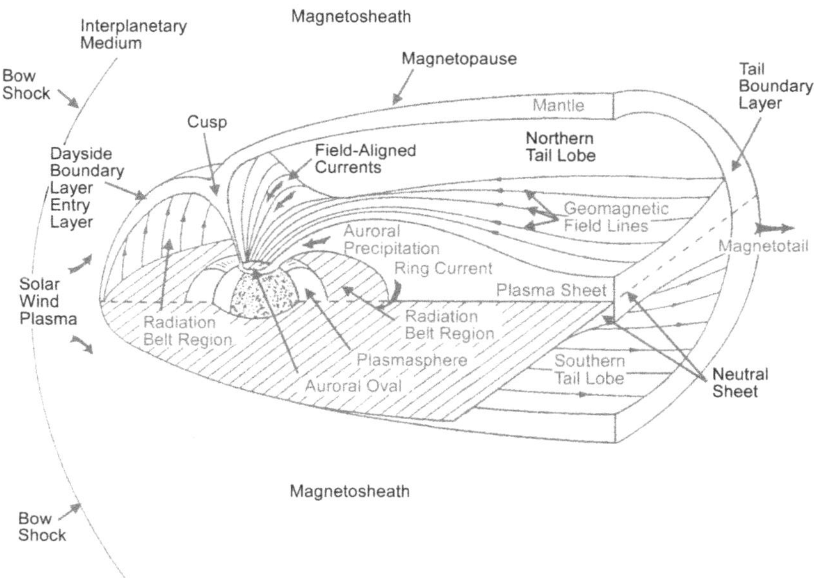

Fig. 3.8. Sectional View of the Earth's Magnetosphere [Jursa 85, Tascione 88]

of forces between the Earth's magnetic field and the solar wind, can, in this case, be interpreted as the body surface. The field lines of the Earth's magnetic field cannot exceed this region. Their upstream boundary lies at a distance from the Earth's center of only between 7 to 10 Earth radii, whereas downstream, the magnetic field stretches into a very elongated structure known as magnetotail. An additional two or three Earth radii in the direction of the Sun above the magnetopause begins the shockwave where the deceleration of the solar plasma starts.

We may summarize the Earth's magnetic field as being caused by two factors: the electric currents inside the Earth and those in the magnetosphere. Both currents contribute 99% and 1%, to the electromagnetic field strength at typical space station altitudes.

3.2.2 The Magnetic Field of the Sun

The Sun as well is surrounded by a magnetic field of a heavily fluctuating intensity. Magnetic energy from local field disturbances in spatially limited regions is accumulated due to the latitude-dependent rotation period of the Sun (25.38 d at the equator, 30.88 d near the poles) and is released in spontaneous relaxation processes in the form of radiation and kinetic energy of emitted plasma. Effects typical of such a process are solar prominences, solar flares or sunspots.

When the Sun is calm, the magnetic field at its surface is about 10^{-4} T and the magnetic momentum about $4.27 \cdot 10^{23}$ Vsm. In the case of sunspots, the magnetic field can reach values of 0.1 to 0.3 T.

Sunspots are very limited regions with a diameter ranging between several hundred kilometers and several thousand kilometers, in which strong local magnetic fields keep gas from escaping from the photosphere. They appear darker than their surroundings, because they are cooler by about 1000 K. They have a lifetime of only a few hours up to a few months and tend to occur in "groups", which in turn mostly occur together with solar flares. Regular observations conducted since 1749 revealed that the occurrence of sunspots happens cyclically, every 11.2 years on the average (cf. Fig. 3.15). The reversal of the Sun's magnetic field at a 22 year-frequency is closely related to the occurrence of sunspots.

The solar corona expels solar wind (a proton-electron plasma) at high velocities so that the field lines of the dipole field are split and carried into interplanetary space. Since the trajectories (the strike lines, to be exact) of the plasma particles have a spiral form due to the rotation of the Sun, the field lines are distorted and nearly form an Archimedean screw.

Close to Earth, the solar activity heavily influences the physical and chemical processes mainly in the atmosphere, but also the solar radiation around a space station. It is not possible to calculate these effects in detail in advance, but its phase can be predicted with the help of its average rotation period of 27 days and the eleven-year solar cycle.

The ULYSSES mission conducted by the European Space Agency ESA is dedicated to research in the field of those complex physical phenomena caused by effects from solar wind, magnetic fields, solar and cosmic radiation, interplanetary dust and gas and the interaction of all those. At the end of July 1995, the ULYSSES probe was the first artificial spacecraft to fly over the north pole of the Sun, and having completed one orbit in April 1998, important measurements that had not been possible ever before were performed. As a consequence, our knowledge of the center of our solar system was considerably increased. For the first time, e.g., the corona could be observed at very high altitudes over a long period of time. As a consequence, conclusions could be drawn about the effects solar radiation has on the terrestrial atmosphere and the ionosphere.

Further information on the magnetic fields of the Sun and the Earth, as well as their impact on space flight systems is provided by [Hallmann 88, Tascione 88, Environment 94, Skrivanek 94].

3.3 Radioactive Radiation

3.3.1 Fundamentals

Corpuscular radiation is a continuous particle flow composed of elementary particles and ions of high energy (or velocity). Together with the very short-wave electromagnetic radiation (X-rays and gamma rays), corpuscular radiation builds up the class of penetrating radiation, i.e. radiation which is capable of penetrating organic and inorganic matter to a certain extent and of giving off its energy to the matter.

Radioactive radiation in near-Earth space is basically composed of six corpuscles (summarized in Table 3.2) and electromagnetic γ-radiation with wavelengths smaller than 10^{-11} m. Some physical fundamentals on radioactivity are given in Fig. 3.9.

Table 3.2. Corpuscles of Radioactive Radiation

	Charge	Rest Mass
Electrons (β-Radiation)	$-e_0$	m_0
Positrons	$+e_0$	m_0
Protons	$+e_0$	$1836.1 \cdot m_0 = m_p$
Neutrons	0	$1838.7 \cdot m_0 = m_n$
Helium Nuclei (α-Radiation)	$2 \cdot e_0$	$7349.6 \cdot m_0$
Heavy Nuclei	$Z \cdot e_0$	$Z \cdot u$

with the Physical Constants:
Elementary Charge e_0 = $1.60219 \cdot 10^{-19}$ C
Electron Rest Mass m_0 = $9.10953 \cdot 10^{-31}$ kg
Atomic Mass Unit u = $1.66057 \cdot 10^{-27}$ kg

3.3.2 Low-Energy Particles of Solar Origin: Solar Wind

The particle flux coming from the Sun is called Solar Cosmic Radiation (SCR) and can be divided into two subcategories: the low-energy, permanent solar wind and the high-energy particles which are released during solar flares.

With respect to radioactivity, energy is usually measured in the unit electron volt (eV), with 1 eV corresponding to an energy level of $1.602 \cdot 10^{-19}$ J.

The energy of one particle is given by the relativistic relation:

$$E = \frac{m}{2} \cdot w^2 \quad \text{where } m = \frac{m_R}{\sqrt{1-\left(\frac{w}{c}\right)^2}}$$

m : Relativistic Mass
w : Particle Velocity
m_R : Rest Mass
c : Speed of Light in a vacuum

For γ radiation, on the other hand, the relation according to the quantum theory is:

$$E = \frac{h \cdot c}{\lambda}$$

h : Planck Constant
λ : Radiation Wavelength

In order to describe the intensity of corpuscular radiation, different terms are used:

Particle Density: $$n \left[\frac{1}{m^3}\right] \tag{3.7}$$

Flux Density: $$J = n \cdot w \left[\frac{1}{m^2 \cdot s}\right] \tag{3.8}$$

Particle Flux: $$\Phi = \int_A J \cdot dA \left[\frac{1}{s}\right] \tag{3.9}$$

Specific Radiation Quantity: $$Q = \int_t J \cdot dt \left[\frac{1}{m^2}\right] \tag{3.10}$$

If the particles have an energy distribution (energy spectrum) not being uniform, the following terms will be used for their description:

Differential Flux Density: $$J_E(E) = \frac{\partial J}{\partial E} \left[\frac{1}{m^2 \cdot s \cdot eV}\right] \tag{3.11}$$

Integral Flux Density: $$J(E > E_0) = \int_{E_0}^{\infty} J_E(E) \cdot dE \left[\frac{1}{m^2 \cdot s}\right] \tag{3.12}$$

Similar differential expressions are defined for the particle flux and the specific radiation quantity.

Fig. 3.9. Physical Fundamentals on Radioactivity

The continuous particle flow of the solar wind is mainly composed of protons (about 99%), α-particles (about 1%) and a few electrons, the particles' energy being about 1 keV. Between the chromosphere (at a temperature of about 10^4 K) and the corona (at about $2 \cdot 10^6$ K), these particles are expelled by the Sun into space with initial velocities of about 1000 km/s; this can be regarded as a kind of evaporation effect. In near-Earth space, the velocities are reduced to an average of 300 to 500 km/s and the proton-electron plasma then has a flux density of about 3 to 40 cm^{-3}, i.e. the flux density fluctuates between 0.09 and $2.0 \cdot 10^9$ $cm^{-2}s^{-1}$. In the course of only a few days, velocity and density may undergo considerable changes due to fluctuations in the solar activity.

3.3.3 High-Energy Particles of Solar Origin: Solar Flares

Highly concentrated, explosive releases of energy within the solar atmosphere which are optically perceived as brief flashes of light from localized areas of the chromosphere are called solar flares. Typically, energies of between 10^{21} and 10^{25} J are released during this process. Although the occurrence of such flares is only statistical (i.e. not predictable), their frequency is closely related with the occurrence of sunspots and, as a consequence, with the 22-year-cycle of the reversal of the Sun's magnetic field. The rather spontaneously occurring events have a duration of 1 to 5 days; the differential flux density of the protons and heavier particles can only be roughly approximated in the form of a model [Environment 94]. The largest solar flares observed up to the present day occurred during the late 1950's and in August of 1972.

The particle flux expelled by the solar flares contains 89% of protons at ultra-high speed with energies around 30 MeV, 10% of α-particles and 1% of heavy nuclei, the so-called HZE-particles, with a high atomic number *Z* and high energy *E* within the range of 10 to 100 MeV, partially up to 1 GeV. The HZE-particles (high *Z*, high *E*) are often called high-LET (linear energy transfer) particles.

3.3.4 Particles of Galactic Origin

Galactic Cosmic Radiation (GCR) consists of extremely high-energy particles from outside the solar system, whose occurrence therefore cannot be correlated with the solar activity. Their origins cannot be precisely indicated, but supernovae and high-energy processes connected with neutron stars and black holes are presumed to be possible sources of origin.

Although the flux density of cosmic particles with 10 particles per $cm^2 \cdot s$ is much smaller than the particle flux density from the Sun, it is permanently present and can barely be shielded. It is composed of protons (85%), α-particles (14%), HZE-particles (in this case, even the presence of iron nuclei was measured!) and electrons and positrons which move nearly at speed of light. Energies are at about 10 GeV: a lower limit of 0.1 GeV and an upper limit of 10^{11} GeV (!) were observed. Especially the lightest particles with atomic numbers of up to 28 have a negative effect on electronic components.

Deflection of GCR particles in the Earth's magnetic field is very low, so the Earth's own atmosphere is the only shield to protect its surface. In a space station orbit, GCR (which is also called *background radiation)* on the average forms 5% to 10% of the total radiation. In this context, the GCR flux in the solar system is also influenced by the solar activity, since with an increase in the density of the solar plasma and in the strength of the magnetic fields occurring with the solar wind, a certain shielding effect takes place.

The differential flux densities of particles exceeding energies of 10 MeV can easily be calculated with the help of simple, but physically unsubstantiated models [Environment 94, Skrivanek 94].

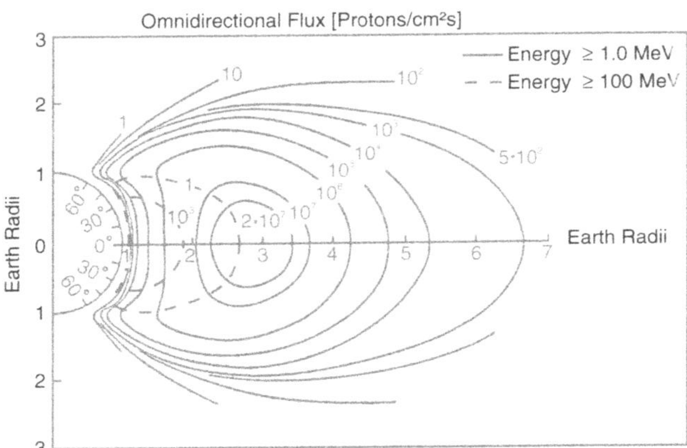

Fig. 3.10. Integral Particle Flux Density of the Protons Trapped in the Earth's Magnetic Field with Energies Exceeding 1 MeV and 100 MeV

3.3.5 Radiation Belts within the Earth's Magnetic Field

Due to the dipole-like structure of the Earth's magnetic field, charged particles (mainly electrons and protons) are trapped in it over the equator in the so-called "Van-Allen Radiation Belts". In this context, the magnetic field has the effect of a magnetic bottle (cf. Sect. 3.2.1), i.e. a kind of trap in which the particles are kept and where they reach life spans of several years. In Fig. 3.10, the particle flux density of protons with energies exceeding 1 MeV and 100 MeV, are shown as a function of the distance from the Earth and its geomagnetic latitude. Figure 3.11, on the other hand, shows the total particle flux density of the electrons as a function of the geographical latitude. This diagram also indicates the two maxima of the flux density, known as *inner* and *outer Van Allen Belt* (VAB) and the South Atlantic Anomaly (SAA). Both are to be considered for space station design. The influence on the particle distribution, due to perturbations in the magnetic field and also due to the solar cycle, grows as the distance from the Earth increases.

In order to calculate the distribution (depending on time and place) of the omnidirectional flux of electrons (within the range of 50 keV–7 MeV) and protons (50 keV–500 MeV), the numerical AP-models from the NASA Goddard Space Flight Center, Greenbelt, Maryland (USA), which were adapted to data actually measured during many satellite missions, are used. The present modeling of the particle flux can only very roughly be adapted to reality, i.e. sunspot period, day/night cycle, etc. For that reason, the US Department of Defense in 1990 launched the specially equipped Combined Release and Radiation Effects Satellite (CRRES) (inclination 18.1°, orbital altitude between 350 km and 33000 km), which in the future will enable better modeling [Skrivanek 94].

The high-energy protons released by solar flares are nearly completely deflected by the Earth's magnetic field, so in low Earth orbits close to the equator and at low

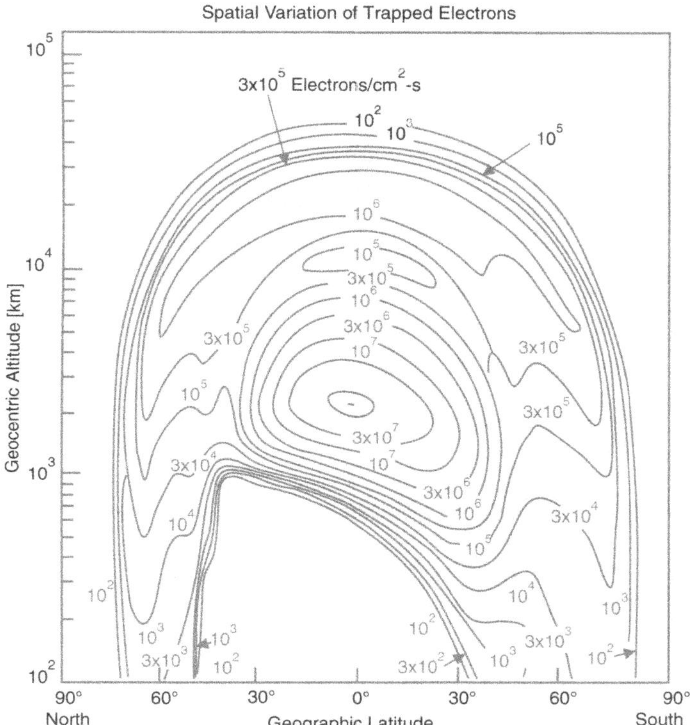

Fig. 3.11. Total Particle Flux of the Electrons Trapped in the Earth's Magnetic Field

inclination, they will be not a large threat, and thus they do not need to be considered among the design criteria. In Earth orbits exceeding 60° inclination or at altitudes of more than 1000 km, however, the flare particles can directly stream into the atmosphere along the field lines. In this case, protective measures especially for crewed space flight are absolutely necessary. Solar flares are also a significant risk to high altitude flying or interplanetary space vehicles. For instance during the Apollo missions, they gave cause for great concern. Fortunately, apart from the flare in August 1972, during the stay in space and on the Moon no threatening flare occurred. The flare in 1972 would, at that time, indeed have been seriously detrimental to the health of any crew in space. Generally, it can be said that solar flares constitute the most significant and most variable part of the radiation load in space, and therefore they are the largest threat and the environmental factor of greatest uncertainty for crewed space flight.

Figure 3.12 shows a summary of the different contributions to corpuscular radiation in space, distinguished according to the particles' origins.

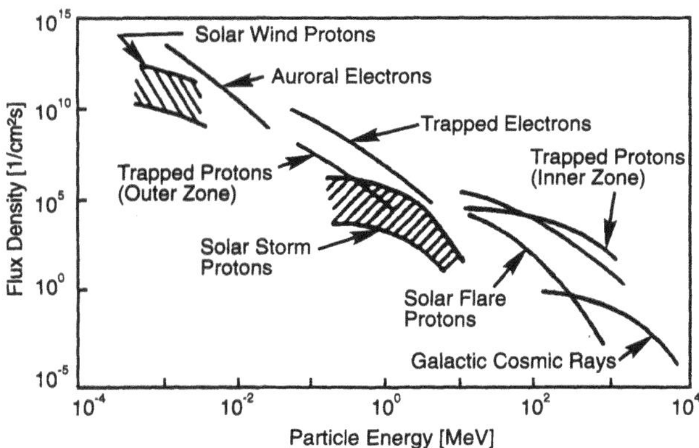

Fig. 3.12. Contributions of Corpuscular Radiation in Space

3.3.6 Radiation Effects on Materials and on the Human Organism

The term *radiation dose* is defined as the amount of energy transferred onto a given mass of irradiated matter. For its description, different quantities and units are used, such as:

Radiation dose D measured in Gy (Gray). The Gray is the official SI unit for radiation dose. One Gy is defined as the amount of radiation energy deposited in 1 J/kg of material which is being irradiated. It is only a unit for energy and does not indicate the kind of energy source or the kind of radiation.

Dose equivalent D_e in Sv (Sievert). Since protons, e.g., are much more detrimental than X-rays of the same energy level, the unit Sievert takes into account the radiation energy and its biological effects. This is achieved by means of the additional factor RBE (Relative Biological Effectiveness) multiplied by the radiation dose:

$$D_e = RBE \cdot D \tag{3.13}$$

The factor RBE for different particles can be seen in Fig. 3.13. For X-rays and gamma rays as well as for electrons, $RBE = 1$ Sv/Gy; for α-particles, on the other hand, according to the most recent knowledge, $RBE \leq 20$ Sv/Gy. The RBE value for other HZE-particles at present is still the subject of discussion, but a value > 10 is assumed safe. For example, virtually every penetration of such a particle through a cell nucleus leads to the permanent damage to the genes or even to the death of the cell.

Historically, the units most often encountered in this context were Roentgen, Curie, rad and rem; nowadays, they are obsolete and should not be used anymore.

When particles from the Van Allen Belt hit a wall (e.g. the outer wall of a space station), first of all bremsstrahlung is released, i.e. X-rays and gamma rays. When the high-energy particles from solar flares or GCR penetrate into such a wall, they lose kinetic energy, but only a few particles are actually stopped. Because of their collision with the wall's material, these few particles cause (apart from bremsstrahl-

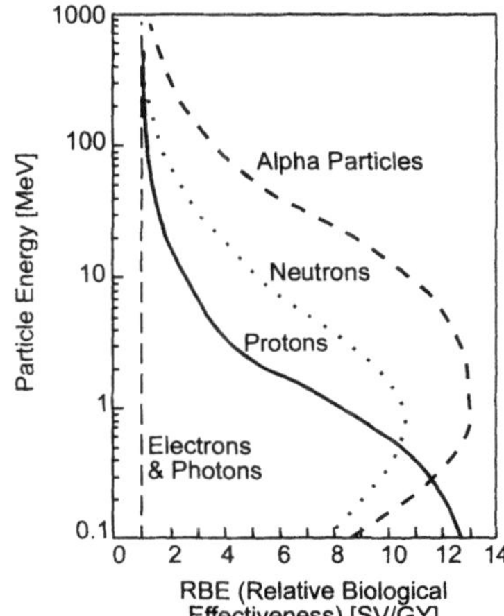

Fig. 3.13. RBE in Sv/Gy for Different Particles as a Function of Particle Energy

ung) veritable secondary showers of particles which considerably increase the total radiation dose. Only extremely thick shields, with area densities of several hundred g/cm^2, could ensure protection comparable to that of the Earth's atmosphere.

Apart from the force effect applied to a station, further effects of ionizing radiation on materials can be single event upsets, latch-ups and burnouts as well as the degradation of materials, for example, if conventional solar cells are considered.

The radiation dose for human beings on Earth amounts to an average of 1.7 mSv/year at sea level. With astronauts, the doses are markedly higher, especially in the case of long-duration missions or frequent stays in space. In this context, Extravehicular Activities (EVAs) play an important role: compared to about 30 g/cm^2 radiation protection inside a space station or transfer vehicle, astronauts during an EVA will be exposed to much higher radiation doses due to the small area density of only 3 g/cm^2 of their EVA space suits.

Figure 3.14 shows values calculated and measured aboard the Space Shuttle indicating the daily radiation dose to which astronauts are exposed at an orbital inclination of 28.5°, summarized as a function of altitude.

In the course of the rapid development of crewed space flight, today more and more persons of different ages and both sexes stay in space for increasingly longer periods of time. Among them, especially mission specialists are relatively frequently assigned for several different missions. This clearly indicates how necessary and important it is to define corresponding guidelines in order to assure crew health.

Particles with a very high level of energy can also penetrate the human body and most of the time, this has no serious consequences. Sometimes, however, in space, the released bremsstrahlung is a serious hazard since it is capable of ionizing the body's atoms. Additionally, the high-energy particles excite adjacent atoms, which in turn emit ionizing radiation. Some long-term consequences could be genetic changes and the development of cancer. Such development, however, always depends on a person's general health, age and sex. An overview of possible short-term consequences for a typical cross-section of the human population is given in Table 3.3.

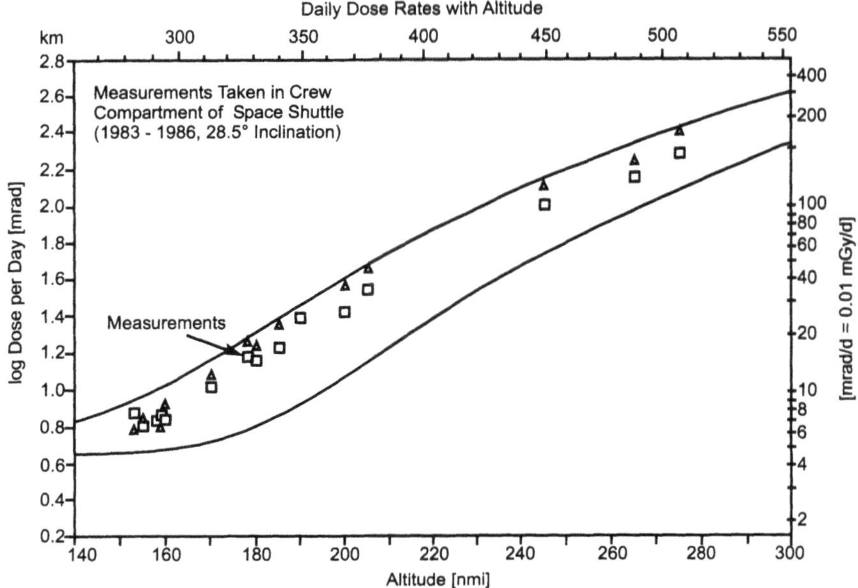

Fig. 3.14. Energy Dose per Day at an Orbit Inclination of 28.5°

Table 3.3. Possible Short-Term Effects Caused by Spontaneously Occurring Radioactive Radiation

Dose [Sv]	Possible Effect
up to 0.5	No obvious effects; possibly minor changes in the blood profile.
0.5–1.0	Nausea for about one day with 10% to 20% of all persons concerned; no serious incapacitation; temporary reduction of lymphocytes.
1.0–2.0	Nausea and first symptoms of radiation sickness with about 50% of all persons involved; reduction of lymphocytes of about 50%; death rate below 5%.
2.0–3.5	Nausea and heavy symptoms of radiation sickness (loss of appetite, diarrhea, light bleeding) with nearly all persons involved; about 5% to 90% deaths within two to six weeks.
3.5–5.5	Heavy symptoms of radiation sickness (fever, bleeding, weight loss) with nearly all persons involved. > 90% deaths within one month, survivors are temporarily incapacitated for a duration of six months.
> 10.0	Presumably no survivors

3.3.7 Protective Measures

Space lacks a protective shielding like the Earth's atmosphere, which repels most of the harmful solar and galactic radiation. For that reason, corresponding protective measures have to be taken for space vehicles in order to guarantee astronaut safety and the undisturbed operation of all instruments.

In orbits at inclinations of 28.5° (Space Shuttle) and 51.6° (ISS), the South Atlantic Anomaly (SAA) is traversed about three to four times per day. On such orbits, 10 h per day will be practically radiation-free. Therefore it is necessary to reduce EVAs as much as possible and not to perform them during passages through or close to the SAA. Additionally, space suits should be outfitted with particular protection for very critical organs like eyes, bone marrow and genitals.

The assembly of large structures in orbit should mainly take place by means of docking maneuvers or robotic systems instead of EVA. If this proves to be impossible, the EVAs should take place in low orbits and the assembled structure should be lifted by a reboost afterwards.

As a protection against the statistically occurring solar flares, the installation of so-called "storm shelters" for the crew is recommended. The shelters provide easy access and sufficient shielding and, due to the different transfer times of the released light and the actually hazardous particles, the lead time of several hours is enough for the crew to take shelter (cf. lower part of Fig. 3.15). Such storm shelters are also possible outside ISS for the safety of astronauts during EVAs. Further precautions include a dosimeter for each crew member as a control instrument and dose predictions during the planning of the operational phase.

Different strategies for the shielding of space vehicles in space are possible, such as the following:

- The outer skin should consist of materials with a low nucleon number (e.g. materials containing hydrogen) to stop charged particles, whereby only a minimum of the resulting energy is released in the form of bremsstrahlung.
- The inner skin, on the contrary, should consist of materials with a high nucleon number (e.g. tantalum or lead), which can absorb the amount of bremsstrahlung nevertheless released.

At the same time and as Fig. 3.16 illustrates, the shielding should have the highest possible area density, since only then a considerable reduction of the radiation dose can be expected.

In the future it should be the task of research in the field of radiation protection to examine not only the shielding effect of different space flight materials, but also the penetration effect of radiation in the human body, the radiation distribution in the inner organs, the RBE of HZE-particles and the influence an interaction of the μg-environment with radiation has on the human organism.

3.4 Electromagnetic Radiation (EMR)

In space station orbits, the frequency range of the electromagnetic radiation present spans the spectrum from long-wave to X-ray. The electromagnetic waves are summarized and categorized in Table 3.4 according to their wavelength and frequency. The main radiation source in our solar system is the Sun. In addition to that, in near-Earth space there are the following: Albedo radiation induced by the Sun, the Earth's infrared (IR) radiation, galactic and natural plasma radiation, and man-made radiofrequency (RF) radiation. The highest energy densities are in the visible and infrared spectrum and they determine the order of magnitude of the solar constant.

Similar to particle radiation, EMR can cause chemical or optical changes in materials and damage to logic circuits and sensors. Secondary effects due to outgassing

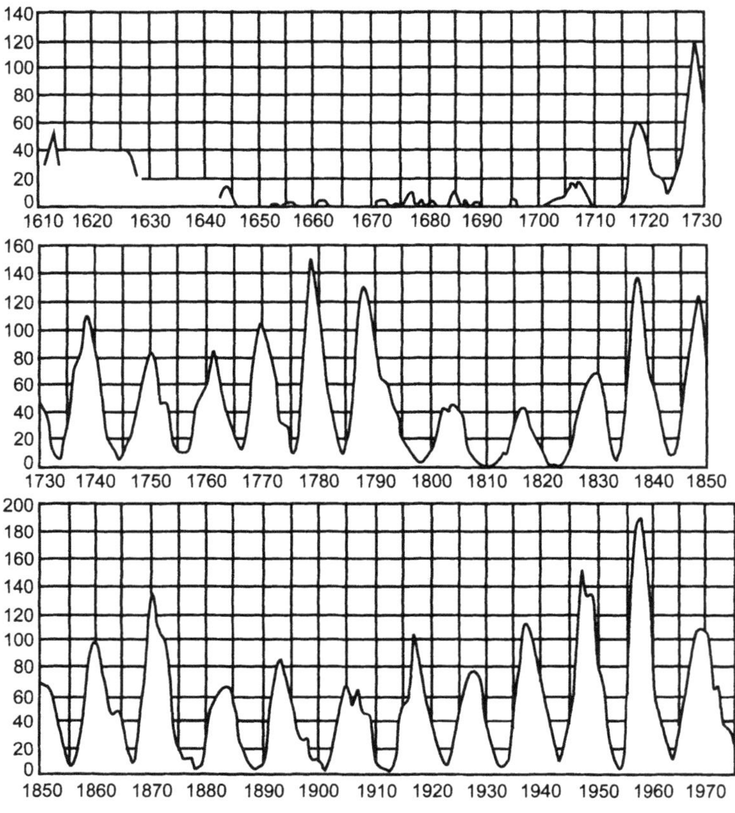

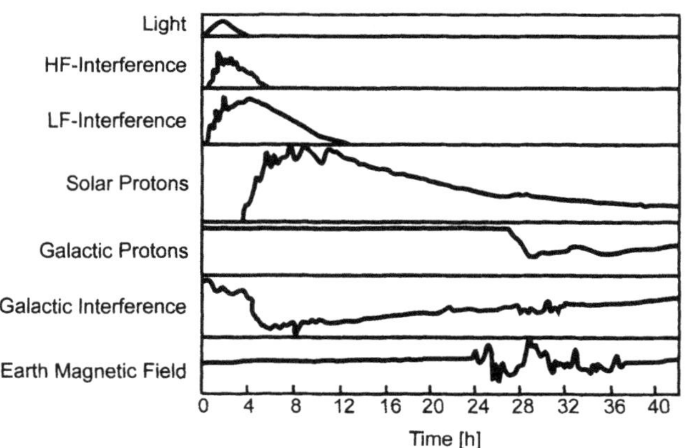

Fig. 3.15. Sunspot Periods Since 1610 (above) and the Course of Effects after a Single Solar Flare (below) in a LEO

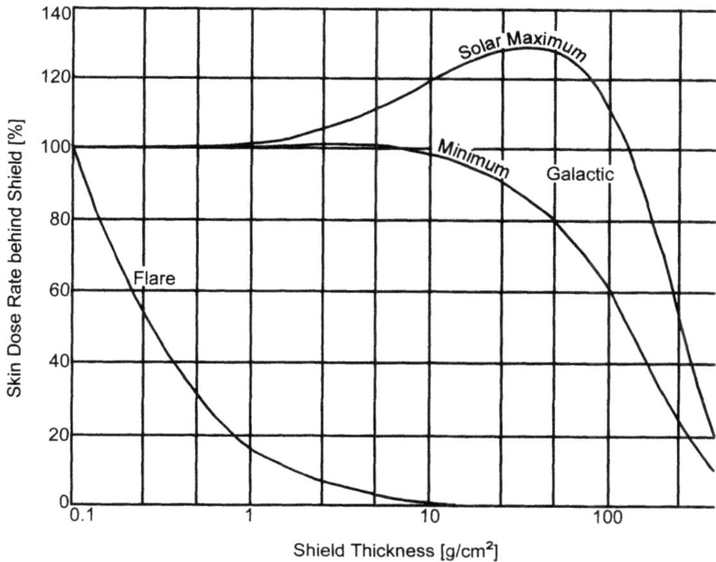

Fig. 3.16. Influence of Area Density of a Shielding on the Radiation Dose behind the Shield

Table 3.4. Classification of Electromagnetic Radiation

Designation	Wavelength λ	Frequency $\nu=c/\lambda$
Gamma (γ)-Radiation	< 0.1 nm	$> 3 \cdot 10^{18}$ Hz
X-Rays	0.1–10 nm	$3 \cdot 10^{18}$–$3 \cdot 10^{16}$ Hz
Ultraviolet (UV) Radiation	10–300 nm	$3 \cdot 10^{16}$–10^{15} Hz
Visible Light	0.3–0.7 μm	10^{15}–$0.43 \cdot 10^{15}$ Hz
Infrared (IR) Radiation	0.7–1000 μm	$0.43 \cdot 10^{15}$–$3 \cdot 10^{11}$ Hz
Microwaves	1–1000 mm	$3 \cdot 10^{11}$–$3 \cdot 10^{8}$ Hz
Short/Medium/Long Wavelengths	> 1 m	$< 3 \cdot 10^{8}$ Hz

and contamination, also referred to as "induced environmental influences", will be addressed separately in Sect. 3.9.

3.4.1 Galactic Radiofrequency (RF) Noise

Galactic RF noise reaching the space station originates mainly from within a range of between 15 MHz and 100 GHz, and is at its strongest in the direction of and perpendicular to the galactic plane. It can disturb communications with the space station in the frequency range of 40–250 MHz.

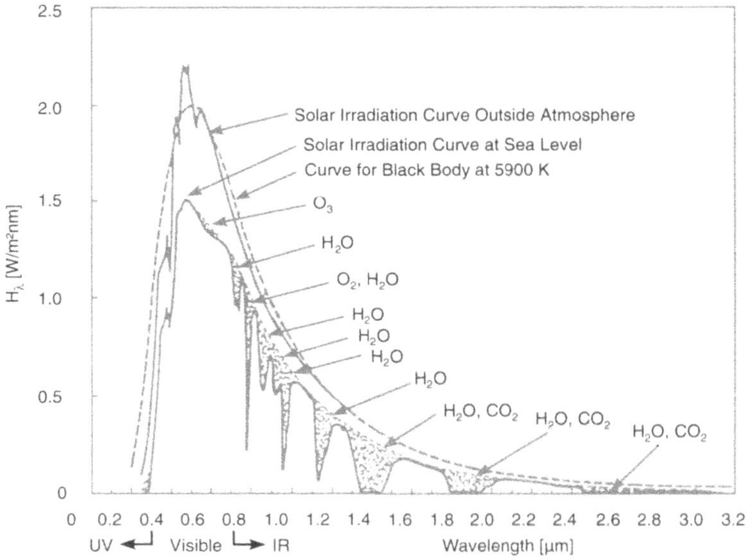

Fig. 3.17. Spectral Radiation Intensity of Solar Radiation

3.4.2 Solar Radiation

Figure 3.17 shows the solar electromagnetic radiation spectrum measured at a distance of 1 AU (=149.5·10^6 km) from the Sun outside the Earth's atmosphere. This spectrum approximately corresponds to that of a black body at T_{Sun} = 5900 K. The solar radiation intensity decreases with increasing distance r to the Sun:

$$M_{Sun}(r) = \sigma \cdot T_{Sun}^4 \cdot \left(\frac{R_{Sun}}{r}\right)^2 \tag{3.14}$$

with the solar radius R_{Sun} and σ, the Stefan-Boltzmann constant. The magnitude of solar radiation at Earth distance is described as a solar constant; it has the value:

$$M_0 = 1371\ (\pm 10)\ W/m^2 \tag{3.15}$$

It is a mean value which e.g. forms the basis for the examination of degradation effects in materials, especially in the case of solar cells. The tolerance is a result of natural fluctuations and inaccurate measurements. Because of the eccentricity of the Earth orbit, the energy flux varies over the course of a year by about 3.3%, with its maximum in the winter solstice (with respect to the northern hemisphere) and its minimum in the summer solstice (beginning of July in northern hemisphere). These seasonal fluctuations can be estimated by means of the following equation:

$$M_{Sun} = M_0 \cdot \left(1.0 + 0.033 \cdot \cos\left(\frac{2 \cdot \pi \cdot n}{365}\right)\right) \tag{3.16}$$

with n being the day of the year beginning with January 1.

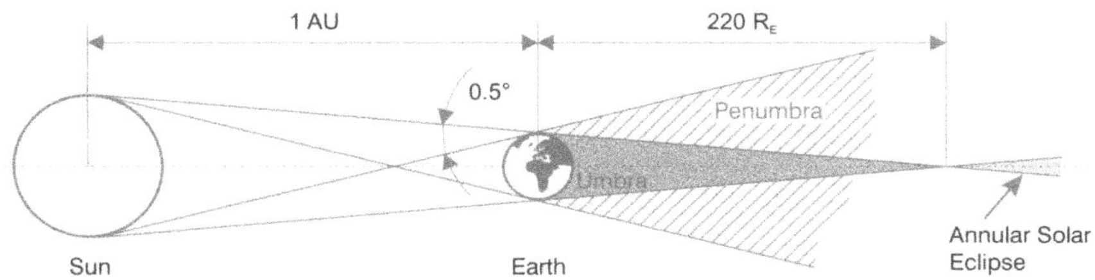

Fig. 3.18. Structure of the Earth's Shadow

Apart from the absorption of certain spectral lines within the Sun's atmosphere, solar radiation is also deflected within the Earth's atmosphere and is absorbed according to the characteristic absorption bands of the atmospheric gases. For this reason, the solar power received on the Earth's surface at sea level is reduced to only 747 W/m^2. The largest portion of the total solar irradiance makes up the visible and long-wave infrared radiation.

The shadowing effect of space vehicles entering the Earth's shadow plays an important role for on-orbit thermal control. Viewed from the Earth, sun beams do not run parallel, but diverge at an angle of 0.5°. The Earth's shadow thus does not have a cylindrical form but is characterized by three shadow zones illustrated in Fig. 3.18. But, since the Earth's umbra stretches into space for up to about 220 Earth radii, the Earth shadow (in a LEO) can be considered more or less cylindrical ($r_{cylinder} = R_E$). For lower orbits, the maximum duration of the shadow phase can be calculated very easily in the special case of a circular orbit in the Earth's ecliptic:

$$\frac{t_E}{t_O} = \frac{1}{\pi} \cdot \arcsin\left(\frac{R_E}{R_E + h}\right) \tag{3.17}$$

In this equation, R_E stands for the Earth radius, h for the flight altitude of the space vehicle, t_E represents the duration of the eclipse and t_O the orbital period, see also Section 5.2.3.

Radiation in the ultraviolet (UV) and X-ray range that strongly varies with the solar activity, accelerates the wear and tear on materials in orbit. The emitted UV-radiation causes direct degradation effects on and within the material. It damages plastics, paint, adhesives, most kinds of glass and the matrix materials in composites. UV-absorbing paints or the coating of irradiated parts with metal foil or metallized plastic films (e.g. aluminized Kapton®) are possible protections. The Sun's X-rays have relatively low energy and intensity, so even the comparatively weak protection offered by a space suit is sufficient. In addition to that, UV-rays and X-rays cause dissociation and ionization processes in the atmosphere which lead to the formation of aggressive atomic oxygen (AO). Atomic oxygen, especially in low orbits, is responsible for further degradation. Moreover, heavily fluctuating thermal loads in low Earth orbits cause material deformation and fatigue, outgassing, degradation, contamination and other changes in material properties.

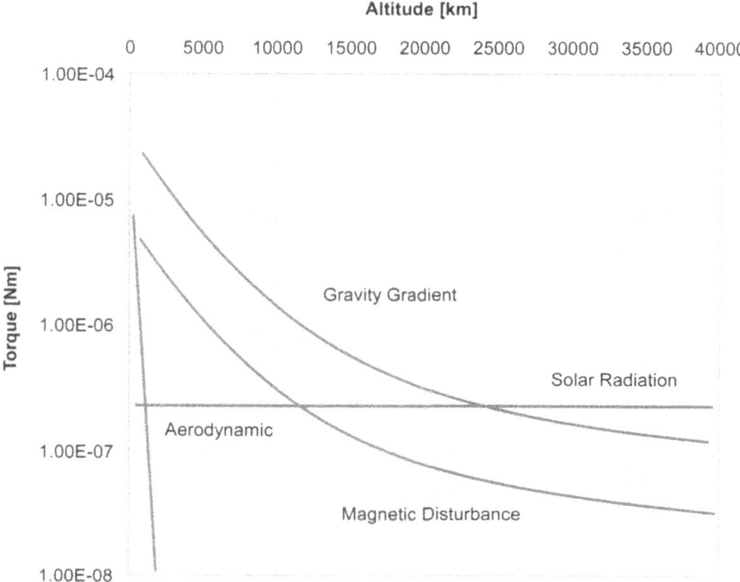

Fig. 3.19. Typical Perturbation Torques as a Function of Orbit Altitude

3.4.3 Solar Radiation Pressure

The quantum theory assigns an impulse to each photon so the radiation impinging on a surface also exerts a pressure. This influences orbital altitude and orientation of space vehicles and at altitudes over 1000 km, these effects exceed those of the residual atmosphere. Especially for geostationary satellites, this effect has to be taken into account. The radiation pressure can also disturb the μg-environment of a space vehicle and thus negatively influence correspondingly sensitive experiments. The radiation pressure applied to a surface in Earth orbit can be calculated according to the following equation:

$$p_s = \frac{4}{3} \cdot \frac{M_0}{c_0} \cdot (1 + r) = 0.46 \cdot 10^{-5} \cdot (1 + r) \cdot \cos^2\Psi \qquad [N/m^2] \qquad (3.18)$$

with the reflection coefficient of the surface $0 \leq r \leq 1$, the speed of light c_0 in a vacuum and the angle Ψ between surface normal and Sun direction. Figure 3.19 illustrates the influence of the disturbances caused by the solar pressure in relation to other perturbations in orbit.

3.4.4 Albedo Radiation

Albedo radiation is the fraction of incident solar radiation which is reflected into space in a diffused manner by the atmosphere or the surface of a celestial body. The diffuse *albedo reflectivity coefficient* ρ_{albedo}, often simply called *albedo*, is strongly dependent on the geography, such as light snow surfaces, dark ocean surfaces and cloud coverage (see Fig. 3.20).

NASA measured the albedo of different orbits with variations in inclinations, altitudes and solar elevation angles (cf. Fig. 3.21), but often, simply the mean value indicated in Table 3.5 is used for design purposes. This value is based on average surface and cloud coverage.

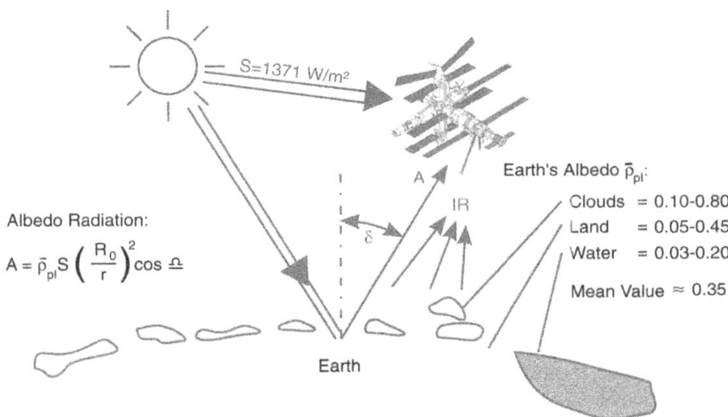

Fig. 3.20. Illustration of Solar, Albedo and Thermal Radiation

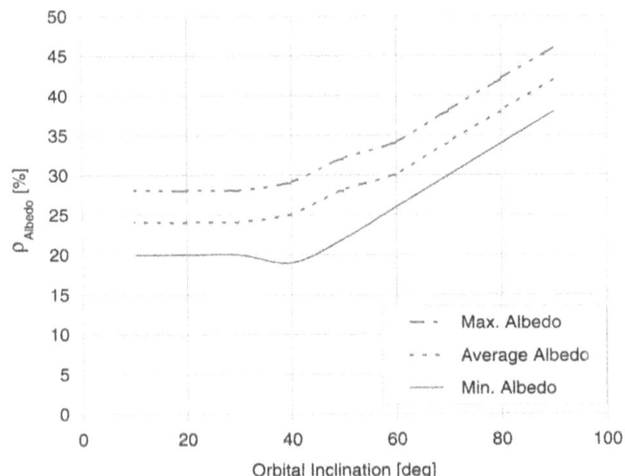

Fig. 3.21. Earth's Albedo for Different Inclinations [Gilmore 94]

Table 3.5. Relevant Thermal Data for the Planets of the Solar System

Celestial Body	Solar Constant $M_{Sun}(r)$ [W/m²]	Mean Albedo Degree of Reflection $\varphi_{Albedo}=1-\alpha_s$ [–]	Mean Planetary IR Radiation Flux M_P [W/m²]	Equivalent Mean Planet Temperature T_P [K]
Mercury	9034	0.53	2139	440
Venus	2588	0.76	155	228
Earth	1371±10	0.35±0.02	240±7	255
Mars	585	0.16	123	215
Jupiter	51	0.73	3.4	88
Saturn	15	0.76	0.9	63
Uranus	3.6	0.93	0.063	32
Neptune	1.5	0.84	0.06	22
Pluto	0.89	0.14	0.191	42
Moon	1353	0.067	316	273

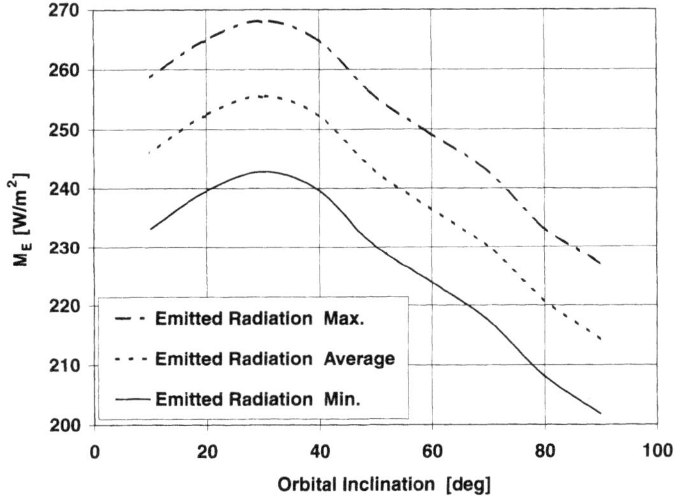

Fig. 3.22. Maximum, Average and Minimum Earth IR Flux as a Function of Orbit Inclination [Gilmore 94]

3.4.5 Thermal Radiation

The Earth absorbs solar radiation with a solar absorbance α_s close to 0.7 and emits radiation in the infrared wavelength range due to its thermal equilibrium temperature. The average radiation flux M_E of the Earth coincides with an equivalent black body radiation of $T = 255$ K. Satellites measured a global annual average radiation flux of $M_E = 237$ W/m² (cf. Fig. 3.22). This corresponds to a radiation temperature of 277 K on the day side and 248 K on the night side of the Earth. As a matter of

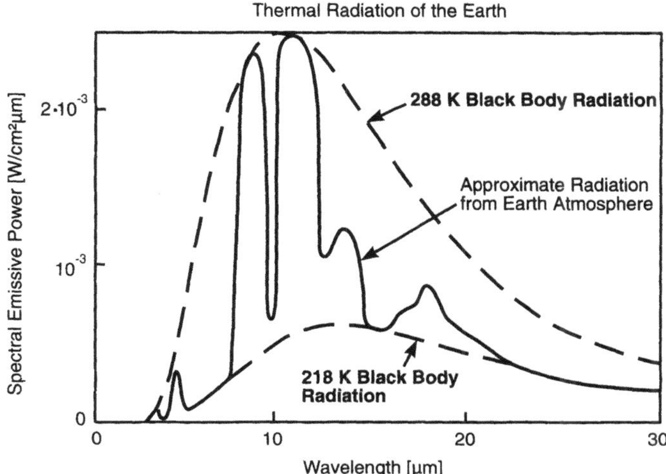

Fig. 3.23. Radiation Emitted by the Earth and Black Body Spectral Intensity Distribution

fact, the Earth-emitted IR radiation has a complicated spectrum with a spectral distribution, which can be approximately indicated as black body radiation at about 218 K (absorption by the atmosphere) with its peak at 288 K (about average Earth temperature) between $\lambda = 8.5$ and 11 μm (cf. Fig. 3.23).

3.5 Natural and Other Radiation Sources

Space station orbits run closely above the altitude of maximum electron densities which were measured with 10^4–10^6 cm^{-3} at 200–300 km altitude. These densities correspond to a maximum electron-plasma (or Langmuir) frequency of 0.9–9 MHz. Below this frequency, EMR cannot propagate through this region of strong electron density without being strongly damped. As a consequence, terrestrial and radiofrequency sources below this region, except for the occurrence of certain conductive plasma waves, cannot reach the space station. In this context, it must be taken into consideration that the maximum electron density and the corresponding altitude strongly fluctuate depending on the solar activity and the local time.

High power fluxes in the radiofrequency range mainly result from man-made noise either on Earth or aboard space vehicles. The transmitter frequencies of radar stations of typically 0.1–5 GHz yield charge densities in locally confined areas of the ionosphere.

Radiation at frequencies below 1–10 MHz originate in natural plasma regions of the magneto-ionosphere system. Such radiation does not depend on plasma temperature and is mainly caused by the interaction of charged particles with the Earth's magnetic field, for example, electron-cyclotron waves at 0.5–30 kHz. Further noise sources can be found in the aurora australis or aurora borealis, or in irregularities of the ionosphere.

Wide-band man-made radiation sources are predominantly emitting below frequencies of 1–10 MHz and are hence not considered a disturbance for high frequency communications outside the ionosphere, i.e. at the altitude of space station orbits. There are, however, a few narrow-band, powerful radar transmitters and other transmitters which can be operated up to the high frequency range of 300 GHz. In the case of perturbations, these have to be identified and faded out by, e.g. electronic circuits or appropriate antenna nulling. Most of the time, the signal strength is a function of the inverse square of the distance transmitter/receiver and hence is helpful in order to determine time and duration of the perturbation (cf. Sect. 12.2).

3.6 The Atmosphere

3.6.1 Composition

The gaseous envelope surrounding the Earth or other planets is called atmosphere. In space flight, its upper boundary is defined at that altitude where the radiation pressure replaces the atmospheric pressure as the dominant perturbing influence at increasing altitudes. For the Earth, this is at an altitude of about 1000 km. The region from 105–750 km is called thermosphere and the one above is known as the exosphere. The schematic structure of our atmosphere is illustrated in Fig. 3.24.

A decisive factor for the planning of missions in LEO, but also for all launches of rockets or other spacecraft, is to have a thorough knowledge of the atmospheric conditions: especially density has a major influence on the aerodynamic drag or on corresponding torques and thus on the attitude of rockets and spacecrafts. Temperature and composition are important characteristics as well, for example in judging material degradation caused by atomic oxygen. For that reason, many semi-empirical models with different levels of accuracy were developed to determine the desired data at any location at a specific point in time. These models are based on systems of differential equations for fluid and thermodynamic processes. Experimentally obtained parameters are used as initial input data in order to solve these differential equations.

But since the atmosphere is subject to the influence of solar radiation and the Earth's magnetic field, its exact condition at a given point depends on several parameters including long-term, short-term and spatial variations. Most of the time, estimations only require a simple model that delivers mean values or minimum and maximum values, respectively.

An example for such a model is the CIRA72 atmosphere. Figure 3.25 shows mean values for the temperature T_∞, the density ρ_∞ and the composition as a function of altitude. In the upper thermosphere and in the exosphere, the temperature only increases asymptotically because of the extremely effective heat conduction. The dotted lines represent the variation range of the data which is primarily caused by the changes in the solar activity. Figure 3.26, on the other hand, shows the influence local time has on the atmosphere density for minimum and maximum solar activity. Density variations $\Delta\rho$ [%] at altitudes between 150 km and 800 km, as they result from different effects and different time scales, are indicated in Table 3.6.

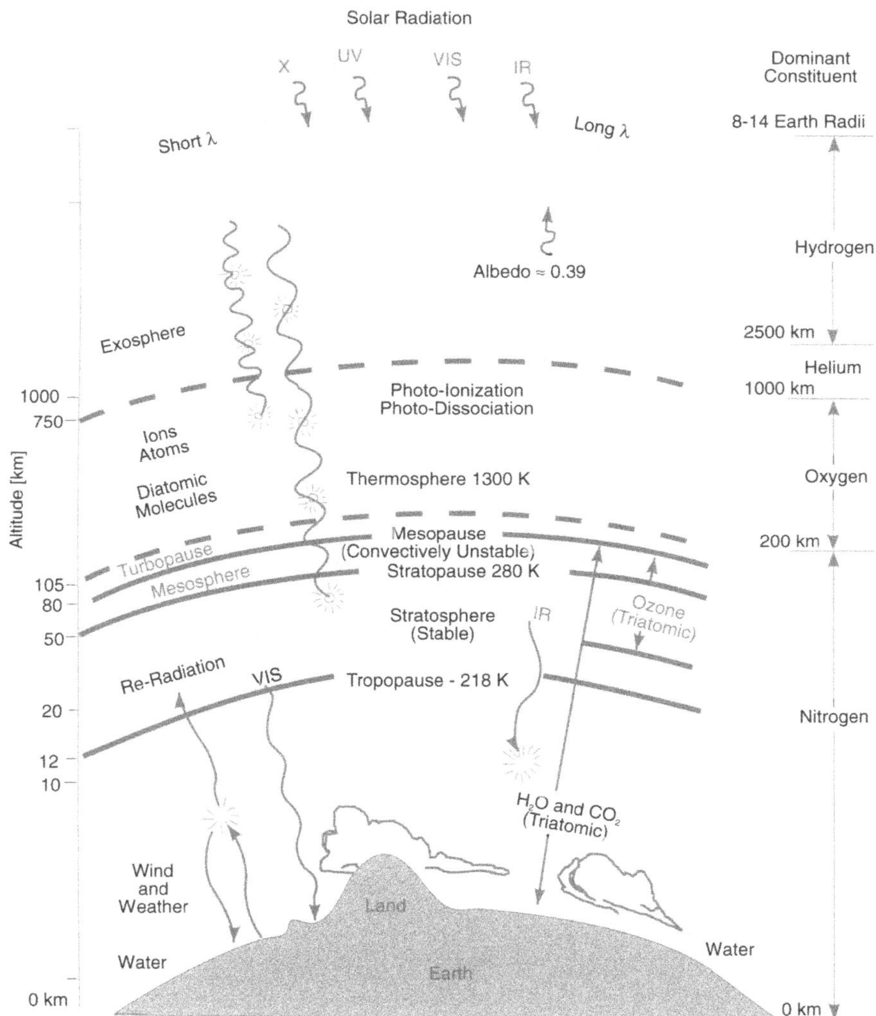

Fig. 3.24. Atmospheric Layers and Composition

Gases in the upper thermosphere are heated by absorption of the extreme ultraviolet (EUV) radiation of the Sun, at lower altitudes by UV radiation. The heating by radiation of course only occurs on the day side of the Earth; conductive and convective transport of heat distribute the energy in the atmosphere only to a small extent. For that reason, a significant temperature gradient exists at the transition zone into the eclipse which increases in the exosphere to a level of over 200 K (day side 1060 K, night side 840 K).

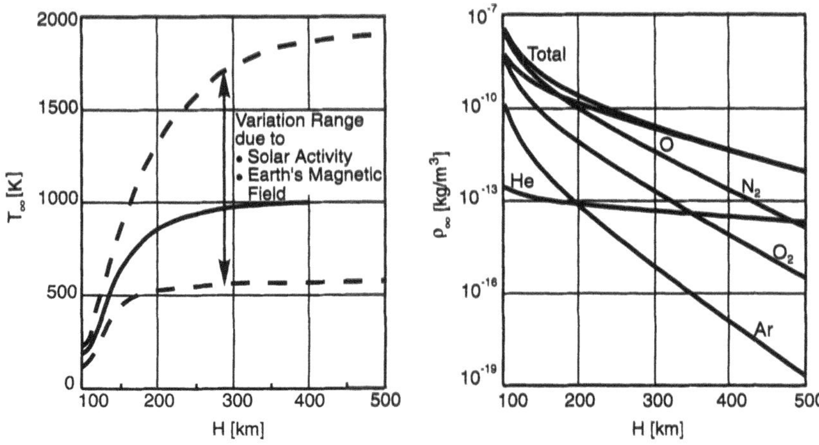

Fig. 3.25. Average Temperature (left), Composition with Average Densities (right)

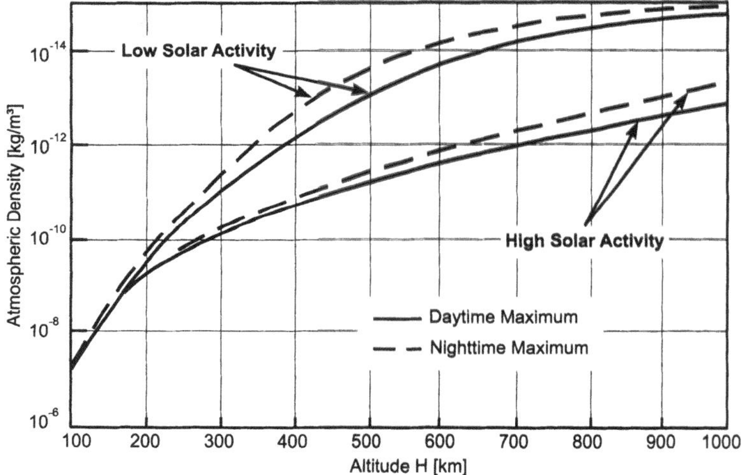

Fig. 3.26. Diurnal Density Variations for High and Low Solar Activity

Unlike the atmosphere considered up to the altitude of the turbopause (at about 100 km altitude) where the composition is almost constant, the atmosphere above this level is dominated by diffusion. Diffusion processes lead, together with gravitational forces, to the effect of light components such as helium and hydrogen (which is produced in lower layers by photo-dissociation of water), accumulating in the higher regions of the atmosphere. At altitudes above 1000 km, these components can even escape slowly from the Earth's gravitational field.

Today, there are highly developed models describing the neutral atmosphere; more detailed information on the subject can be found in [Skrivanek 94]. However,

Table 3.6. Density Changes in the Thermosphere for Different Altitudes and their Time Scales

	Δρ [%]				
Effect	**150 km**	**200 km**	**400 km**	**800 km**	**Time Scale**
1. Flux (Solar Cycle)	25	110	1165	3800	Years
2. Flux (Daily)	0	1	5	15	Days
3. Geomagnetic Activity	25	35	60	100	Hours
4. Local Time	10	25	115	230	Hours
5. Semi-Annual	15	15	50	80	Months
6. Latitude	10	15	60	90	Months
7. Longitude	2	2	5	15	Days

such models are relatively inexact especially for space flight application, since the principal influencing factors stated in Table 3.6 are insufficient. Other factors, such as sporadic factors which may be influenced, for example, by gravitational forces, cannot be reproduced as a model. Some effects of the empirical models can be described in more detail by the factors $F_{10.7}$ and K_p, which correlate with the electromagnetic and auroral heating. The $F_{10.7}$ radiation, measured at 10.7 cm wavelength, is caused by other than EUV mechanisms in the solar atmosphere and is not absorbed by the Earth's atmosphere; it is an approximate indicator for the solar activity. The geomagnetic index K_p, and a further index a_p, include the global influence by changes in geomagnetism, measured by a network of magnetometer stations on the Earth's surface. The models mostly used at present J70, MSISE90, GRAM and the latest one, VSH, vary only from 10 to 15%. They differ from each other mainly in their range of validity at lower altitudes and the user-friendliness of their interfaces. Figure 3.27 [Skrivanek 94] illustrates the standard deviations of the different models as compared to data gauged by electrostatic triaxial acceleration measurement aboard satellites at an altitude of 250 km, with an accuracy of about 1%. At higher altitudes, model errors increase: it is reported, for instance, that an error with a 25% standard deviation occurred for the typical altitude of space stations (400 km). These deviations are, among others, due to wind velocities not incorporated in the models which are also present at high altitudes (previous measurements already indicated wind velocities of 1.4 km/s at $K_p = 9$). The deviations, however, may be reduced to below 5% by measuring the drag coefficient.

The influences the atmosphere has on space vehicles can mainly be seen in the following points:

- At relative velocities of about 8 km/s, a significant impulse and energy exchange takes place between ambient flow and spacecraft. The applied drag is described by means of the drag coefficient C_D which, as a basis for simple estimations, can be set to about 2.2–2.5, where the effective surface is the space vehicle's surface normal to the flight direction. Additionally, the aerodynamic forces also cause

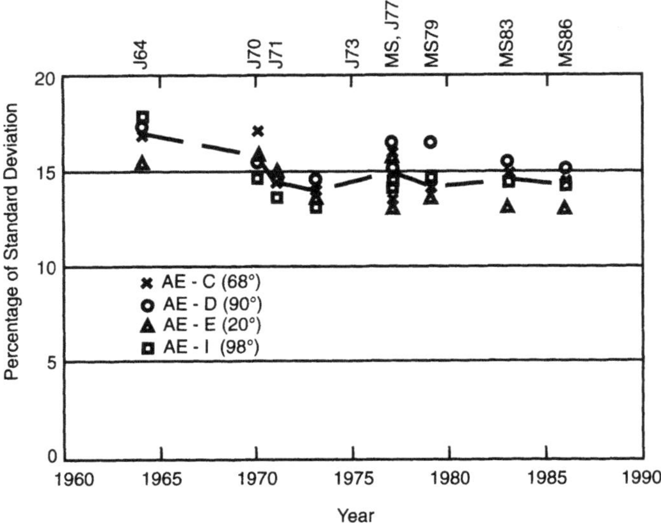

Fig. 3.27. Statistical Comparison of Empirical Models for the Neutral Atmosphere

torques, which can change the attitude of a space vehicle. Both influences must be balanced by active attitude and orbit control systems.

- Gas particles that impact the surface of a space vehicle can have mechanically or chemically erosive effects, especially atomic oxygen.
- Sometimes, gas particles in the atmosphere collide with the gases emitted by the vehicle. This can result in contamination of the vehicle surface (for example, optical glasses).
- In very low orbits, aerodynamic heating plays an important role. At an altitude of about 150 km, the aerodynamic heating reaches the order of magnitude of the solar constant and quickly decreases with increasing altitudes (see Fig. 3.28). At 300 km, this portion drops to about 1% of the solar radiation.

Events induced by the space station itself can, at least temporarily, have considerable effects on the immediate environment of the station. These effects will be described in Sect. 3.9, and Chapter 6 ("Attitude and Orbit Control System") will include a description of the atmosphere with respect to the design of a space vehicle.

3.6.2 Atomic Oxygen (AO)

At high altitudes, atomic oxygen forms a large part of the Earth's atmosphere and above 150 km, it is even its main constituent (66% at 200 km altitude, 90% at 500 km). Despite its relatively low particle density at high altitudes, it causes significant erosive effects due to its high chemical reactivity. This influence was discovered during the first US Space Shuttle missions with the unexpectedly occurring phenomenon of the so-called *Shuttle Glow*, i.e. a shine on the surfaces directly exposed to the ambient flow of the residual atmosphere. This shimmer occurs as a re-

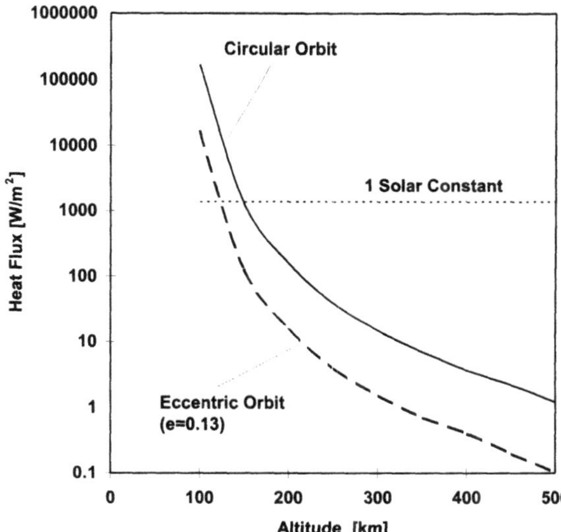

Fig. 3.28. Aerodynamic Heating in Low Earth Orbits

sult of chemical reactions between the surfaces and the atomic oxygen. At the same time, severe erosion effects of certain materials observed after the mission, for example weight loss of 35% within three days with Mylar® foils. Further experiments, for example aboard the Long Duration Exposure Facility (LDEF), the EURECA-1 platform and during the D2 mission confirmed the lasting influence atomic oxygen has on all surfaces exposed to flight direction [Ham 87, WPF 95].

The effects of atomic oxygen depend on the altitude but also on the solar activity, since atomic oxygen results mainly from photo-dissociation. Figure 3.29 shows the oxygen atomic flux as a function of altitude for a duration of one year: for the orientation of surfaces in ram exposure and for solar-viewing surfaces. Erosion rates caused by atomic oxygen can become very high, so coatings in the μm range can easily be eroded away after only a few days. A flow of 10^{22} atoms/cm^2 over a period of 11 years (one solar cycle) forms the basis of assumptions made for the International Space Station. Under these circumstances, a supporting truss made of a composite based on graphite, would within 15 years (the projected design life of the station), lose more than 30% of its initial wall strength of 1.25 mm. Figure 3.29 also shows the surface recession per year for a typical loss rate (R_e) of $3.0 \cdot 10^{-24}$ cm^3/oxygen atom. Apart from erosion, also recombination of oxygen atoms (under release of bonding energy) and a reaction with the wall (producing non-volatile constituents) occurs.

Most metals are usually not damaged by atomic oxygen; sometimes they even build up protective oxide layers, as in the case of aluminum. Exceptions, however, are silver (used for electrical contacts) and osmium (used for the vapor-deposited surfaces of optical instruments). On the other hand, many polymers that are used in space flight as foils, adhesives, paints or windows are subject to degradation. Also fibrous composites, after experiments aboard EURECA and D-2, partially showed severe damage of resin and matrix; sometimes the fibers even became exposed or porous.

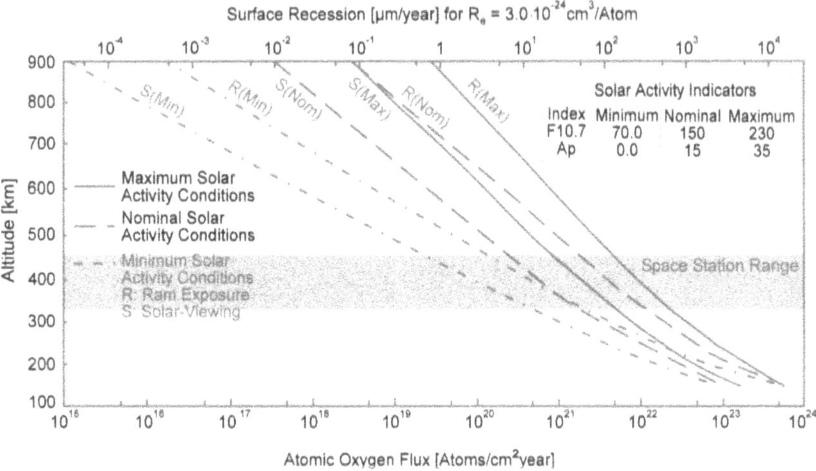

Fig. 3.29. Flux Density of Oxygen Atoms over One Year

Table 3.7. Loss Rates for Different Materials

Material	R_e [10^{-24} cm^3/O-Atom]
Polyethylene	3.7
Kapton	3.0
Mylar	2.2
Polyamide	2.2
Teflon	< 0.02
Material on Silicone Basis	< 0.1

The erosive effect of oxygen atoms can be expressed in loss rates. These rates indicate which volume R_e (Reaction Efficiency) of a material is eroded per impacting atom. Values of loss rates R_e obtained by means of experiments for different materials are summarized in Table 3.7.

Various coatings may be taken into consideration as protective measures against atomic oxygen erosion, e.g. surface treatments such as vapor-deposited gold for silver contacts or resistant coatings for fibrous composites. The long-term development objectives are resistant polymers [Dueber 93].

The erosion efficiency η is defined as the relation of R_e to the atomic volume of typical wall materials with $V_A \approx 3.0 \cdot 10^{-24}$ cm^3 as follows:

$$\eta = \frac{R_e}{V_A} \tag{3.19}$$

The erosion rate r in m/s amounts to:

$$r = \frac{N_O \cdot u \cdot \eta}{N_W} = \frac{\Phi_O \cdot \eta}{N_W} \tag{3.20}$$

N_O: Particle Density of Atomic Oxygen
N_W: Particle Density of Wall Material
u: Orbital Velocity
Φ_O: Flux Density of Oxygen Atoms

3.7 The Ionosphere

Solar radiation contains sufficient energy at short wavelengths to cause considerable photo-ionization in the upper Earth atmosphere. This is how results the *ionosphere*; it is a partly ionized layer mainly located between 50 and 200 km and extending up to an upper boundary of about 2000 km from the Earth's surface. The ionosphere only constitutes a small part of the atmospheric density. Although it extends to a maximum on the day side, it never exceeds 1% of the neutral gas density. The special importance of this conductive layer lies in its capability to reflect impacting radio waves and thus to make possible their distribution around the globe. This phenomenon was demonstrated for the first time by G. Marconi in 1901. Today, this property is the fundamental basis upon which research on the ionosphere is put. The ionosphere does not allow a radio link between a space station and the Earth below frequencies of 100 MHz. It can notably influence experiments and exposed parts of a station negatively, but it can also affect, for example, electromagnetic tethers in a positive way.

Since the 1920's it has been known that the condition of the ionosphere strongly depends on the course of the field lines of the Earth's magnetic field and hence on the geomagnetic latitude. We distinguish three regions of high, mean, and low latitude. Additionally, they are dependent on day, time, and seasons.

3.7.1 Ionosphere Models

The classic ionosphere model which can be attributed partly to S. Chapman, describes the ionosphere's characteristics at mean geomagnetic latitudes. Four different layers can be defined according to their electron concentration, designated as D (50–90 km altitude), E (90–160 km altitude), F_1 and F_2 (160–900 km altitude). Over the F-layer up to an altitude of 1200 km, we define the *upper ionosphere* and above that one the *plasmasphere*. This stratification is mainly caused by the following three reasons:

- Depending on their absorption characteristics, received solar energy is absorbed in different layers.
- Recombination of electrons and positive ions depends on the atmospheric density and this, in turn, on the altitude.
- The composition of the atmosphere changes with the altitude.

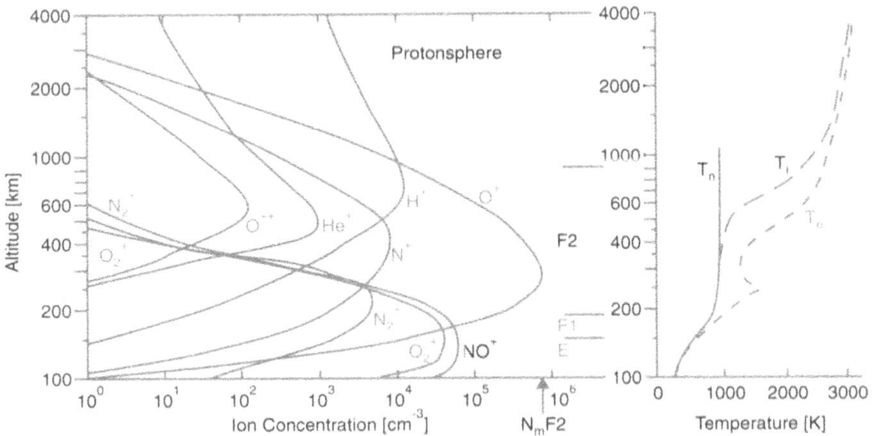

Fig. 3.30. Particle and Temperature Profiles in the Ionosphere

Except for the classic atmosphere model, there are other models such as:

- Empirical models, the best known of which is the International Reference Ionosphere (IRI) and the latest of which is the IRI-94
- Physical, time-dependent models (USU Time-Dependent Model of the Global Atmosphere)
- Models coupled with the thermosphere (NCAR Global Ionosphere-Thermosphere Model)
- Further models, partly very complex ones from the numerical point of view; with these models, a high-performance computer solves conservation equations (mass, impulse, energy, species) coupled with equations from the reaction kinetics and transport processes as a function of the different influencing factors [Skrivanek 94].

Figure 3.30 shows representatively the distribution of particle concentration and the temperature of electrons, ions and neutral particles as a function of altitude.

3.7.2 Variations in the Ionosphere

Only a few minutes after the occurrence of strong solar flares, the electron density in the D and lower E layers increases strongly. As a consequence, high frequency radio waves, which normally would be reflected by higher layers, are absorbed within these lower layers. These interruptions of communication links are called *Sudden Ionospheric Disturbances (SID)* and last for about one hour. SIDs are presumably caused by the solar X-rays coming from solar flares and penetrating the lower layers of the ionosphere.

Another kind of communication disturbances are the so-called *Ionospheric Storms*. Unlike SIDs, they can last for a duration of up to several days. Within the group of Ionospheric Storms, we distinguish the *Polar-Cap Absorption (PCA)* and the *Geomagnetically Induced Storm (GIS)*. In the case of PCA, high-energy protons, mainly originating from the solar flares, penetrate along the field lines of the Earth's magnetic field into the polar regions of the lower ionosphere layers (55–90 km) where they cause an increase in electron density due to ionization processes. As a consequence, communication is considerably impaired. The second phenomenon, the GIS, occurs about 20 hours after the actual solar flares. Firstly, low-energy electrons and protons penetrate along the field lines into the lower ionosphere layers, causing an increase in electron density (this effect is also referred to as an *Auroral Substorm*). Secondly, due to the enormous energy transfer to the Earth's magnetic field after the occurrence of solar flares, nitrogen and oxygen from the thermosphere (100–750 km) rise into the F-layer. Consequently, the electron density in these F-layers is considerably reduced as a result of the aforementioned recombination mechanism. The resulting concentration gradient causes considerable turbulence and perturbations in the ionosphere. Their effects can last up to one month at solar minimum.

3.7.3 Behavior of Radio Waves in the Ionosphere

In the ionosphere, charged particles can take away energy from impacting electromagnetic waves and thus weaken or even totally absorb them. However, a wave that moves through a region of a constantly varying electron density changes its direction and that, under certain circumstances, can lead to its reflection. The reflection of radio waves is hence based on an effect similar to the refraction of light in transition to an optically thinner medium, i.e. the beam angle increases with respect to the local vertical of the interface. Continuous refraction on layers with different electron densities causes the diversion of the beam path.

In principle, free electrons take up the energy of radio waves and reflect them at the same frequency. If, however, the density of the object with which they collide is high (e.g. in the D-layer), collisions between electrons and neutral particles will take place and most of the energy of the electrons will be transferred to the neutral particles. This energy appears in the form of heat (i.e. statistically distributed kinetic energy) and is lost for the signal transmission.

The ionosphere's dispersive properties also decelerate radio waves as a function of their frequency. This effect is often made use of, for example, in the case of satellite and space station supported navigation systems in order to compensate related errors by means of multi-frequent transmission techniques. The International Space Station will support the navigational and time-synchronizing signal techniques in various ways in order to synchronize clocks, computer networks, etc. For that reason, these ionospheric effects will be subject to future investigations by space station experiments.

3.8 Solid Matter

3.8.1 Meteoroids

A visible sign of the presence of cosmic matter in space is the so-called *Zodiac light*, which can be observed shortly before sunrise and shortly after sunset. Zodiac light is sunlight scattered within the interplanetary matter close to the ecliptical plane. Solid matter originating in deep space and which does not belong to planets or asteroids, is divided into the following categories:

- *Comets* consist of a solid core surrounded by the so-called coma (gaseous envelope), and a tail resulting from the influence of the solar wind; the tail appears fluorescent in the sunlight.
- *Meteoroids* are solid, non-fluorescent bodies in space.
- *Micrometeoroids* are meteoroids with a mass ≤ 1 g. They are important, because they occur much more frequently than meteoroids do.
- *Meteors* are meteoroids penetrating the atmosphere which partially or entirely burn up on entry. Meteors often occur in the form of periodic showers.
- Meteorites are fragments or residual pieces of meteors that impacted a planet's surface.

The latest estimations say that about 4000 tons of such matter collide with the Earth per year. We can identify the following sources of meteoroids:

- Disintegrated comets (especially causing showers of meteors)
- Asteroids, possibly also due to collisions within asteroid belts
- Release of dust particles and pieces of broken rock on the Moon, caused by the impact of primary meteoroids
- Residual matter from the birth of the solar system
- Interstellar dust grains moving through the solar system

Velocities observed in meteoroids range from 11 to 82 km/s (Fig. 3.31). The micrometeoroid density at a distance of 1 AU from the Sun (i.e. in the Earth orbit) is about $9.6 \cdot 10^{-20}$ kg/m^3; for most of the particles, masses are between 10^{-7} and 10^{-9} kg and the diameter on an order of magnitude of 0.01 cm. The mean density of the particles is about 0.5 g/cm^3. Satellite measurements came to the estimated value of 200 kg total mass below 2000 km altitude, distributed over the entire surface of the Earth [Debris 95].

A meteoroid storm named Leonids nearing Earth November 17–18, 1998 represented one of the largest meteoroid threat to a spacecraft in history. The particle shower, from the tail of the comet 55P/Temple-Tuttle, generated far more high-speed particles than any previous natural event since 1965. The closing velocities between spacecraft in orbit and incoming meteoroids were about 70 km/s – enough for a particle one-sixth of a millimeter in diameter to generate the same impact as a 22-cal. bullet [AWST Oct. 5, 98]. As a precaution, the world's satellite operators were gearing to turn solar arrays and other vital parts away from the incoming Leonid particle shower; they also powered down many nonessential systems and kept

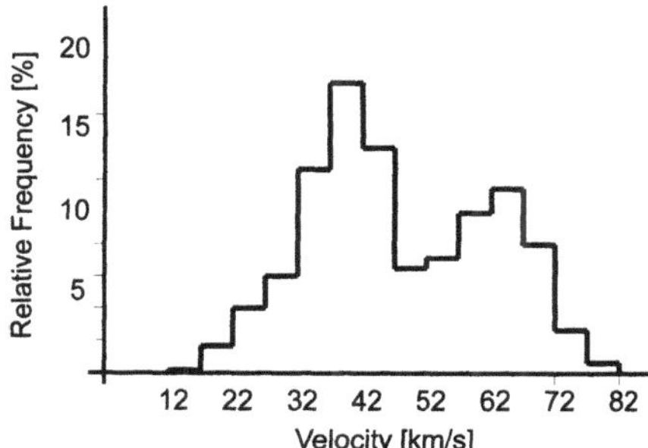

Fig. 3.31. Velocity Distribution of 11000 Meteoroids

computer commands and maneuvers at a minimum. NASA decided that no shuttle mission would be aloft during the period, and the November 20, 1998 launch date for the first element of the ISS, FGB, was specially chosen to avoid the meteoroid risk. It was even considered to ask the crew of the Mir space station to go into the Soyuz capsule during periods when Mir is on the side of the Earth facing into the storm. Fortunately, no severe impact on any operational spacecraft was observed.

The flux density, i.e. the number of micrometeoroids passing through a certain area each second, is independent of the direction, but close to the Earth, shading effects or gravitational influences may occur. The particle flux was measured by radar observation (only possible for particles exceeding a diameter of 4 cm), high-altitude sounding-rockets (TEXUS/MAXUS), and satellites (LDEF, SOLAR MAX, PEGASUS). These observations indicated that the total particle flux is composed of a statistically fluctuating part (sporadic flux) and a second part fluctuating on a yearly basis (showers of meteoroids).

3.8.2 Sporadic Flux

Different models exist for the flux that occurs sporadically with respect to time and space. Some of them are presented in Fig. 3.32. In this context, the integral flux density $F(m > m_0)$ is defined as the number of particles impacting per area and time with a mass $m > m_0$. The flux density is plotted over the lower mass limit m_0. Particles with hyperbolic trajectories that are capable of leaving the solar system are called β-meteoroids.

The so characterized natural environment in the form of meteoroid streams around the Earth remains stable for a long period of time and can hence serve as a means of comparison for the increasing number of artificial objects in near-Earth space.

3.8.3 Showers of Meteoroids

Contrary to sporadically occurring meteoroids, there are groups of meteoroids that orbit the Sun on a highly eccentric trajectory. Generally, these are remains of disintegrated comets. The intensity of meteoroid showers can be described by means of a time-dependent flux density value *z*, which varies over the year between 0 and 22, the average annual value being about 2.

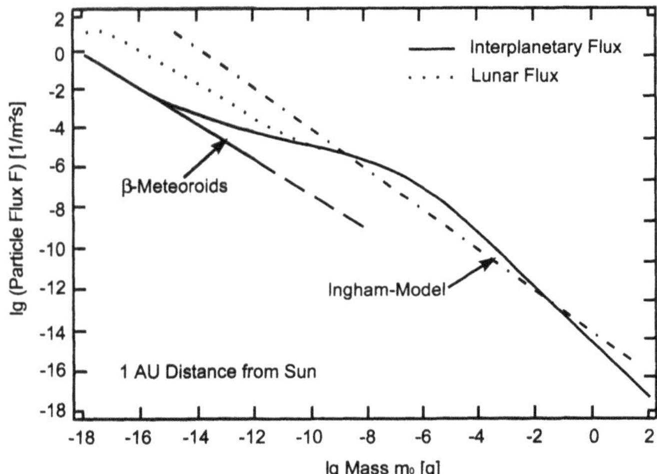

Fig. 3.32. ESA Meteoroid Model (1988)

As a result of the precession of the direction vectors of the space station's orbital plane and the equatorial plane with respect to the ecliptic plane, a potential collision with meteoroids can be regarded as omnidirectional from the design point of view. More exact calculations have to take the velocity and the relative orientation of space stations, showers of meteoroids, shading effects of the Earth, and the focussing effect of the gravitational field into account. The hazard for crew and equipment in a LEO is small compared to that presented by the man-made objects over 1 mm in diameter.

Between the differential flux density $F_m(m)$ and the overall flux density $F(m > m_0)$ exists the relation:

$$F(m > m_0) = \int_{m_0}^{\infty} F_m(m) \cdot dm = \frac{L(m > m_0)}{A \cdot \Delta t} \tag{3.21}$$

$L(m > m_0)$: Number of Impacting Particles with a Mass $m > m_0$
A : Reference Area
Δt : Measured Period of Time

The *Ingham* model gives a good approximation for the overall flux density F:

$$F(m > m_0) = (1 + z) \cdot \frac{C}{m_0} \tag{3.22}$$

$$\text{where } C = 6 \cdot 10^{-15} \frac{g}{m^2 \cdot s} \tag{3.23}$$

In this equation, the occurrence of sporadic meteoroids ($z = 0$) and showers of meteoroids is described by means of the flux density value z. The effect of this "meteoroid bombardment" onto the outer wall of a space vehicle can be described as follows:

If the structure withstands an impact of a mass $m < m_{kr}$, the mean number L of impacts on area A in the time Δt is given by:

$$L = F(m > m_{kr}) \cdot A \cdot \Delta t \tag{3.24}$$

Since in practice generally $L \ll 1$, and the occurrence of the particles with $m > m_{kr}$ being statistically distributed over the time, the probability that exactly N penetrations will occur, may be computed using the Poisson distribution:

$$p(N) = \frac{L^N}{N!}\exp(-L) \tag{3.25}$$

3.8.4 Space Debris

Any scrap of artificial, man-made origin moving in space is called space debris. A more detailed definition groups the following items under the term "debris":

- Uncontrolled payloads
- Operational debris and scrap (caused by normal space activities)
- Debris from explosions and collisions
- Microparticles (e.g. aluminum from solid boosters, paint chips, etc.)

3.8.5 Occurrence of Space Debris

Since the launch of Sputnik 1 in 1957, the US monitoring network NORAD observed and cataloged around 20000 objects of diameters exceeding 10 cm. Today, about 7500 of them are still in Earth orbits, the major part in LEOs (5747 objects in 1995). The remaining 13000 objects burned up or, in a few cases, payloads were returned to Earth. In addition to that, some 10000 objects > 1 cm and several 100000 objects > 1 mm have to be taken into consideration. The number of particles will continue to increase due to fragmentation of the debris particles as they collide with one another (the resulting fragments are called secondary debris), see Fig. 3.33.

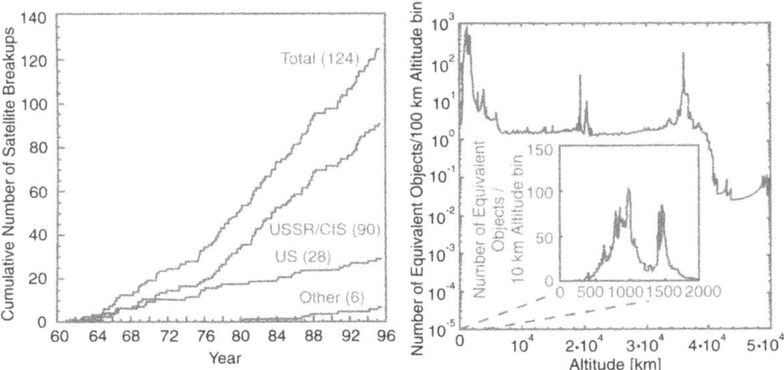

Fig. 3.33. Total Number of All Objects Cataloged in Earth Orbits and their Distribution as a Function of Altitude

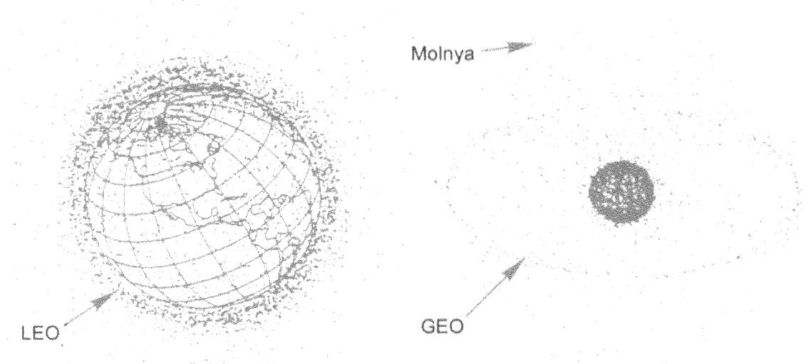

Fig. 3.34. All Objects > 10 cm in Diameter on January 1, 1987

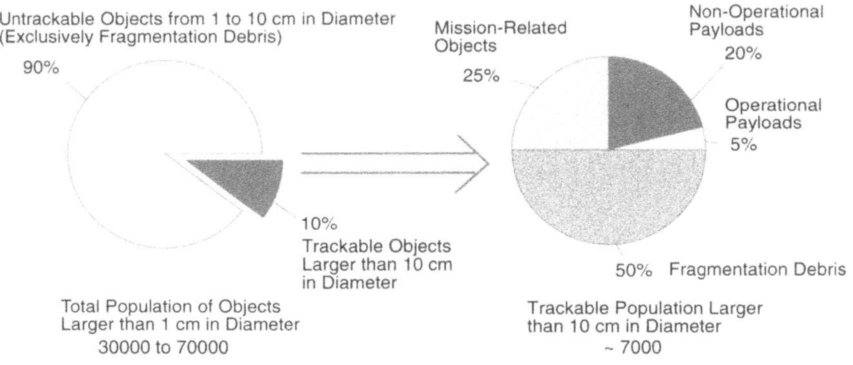

Fig. 3.35. Distribution by Percentage of All Objects > 1 cm and > 10 cm

Figure 3.34 shows an image of all objects registered in 1987, whereas Fig. 3.35 presents today's way of classifying objects into different categories. It is striking that of all objects > 10 cm, only 5% are active payloads.

Most debris objects have a nearly circular orbit. The lifetime of objects < 1 g at 500 km altitude or less is only a few years at the most, with high solar activity for only a few months. In geostationary orbits, on the other hand, their lifetime is practically unlimited. Figure 3.36 shows the orbital lifetime of objects as a function of orbit altitude and the influence of the solar activity for typical space station altitudes. This orbital lifetime is strongly dependent on the atmospheric density and the *ballistic coefficient* ≈ *mass/*(C_D·*area)*, i.e. a particle of aluminum foil, at the same initial altitude, will descend faster than the steel balls of a ball bearing.

In the case of elliptical orbits, the altitude of the perigee is of course a decisive factor. A fast lowering of the apogee as a consequence of the braking at perigee altitudes leads to a reduction of the eccentricity. The resulting "cleansing effect" below 600 km cannot, however, avoid secondary debris which results from collisions of debris objects with one another and their subsequent fragmentation.

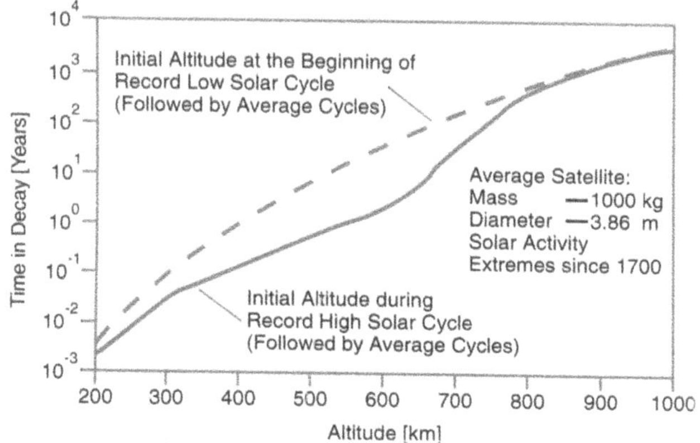

Fig. 3.36. Orbital Lifetime of Objects as a Function of Altitude and Solar Activity

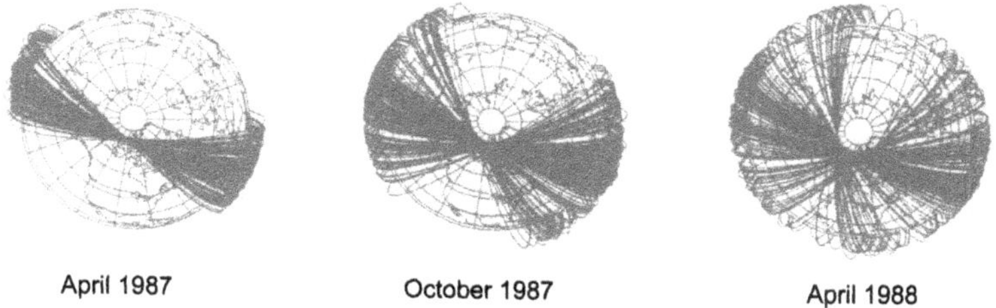

Fig. 3.37. The Dynamics of the Ariane-V16 Debris Cloud

Whereas the mean impact velocity in LEO is about 11 km/s, in GEO it is only about 100 to 600 m/s (for meteoroids about 20 km/s); the density is between 2.7 and 8 g/cm^3 since the objects are mostly metal. Due to its origins, the flux of the artificial debris particles depends on the direction and, additionally, on orbit altitude and inclination.

The trajectories of the debris particles may be perturbed and changed by the following four main influences which can have different effects depending on the corresponding orbit and object parameters:

- Atmospheric braking (LEO, GTO)
- Inhomogeneity of the Earth's gravitational field
- Solar radiation pressure (GEO, but also in the case of small mass/area ratio)
- Gravitation of the Sun and the Moon (GTO, GEO)

Figure 3.37 illustrates the influence such perturbations have; it shows especially the influence of the Earth's oblateness. In this figure, the distribution of the trajectories of debris caused by the destroyed upper stage of the Ariane V16 (exploded in 1986) is shown.

3.8.6 The Development of Space Debris-Caused Risk

Debris particles may be divided into the following three groups according to their size and possible effects:

- Particles < 0.01 cm cause surface erosions on layers of paint, plastic and metal parts.
- Particles of 0.01–1 cm cause severe damage which, according to the degree of shielding, can be dangerous for a spacecraft and its mission.
- Objects > 1 cm can cause catastrophic damage.

Further adverse effects are the disturbances of observation by debris particles obstructing the view. Radioastronomy is also restricted by direct reflections from such particles.

The space flight activities of the past decades have produced much more debris particles at space station altitudes than were eliminated by the natural cleansing effect resulting from the decelerative effect of the atmosphere. The result of worldwide launches (about 100 per year, from 1981 to 1994 their number decreased from 123 to 93, with a maximum of 129 in 1984 and a minimum of 79 in 1993) is that the number of the cataloged, i.e. large objects, increased by 200–300 per year [Debris 95]. The effect of high solar activity was observed toward the end of the 1970's and at the beginning of the 1990's, when the number of cataloged objects increased marginally, which was less than usual.

The future projection is dependent on the sources and sinks of debris, and those, in turn, depend on models for launch, fragmentation, distribution, lifetime, and possibilities for observation. The models developed by NASA are supported by fragmentation models and observational data derived from laboratory experiments. In simplified form, they are used as engineering models for prediction. That such a prediction is uncertain, above all, is because of the inaccuracy of the fragmentation models and, to a lesser degree, because of the atmosphere or transport model. Such a prediction model was derived from the NASA model EVOLVE and has been used since 1991. The cumulative impacts per m^2 and per year presented in Fig. 3.38 and Fig. 3.39, result from the assumption of the number of launches and type of payloads during the past 10 years, average solar cycles and additional efforts to avoid explosions. These values are valid for altitudes of 400 km and 1000 km, and 400 km and 800 km, respectively. As can be seen, the cumulative collisional cross section rather increases, despite avoiding explosions (launcher upper stages, batteries, etc.) and also despite the extensive efforts to avoid new debris caused by collision and fragmentation mechanisms. This increase is especially valid when we think of the fact that several of the planned Personal Satellite Communications Systems such as Iridium, Globalstar, ICO and Teledesic in some cases comprise hundreds of individual satellites.

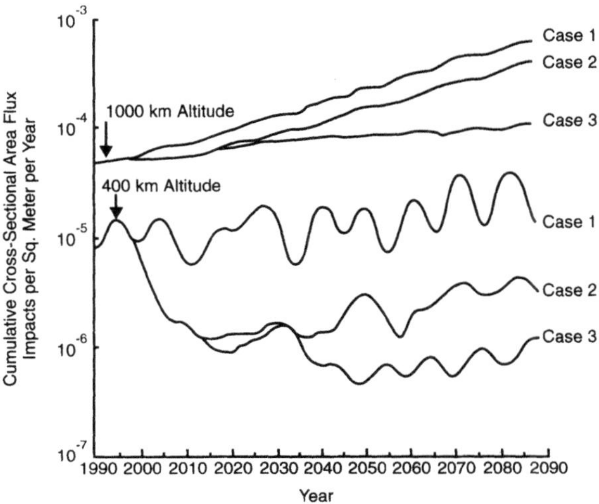

Fig. 3.38. Probable Development of the Cumulative Collision Rate per m^2 and per Year in the Period 1990–2090, Chain Reactions not Included

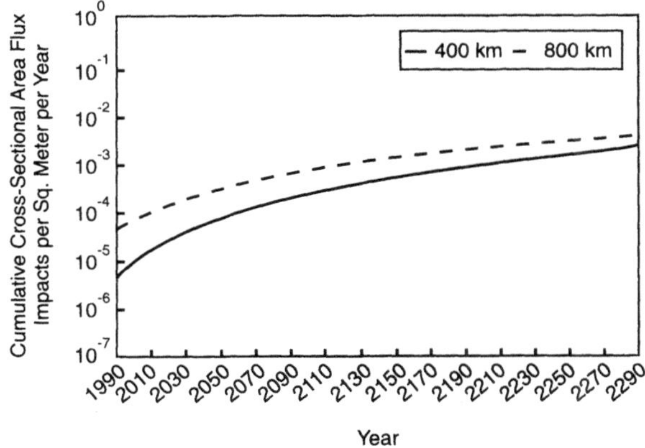

Fig. 3.39. Probable Development of the Cumulative Collision Rate per m^2 and Year in the Period 1990–2290, Chain Reactions not Included

Fig. 3.40. Possible Designs for Wall Protection [Hallmann 88]

3.8.7 Protection against Space Debris and Implications for Space Stations

In order to protect especially crewed space vehicles from catastrophic or even fatal consequences caused by debris impact, various protection mechanisms have been discussed and partially implemented. Certain parts of a space vehicle, such as solar arrays or antennas, can practically not be protected against an impact. In such a case, the systems have to be designed in such way that they include redundancies. A corresponding design can keep the cross-section of the vehicle in flight direction as small as possible, since most of the debris impacts can be expected to come in the local horizontal plane (debris plane, cf. Fig. 9.10). Evasive maneuvers as they have been carried out with the Space Shuttle are another possibility; this option, however, will be less practicable in the case of a space station. Finally, protective layers can be integrated into the wall itself.

Due to weight penalties, sufficiently dimensioned wall strengths are usually not feasible. For that reason it is a good solution to design a dual wall with an intermediate layer which prevents the particles from penetrating the inner wall. Multi-wall shields are very promising where the particles go through a thin outer wall and the tiny fragments resulting from this interaction are stopped by a second wall (Fig. 3.40).

Also an improved prediction of debris distribution may contribute to the protection of vehicle and crew. Therefore, the improvement of prediction models is constantly being worked on, and space vehicles and components that had previously been in space (e.g. the Long Duration Exposure Platform LDEF, the platform EURECA or the solar panel which had been exchanged at the Hubble Space Telescope) now deliver valuable contributions to it.

To date, intended or unintended explosions in space have been the main source of space debris. In the future, however, the increasing number of particles and the

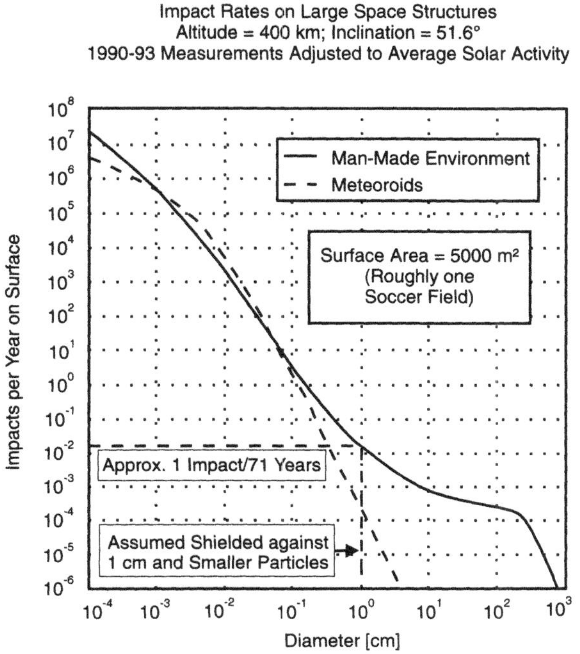

Fig. 3.41. Predicted Impact Frequency of Debris and Meteoroids for ISS between 1994 and 2030

collisions between them will have to be given more attention. According to recent calculations, there is the danger of a chain reaction setting in, when double or triple the amount of objects existing at present is reached. Such a reaction would make certain orbit altitudes unusable for a long period of time. Figure 3.41 shows the cumulative impact frequency predicted for the period 1994–2030 for a space station at an altitude of 400 km, with 51.6° inclination and an area of 5000 m^2. Whenever a spacecraft collides with a large object, there is high probability that this object will be a debris particle. As far as collisions including objects smaller than 0.1 cm are concerned, there is an equal probability that the collision will be caused by a meteoroid.

The implication of the growing hazard caused by space debris on the design and operation of a space station is manifold. The International Space Station, for example, has been designed such that the critical areas at the front part (= in flight direction, see Fig. 9.10) can withstand the particles which are most likely to occur: objects with 1.4 cm diameter or smaller, i.e. 99.8% of the debris population. The result of an analysis presented in Fig. 3.41 predicts approximately one impact of a particle with 1.0 cm diameter or smaller over a period of 71 years. A particle of 1.4 cm or larger does not, however, necessarily lead to a catastrophic failure for the space station.

The ISS strategy for collision avoidance with large objects is based on prediction and, if necessary, swerving, as has been the norm with the Space Shuttle and the space station Mir. Compared to the Apollo program (0.01–0.05 probability of penetration per mission), these strategies have led to an acceptable degree of risk.

More frequently, impacts of particles which are too small for penetration or structural damage occur. Most of the particles are the size of a grain of sand. These, however, can impair the vehicle surface or the sensitive solar cell surfaces and hence their functions. This kind of damage has to be taken into consideration and repaired by means of routine measures.

Apart from the aforementioned protective measures, the positions of the air locks are determined at places with lower impact risk, and internal structures such as racks, facilities, etc. are installed in places with higher impact risks. Various repair measures and operational procedures for periods with high particle fluxes are being prepared and additional debris–protective surfaces can be installed any time. With respect to the surface of an EVA suit and to the duration of an EVA, the hazard for astronauts working outside the station is relatively low; nevertheless the number of EVAs should be as small as possible.

All investigations addressing the long-term evolution of orbital debris conclude that, without changes to the way space missions are performed, regions of near Earth space will become so cluttered by debris that routine operations will not be possible. The options available to decrease the growth of orbital debris depend greatly on the altitude of the mission, design of hardware, and the commitment of the international spacefaring community.

The amount of debris can be controlled in one of two ways: debris prevention or debris removal. Table 3.8 shows individual techniques under each of these categories. Several of these techniques are already practiced by space users at this time. The fact that some debris minimization techniques are already being used voluntarily bodes well for the future, but is not clear at this time which of the methods are most effective and how to measure the cost-benefit tradeoffs for each. Continued research is required in this area. Identification of realistic and effective methods is the most important issue.

Further important information on strategies to avoid damage caused by debris, international agreements, observational techniques, particular models for the description of debris environment in LEO, suggestions for future research activities, and finally general recommendations, etc. can be found in [Debris 95, IAA Debris 95].

Table 3.8. Methods to Reduce Debris Population

Prevention	Removal
Design and Operations	Retrieval
Expulsion of Residual Propellants and Pressurants	Propulsive Maneuvers (Deorbit)
Battery Safety (Vent or Fuse)	Drag Augmentation
Retention of Covers and Separation Devices	Solar Sail
Propulsive Maneuvers (Reorbit)	Tether Sweeping Laser

3.9 Induced Environment – Contamination

The previous sections in this chapter on "Orbital Environment" have been dealing with the *natural environment* of space stations and/or vehicles. The so-called (self-) *induced environment*, on the other hand, also has an influence on the operation of a space station or on the experiments to be performed; that influence must not be ignored [ESA Guide 96]. During normal operation, the space station itself or other space vehicles cause contamination in their immediate environment. Such contamination may be caused, for example, by venting of air locks, outgassing of materials, operating of thrusters, or by particulate matter resulting from the weathering of materials which are exposed to the space environment. During its operation, the International Space Station is intended to have so-called "quiescent periods" (by analogy with "microgravity periods"), nominally of 30 days duration during which no active contamination will take place, and so-called "non-quiescent periods", when active contamination will take place. As a consequence, it may be necessary to stop the operation of sensitive external payloads during these non-quiescent periods.

The induced environment can be characterized as follows:

Molecular Deposition (MD). Material outgassed from the external surfaces will form the major part of the molecular deposition during the quiescent periods. This can be reduced by advance vacuum-baking of the external materials (e.g. silicone sealants). During the non-quiescent periods, molecular deposition is mainly caused by:

- Operation of the AOCS thrusters of the Space Shuttle close to the space station
- Dumping of Space Shuttle waste water
- Dumping of condensed water from the space station via vents
- Operation of ACS and RCS thrusters aboard the space station

Further vehicles visiting the space station will be the Russian Soyuz and Progress capsules as well as the European ATV. According to present estimations, the molecular depositions of the above mentioned contributors will presumably exceed space station requirements during the quiescent periods.

As an example, it can be shown how the operation of a thruster may influence the space station environment. Concretely, the influence that a thruster of a Progress-M capsule (docked at the Russian Service Module) has on the environment of the Payload Attached Structure was examined with the help of a simulation called "Direct Simulation Monte-Carlo" (DSMC). The Progress-M transporter is located in flight direction at the rear end of ISS, whereas the Payload Attached Structure is located at the port side at the end of the truss. Figure 3.42 shows the isolines of the particle densities during the operation of the thruster. The particle density in the immediate vicinity of the Payload Attach Structure (PAS) reaches 10^{21} 1/m^3 and is thus approximately seven times higher than the natural particle density [Depth 96].

Molecular Column Density (MCD). The molecular density depends on the intervals and durations of the quiescent periods. Except for the operation of the Space

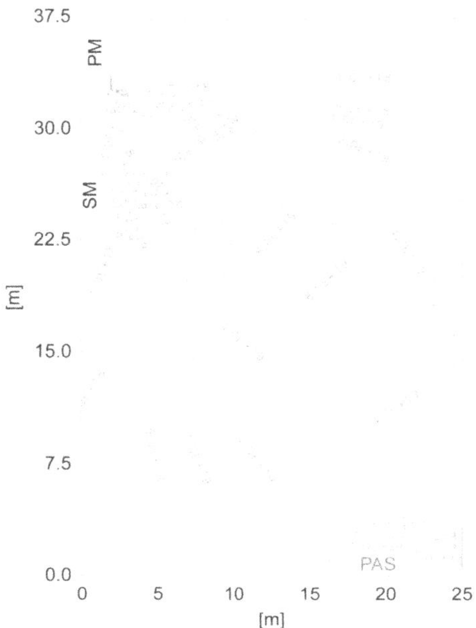

Fig. 3.42. Influence of a Thruster on the ISS Environment: Isolines of Particle Densities [Depth 96] (PM: Progress-M)

Shuttle, venting of laboratory modules will lead to short, unplanned, non-quiescent periods, and the venting of condensate water will cause longer non-quiescent periods once every few days. The influence caused by the venting of CO_2 by the Life Support System (ECLSS) and by leaks in the modules is still being examined, but it is presumed that the predicted CO_2 concentrations in the space station environment can lead to unplanned non-quiescent periods.

Particle Release. The release of particles from the outer surfaces of the space station during the quiescent periods is not considered a problem.

The above described contamination can influence the space station's operation and therefore can also directly or indirectly influence the payloads in different ways. Molecular or particulate depositions, in combination with ultraviolet radiation and atomic oxygen, may influence the thermo-optical properties of surfaces. This, in turn, can reduce the efficiency of the Thermal Control System. A high molecular density that may limit the field of vision of optical payloads or that leads up to depositions on optical surfaces, may reduce the optical quality, e.g. by absorption and/or scattering of radiation, reduction of signal strength, increase in background noise, and/or interferences. Contamination also can influence the power of solar generators by attenuation or absorption of certain wavelength ranges of electromagnetic radiation which is intended to reach the solar cells.

Measures for minimizing contamination have to be carried out before launch (such as cleaning of surfaces, selection of appropriate materials, suitable configuration and arrangement of valves) as well as during space station operation (such as timing of venting in order to ensure reliable prediction of quiescent and non-quiescent periods). This applies especially to the emission of water that may influence observation within the IR range.

Other, more specific types of contamination can become important for individual payloads, such as the following:

- Plasma wake: The variation of plasma density from the ram to the wake side of the space station
- Neutral wake: The variation of neutral atmosphere density
- Plasma waves induced by the space station's motions
- Vehicle and station glow on the ram or forward side
- Electrostatic interaction between the space vehicle surface and the plasma environment leading up to charging effects that determine the electric potential of the station
- Variation of the plasma density and generation of electrical noise due to static charging of vehicles
- Emission of conducting and radiating electromagnetic interferences
- Visible light generated by the space station and its reflections
- Induced electric potential caused by the motion of the station and by the Earth's magnetic field
- Vibrations or movements influencing external payloads
- Intended influence on the environment by experiments

Electrostatic charging of the vehicle surface due to interaction with the plasma environment plays an important role for the space station. For example, on May 5, 1995 there was a malfunction in the F13 satellite belonging to the Defense Meteorological Satellite Program (DMSP) on a Sun-synchronous, near-Earth orbit (= 840 km). From parallel measurements of the energy levels and electron densities, for the first time conclusions were able to be drawn about the processes during electrostatic charging in LEOs [Tribble 95]. Vehicles in LEOs traverse the ionospheric plasma (cf. Sect. 3.7) which is composed of oxygen atoms and, in less than 1%, of free electrons, ions, and ionized molecules within the range of only few keV, resulting from UV radiation. Apart from naturally resulting ions, ions can also be mechanically produced by collision with high-energy particle fluxes or by photoionization from the vehicle's surface. The thermal velocity of the charged particles (which, for the most part, consists of O_2^+-ions) on such trajectories is lower than the vehicle velocity, which, in turn, is lower than the electron thermal velocity. For that reason, the spacecraft collects electrons on all surfaces, but collects ions only on the surface of the ram side. Therefore, negative electric potentials typically within the Volt range (1 Volt negative is a typical LEO floating potential for a spacecraft powered by fuel cells such as the Space Shuttle) can occur on non–grounded surfaces in the wake (backward) vector direction. This negative potential called "floating potential" is not critical for most of the space vehicles. Since vehicles normally have surface materials with different conductivity, the occurrence of large potential differences may lead to sudden discharges in combination with arcing between or through materials. Such arcs, for example between the coverslide of the solar cells (a dielectric) and a nearby solar cell interconnect (a conductor) may melt these conductor lines and lead to a loss of electrical energy for this cell string. In the case of thermal insulation, which mostly consists of dielectic materials, such arcing causes micron-sized channels in the material. As a consequence, the thermo-optical properties of the surface are changed, and that, in the long run, leads to breakdown of the thermal insulation. Additionally, the environment will be contaminated. The arcing can also perturb sensitive onboard electronics due to simultaneously occurring electromagnetic interferences (EMI). Such damage, for example, in the life support system of crewed space systems, may easily cause catastrophic failures. In the case of the International Space Station with its negatively grounded solar arrays (-160 V), a floating potential of as much as -140 V is assumed [Tribble 95]. In order to avoid the occurrence of arcing e.g. between ISS and a docking Shuttle, a plasma conductor is installed to reduce the electric potential to -40 V.

The effects solar flares have on the ionosphere (cf. Sect. 3.3.3), which have also been discovered in relation to the malfunction of the DMSP-F13, pose a particular problem. The satellite was in a zone of very low plasma density ($< 10^4$ cm^{-3}) in the Earth umbra, when the sensors within seven minutes measured an increase of the integral flux density of high-energy electrons (31 keV and more) to over 108 electrons $cm^{-2}s^{-1}sr^{1}$. Thus, the energy density of the particles impacting the satellite was markedly larger than that of the free, thermal ions. The electron flow was 50 times higher than the nominal value, and the potential at the vehicle's surface increased to over -459 V when the potential discharged itself over the entire satellite structure and caused the breakdown of a control computer. The consequence for vehicles in LEOs is that, in the case of all thermal coatings designed as MLIs (Multi-Layer Insulation), the bottom side of the top layer needs to be directly grounded in order to achieve larger time constants: Due to the high dielectric constant of the Teflon® used for such coatings, the time in which the vehicle can cross such an anomaly is not sufficient for a discharging arc to cause a breakdown and the charging can be conducted away by the bonding.

If, in concluding this chapter, we look at the subject from the experimental point of view, the contamination situation for ISS seems to be extremely complicated. Nevertheless, by being precisely familiar with the induced environment, its variations over time as well as the direct and indirect effects, experimenters may design payloads and their operation such that not only the contamination hazard is minimized, but contamination originating from the payload can also be detected and perhaps even avoided.

In order to achieve better modeling of both the natural and the induced environment of the International Space Station, R. Bertrand et al. developed a Space Station Design Workshop (SSDW) on the basis of earlier works by F. Pohlemann. In the framework of the SSDW, the most important physical models were integrated into a coherent work environment (cf. Sect. 9.3.3). The example of an environment induced by thrusters, shown in Fig. 3.42, proved that available numerical-theoretical transport models for an LEO environment (e.g. Direct Simulation Monte Carlo models) can be incorporated into the SSDW models and yield important results.

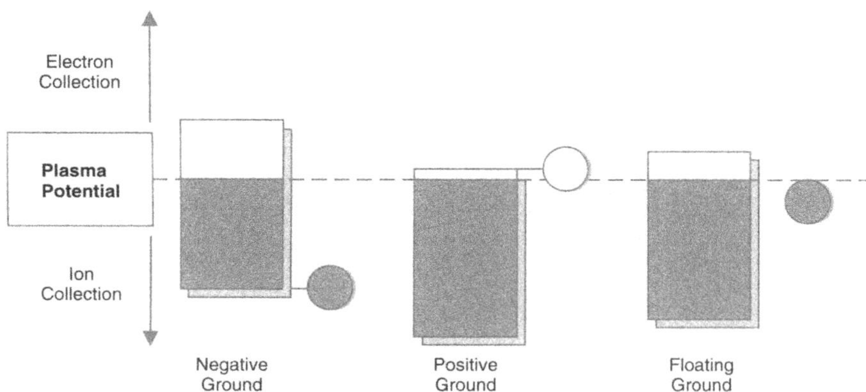

Fig. 3.43. LEO Spacecraft Charging

For a more general application which is not limited to the LEO environment, the European Cooperation for Space Standardization (ECSS) is producing a system of standards covering all aspects of space mission execution. There is a Space Environment Standard, ECSS-E-10-04, within the Systems Engineering Branch of ESA/ ESTEC ("E-10"). As a "level 3" standard, it aims to provide authoritative information which will help to make the design, development and execution of missions more efficient. In this engineering process, it is important that environments and their effects are properly understood and taken into account. In addition, use of a common standard ensures visibility of practices among collaborating research companies or organizations.

The standard provides information on the various aspects of the space environment and its effects. Where possible it identifies recognized standard environmental models and methods for assessing environmental effects. The environmental aspects covered are the following: gravitation, geomagnetic field, solar and Earth electromagnetic radiation and indices, atmospheres, plasmas, energetic particle radiation, particulates, and contamination. In some of these areas, well-established internationally recognized standard models exist, such as the International Reference Ionosphere, the International Geomagnetic Reference Field, and the COSPAR International Reference Atmosphere. In others, widely used models and methods have become de-facto standards. Problems caused by a lack of internationally accepted standards and by models which had shortcomings were identified. As a result, the Standard is expected to continually evolve as these problems are resolved. Coordination with CEN, COSPAR, ISO and other organizations is underway. An "active" WWW-based version is under development within Spenvis (the Space Environment Information System: http://www.spenvis.oma.be/) [Daly 98].

The Space Environment Standard ECSS-E-10-04 is currently being prepared by a working group under the responsibility of the above mentioned ESTEC Branch.

4 Environmental Control and Life Support System

The Environmental Control and Life Support System (ECLSS) is a subsystem typical of crewed space vehicles which provides all the necessary conditions in order to make life in space possible. In its first section, this chapter places the focus on human life and the requirements which should be fulfilled in order to establish a human presence in space. The second section will discuss all tasks of a life support system resulting from these requirements. Moreover, different ways of classifying life support systems will be presented; available life support methods and technologies will be described and also some examples of life support system design will be given. The chapter will be concluded with an outlook on bioregenerative life support systems and how the subsystem "life support" is integrated in the overall system "space station".

4.1 ECLSS: Environmental Protection for the Crew

The crew aboard a space station has to perform a wide range of tasks with the utmost degree of accuracy and reliability. This can only be achieved if they are provided with a physiologically and psychologically tolerable environment. Moreover, human beings need sufficient amounts of energy, and particularly in the case of long-duration missions like duty aboard a space station, they have to maintain good health.

From the point of view of system design, a human body can be regarded as a "black box". Some input flows enter this black box, output flows leave it, and additionally, heat is released (cf. Fig. 4.1). The task of a space vehicle's life support system is, to keep the balance between these two flows. Depending on their order of importance, different boundary conditions have to be heeded; they will be addressed in the following sections.

4.1.1 Physiological Boundary Conditions

Physiological boundary conditions are direct short-term requirements for the survival of a human organism. Firstly, they determine the environment directly surrounding the human body. In this context, the following characteristics are of primary importance:

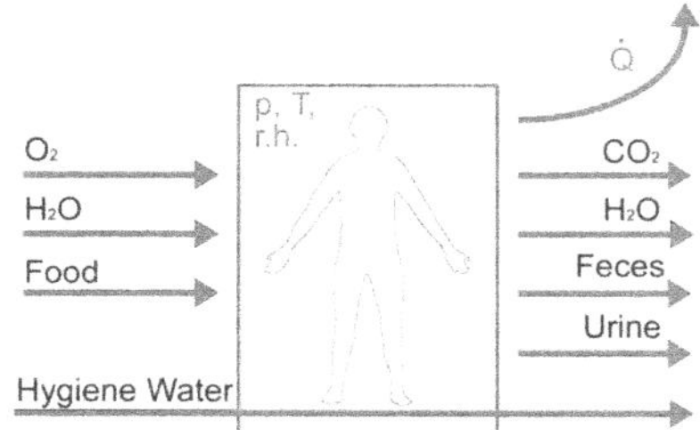

Fig. 4.1. The Subsystem "Human Being"

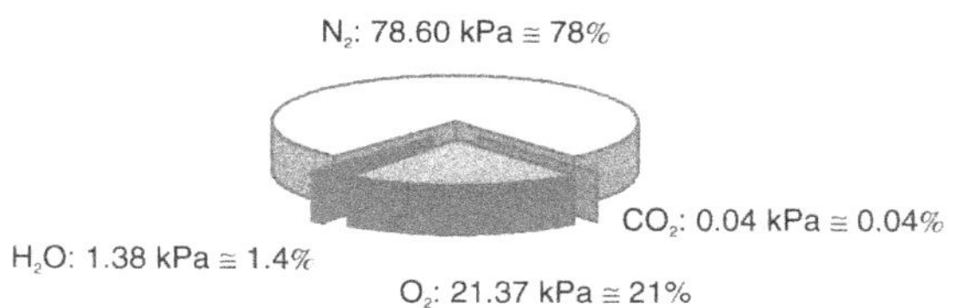

Fig. 4.2. Normal Atmosphere Composition

Gas Pressure and Composition. The normal atmosphere at sea level has a mean total pressure of p = 1013.6 hPa. It is a mixture of mainly nitrogen and oxygen (cf. Fig. 4.2).

The total pressure of an artificial cabin atmosphere can vary as long as it is kept within two defined boundaries: the oxygen partial pressure must be around 220 hPa, and the content of carbon dioxide must be kept well below 1% by vol. The nominal carbon dioxide value for ESA's Spacelab was 6.7 hPa at a total pressure of 1013 hPa, which corresponds to a volumetric fraction of 0.66%.

Discounting the safety concerns such as increased fire hazard, the total cabin pressure theoretically can be lowered until it reaches the necessary oxygen partial pressure. This reduction of the total pressure, however, requires that the oxygen volumetric fraction be increased. Earlier US missions used this method with oxygen fractions of nearly 100% (cf. Fig. 4.3). This strategy was abandoned in 1967 after a tragic accident in which three astronauts lost their lives during ground testing for the Apollo program: the interior of the command capsule, containing a pure oxygen atmosphere, was completely burned up due to a short circuit.

The minimum value for human existence in a pure oxygen atmosphere is p_{min} = 186.2 hPa. Toxic symptoms occur when human beings are exposed to oxygen partial pressures of more than 700 hPa over several days. The first signs of intoxication are, for example, cough and hyperoxia.

As Fig. 4.2 shows, the carbon dioxide content in the Earth's atmosphere is about 0.04% by volume. An increase in the carbon dioxide concentration to values of 1–4% by vol. leads to a higher metabolic rate in the human body, and every value exceeding this level will immediately cause symptoms of intoxication.

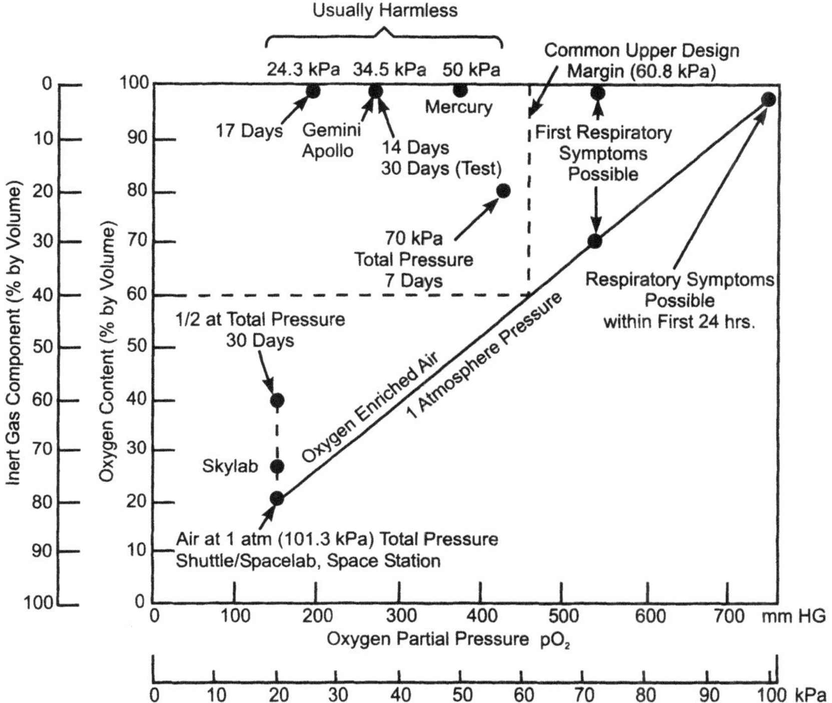

Fig. 4.3. Air Composition of Selected ECLSS [Hallmann 88]

The Space Shuttle cabin has a normal atmospheric composition, as it will also be used for the International Space Station. Based on a fixed oxygen partial pressure, the total pressure will be controlled at 1013 hPa with the help of nitrogen. Carbon dioxide will be removed by a special filtration unit.

Temperature. The onboard temperature influences the physiological as well as the psychological well-being. However high or low the crew members perceive the temperature inside the cabin to be, depends on various factors such as humidity, physical activity, air circulation, clothing, etc. For that reason, the crew should be able to control the temperature locally. Literature on this subject suggests optimal skin temperatures of 32.8 ± 3°C, corresponding to an air temperature of 18–21°C. In Table 4.1, some design values of temperatures aboard earlier missions are given.

Humidity. For technical safety, i.e. in order to avoid condensed water entering the onboard systems or biofilms growing on free surfaces, the relative humidity is usually controlled between 25% and 75%. Specifications often indicate the humidity by means of the so-called "dew point". The dew point is the temperature at which vapor begins to condense after an isobaric cooling. With the help of the equation of the

Table 4.1. Cabin Atmosphere Aboard Different Space Vehicles

Space Vehicle	T [°C]	Rel. Humidity [%]
Mercury	32–38[a]	
Gemini	15.5–27	
Apollo	21.2–26.8	40–70
Skylab	14–32[b]	
Space Shuttle	18–27	30–70
Spacelab	18–27	30–70

[a] Additional Cooled Pressurized Suit
[b] Depending on Wall Temperature

steam pressure curve (Clausius-Clapeyron equation), an approximate conversion can be achieved (cf. Eq. 4.1):

$$\frac{1}{T_{DP}} = \frac{1}{T_0} - \frac{R}{r}\ln\varphi \tag{4.1}$$

T_{DP} : Dew Point Temperature [K]
T_0 : Onboard Temperature [K]
R : Specific Gas Constant for Water = 462 J/kg K
r : Evaporation Enthalpie for Water = 2454 kJ/kg
φ : Relative Humidity

For an exact conversion, a steam table is necessary. As a physiological parameter, however, the absolute humidity is more important than the relative humidity. Its standard value can be indicated with a water vapor partial pressure of 13 hPa. When the absolute humidity is too low, symptoms we know from our terrestrial winter such as dry skin and chapped lips occur.

4.1.2 Metabolic Boundary Conditions

When mission duration exceeds a few hours, the most important task of the life support system is to maintain the mass flows (cf. Fig. 4.1). As indicated by the term "*metabolic* boundary conditions", these mass flows are mainly determined by human metabolism. Table 4.2 shows the mass flows for the average adult person per day, as specified for ISS.

The average heat production per person per day is about 11.83 MJ. According to crew activity, the heat flow obviously varies from the average value of 137 W. Consumption of potable water per crew member per day is slightly over 3.5 kg, 2.35 kg of which are consumed in the form of beverages and the rest being used for oral hygiene purposes and food rehydration. Depending on how much comfort the crew is granted, the amount of hygiene water indicated in literature on this subject varies

Table 4.2. Human Mass Flows

Inputs [kg/Person-Day]		Outputs [kg/Person-Day]	
O_2	0.84	$CO_{2\,out}$	1.00
H_2O_{in}	3.55	H_2O_{out}	1.83
Food (dry)	0.64	Urine Feces	1.63 0.12
Hygiene Water	6.80	Waste Water	6.80
Σ_{in}	11.83	Σ_{out}	11.38

from 1.15 to 20 kg; with increasing allowance, when including washing machines or dishwashers for long-duration missions.

Aboard the International Space Station, each crew member will be provided 2.8 kg of water on average for food rehydration, potable water and oral hygiene; the amount of hygiene water is specified as 6.8 kg/person-day [SSP 41000 D].

4.1.3 Additional Boundary Conditions

Additional factors have to be taken into consideration to ensure a safe, healthy and pleasant environment aboard a space station, for example:

- *Preventing injuries.* Offer radiation protection and noise reduction, limit maximum accelerations and gravity (not applicable if medical countermeasures are taken)
- *Keeping the psychological balance.* Maximize available space, provide visual privacy and Earth viewing opportunities, balance illumination and colors, etc.

Even though these factors are of a certain importance for the success of a mission, they do not fall under the heading of the tasks of a life support system. For this reason, they will not be explained in detail in this chapter, but in Chapter 11 on "Human Factors".

4.2 Tasks of an ECLSS

4.2.1 Overview and Classification

On the basis of the boundary conditions mentioned before, five main tasks of a life support system can be distinguished:

- Atmosphere management
- Water management
- Food production
- Waste management
- Crew safety (i.e. fire detection and suppression as well as radiation protection)

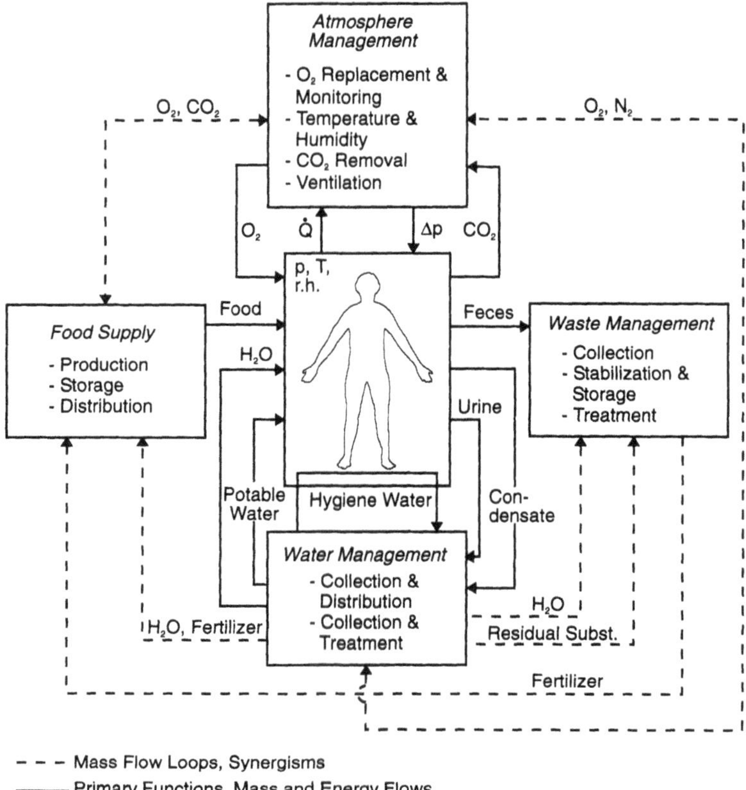

Fig. 4.4. Tasks and Interfaces of a Life Support System

Occasionally, systems for Extravehicular Activities (EVA) are included as a part of an ECLSS. In fact, EVA suits are pressurized suits with an integrated open life support system, designed for an operating period of 6 to 21 hours. This chapter, however, will not address them separately.

Further human requirements are found in the field of hygiene, clothing, food preparation and medical care. All these are summed up in the notion of "Crew Systems" and do not form part of a life support system either.

In Fig. 4.4, the different tasks and functions of an ECLSS are presented, together with their interfaces to the subsystem "human being". The solid arrows stand for the process flows which ensure and control the required environmental and living conditions, i.e. the physiological and metabolic boundary conditions. The dashed arrows represent possible cross connections as they occur in the realization of regenerative process cycles (recycling). When these connections are missing, the resulting system is called an open life support system. In an *open life support system*, all consumables (e.g. oxygen, water, etc.) are to be resupplied and all waste has to be disposed of. If, on the other hand, the process cycles are closed and resupplies are not necessary, the system is called a *closed life support system*. Between these two extremes, we can think of any possible degree of closure in a *regenerative life support system*. In order to achieve an optimal design for a fixed mission duration, the hardware requirements for regenerative systems and those for resupply have to be weighed against one another carefully. In the following sections, this will only be addressed in the form of examples, but the Chapter 10 "Synergisms" will treat the subject in detail.

The task "crew safety" does not appear in Fig. 4.4 since its functions relate to the whole life support system and cannot be represented by means of the mass and energy flows of the subsystem "human being".

Besides describing such systems as "open", "regenerative" or "closed", there is yet another way of classifying life support systems: according to the kind of process involved, we can distinguish between physico-chemical and biological life support systems. The latter uses biological processes or components instead of physical or chemical ones for its functions.

This chapter shall introduce mainly physico-chemical regenerative processes, while Section 4.3 will offer an overview of bioregenerative components and systems.

4.2.2 Atmosphere Management

Atmosphere management comprises the following functions:

- Cabin ventilation
- Temperature and humidity control
- Atmospheric composition and pressure
- Air filtration (removal of particles and trace contaminants)
- Generation of O_2 and N_2
- CO_2 removal
- Regenerative functions

Often, the functions of atmospheric composition and pressure, CO_2 removal, air filtration, oxygen generation and regeneration are summed up with the notion of "Atmosphere Revitalization". As for its technical realization, it can be practical to include cabin ventilation, atmospheric composition and pressure, CO_2 removal and removal of particles and trace contaminants in one unit. Due to the lack of gravity and, as a consequence, natural air convection, the cabin air must be circulated by fans to avoid local enrichment of CO_2 and heat. The functions mentioned above, and, according to the given requirements, regenerative components as well, can be integrated into such an air loop.

Figure 4.5 shows such a loop as it was designed for Spacelab; this design is also typical for space stations.

Aboard Spacelab, the cabin air enters the system through a central fan and the CO_2 is removed by a carbon dioxide filter. The air is moved by fans at a rate of 700 m^3/h, i.e. the atmosphere is exchanged about 10 times per hour. Crews appreciate the possibility of regulating the air speed locally: the comfort range is 8–20 cm/s.

After the air stream is moved through the carbon dioxide filter, it is divided, and one part is separated for trace contaminant removal. The Trace Contaminant Control (TCC) system consists of a cascade of catalytic cartridges, particle filters and reac-

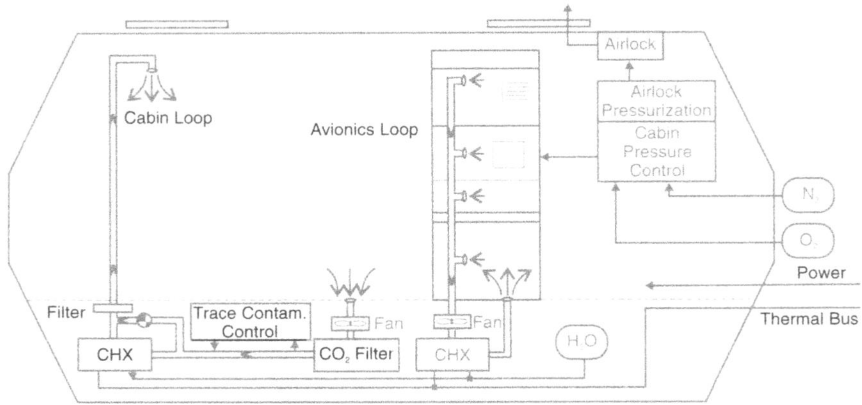

Fig. 4.5. Air Loop of the Spacelab

tors which oxidize the contaminants. The separated air stream is reintroduced into the other part, and then the complete stream enters a condensing heat exchanger (CHX) which provides temperature and humidity control. Inside the CHX, the air is cooled below its dew point. Since water droplets move with the air flow in microgravity, one method for phase separation is to pass the air stream through a centrifuge or vortex flow. In order to reach the required exit temperature, part of the (warm) air stream is led around the CHX by means of a variable bypass and back to the main stream.

Before blowing the air back into the cabin, activated charcoal filters remove contaminants or odors resulting from either the cabin or technical systems.

Usually, the total pressure and composition of the cabin atmosphere are regulated jointly, but separated from the air loop described above. One possible method: oxygen flows into the cabin with a mass flow determined by the metabolic rate while as much nitrogen as needed is added to achieve a given oxygen partial pressure.

The regenerated air reenters the cabin through cabin diffusers. In addition to the air loop for the pressurized cabin ("cabin loop") in most cases, there is a second loop for experiments and subsystems called the "avionics loop". All resources necessary for operation (for example: oxygen, nitrogen, cooling, electrical power, etc.) are provided with the help of centralized infrastructure systems or, as it was the case with Spacelab, by the US Space Shuttle.

Air Filtration

The cabin atmosphere of a space vehicle is constantly polluted by trace contaminants resulting from subsystem and payload operations and from outgassing of materials. Especially on long-duration missions, it is necessary to filter or decompose these trace contaminants.

Trace Contaminant Monitoring has the difficult task of surveying many chemical

Table 4.3. Possibilities for Trace Contamination Removal [Tan 93]

Contamination Group	Removable by Means of		
	Activated Charcoal	**Chemisorption**	**Catalytic Oxidation**
1 Alcohols	X		X
2 Aldehydes	X	X	X
3 Aromatics	X		X
4 Esters	X		X
5 Ethers	X		X
6a Halocarbons, low b.p.[a]		Non-Removable	
6b Halocarbons, high b.p.[a]	X		
7a Hydrocarbons, low b.p.[a]			X
7b Hydrocarbons, high b.p.[a]	X		X
8 Ketones	X		X
9a Ammonia		X	
9b Acetonitrile	X		X
9c Carbon Monoxide			X
9d Dimethyl Sulfide		X	
9e Hydrogen Sulfide		X	
9f Hydrogen			X
9g Ozone		X	X
9h Sulfur Dioxide		X	
9i Nitrogen Dioxide		X	
9j Nitrogen Monoxide		X	

[a] b.p.: boiling point

substances with a high degree of accuracy. For this reason, highly sensitive devices such as instruments for gas chromatography/mass spectrometry are used, allowing measurements with a degree of accuracy of up to 1 $\mu g/m^3$. For every substance, there exists a so-called Spacecraft Maximum Allowable Concentration (SMAC) value [ESA PSS-03-401]. When those SMAC values are adhered to, there will be no negative consequences for the crew.

For Trace Contaminant Control, i.e. reduction of the trace contaminants in the cabin air, different methods are applicable (cf. Table 4.3): First of all, dust and aerosols are removed by a particle filter. Activated charcoal beds are used to bind contaminants with a higher molecular weight. A catalytic oxidizer can convert contaminants which cannot be directly adsorbed into carbon dioxide, water, nitrogen compounds, sulfur compounds or halogen compounds. Those can finally be removed by chemical adsorption beds together with other, non-oxidable substances.

Generation of O_2 and N_2

Oxygen and nitrogen have to be permanently fed into the cabin for several reasons:

- Oxygen is consumed by crew respiration.
- Air locks and pressurized modules always have a certain loss of oxygen and nitrogen due to leakage and thus these gases have to be replenished. For example, the nitrogen and oxygen losses due to leakage aboard the Space Station Freedom with four pressurized modules, six nodes and some logistics modules were estimated at 2.8 kg per day [Wydeven 88].

Oxygen and nitrogen can generally be stored in three different ways:

- High pressure storage
- Cryogenic storage
- Storage in the form of chemical compounds

From the technological point of view, high pressure storage is the simplest method and also, the most reliable. The gas is stored at ambient temperature and at a pressure of several hundred atmospheres. High pressure tanks, however, have a relatively high structural mass, and they present a certain risk.

With cryogenic storage, considerably higher storage densities and reduced tank masses can be achieved. For example, the density of liquid oxygen is three times the density of a gas stored under high pressure. Moreover, tanks for cryogenic storage only have a structural mass of about 0.25 kg per kilogram O_2. What is problematic about this method of storage is that a portion of the liquid gas is vaporized by inevitable ambient heating, which causes partial evaporation of the liquid gas and thus results in two phases occurring inside the tank. These problems can be avoided by storing the gases in their supercritical state, i.e. at temperatures and pressures above the critical point. In this case, the gas is present in a homogeneous phase, so no components for phase separation are necessary. Moreover, application of any external pressurization of the tanks is not required since the gas can be evacuated by a simple increase in pressure due to heating.

Table 4.4 shows some characteristic values taking oxygen storage as an example. The tank masses indicated do not account for any overhead for filling/emptying, pressurization, phase separation, and pressure control. If they were taken into consideration, supercritical storage would be a suitable method to opt for.

The third possibility for oxygen and nitrogen storage aboard a space vehicle are chemical compounds. A possible substance for nitrogen generation is hydrazine, which, for propulsion purposes, is usually stored aboard anyway. The catalytic dissociation of hydrazine takes place in two steps, yielding nitrogen and hydrogen:

$$\begin{aligned} 3N_2H_4 &\rightarrow N_2 + 4NH_3 + E_{therm} \\ 4NH_3 + E_{therm} &\rightarrow 2N_2 + 6H_2 \end{aligned} \qquad (4.2)$$

Oxygen, on the other hand, can be produced from water, oxides, superoxides of alkali metals, chlorates, perchlorates, etc.

Table 4.4. Methods for Oxygen Storage

Method of Storage	Oxygen Density [kg/m^3]	Tank Mass [kg/kg Filling]
High Pressure	about 300	2.0
Cryogenic	1140	0.25
Supercritical	430	0.26 to 0.7
Oxygen Candles	> 300	2.5[a]
Water[b]	about 880	< 0.24

[a] kg $KCLO_3$ / kg O_2 [b] for Comparison

Oxygen Candles. Aboard the Russian space station Mir, oxygen can be obtained from so-called "oxygen candles". Those oxygen candles are cylindrical cartridges (diameter about 12 cm, length about 25 cm) filled with about 2.2 kg of $LiClO_4$. When ignited with a small amount of solid fuel, such a "candle" releases about 600 liters of oxygen, and it takes a cartridge about 5 to 20 minutes to decompose. Oxygen candles are a robust and reliable technology, even though the exothermal decomposition reaction implies some risk. For that reason, they will be used aboard ISS, referred to there as "Solid-Fuel Oxygen Generator" (SFOG). Yet, as a classic example for a method intended only to be used once, it is not suitable for closing process flows. This is the reason why they will only be used during early phases of operation of ISS; after completion of assembly, oxygen generation will be effected by electrolyzers while the oxygen candles will only be used in case of emergency.

Electrolysis Cells. Water as a basic substance has the following advantages from a process engineering and logistical point of view:

- Chemically inert behavior
- Relatively high density compared to gases
- Safe and easy handling

This results in mass savings due to smaller, simple tanks, pipes, pumps, etc. All these advantages make water the top priority candidate for oxygen generation, which can easily be combined with other, water-producing subsystems. Water is electrolyzed according to the following equation:

$$\begin{aligned} &2H_2O + 2e^- \rightarrow H_2 + 2OH^- && \text{(Cathode, "-" Pole)} \\ &2OH^- \rightarrow \frac{1}{2}O_2 + H_2O + 2e^- && \text{(Anode, "+" Pole)} \end{aligned} \tag{4.3}$$

The different technical processes for water electrolysis can either be distinguished by the state of the electrolyte used (liquid or gaseous), or by the temperature level (low-temperatures up to about 395 K and high-temperatures up to about 1200 K) as well as by the current density inside the cell (about 100 to 1000 mA/cm^2). In principle, better efficiencies can be achieved (up to 96%) at higher temperatures and currents, however, those approaches would require higher system masses in order to

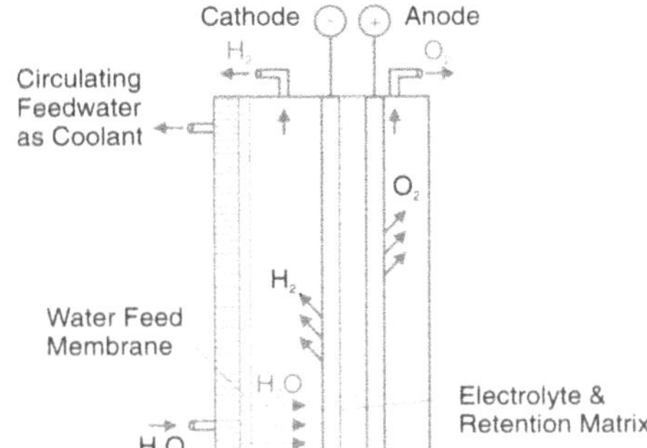

Fig. 4.6. Static Feed Electrolyzer (SFWE) [Wydeven 88]

ensure evaporation, heating and cooling. Low-temperature electrolyzers achieve efficiencies up to 80% with typical specific system masses of 3 kg/kW. An example of the technical realization of electrolyzers is the "Static Feed Water Electrolyzer" (SFWE), as shown in Fig. 4.6.

In the SFWE, the water circulates through a chamber which is separated from the electrolysis cell by a water-permeable membrane (cf. Fig. 4.6). It passes into the chamber in which the actual electrolysis takes place by diffusing through the hydrogen vapor space. The electrolyte is held in its chamber by gas- and water-permeable membranes (normally made from asbestos). At the same time, these membranes serve as electrodes. The produced heat is carried off by the circulating water stream. The advantage of an SFWE cell: contamination of the electrodes by polluted water is impossible; this is very important when waste water is to be used. In order to electrolyze 1 kg of water, an average of 2350 Wh of electrical power is necessary. Different process rates can be achieved either by varying the cell surface area or by serializing or stacking of several cell modules.

CO_2 Removal

In order to remove carbon dioxide produced by the crew, a number of different methods may be implemented:

Lithium Hydroxide (LiOH). This method is the simplest of all and is based on lithium hydroxide, which, when combined chemically with CO_2, forms lithium carbonate and water. The resulting mass flows can be determined easily by looking at the chemical equation and the molar masses:

$$\begin{gathered} 2\text{LiOH} + \text{CO}_2 \rightarrow \text{Li}_2\text{CO}_3 + \text{H}_2\text{O} \\ 47.90\text{ g} + 44.01\text{ g} \rightarrow 73.89\text{ g} + 18.02\text{ g} \end{gathered} \tag{4.4}$$

Thus, for one crew member producing 1 kg of CO_2 per day, 1.088 kg/d of LiOH are needed:

$$1088.4\text{ g} + 1000.0\text{ g} \rightarrow 1678.9\text{ g} + 409.5\text{ g} \tag{4.5}$$

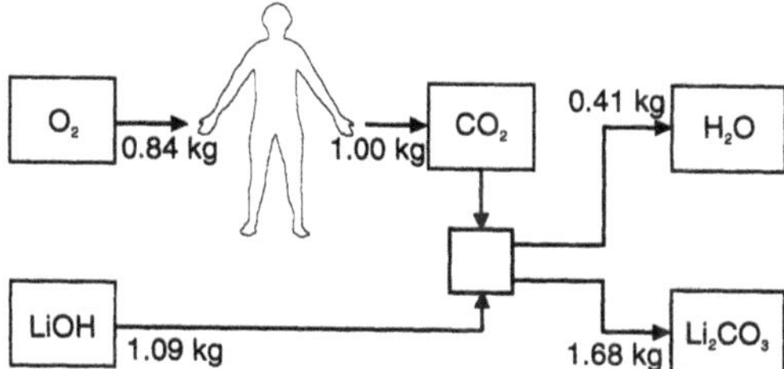

Fig. 4.7. Block Diagram of an LiOH Filter

This process can also be shown with the help of the following block diagram in Fig. 4.7.

In the technical realization of the LiOH process, the cabin air is pumped through metal cartridges filled with LiOH granulate. Depending on size, cartridges are manually exchanged once or twice per day.

The advantage of this method is that it is rather uncomplicated in design and use, and very few failures have occurred up to the present day. Its disadvantage is that the oxygen contained in the metabolic CO_2 is chemically bound and therefore not available for further usage. Additionally, as it is characteristic of so-called "open loops", there are relatively high mass flows for resupply and disposal. Therefore, this method is only appropriate for short missions (Space Shuttle, Spacelab); long duration missions will require regenerative processes.

Electrochemical Depolarized CO_2 Concentration (EDC). This method is based on the principles of the fuel cell reaction (cf. Fig. 4.8). In a fuel cell, water is oxidized into hydroxyl ions by a cold oxidation (which takes place at the cathode) by taking in electrons. These migrate through the electric field to the anode where they recombine with hydrogen ions and thus form water. Additionally, electrical and thermal energy are released.

The EDC reaction now involves carbon dioxide on the cathode side: the hydroxyl ions and the dissolved carbon dioxide form, apart from water, carbonate ions, which, in place of the hydroxyl ions, migrate from the cathode to the anode (cf.

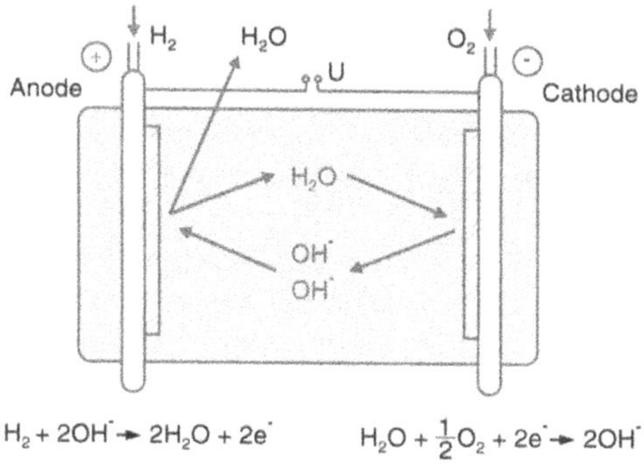

Fig. 4.8. Principle of a Fuel Cell

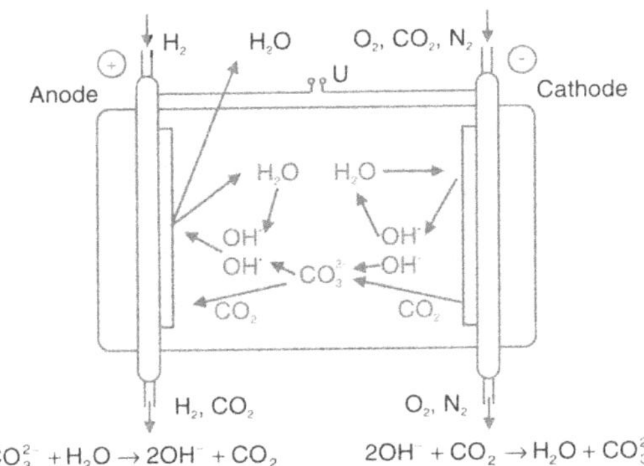

Fig. 4.9. Principle of the Electrochemical Depolarized CO_2 Concentration (EDC)

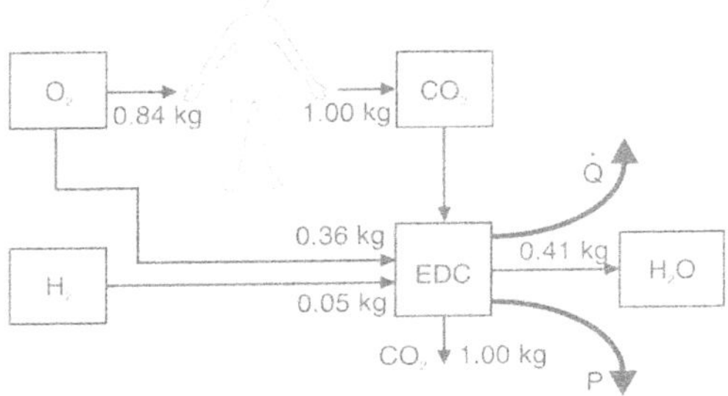

Fig. 4.10. Block Diagram of an EDC Cell for CO_2 Removal

Fig. 4.9). Due to a shift in pH, the carbon dioxide is released from the solution at the anode side. The reaction taking place at the anode is identical to the usual fuel cell reaction.

$$CO_2 + H_2 + \frac{1}{2}O_2 \rightarrow CO_2 + H_2O + E_{therm} + E_{electr} \tag{4.6}$$

Thus, carbon dioxide originating from the cabin loop will be concentrated inside the anode's vapor space. Three steps are necessary for this process: hydrogen (directly fed) and oxygen (indirectly fed since it is contained in the cabin air) are added, and then process heat has to be carried off. In turn, the process releases DC power. This net formula may again be presented as mass turnover per person/day in the form of a block diagram (cf. Fig. 4.10).

When comparing the EDC to the LiOH process, it is obvious that the resulting resupply volume is considerably smaller. On the other hand, the hardware requirements for an EDC filter are markedly more complex, compared to the relatively simple LiOH cartridges. As for subsystem design, engineers have to decide which concept is preferred for every mission, each with its unique and special characteristics. This problem of comparing and evaluating concepts is typical for systems engineering problems and will be discussed in detail in Chapter 10 "Synergisms". An

One possibility for a system-level comparison is the description of all pros and cons of a design alternative with the help of the criterion "system mass" (cf. Chapter 10). In this context, we will distinguish as follows:

- Fixed system mass: Mass of hardware necessary to perform the vital functions of a system; in case of the EDC cell, all hardware including fans, piping, etc.
- Resupply mass: Masses which permanently have to be resupplied during operation; in our example gases, expendables such as gases, filters, spare bulbs or heaters for maintenance.
- Synergistic mass: System mass differences resulting from interaction with other subsystems or the overall system; for example, the EDC unit yields electrical power, and as a consequence, the space station's electrical power supply system may be smaller than initially intended. On the other hand, for the EDC process, heat must be provided, so the thermal control system must be larger. These influences are typically described with the help of parametric masses, for example, kg/kW.

When adding up all three system masses, the result is the so-called "equivalent system mass" which is a theoretical reference value for evaluating the alternatives with the help of the system mass.
It is interesting to look at the following comparison of methods for CO_2 removal, LiOH and EDC, based on a four-person crew:

	LiOH	**EDC**
Fixed system mass:	We may assume that for both LiOH and EDC, the necessary amount of fans, supply lines, etc. remains stable. Additional fixed system masses only come about in the case of EDC; the value beneath was found in [Heppner 83]	
	$m_{fix,LiOH} = 0$	$m_{fix,EDC} = 39.1$ kg
Resupply mass:	LiOH granulate (cf. Fig. 4.7): 4.35 kg/d metal cartridges: 50% of LiOH mass	oxygen (cf. Fig. 4.10): 1.44 kg/d hydrogen: 0.182 kg/d
	$m_{LiOH} = 6.52$ kg/d	$m_{EDC} = 1.63$ kg/d
Synergistic mass:	Assumed are the following specific masses: For the thermal control system: $\mu_{therm} = 200$ kg/kW For the electrical power supply system: $\mu_{electr} = 268$ kg/kW LiOH is a thermally and electrically neutral process, i.e. neither heat nor power must be provided for or will result from it. For the EDC, literature indicates an output of 119 W of electrical power and an input of 173 W of thermal power. The resulting synergistic masses are:	
	$m_{syn,LiOH} = 0$	$m_{syn,EDC} = 0.173\ \text{kW}\cdot 200\ \text{kg/kW} - 0.119\ \text{kW}\cdot 268\ \text{kg/kW} = 2.7\ \text{kg}$

With the help of these results, we can formulate the mass balance for the point where both design alternatives reach identical system masses. The time of the so-called "break-even point" is determined as follows:

$$t_{\text{break-even}} = \frac{m_{\text{fix,EDC}} + m_{\text{syn,EDC}} - (m_{\text{fix,LiOH}} + m_{\text{syn,LiOH}})}{\dot{m}_{\text{LiOH}} + \dot{m}_{\text{EDC}}} = 8.5\ \text{d}$$

We can see that the LiOH process is more suitable for short-term missions, whereas for long-duration missions applying the EDC method is more appropriate. In case of the EDC, this comparison ignored the effort necessary to provide hydrogen and oxygen, but even if it were taken into consideration, the result would not be influenced. On the contrary, the EDC cell's results would be even better because a new factor could be applied: the recovery of oxygen.

The reliability of such system balances depends, of course, on the basic hypotheses and simplifications assumed. This must be considered when viewing the results. At the same time, the equivalent system mass is only one of many possible criteria for comparison. For that reason, system balances are covered in detail in Chapter 10, "Synergisms"

Fig. 4.11. Comparison of LiOH and EDC systems for CO_2 removal

example of system-level comparison under the criterion of the so-called "equivalent system mass" is presented in Fig. 4.11.

Molecular Sieves. The oldest and therefore the most mature method of regenerative CO_2 removal is the use of molecular sieves. Initially operated on the US Skylab mission, they will also be used in the life support system of ISS.

Molecular sieve systems make use of the ability of synthetic zeolite or aluminosilicate compounds to adsorb CO_2. The CO_2 is removed from the cabin air by pumping the air through adsorption beds containing the zeolite material. Once the adsorption bed is saturated, the carbon dioxide can be desorbed with the help of heat and vacuum.

Figure 4.12 shows a schematic of a molecular sieve system. In order to ensure continuous operation, two CO_2 adsorption beds are mounted, operating alternately in either adsorption or desorption mode. Since water vapor contained in the cabin air would also be adsorbed into the CO_2 adsorption bed, a desiccant bed placed before the adsorption bed provides for selective water adsorption. Again, two desiccant beds in alternating mode ensure continuous operation of the system. Desorption of water is effected by means of the dry air stream from which the CO_2 was removed, so no water will be lost throughout the whole filtering process.

The four-bed system displayed in Fig. 4.12 is very reliable due to its sturdy nature and its advanced technological development. However, for heat desorption of the carbon dioxide beds, relatively large amounts of electrical power are needed (on average, 587 W for a four-person ECLSS). If desorption takes place by venting the filtered CO_2 to space vacuum, it will, of course, be lost to the space station's mass balance.

If engineers want to reduce system mass, system volume, and electrical power, and at the same time use identical hardware components, they will have to accept high loss rates for CO_2, water and residual air. For example, in a two-bed molecular sieve as presented in Fig. 4.13, the adsorbing beds contain desiccants and CO_2-adsorbents, so each bed can adsorb water and carbon dioxide. Desorption is effected by venting water, CO_2 and the residual air to space vacuum.

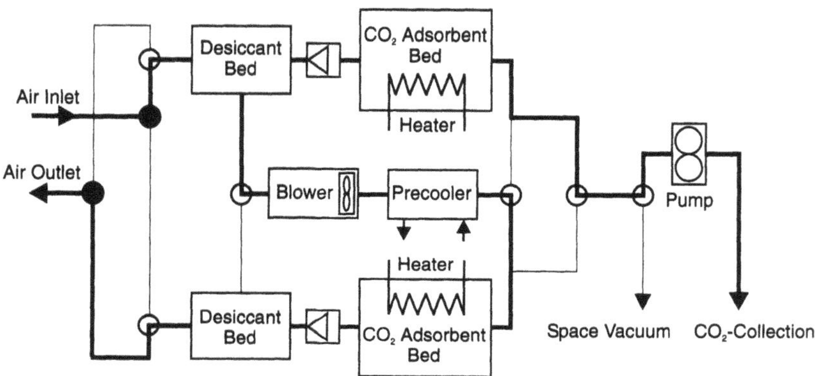

Fig. 4.12. Four-Bed Molecular Sieve (4BMS)

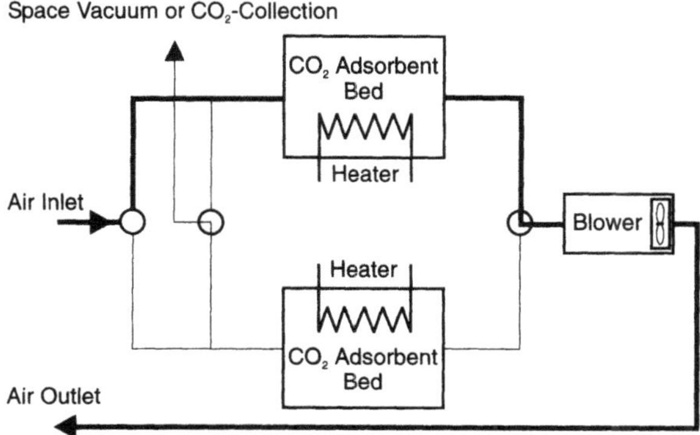

Fig. 4.13. Two-Bed Molecular Sieve (2BMS)

In its US segment, ISS will contain a four-bed molecular sieve for CO_2 removal which has a specified mass flow rate for air of 23 kg/h, and one half-cycle duration (adsorption/desorption) will take three hours. In the station's first phase of operation, desorption of the zeolite beds will be effected by exposing them to space vacuum. This filter system, however, is prepared for being upgraded for processing the carbon dioxide in the future.

Solid Amine Water Desorption (SAWD). Like molecular sieves, the SAWD system is a "passive" adsorption system. The cabin air is drawn through a granulated amine resin. With H_2O acting as a catalyst, the resin forms a hydrated amine which in turn reacts with the CO_2 forming a bicarbonate. This process is presented in the following equations:

$$\begin{aligned} \text{Amine} + H_2O &\rightarrow \text{Aminehydrate} \\ \text{Aminehydrate} + CO_2 &\rightarrow \text{Amine} - H_2CO \end{aligned} \tag{4.7}$$

Regeneration is possible by flooding the amine bed with steam; its heat will break the bicarbonate bond, as the following equation shows:

$$\text{Amine} - H_2CO_3 + \text{Steam} + E_{\text{therm}} \rightarrow CO_2 + H_2O + \text{Amine} \tag{4.8}$$

The principle of the SAWD is presented in Fig. 4.14. Desaturation of the amine bed by desorption with steam also regulates the amount of water the resin needs for optimum absorption. Different kinds of resin materials are available, e.g. Amberlyte-IRA 45, Mitsubishi Diaion WA 21 or Bayer Levatit.

Similar to a 2BMS, two amine beds are mounted for alternating adsorption and desorption mode, but the bed's CO_2 desorption does not take place by using vacuum but water vapor instead. The water vapor remaining in the amine bed after desorption is vented during the next adsorption mode by the cabin air entering the bed. At the system output, the water vapor is collected by a condensing heat exchanger, which is not included in the above diagram since, alternatively, the heat exchanger of the air conditioning system could be used here as well.

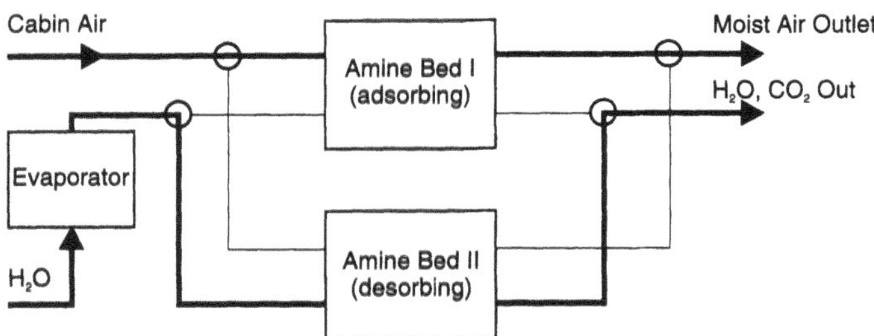

Fig. 4.14. Solid Amine Water Desorption (SAWD) Filter

Thus, an SAWD filtering system can also be used for a fully regenerative CO_2 removal without loss of mass. A high level of electrical power is needed to convert water into steam for desorption (about 460 W for a three-person SAWD system). For a system-level comparison, the condensing heat exchangers downstream of the SAWD filtration unit with their system mass and their resource needs also have to be taken into consideration.

Additional Possibilities for Filtration of CO_2. CO_2 filtering by means of LiOH, the EDC system, molecular sieves or the SAWD system are the "classic" physico-chemical processes for removal of CO_2 from the cabin air. Other possible processes include:

- Semipermeable membranes which remove carbon dioxide by osmosis
- Electroactive carriers which are able to selectively transport CO_2 through a membrane
- Metal oxides which serve as adsorption material for CO_2
- Chemical washout techniques (e.g. $NaOH/N_2SO_4$), combined with bipolar membranes for regeneration of the chemicals

Unfortunately, the use of these processes is still in the distant future since their (degree of) selectivity and efficiency is relatively low, and the processes are not sufficiently technologically advanced. All processes involving membranes surely hold a huge potential for lightweight, reliable filtration units.

Regenerative Functions for Atmosphere Management

One possibility of reducing logistics mass by closing process flows is to recover the oxygen bound in the metabolic carbon dioxide. This problem offers its own solution since, in case of regenerative filtering processes, carbon dioxide is the resulting process gas.

Sabatier Process. CO_2 and hydrogen are fed through a reactor, where they split into methane and water under presence of a catalyst (e.g. a ruthenium catalyst on an aluminum substrate). This process is presented in the following equation:

$$4H_2 + CO_2 \rightarrow 2H_2O + CH_4 \tag{4.9}$$

Fig. 4.15. Sabatier Reactor

The process is exothermic only after pre-heating to 370°C. After pre-heating, the temperature steadily increases until it is limited to 600°C by an endothermic reverse reaction. Normally, the reactor will be equipped with a heat exchanger for the exhaust gas. In practice, it is possible to achieve process efficiencies of over 99%. In this context, the ratio of the molecular mixture of hydrogen to carbon dioxide can vary between 2:1 and 5:1 (stoichiometric: 4:1) without impairing the reaction's efficiency. This allows the Sabatier reactor (cf. Fig. 4.15) to be easily connected with other components of the life support system. The combination of an EDC filter with a Sabatier reactor is especially interesting since in the EDC process, a mixture of hydrogen and carbon dioxide is released on the reactor's anode side (which in turn can be used for the Sabatier reaction).

After condensation, the water vapor produced in the Sabatier reactor may be fed into an electrolyzer where it is divided into molecular hydrogen and oxygen (cf. section on "Generation of O_2 and N_2", page 118). Also, the methane obtained during the Sabatier process may be put to use again: for example, if the spacecraft in question includes resistojet or arcjet thrusters, the methane may be used as a propellant for attitude and orbit control. However, in order to recover the hydrogen bound in methane for other use, a pyrolytic process is necessary.

Carbon Formation Unit (CFU). When heated, a pyrolytic decomposition of methane (into hydrogen and elemental carbon) is achieved at a temperature of about 960°C, as shown in the following equation:

$$CH_4 + E_{therm} \rightarrow C + 2H_2 \tag{4.10}$$

Such a "Carbon Formation Unit" (CFU) may consist, for example, of a quartz glass cone filled with quartz wool in which the pyrolysis takes place. The carbon obtained from such a reaction may be regularly removed from the reactor in form of a massive, cone-shaped rod.

Bosch Process. This process may alternatively be used instead of combining a Sabatier reactor and a CFU. The Bosch process is similar to the Sabatier process, but carbon is directly obtained according to the following equation:

$$2H_2 + CO_2 \rightarrow 2H_2O + C \tag{4.11}$$

This reaction needs a catalyst (iron), too, and an even higher temperature level of 530–730°C. Compared to the Sabatier process, the Bosch process has the following disadvantages:

- Relatively low efficiency of under 10%
- High process temperatures
- High maintenance requirements since the catalyst bed must be regularly exchanged

On the other hand, when comparing these two processes, the Sabatier unit's higher system mass (due to additional components like the CFU) has to be taken into account as well.

In conclusion, an example of a closed oxygen loop shall be presented in the form of a block diagram (cf. Fig. 4.16) which illustrates the figures relating to the mass turnover per person/day.

The system presented removes CO_2 from the cabin air by means of an EDC and feeds it into a Sabatier Reactor (SR). The water escaping from the EDC and the SR is split up into oxygen and hydrogen by an electrolyzer (EL). The oxygen is recirculated to the cabin and the hydrogen can again be used by feeding it into the EDC

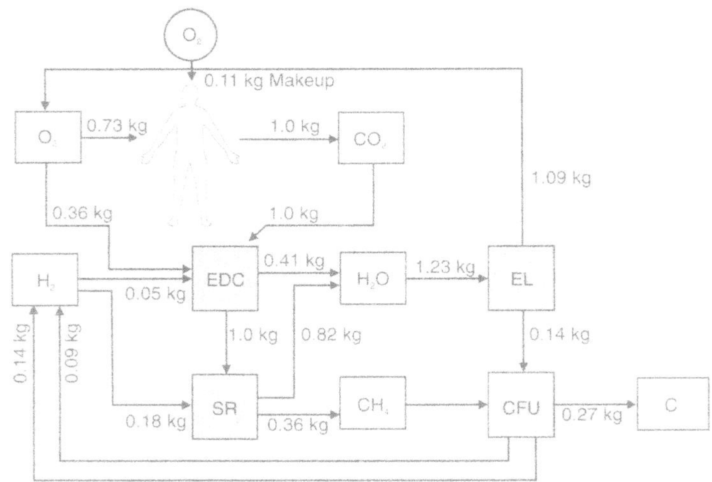

Fig. 4.16. Example of a Regenerative Oxygen Loop for Atmosphere Management (Mass Flows indicated in kg/day)

and SR. To this hydrogen flow, the additional hydrogen obtained in the CFU (where it was separated from the methane) is added.

The example shown in Fig. 4.16 does not take process efficiencies into consideration; in reality, incomplete conversion inside SR or CFU, or incomplete CO_2 removal within the EDC lead up to correspondingly smaller mass flows.

The block diagram also shows that only 0.73 kg per person-day of oxygen can be provided, even if O_2 is completely regenerated. The metabolic requirement, however, comes to 0.84 kg per person-day. Having a closer look at this calculation, there is an apparent daily "loss" of oxygen within the human body of about 110 g/day, which has to be replenished from "outside". Together with the oxygen atoms taken from the water consumed by a person, the oxygen inhaled is used within the body for purposes such as cell production, support of the immune system and oxidation of contaminants. Considering the context of a long-term balance, it does leave the human body, but it cannot be recovered from the fluid loops described above.

4.2.3 Water Management

Water aboard a space station is not only needed for consumption by the crew, but also for payloads, for washing or showers, and in some cases for food production, washing machines and dishwashers as well. Of all subsystems of an ECLSS, the water consumption by the crew covers the largest part of the mass flow requirements of a life support system. Table 4.5 shows an estimate of the expected water requirements for a space station with a crew of four.

Water Supply

A life support system or ECLSS has to provide for water at the required rate and at the temperature requested. For example, the galley water dispenser is designed by NASA to provide for 2.8–5.1 kg of water per person-day at the following temperatures:

cold:	4.0°C
ambient temperature:	21.0°C
hot:	65.5°C

If water produced by fuel cells is used, it will at first be within the upper temperature range, but usually will cool down to ambient temperature during storage. For this reason, the galley needs both, a connection to the space station's cooling loops as well as a heater for warm water.

Water resulting from a fuel cell reaction is virtually sterile and therefore not of hygienic concern, but in order to kill germs in the storage tanks, the potable water is treated all the same. For example, in the US Space Shuttle this is done by adding a light supplement of iodine applied with a self-regulating ion exchanger. Less restricted contamination levels are accepted for washing water aboard a space station; on the other hand, much higher quantities are needed.

Apart from absolutely necessary hygiene systems for the crew, some thought has been given to a washing machine and a dishwasher to form part of the equipment of a space station. By using them, resupply requirements in the form of fresh clothing and weight resulting from the individual single portion packages of food can be reduced. Such appliances have not been used yet, but in an advanced phase of the assembly of the International Space Station, a washing machine will pay off, since the need for clothing per astronaut/year is about 600 kg [Boeing 96].

Water Disposal and Treatment

The US Space Shuttle regularly empties its waste water tanks into space. When looking at the water requirements of a space station (cf. Table 4.5), however, it is obvious that in this case, such an open system would result in unrealistic resupply requirements. Closing the water loop, therefore, is a prerequisite for the efficient operation of a space station and the most logical starting point on the way to overall "loop closure".

The first step is fairly simple: water vapor metabolically produced by the crew must not only be carried off from the heat exchanger of the air conditioning unit in the form of condensate, but recycled into potable water. This method was and still is applied aboard Soviet/Russian space stations. Of course, contaminants must be carefully removed and microbe growth inhibited.

When processing all other waste waters which are produced, the water molecules have to be separated from the dissolved and dispersed contaminants. In the case of urine, those contaminants are mainly hydrocarbons and ammonia, and in the case of washing water, there are soaps (sodium hydroxide), sodium, chlorides and lactic acids.

Water quality is basically judged with the help of the following parameters: Total Organic Carbon Content (TOC), pH, conductivity, transparency, different ion concentrations (e.g. Cl^-, Na^+, NO_3^-, SO_4^{2-}, NH_4^+) and microbiological purity.

For contaminant separation, different basic principles can be applied:

- Filtration
- Phase change and removal of contaminants
- Osmosis
- Oxidation

Table 4.5. Expected Water Requirements for a Four-Person Crew aboard a Space Station [Gustavino 94]

	Waste Water [kg/day]
Condensate	9.4
Shower/Hand Washing	27.3
Washing Machine	50.0
Urine Condensate	8.8
EVA Waste Water	0.45
Total	about 96

All four methods have in common: Apart from the recycled water, there will be a residual fraction of highly concentrated waste water, the so-called "sludge". The water contained in the sludge must be replaced by resupply. At the same time, the concentrate has to be chemically stabilized for intermediate storage. Finally, regardless of the technology used, the whole processing unit has to be periodically sterilized in order to fight the propagation of biofilms and bacteria.

Before presenting an entire water loop, the different methods of filtration shall be introduced:

Multifiltration (MF). In this process, waste water is purified by passing it through three different steps of filtration:

1. Suspended particles are removed by a series of mechanical filters with gradually decreasing pore size.
2. Organic contaminants are removed by an activated charcoal filter.
3. Inorganic salts are removed by cation and anion exchange resin beds.

In order to reduce filter resupply, several filtration cartridges containing activated charcoal and ion exchange resins are connected in series one after another. If the required water quality cannot be achieved anymore due to spent filtration material, the first cartridge at the upstream end of the filtration cascade will be removed and a new cartridge inserted at the downstream end.

Multifiltration offers a good degree of purification by means of relatively simple technology. For instance, water that passed through such a multifiltration cascade can be used as potable water after the residual volatile organic substances have been removed. Yet, the spent cartridges cannot be regenerated, so new ones must be continually resupplied from Earth. MF is successfully applied aboard the Russian space station Mir and has also been chosen for water treatment aboard ISS (cf. Fig. 4.20).

Vapor Compression Distillation (VCD). As the name shows, this process involves a phase change. The waste water is evaporated on the inside of a centrifuge and the then "purified" water vapor is compressed (cf. Fig. 4.17). As a consequence, the water vapor's saturation temperature is increased so that it can recondense in the centrifuge's cylindrical slit in direct thermal contact with the evaporation zone. A VCD unit only needs electrical power for operating the compressor and compensating thermal and mechanical losses. The waste water is circulated through the inner centrifuge, until the content of solid matter reaches the value specified. Nowadays, VCD filters achieve water recovery efficiency rates of over 96%.

Gases contained in the waste water which cannot be condensed (like carbon dioxide or nitrogen), however, pose technical problems. They cause an increase in the system pressure resulting in a decreasing evaporation efficiency. This problem may be solved by periodical evacuation of the unit or by removing these gases by special pre-treatment of the waste water. Additionally, before installing a VCD unit, the effects of the rotating centrifuge on the station's microgravity environment must be examined carefully.

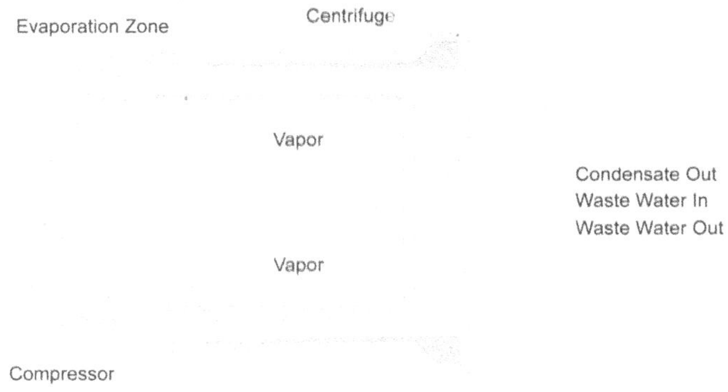

Fig. 4.17. Vapor Compression Distillation (VCD)

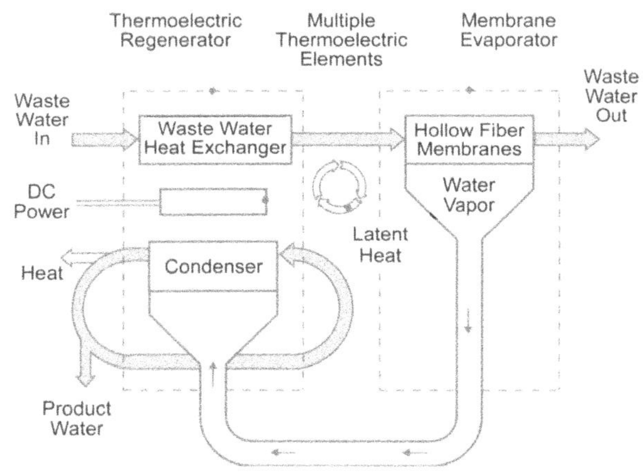

Fig. 4.18. TIMES Process [Wydeven 88]

Thermoelectric Integrated Membrane Evaporation System (TIMES). TIMES is an evaporation process as well, but unlike VCD it includes a membrane for phase separation, see Fig. 4.18.

The waste water is passed through a membrane tube whose surrounding pressure is below evaporation pressure ($p_a < p_{evap}$ H_2O). The water molecules pass through the membrane and evaporate; so the slowly circulating waste water inside the tube becomes more and more concentrated. The resulting water vapor is finally reliquified in a condenser. When compared to VCD, TIMES requires double the amount of heat per kg water (202 Wh/kg), although part of the latent heat of the vapor can be reused with the help of thermoelectrical elements inside the condenser. As a consequence, TIMES requires active thermal control to carry off the fraction of the condensation heat which cannot be recycled thermoelectrically.

Since both the VCD and the TIMES process involve phase changes, the attainable water quality depends on the amount of volatile gases (hydrocarbons and ammonia) contained in the waste water. These gases evaporate and recondense with the water. Consequently, in both cases, pre-treatment of the waste water with acid is necessary. As ammonia, for instance, is transformed into a salt, the pH decreases and thus prevents a decomposition of lactic acids into ammonia (and other components). Before being used as potable water, however, separation of the residual gases is absolutely necessary.

Vapor Phase Catalytic Ammonia Removal (VPCAR). As an alternative to pre-treatment of the waste water, this hybrid evaporation-oxidation process can be applied. A membrane evaporator (similar to TIMES) is followed by two catalytic beds which oxidize the ammonia in its vapor phase according to the following equation:

$$\begin{aligned} 8NH_3 + 8O_2 &\rightarrow 4N_2O + 12H_2O \qquad (\text{for } T = 250°C) \\ 4N_2O &\rightarrow 4N_2 + 2O_2 \qquad (\text{for } T = 450°C) \end{aligned} \tag{4.12}$$

At the same time, hydrocarbons are decomposed according to the following equation:

$$C_xH_y + \left(x + \frac{y}{4}\right)O_2 \rightarrow x\,CO_2 + \frac{y}{2}H_2O \tag{4.13}$$

Generally speaking, the VPCAR system is fed with urine and oxygen, producing nitrogen, water and carbon dioxide which, in turn, can all be used in other components of the ECLSS.

When compared with VCD and TIMES, VPCAR achieved the highest water quality with regard to ammonia content, conductivity and organic residual substances. As to electrical power, it requires about 217 Wh per kg of purified water; this is also approximately what is necessary for TIMES.

Reverse Osmosis (RO). This method is based on the principle of osmosis as shown below and does not require phase change of waste water.

Osmosis takes place when two liquids of different contaminant concentrations are separated by a semipermeable membrane: the system will begin to compensate for the concentration difference by H_2O diffusing from the section of low concentration to the section of high concentration. Diffusion will continue until the so-called "osmotic pressure" is reached in the section of high concentration, in our example, (cf. Fig. 4.19, center) in the right cell. If the pressure on this side is increased artificially, H_2O will diffuse back through the membrane to the side of lower concentration. This effect is called "reverse osmosis". In very general terms, it is a filtration process on the molecular level.

For our application, it is possible, for example, to circulate waste water under pressure in a system of membrane pipes until a prespecified amount of H_2O has diffused back. Reverse osmosis systems normally operate at pressures of 6900–55000 hPa and boast low power requirements of only about 10 Wh per kg water. Still, the realization of complete water recovery by reverse osmosis is difficult since the osmotic pressure rises sharply with increasing waste water contaminant concentration.

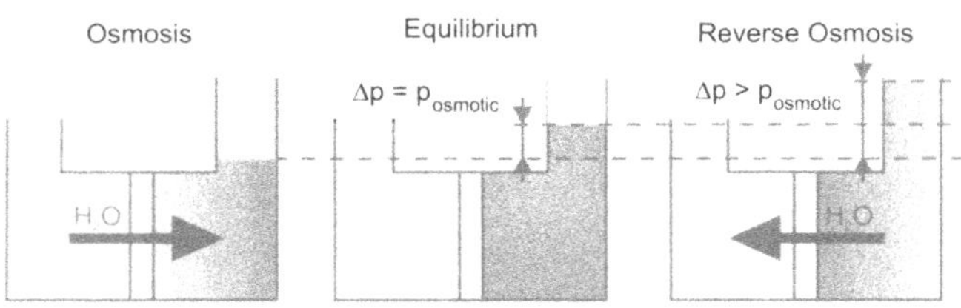

Fig. 4.19. Reverse Osmosis Process

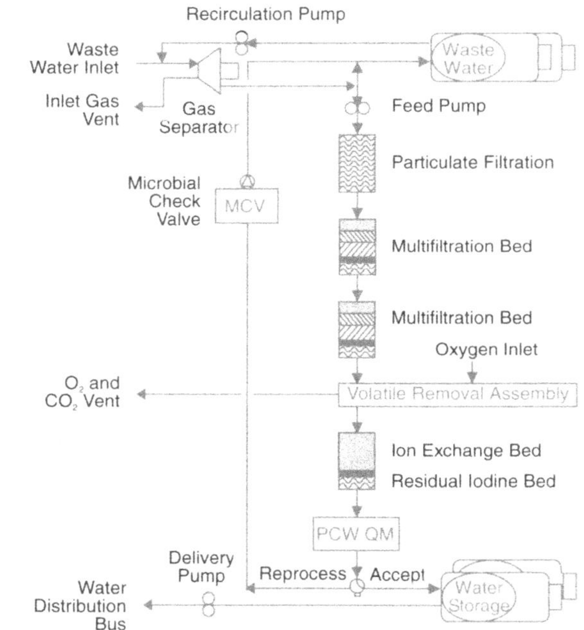

Fig. 4.20. Schema of the Water Processor aboard ISS

The Complete Water Loop. Figure 4.20 presents a scheme of all systems aboard the International Space Station involved in waste water processing. All waste water resulting from cabin loop condensers, hygiene systems and experiments directly enters the Water Processor (WP) whereas urine is pre-treated in a VCD unit before being processed in the WP.

The order of steps through which the waste water passes in the WP is as follows:

- Particulate filter
- Multifiltration bed
- Volatile removal assembly (removes dissolved gases)
- Ion exchange bed

At the outlet of this ion exchange bed, the water is treated with iodine for further sterilization. The purification efficiency is monitored in a control unit called "Process Control Water Quality Monitor" (PCWQM). If the specified quality standards are not met, the filtered water is led back into the waste water tank through a microbial check valve for re-processing.

If we wanted to group the different systems of a water loop, different philosophies would be applicable. A uniform solution is the example mentioned above: the waste water resulting from cabin loop condensers, hygiene systems and experiments is processed into the quality of potable water. Such a solution allows for the centralization of water distribution and processing, as it is shown in Fig. 4.21.

The argument against this integral solution is the higher total quantity of waste water to be treated, of which, in the final count, only a small quantity actually must have the quality of potable water. A possible alternative would be to use the processed metabolic crew output as hygiene water (cf. Fig. 4.22). The crew is provided with fresh water from tanks or fuel cells; at the same rate, concentrated waste water is removed from the system and disposed of.

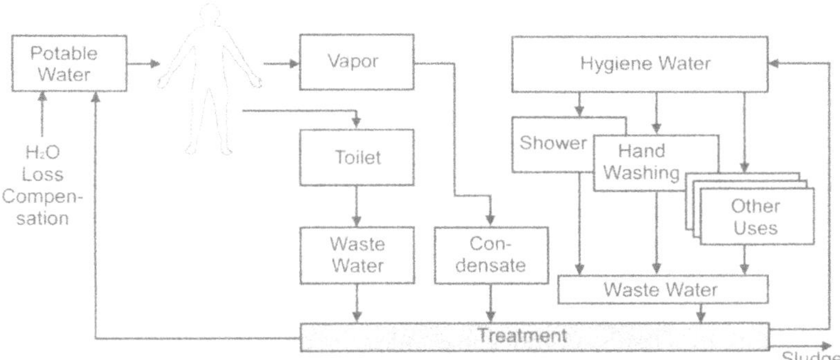

Fig. 4.21. Centralized Concept of Water Treatment

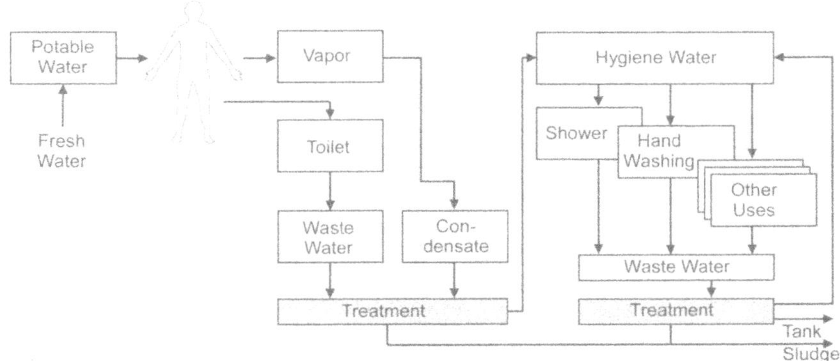

Fig. 4.22. Serial Concept of Water Treatment

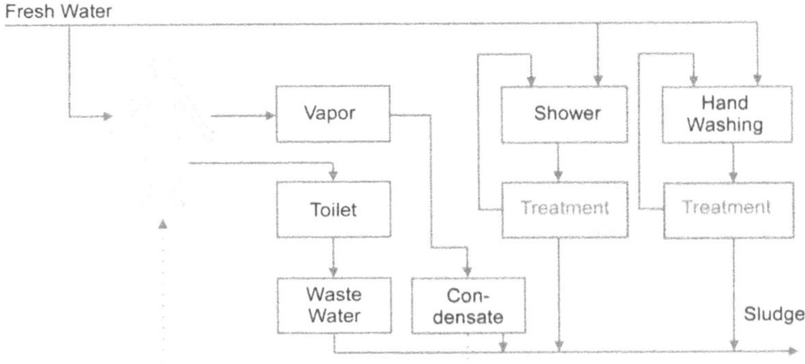

Fig. 4.23. Decentralized Concept of Water Treatment

Except for this serial solution, it is also possible to operate all parts of the system separately from each other, with every system processing its own waste water (cf. Fig. 4.23). In this case, the centralized water supply mentioned above could also be applied. Moreover, mutual contamination of subsystems can be avoided. An argument against this concept is, however, the "system mass" criterion, since there would be a high redundancy required of the processing hardware.

In order to structure the water loop, further alternatives are available through the use of possible synergisms with other subsystems. For example, a regenerative fuel cell used for energy storage in the electrical power supply system could simultaneously produce potable water for the crew. These possible options are treated in detail in Chapter 10 "Synergisms".

In the end, only analysis of the actual situation will show which concept will suit the purposes best. Apart from the resulting equivalent system mass and total quantity considerations mentioned above, the following aspects should be taken into account:

- Size of the station
- Intentional redundancies to increase reliability
- Possibilities of operation in the different stages of assembly

4.2.4 Waste Management

Solid waste aboard a space station results from the following:

- Metabolic output of the crew
- Food preparation
- Payload operation, e.g. packaging or expendable parts
- Subsystem operation (e.g. residual substances from water processing)

Firstly, the waste is collected and, if necessary, separated. According to its composition and the applied concepts for disposal, it is shredded, compressed, chemically and biologically stabilized, and then stored.

Dry waste only requires an adequate place for storage. Differently, wet waste resulting from hygiene systems, galley (food leftovers), or water processing, has to be chemically or biologically stabilized to prevent any contamination of the habitat by decomposition products or microbes. Possible methods of stabilization are heat treatment, drying, freezing, storage at extreme pH levels or treatment with metallic or organic toxins.

Also, on the waste management side, there have been ideas on how to recover resources by oxidation of waste. Possible physico-chemical processes are incineration (similar to the terrestrial waste incineration), the "wet oxidation" of waste dissolved in water or suspended waste with higher temperature and pressure levels, or the so-called "Super Critical Wet Oxidation" (SCWO), an oxidation involving water in the supercritical state (over 922 K and 22 MPa). The latter is of particular interest, since the process takes place in a homogeneous phase (μg !), and it allows nearly complete decomposition of any waste of any consistency in processing times of under five minutes. All three methods require considerable amounts of oxygen as the agent for oxidization; they mainly produce carbon dioxide and water. Sometimes, sulfur and nitrogen oxides occur as reaction products as well. In principle, these products can more or less easily be reintroduced into the space station's process cycles.

When taking a closer look at these methods, it must be taken into account that solid waste produced by the crew is only a small part of the total mass balance aboard a space station (the CO_2 production rate alone is four times as high). As long as the significant amounts of reusable wastes and/or the station's size are not large enough to produce corresponding quantities, the hardware requirements for waste processing are not worth the while. So for the time being, the International Space Station's waste management concept will consist of either returning all waste to Earth or disposing of it, like on Mir, by loading it into expendable resupply vehicles which are then burned up through a controlled reentry into the Earth's atmosphere.

When considering bioregenerative systems or concepts (such as providing food by growing plants), a waste converter which decomposes dry and wet waste into basic chemical substances (like carbon dioxide and water) will be absolutely necessary. At the same time, the use of such a concept could, e.g. by means of photosynthesis, reduce the problem of the enormous amount of oxygen required for waste oxidation.

4.2.5 Food Supply

The amount of food indicated in Sect. 4.1.2 must be able to provide the guaranteed daily energy requirement for each crew member. On the average, 12.6 MJ per day, i.e. 146 W, are sufficient for medium to extreme physical stress of a person (25 years old, male, 75 kg weight); they contain the following:

56%	carbohydrates	0.42 kg	
27%	fat	0.09 kg	
17%	protein	0.13 kg	
		0.64 kg	dry mass

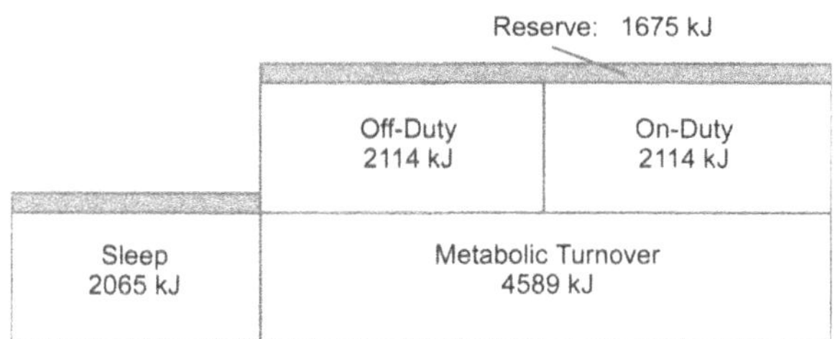

Fig. 4.24. Daily Energy Requirements for One Crew Member

This minimum supply is completed by vitamins, trace elements, about 1.1 kg of water contained in food, and additional to that, potable water (cf. Table 4.2 and Table 4.6). In addition to providing energy, the food components must also help to balance the known losses of calcium and presumably, also losses of blood-corpuscles and muscle tissue caused by weightlessness. Figure 4.24 demonstrates the way in which energy provided by food is consumed during a routine working day (not considering EVAs).

In order to save crew time in orbit, most food is – and will continue to be in the foreseeable future – prepared on ground, and is provided aboard the space station in the form of the following:

- Fresh food
- Cooled or frozen food in natural form
- Dehydrated prepared meals
- Conserved food in packages and cans
- Instant beverages

The daily menu will be drawn up, while considering the timeline, during the mission in coordination with the crew. At this point, however, we leave the domain of ECLSS and enter the field of "Crew Systems".

4.2.6 Crew Safety

Systems for crew safety are generally considered to be a part of the life support subsystem. The highest safety risks for the crew result from:

- Radiation in orbit (mainly during periods of the so-called "solar flares")
- Collisions with space debris
- Contamination of the habitat due to fire

The first two risks can only be mitigated by appropriate structural design (shielding, meteoroid shields) or operative measures such as collision avoidance maneuvers or mission abort. Fire detection and suppression, though, are clearly included in the field of an ECLSS due to the technical processes that are applied.

Fire Detection and Suppression. Traditionally it is assumed that combustion processes under microgravity are slower and take place at lower temperatures than those on the ground under normal gravity. This is due to the missing thermal convection forces and the resulting reduction in oxygen supply. However, as happened during a malfunction of an oxygen candle aboard Mir in February 1997, a fire may occur.

During the Columbus Program [Rygh 93], ESA conducted experiments with fire during parabolic flights showing that the aforementioned assumption is not generally true. When burning certain solid substances (e.g. paper, clothing, wood), the combustion extinguished itself, but in the case of hydrocarbons, the oxidation proceeded, even when no flames or smoke were visible. For this reason, a system for fire detection is vital for a space station. The most simple method is to install sensors inside the air loops, where they can detect smoke either by the interaction of smoke particles with ionized air molecules or by optical obscuration or scattering detectors. An uncontrolled combustion can also be detected by the amount of CO_2 contained in the air. To ensure very reliable fire detection, the cabin must additionally be monitored with the help of optical sensors for infrared, UV and visible signatures typical of flames.

Fire suppression also can make use of the air loops in order to successfully spread extinguishing agents. For example, aboard the European Spacelab, the gaseous extinguishing agent Halon 1301 can be fed from a central tank into the experiment racks. Additionally, an outlet for the venting of the pressurized modules to space vacuum must be provided.

Aboard ISS, mainly portable CO_2 fire-extinguishers will be used. The experiment racks will have so-called "Fire Suppression Ports", through which a gaseous extinguishing agent may be fed if necessary. Apart from that, there are a number of design requirements intended to reliably suppress fire [SSP 41000 D]: For the pressure control system, it is required that, in case of a fire, the volumetric fraction of oxygen can be reduced to below 10.5% within just 60 seconds. A further requirement is that the oxygen partial pressure be reduced to below 70 hPa (1 psia). This emergency depressurization can be initiated by the crew aboard the station as well as by ground controllers.

4.3 Outlook on Bioregenerative ECLSS

The previous sections showed how to partly close air and water loops involving physico-chemical processes. As for carbon, however, those systems always remain open. Carbon enters the loop contained in food, and at the end, it is present in the elemental form (Bosch process, CFU) or as waste. During long-duration missions, resupply either becomes too voluminous or, because of the enormous distances, e.g. in case of Lunar or Martian missions, even impossible.

Closing the carbon loop requires a life support system which can regenerate metabolic products and produce food: a Biological Life Support System (BLSS). The BLSS using photosynthesis and microbiological processes is able to close the loop as it is presented in Fig. 4.25. Photosynthesis allows regeneration of oxygen from

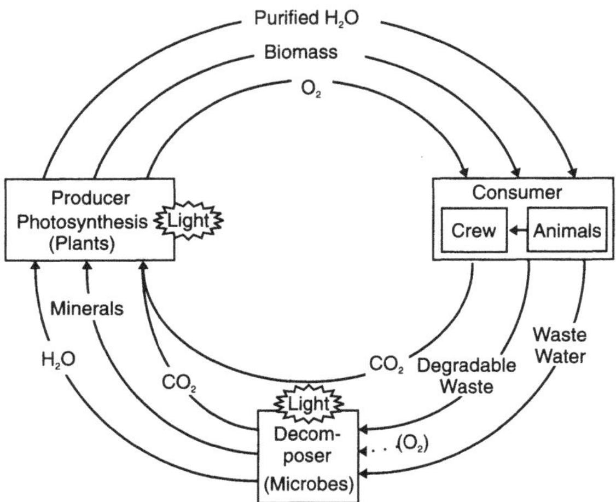

Fig. 4.25. Biological Life Support System (BLSS)

carbon dioxide on one hand, and on the other, it allows the production of biomass as the basic substance for food as shown in the following equation:

$$CO_2 + H_2O + \text{Light} + \text{Nutrients} \rightarrow O_2 + \text{Biomass} + \text{Heat} \tag{4.14}$$

Additionally, organic and biodegradable waste could be decomposed into the chemical basic substances CO_2, water and minerals, thus closing the carbon loop.

When it comes to the realization of a BLSS, the first thing to do is to choose the appropriate producers (plants and algae), consumers (crew and animals) and decomposers (microorganisms). They must be able to perform all functions concerning biological analysis and synthesis, and live together in a stable ecosystem. Special problems occurring in this context are for instance:

- The complexity of processes in an ecosystem with various kinds of feedback loops. Even when compared with lessons learned on Earth, these processes are only partially understood.
- The buffering ability of single substances, which is considerably limited due to the ecosystem's size. On Earth, the size of the ecosystem allows compensation between different growth periods and mass flows of, for example, water, CO_2 or O_2. An artificial biosphere with its small dimensions is very sensitive to changes in the biological activity of an organism. The stability of the entire ecosystem has to be ensured by a monitoring and control system.

Moreover, the specific environmental conditions of space such as microgravity and radiation must be considered; these are relatively easy to handle with physico-chemical systems and units, but are not yet sufficiently examined as to their impact on live organisms.

Yet, when compared to physico-chemical systems, there are considerable advantages as well:

- By closing the carbon loop, an almost closed ECLSS is achieved. Resupply would only be necessary to balance losses due to leakage.
- In some tasks of a life support system, component tasks can be accomplished more or less in the form of synergistic effects. For example:
  Atmosphere Management. Volatile hydrocarbons and other trace contaminants could serve as intake for various microorganisms which thus form a biological "air filtration unit".
  Water Management. Water evaporated by plants is very pure, so it would be possible to save the most energy-consuming processes used for physico-chemical water filtration.
  Waste Management. The amount of non-regenerable waste could be drastically reduced by using biodegradable material.
- Last, but not least: a BLSS also would have positive effects on the crew, since during long-duration missions, a miniature biosphere is more easily accepted than an environment which is maintained with the help of physico-chemical processes. Moreover, the quality of the crew's menu could be considerably improved by offering fresh fruit and vegetables, and perhaps even meat.

In the near future, however, a complete bioregenerative life support system is not very likely to be brought about. Apart from the above mentioned problems, this improbability is also due to the high system mass of a BLSS. For such a system to provide advantages, a space station of large size – or a mission of sufficiently long duration – is necessary.

At present, research in the field of biological life support systems takes place by developing individual biological components and fundamental studies on the basis of small, defined biospheres. The latter even exceeds the boundaries of life support systems. However, it is of considerable interest for the understanding of the terrestrial ecosystem as well: compared to the Earth, mass turnover takes place much more quickly in small biospheres, and individual phenomena are more observable.

Biosphere Research. A good example for a nearly closed biosphere (except for energy) is the project Biosphere 2, Arizona, USA. A volume of about 180000 m^3 offers space for a crew of eight, together with seven biomes, representing the different vegetation zones on Earth, are contained inside: rain forest, savannah, ocean, desert, marsh, agriculture and human habitat. The project's aim is to improve the understanding of the laws governing a biosphere and especially of all mechanisms developed for self-regulation. During one of the experiments, when humans were sealed in Biosphere 2 in 1993, the experiment had to be aborted, since the content of oxygen fell to under 14%. At the same time, the amount of carbon dioxide had risen to a value 10 times higher than normal due to chemical reactions in the concrete foundations of the biosphere building. These problems illustrate impressively the complexity mentioned in the beginning which has to be taken into account when defining and implementing small biospheres. They also show that the understanding of related chemical, physical and biological processes still is rather limited.

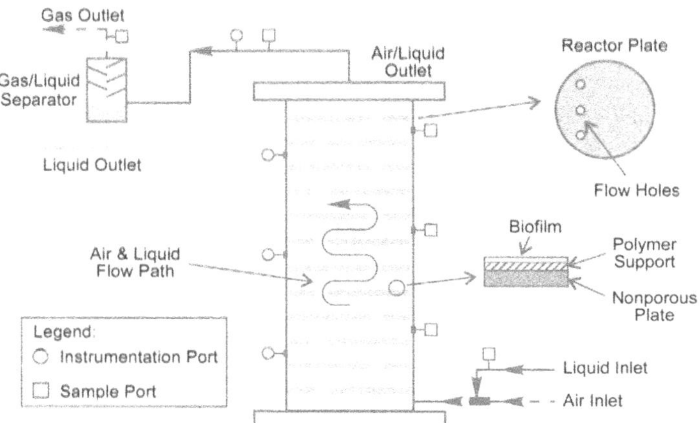

Fig. 4.26. Scheme of an Immobilized Cell Bioreactor [Kumagai 94]

Further models of biospheres containing algae and microorganisms, but higher plants and humans as well, were scientifically examined in Krasnoyarsk, Siberia. The system is called BIOS-3; it is located underground and is hermetically sealed. Several times, a crew of two to three have lived in it for a duration of five to six months [Eckart 96, Melesheko 93]. During this time, only electrical power was supplied from external sources.

Individual Biological Components. Figure 4.26 shows the so-called "Immobilized Cell Bioreactor" tested as a laboratory model at the NASA Johnson Space Center – an example of component design for water reprocessing. Microorganisms are immobilized on a biofilm applied onto carrier plates. When the waste water thus flows over these plates, up to 98.9% of the organic contaminants (TOC) are retained.

The possible concept of growing more complex plants can be illustrated with the help of another example: the so-called "Salad Machine" (Fig. 4.27), [Eckart 96, McElroy 90]. When integrated into a standardized double rack of ISS, this system could supply other vegetables (e.g. lettuce, carrots, green beans, cucumbers) on an area of 2.8 m^2 and thus provide fresh produce for a crew of four, three times per week.

The Salad Machine was planned by the NASA Ames Research Center to be a part of a CELSS (Controlled Ecological Life Support System), and it could be the first step towards bioregenerative components in a life support system for space flight application. Such a system would have to consist of both biological and physico-chemical processes with the key drivers determining process selection being reliability, compatibility, low energy consumption and low system mass.

This section is basically intended to introduce the problem of bioregenerative life support systems. Looking at them in detail would by far exceed the framework and purpose of this chapter. For more detailed treatment, we recommend [Eckart 96] and [Eckart 97].

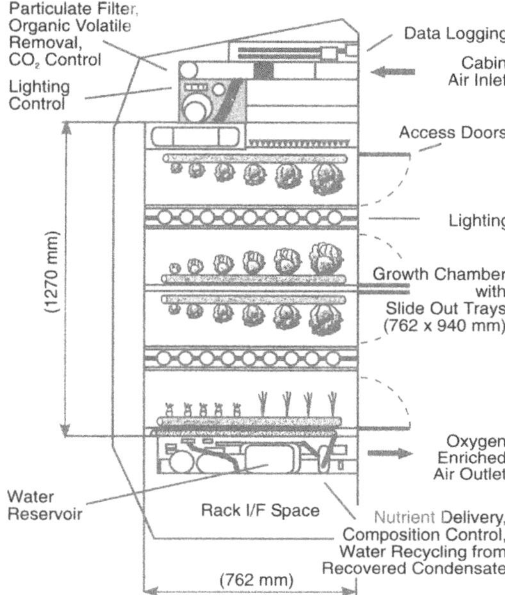

Fig. 4.27. Salad Machine Design Concept [Eckart 96]

4.4 Summary

As mentioned in the previous sections, the task of the entire ECLSS of a space station is to meet the physiological, metabolic and hygienic needs of human beings. Table 4.6 summarizes the mass flows involved (cf. Section 4.1.2) for a crew of six (which is the number of astronauts projected for the completely assembled International Space Station).

Table 4.6. Net Resupply Requirements for a Six-Person Crew

Supplies	**Daily Requirements** [kg/Astronaut/Day]	**Yearly Requirements**[a] (6 Astronauts) [kg/Year]
O_2 (Metabolic)	0.84	1836
H_2O (Potable Water, Oral Hygiene, Food Water Content)	3.55	7759
Food (Dry)	0.64	1399
Hygiene Water	6.80[b]	14861
Clothing	1.64[b]	3584
LiOH-Filter	1.09	2382
Leakage Losses		250[b]
Total	14.56 kg/d	32071 kg/y

[a] 364.25 Days/Year [b] Value for ISS-USOS

With an open ECLSS, the above corresponds to the system's net amount of resupply. The mass of approximately 5345 kg/y per person could be reduced by gradually closing loops; at the same time, resupply requirements for consumables and spare parts for operation of regenerative systems (filter, valves, etc.) would decrease. The current state-of-the-art technology suggests the following order for the closure of loops in physico-chemical systems:

- Recycling of waste water
- Regenerative CO_2 filtration
- Oxygen recovery

Figure 4.28 shows the possible reduction of the net amount of resupply. The level indicating 100% corresponds to the completely open life support system of the Apollo program.

From the processes presented in this chapter, different components can be combined to form complete ECLS units. Apart from the minimization of resupply mass mentioned above, further qualitative criteria are of importance, such as safety, reliability, technological readiness, etc. (cf. Table 4.7).

When comparing different systems, their hidden technical disadvantages, as well as the requirements occurring during the phase of operation, have to be taken into consideration.

An adequate criterion of comparison is the required total system mass accumulated over the life-cycle. Related literature offers extensive information on the subject. However, in each case, careful examination must be made as to whether or not the values assumed for calculating the so-called "equivalent system mass" are comparable and compatible.

An example of a system-level comparison on the basis of equivalent masses is illustrated in Fig. 4.29 weighing the Sabatier process against the Bosch process for air regeneration. If the entire ECLSS was included in the comparison, the ideal case,

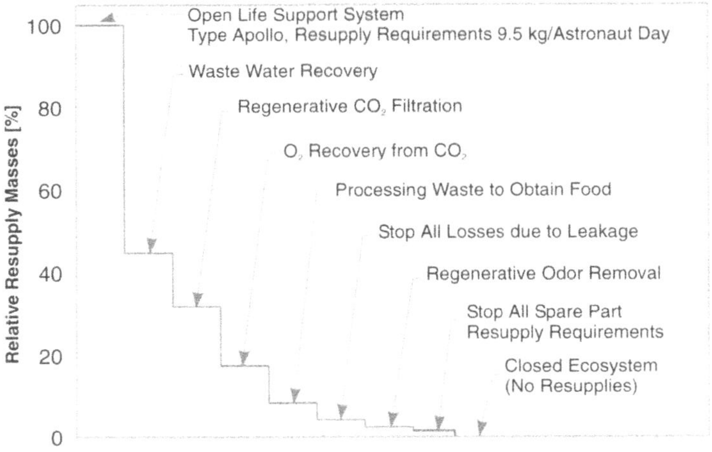

Fig. 4.28. Possible Steps for Closing Loops in Life Support Systems [Hallmann 88]

Table 4.7. Criteria of Assessment for an ECLSS

Qualitative Criteria	Quantitative Criteria
Safety	System Mass
Reliability	System Volume
Current State of Development	Resupply Mass
Growth Potential	Development Costs
Compatibility	Hardware Costs
Degree of Maintainability	

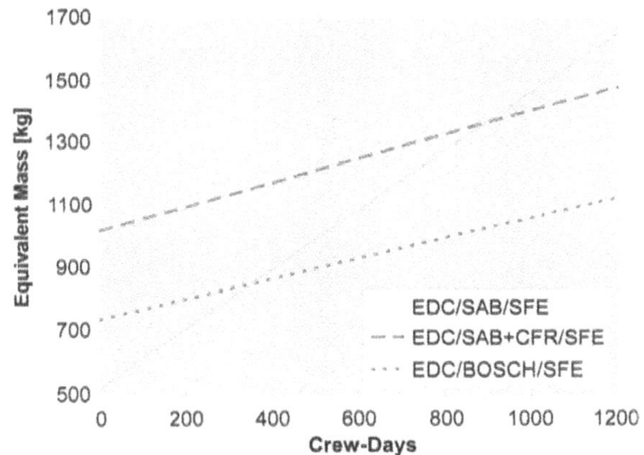

Fig. 4.29. Comparison on the Basis of Equivalent System Masses

a completely closed ECLSS, would be represented by a horizontal line. This can only be realized with a bioregenerative strategy; those are already subjects for research in the USA (CELSS) and Russia (BIOS). A completely closed biological ECLSS requires a considerably higher mass than the often published concept of a "space station including greenhouse". Thus, the only functioning example of a closed biological life support system known to the present day remains our planet Earth.

5 Power and Thermal Control System

Space stations represent the largest kind of infrastructures in space. As they are used intensively over long periods of time, it is not surprising that they are also the largest power consumers in space. Due to the current mode of power generation by means of solar generators, the large collector surfaces are the dominant components of a station. Skylab had a total solar array area of 216 m^2, providing enough energy for the in-orbit residence of three astronauts and the operation of power consuming equipment.

Likewise, the early Russian Salyut space stations had remarkable solar generators installed. The total area of three solar wings on Salyut 6 was approximately 60 m^2, while Salyut 7 possessed a forth solar wing after an extravehicular installation performed by the astronauts. Large solar arrays are the most dominant visual feature of the International Space Station (ISS). The ISS-arrays will have a total area of about 3000 m^2 when completed (around 2004). Table 2.10 shows the total electrical power, generated for the operational use of the above-mentioned space stations. Figure 5.1 puts this electrical power in relation to the stations' total mass. State-of-the-art power demand is approximately 250 W for each ton of station mass, while a maximum of 45% is available for actual use (payload, i.e. experiments, etc.).

A typical task of the thermal control system (TCS) is the removal of waste energy that is produced by external heat loads (solar, albedo, Earth IR radiation), and the internal heat dissipation of energy converters, computers, instruments, payloads and finally, by human beings. Due to the high level of installed electrical power, the long period of operation, the large physical extension, the frequently changing mission profile and payload requirements, the thermal control system of space stations plays a role equally important to that of the power system. During some phases of the mission, up to 90% of the captured electrical power in a space station must be thermally rejected. For example, in the case of Skylab during inactive phases, nearly all the electrical power available for experiments was directly rejected by shunts. In the case of EURECA (cf. Sect. 8.2), such shunts were also used for controlled deep discharge of batteries (2.3 kW) in order to increase their lifetime. In the following sections of this chapter power systems will be addressed, followed by a discussion of thermal control systems. Since fundamentals will not be described in detail, the following introductory literature on the subject of power and thermal control technology is recommended: [Gilmore 94, Kurtz 89, Krüger 97, NASA 8105, Sorensen, Woodcock 86].

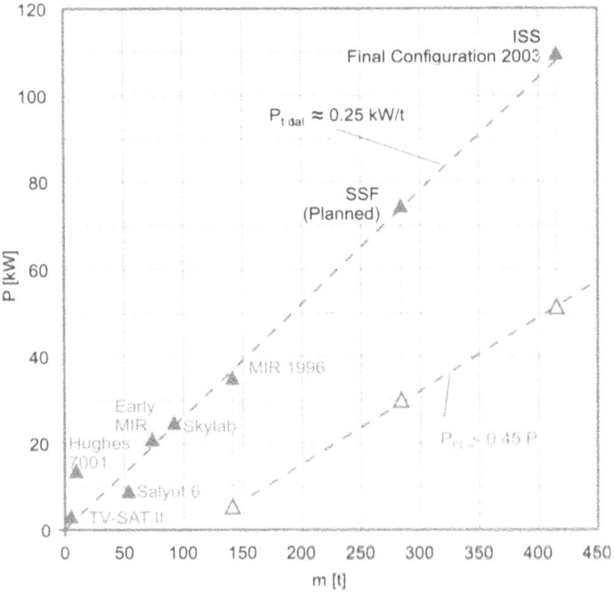

Fig. 5.1. Total Electrical Power of Space Stations (Averaged over One Orbit) and Electrical Power Effectively Available for Experiments (Payloads)

5.1 Power Supplies

Since space stations will be operated over long periods of time, naturally the following question arises: which energy sources (besides the prevailing but unfortunately life-limited photovoltaic units) are suitable? As shown in Fig. 5.2, solar dynamic and nuclear power plants must be considered, especially when in the high power range. Fuel cells are not relevant due to their high replenishment demand for fuel, unless they are operated regeneratively in connection with the life support system.

5.1.1 Characteristics of Space Stations

Large orbital systems like space stations have a high power consumption for the maintenance of both basic space station functions: "housekeeping" and "mission or utilization tasks" such as experiments, manipulators, and lighting for EVA in Earth's shadow [Woodcock 86]. A third category is the emergency power supply system. This is specified separately since it is operated independently of the other two categories and is designed to be exceptionally robust. Energy consumption determines the dimensions of the generators and converters, while the distribution network is designed to transport high power to the end consumers with little losses.

As in the case of satellites, space station elements that carry solar arrays and payloads have general alignment requirements. Thus, at most attitudes rigidly mounted

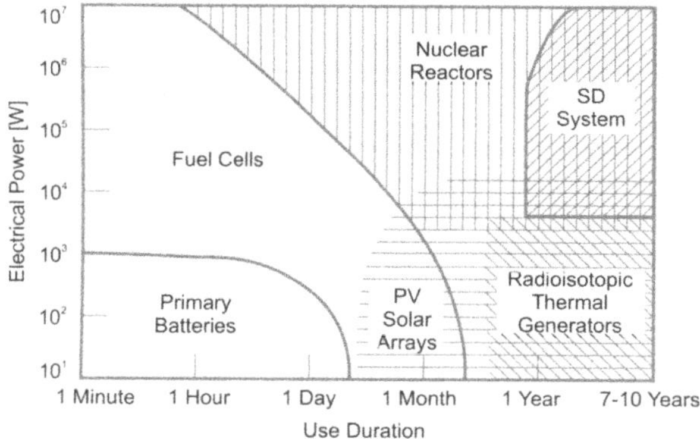

Fig. 5.2. Electrical Power as a Function of Operating Duration and Energy Sources of Space Systems

solar arrays would not be in their most favourable orientation. Consequently, tracking in the direction of the Sun has to be performed for both the orbit movement (α-angle) and with respect to the declination (β-angle). The needed mechanisms represent a non-trivial task especially on a space station where large power levels are required and thus demand large generators.

When conducting crewed space flight, the safety of the crew takes priority over the protection of orbital systems. Thus, the safety regulations for power systems are required to cope with catastrophic failures. It must be assured that the crew can survive or at least leave the station safely in case of a power system failure. Consequently, an emergency power supply system is a must.

These requirements result in redundant distribution network systems with individual sections which, in case of a short circuit, could be electrically isolated. In addition, allotment of the power supply for different technologies (e.g., photovoltaic systems with batteries and solar dynamic systems) increases safety. Any risk of danger during normal operation must be excluded. Examples of such risks are mechanical collisions between the rotating collectors and a manipulator arm or excessively high emissions from a nuclear reactor.

At present, space stations have a planned design life of 15 to 30 years. This is approximately twice the projected service life for present and future communications satellites in geostationary orbit. Unlike a satellite, however, a defective or degraded power supply can be replaced aboard an operational space station. Consequently, systems should be divisible into easily manageable modules, replaceable in orbit (Orbital Replaceable Units, ORUs) with no danger to the crew, and disposable in case defective.

Most concepts incorporate a space station expansion, as Russia demonstrated with their operating station Mir. Thus, power supply systems must possess the capability of having their performance gradually expanded. System elements which cannot be improved by inserting additional components must be sized for future loads in their original design.

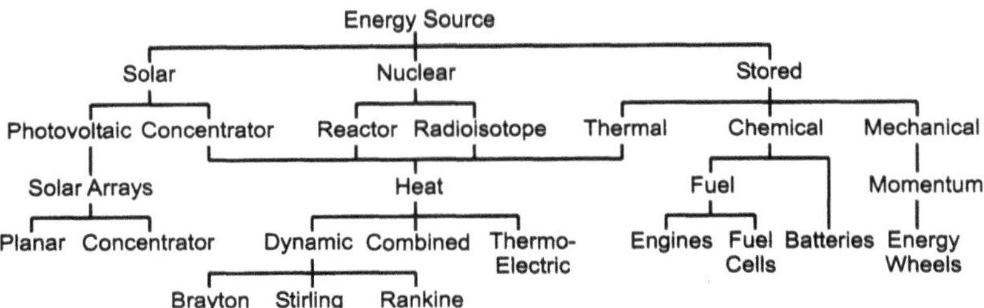

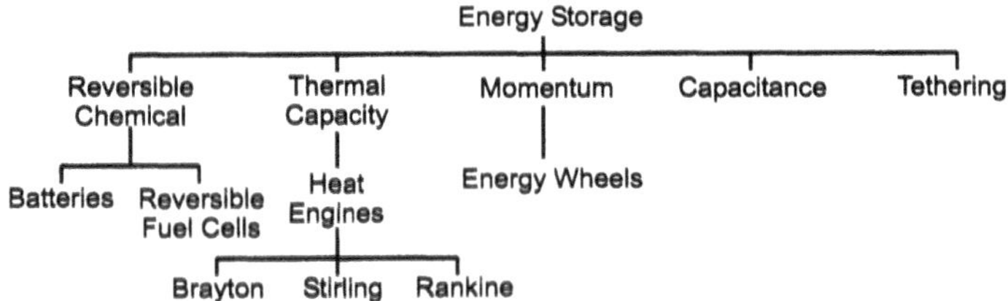

Fig. 5.3. Choices for Energy Sources and Storage Systems for Space Stations in Low Earth Orbit (LEO) [Sorensen 84]

5.1.2 Energy Sources and Storage Systems

Figure 5.3 shows energy sources and storage systems suitable for low orbit applications, while Table 5.1 and Table 5.2 list the established technical solutions. Converters transform the primary energy into the electrical energy needed aboard a spacecraft.

Depending on the conversion process, the energy processed can be stored not only electrically but also in more beneficial energy forms. Conventional fuel cells and chemical batteries are not suitable for long-term applications such as for space stations due to their high mass. For a space station as large as the International Space Station, the energy supply provided by the best non-nuclear fuel cells (H_2/O_2-cells with a mass-specific energy output of 2200 Wh/kg) requires approximately 36 metric tons of H_2/O_2 for 30 days. This resupply demand would put a strain on all available service flights.

The remaining options clearly yield different masses due to their design and conversion choices. For example, Table 5.3 lists results of a study [CNES 89] comparing several systems for a 20 kW uncrewed system in a Sun-synchronous orbit at 1000 km altitude and a design life of 7 years. A tether generator case [Martinez 88] applied to a 300 km orbit completes the data.

The data analyzed for the uncrewed spacecraft supports the radioisotope-driven Brayton process as well as both the nuclear-thermal and the advanced photovoltaic solution. A possible explanation for the unfavorable results of the tether option may be the fact that the loss of orbital energy must be compensated for by a propulsion

Table 5.1. Technical Solutions of Various Energy Sources

Energy Source	Technical Solution
Nuclear Decay of Propellant	Radioisotope Battery, Nuclear Reactor
Chemical Reaction of Propellant	Battery, Fuel Cell
Solar Radiation	Photovoltaic or Solar Dynamic Generator
Kinetic Orbital Energy	Electrodynamic Tether

Table 5.2. Energy Conversion

Conversion		Converter
Direct Energy Conversion (Single Stage)	Chemical – Electrical	Fuel Cells, Batteries
	Radiation – Electrical	Photovoltaic Cells
	Kinetic – Electrical	Electrodynamic Tether in Earth's Magnetic Field
Conversion with Intermediate Stages (Multiple Stages)	Nuclear – Thermal – Electrical	Radioisotope Batteries
	Nuclear – Thermal – Mechanical – Electrical	Reactor or Radioisotope Battery and Turbine
	Radiation – Thermal –Mechanical – Electrical	Solar Dynamic Generator and Turbine

Table 5.3. Specific Mass of Different Power Plants in the 20 kW Category for a Design Life of 7 Years

Technology	Example	Power [kW]	Specific Mass [kg/kW]
Photovoltaic Cells	• Si & NiCd Battery	25	188
	• GaAs & NiH_2 Battery	25	111
Solar Dynamic System	• Brayton Cycle	25	271
Radioisotope Generator	• Thermoelectrical	20	187–227
	• Brayton	20	100
Nuclear Brayton	• UO_2/Na/SS	20	110
	• UO_2/HeXe/HRA	20	105
	• UN/Li/Mo-Re	20	100
Tether	• Al, 20 km, h=0.6, NiH_2 Battery, I_{sp}=447 s	20	>300

system. In seven years time, a considerable amount of propellant is expended. It also could be that, at the time the study was conducted, too little was known about tethers and very unfavorable assumptions were made.

One must consider that other characteristics, besides the most adequate converter choices, play an important role in the conception of any power system. These are the following:

Earth's Shadow: The design of solar-driven power systems depends strongly on their orbit. This determines the dimensions of the converter based on the time spent in the Earth's shadow (see Sect. 5.2.3). Space stations flying in orbits below 600 km or at an inclination of less than 60°, need approximately one third of the orbital period in order to cross the shadow. During this period, the energy stored previously has to maintain the electrical power supply for systems and payloads. Nuclear reactors and radioisotope batteries work independently of solar irradiation and thus save battery mass (storage), while solar dynamic systems possess a high thermal energy storage efficiency.

Orbit Control: The larger the total area of incidence of the space station the more the station's velocity is reduced due to the residual atmosphere. Photovoltaic systems are less suitable in this particular case, since the solar array surface increases linearly with the power output. Solar dynamic systems are considerably more surface-efficient, while nuclear energy sources with their compact architecture produce the smallest atmospheric drag.

Attitude Control: Solar generators with large areas pose a problem for missions requiring precise alignment because their long flexible structures are susceptible to vibrations. In addition, the space station's center of pressure drifts due to the continuous Sun-tracking of the collectors. For example, low-frequency disruptive torques are generated which the attitude control system must compensate for.

Compact reactors and radioisotope batteries have better drag characteristics. However, radiation shields will be necessary at a certain distance from the crewed parts of the station, especially if reactors are used. Moreover, the structures used for those radiation shields could induce perturbation torques due to gravity gradients (as long as they are not used for attitude stabilization).

Safety during Launch and Operation: Photovoltaic and solar dynamic systems generally do not cause safety problems at launch, while the risk of a launch abort or a launch failure must be thoroughly investigated for generators with radioactive materials. A radioisotope source working in a static or dynamic mode would need 200–650 kg of plutonium to generate 20 kW. Since the smallest amount of this highly toxic material causes severe environmental damage, the launch of such a mass is problematic since public acceptance is less certain in the near future.

Control elements could keep nuclear reactors subcritical during launch. Once in orbit, the reactors could be started. A more significant problem is to ensure that the operating or burned-out reactor does not enter Earth's atmosphere before its radioactivity falls below acceptable levels. A typical reactor (operating over 10 years with 1 MW output power) needs approximately 300 years to reach this standard value and requires a minimum orbit of approximately 800 km altitude (Fig. 3.36). Thus, sufficient protection of the Earth environment from space stations on orbits lower than 800 km is compromised by the natural orbit decay. The same applies for the radioisotope sources where most of the plutonium does decay further. However, it should not be discounted that nuclear reactors will be accepted in the future above the minimum orbit altitude of 800 to 1000 km.

Furthermore, it is clear that it is necessary to consider the safety of the crew. An adequately sized radiation shield would represent a large percentage of the station mass. Or, another option is to separate the reactor from the main part of the station by a certain distance which would exceed rational structural dimensions (perhaps tethers as a solution might be used).

Generally, photovoltaic systems pose no safety problems. While using NiCd batteries, a control unit must protect the battery from total discharge which as a consequence might cause the batteries to explode. All dynamic systems operate on a thermodynamic cycle whose high-temperature working fluid will pose a hazard if the components of the system fail.

Reliability and Lifetime: Parallel connections yield redundancy for photovoltaic systems and thermoelectric radioisotope batteries. Due to mass limitations, dynamic operating systems depend on single energy sources (reactor, concentrator) whose failure must be avoided. In addition, dynamic systems contain moving parts in their turbines that undergo intrinsic wear. Components of nuclear systems running at their operating point are less stressed compared to solar systems operating at cyclically varying power levels and requiring mechanical tracking of generators.

Since photovoltaic cells degrade over time, their design must incorporate oversizing or replacement. Certainly, parts subject to wear of a thermodynamic process (e.g. turbine and compressor) cannot operate for the lifetime of a space station. Maintenance work must be scheduled at proper intervals.

Thermal Management: With respect to thermal management, thermodynamic concepts present a larger problem. Heat-dissipating radiators have to be placed in such a way so they neither block optically nor thermally other station sections. Solar dynamic systems need more heat rejection capability than photovoltaic systems.

Deployment: Both the telescopic boom of a nuclear generator and the deployable structure of a photovoltaic generator represent known technologies. The only additional obstacle to using the latter option is the fact that the station itself has such large dimensions. The precisely formed concentrator mirrors of a solar dynamic collector are a structural element which has no parallel to date. A folding mechanism or on-site construction might be considered, while maintaining the shape could profit from lessons learned in antenna design and construction.

In summary, the use of nuclear power plants in space is doubtful especially because of the strict safety regulations of crewed spacecraft and the safety requirements of Earth's population. In spite of their distinct advantages in other areas, their use on space stations remains out of the question for the foreseeable future. Photovoltaic and solar dynamic power systems are and remain in the future the serious relevant options. The latter promises several advantages although these converters have not been tested in orbit except for a few individual components. Nevertheless, they have a significantly higher basic mass compared with photovoltaic options. Hence, their use is only worthwhile at a certain minimum size of approximately 7 kW; however, this aspect suits applications on a space station. As a part of a German study [SDR 87], a parametric investigation compared different solar energy sources for a

Table 5.4. Mass for 25 kW Modules (User Power at End of Life) [SDR 87]
PV1: Highly Developed Si Cells with Bifacial Cell Technology (17%)
PV2: GaAs Cells (20%), Both with Flexible Blanket
PV3: Mini-Cassegrain Panels with GaAs Micro Cells, 100 Times Concentration on Additional Base Structure

	PV0	PV1	PV2	PV3
Blanket/Panel	950	480	1000	2500
Structure	760	400	800	330
Storage battery	1500	1220	1220	1220
PCS/Cabling	750	430	430	430
$\Sigma M^{4)}$	**3960**	**2530**	**3450**	**4480**
η_{Cell}	0.12	0.17	0.20	0.24
η_{Total}	0.057	0.08	0.09	0.135
η_{Orbit}	0.034	0.048	0.052	0.081
Surface$^{3)}$ [m^2]	**550**	**380**	**350**	**222**

	SD1	SD2	SD3
Collector	310	270	270
Mechanism	160	160	160
Receiver/ Thermal Storage	940$^{1)}$	580$^{2)}$	655
Machine/Generator	250	500	250
Radiator	1230	980	1500
PCS/Cabling	300	300	400
$\Sigma M^{4)}$	**3200**	**2800**	**3235**
$\eta_{Machine}$	0.28	0.38	0.34
η_{Total}	0.22	0.30	0.27
η_{Orbit}	0.136	0.175	0.155
Surface$^{3)}$ [m^2]	**133**	**104**	**117**

PV0 – "state-of-the-art" (1987)
PV1 – advanced Silicon
PV2 – Planar GaAs
PV3 – GaAs Cassegrain

SD1 – Organic Rankine
SD2 – Brayton
SD3 – Stirling

1) LiOH + LiF
2) LiF
3) Sun-Facing
4) all Masses in kg

25 kW space station. The assessment included the mass data listed in Table 5.4 and both a technological and programmatic rating. In the end, advanced silicon cells and the solar dynamic Brayton process were found to be the best solutions.

5.2 Technology

After the previous sections have outlined and defined options for power generation, the subsequent sections will examine a few actual examples. The main options, photovoltaic and solardynamic systems, are described, followed by a system comparison of the two.

5.2.1 Photovoltaic Solar Generators

Skylab employed two parallel photovoltaic generator systems operating separately from each other [Belew 73]. One system (cf. Fig. 2.7) consisted of two solar wings mounted on either side of the main module, the Orbital Workshop (OWS), while the other had four panels arranged like a windmill on the Apollo Telescope Mount (ATM). Each system had a total surface area of 108 m^2 and a maximum generator output power of 12 kW. After deducting losses and power for battery charging, each system provided a constant load power of 3.8 kW. It can be noted from the figurs that generator power here is in excess of three times the delivered load power. Higher efficiencies of todays photovoltaic systems brought this factor down to two.

The solar wings on the OWS had crossbeams 12 m in length which were stowed alongside the vehicle at launch, and once in orbit, were deployed by 90° from their original position. Then three solar wings, each consisting of 10 panels, were deployed from each crossbeam. The four solar wings on the ATM system extended to 13 m in length via a crossbar control mechanism. Because of having lost the meteoroid shield during launch, one of the wings on the OWS was completely severed off, leaving Skylab operating at only 75% of its total generator area which provided 18 kW of power (Fig. 2.7 and Table 2.10). The available electrical power of 5.7 kW for the three astronauts, and a total interior volume of 354 m^3, certainly represents the quintessential lower power limit.

Starting with Salyut 4, Russian space stations had three solar wings, while Salyut 7 possessed four wings after subsequently having installed an additional wing in 1984. This increased the total area to approximately 80 m^2. In addition, the Soyuz capsule had two side wings and an area covered by solar cells on the cylindrical mid-section above the propellant tanks (Figs. 2.18 and 2.19). The total volume of the Salyut 7 complex was approximately 165 m^3 corresponding to the Space Shuttle/Spacelab and Mir's basic station volume. By the end of 1989, normally two to three astronauts were living and working aboard Mir and were occasionally accompanied by an additional 2–3 astronauts for short periods of time. Thus, with about the same volume and crew size, the primary power needed for the three technologically very different complexes was in the order of 10 kW, and less than half of that was available for experiments.

Two silicon solar wings each 25.5 m x 3.9 m in size were planned for the European Columbus Free Flying Laboratory (CFFL). The wings had to be retractable since the CFFL was to be maintained at the space station. Its structure consisted of an Extendable/Retractable Mast (ERM) and a pad with solar cells unwound from a drum positioned at the tip of the mast (Fig. 5.4). Together, the solar panels were intended to have generated 21 kW at the beginning of the mission and approximately 19 kW after 8 years. It was intended then that the wings would be exchanged in orbit [Leisten 88, Fachinetti 89, Longhurst 89].

For the Space Station Freedom (SSF), a power of 75 kW was planned, to be produced by four generators each consisting of a pylon and two silicon solar wings measuring 5 m x 33 m [Haas 89, SSF 88, Glines]. A beta gimbal assembly connects two solar wings in a manner opposite to one another to the main truss of the station. The outer segments of the truss are pivoted to allow alpha tracking of both the gen-

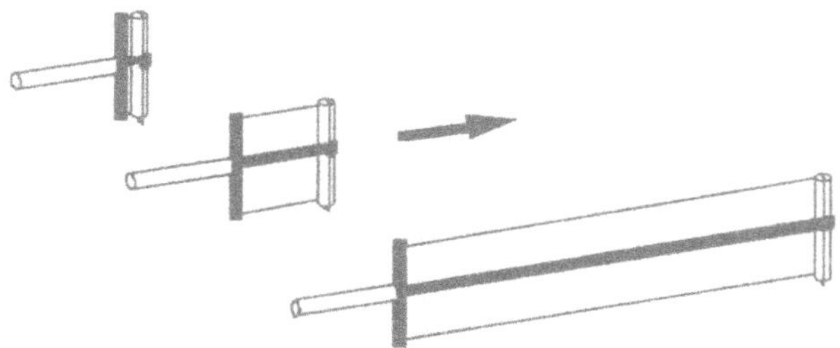

Fig. 5.4. CFFL Solar Wings and Deployment Mechanism

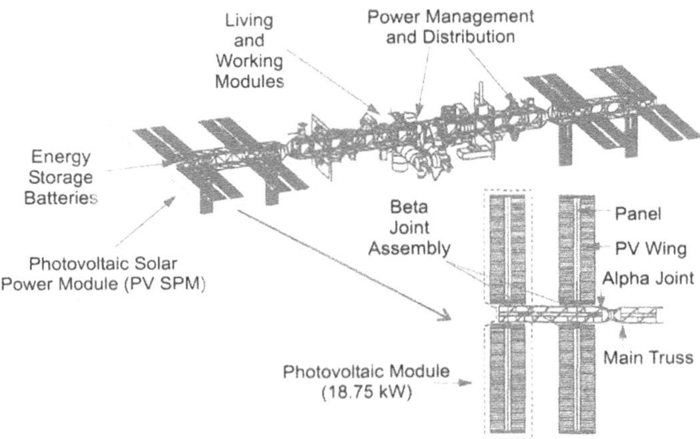

Fig. 5.5. Components of the SSF/ISS Power Supply System

erators and the thermal system radiators (see Fig. 5.5). Similar to CFFL, the solar wings on SSF were planned to be retractable for stowage during transportation. All the power subsystems developed for the SSF will be used on the United States On-Orbit Segment (USOS) of the ISS.

As demonstrated in 1984 with the OAST-1 experiment aboard Space Shuttle flight STS 41-D, the solar cell generators of ISS can be unfolded with a retractable pylon. The dimensions of the entire outer truss incorporating the mechanism for the α-rotary joint do not allow for unfolding or opening and hence must be assembled. Figure 5.6 shows the concept for assembling the generator system from the Space Shuttle. The folded solar wings, the radiator, and the cables are installed after completing a part of the truss. The generators are extended after attaching the alpha-joint onto the truss. Figure 5.7 illustrates the solar cell cross section of an US generator.

The Russian ISS solar cell generators are installed on a pylon that can be rotated in order to adjust the beta angle. They are located above the plane of the US solar cell units, i.e. further in the z-direction. Figure 5.8 shows the panel arrangement on the Science Power Platform (SPP) during an early ISS assembly phase.

Photovoltaic generators are reliable components whose technology is highly developed. However, the dimensions of conventional silicon solar arrays necessary for the performance of space stations border on the tolerable level of atmospheric drag. In this context, improvements are continually being investigated.

Other semiconductor materials besides silicon are suitable for solar cell manufacturing, in particular gallium arsenide (GaAs). GaAs cells demonstrate a greater efficiency even at high temperatures since the junction band gap is well adapted to the frequency spectrum of the Sun. However, they are more expensive than Si cells. Future developments in both multi-layer cells (such as GaAs-GaSb-cells) and Fresnel lenses (improving the optical efficiency) promise an efficiency of 30% or more. This is achieved today with only a few laboratory cells under clean-room conditions. Since GaAs cells have a high efficiency even at high photon intensities and, consequently, high temperatures, they are able to handle concentrated sunlight. Figure 5.9

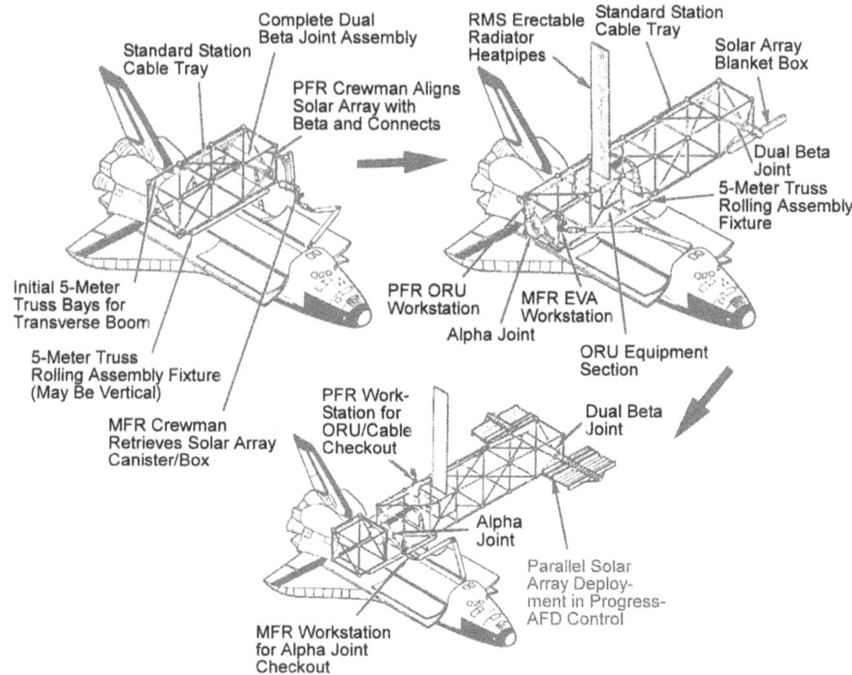

Fig. 5.6. Assembly Concept of SSF and ISS Generators

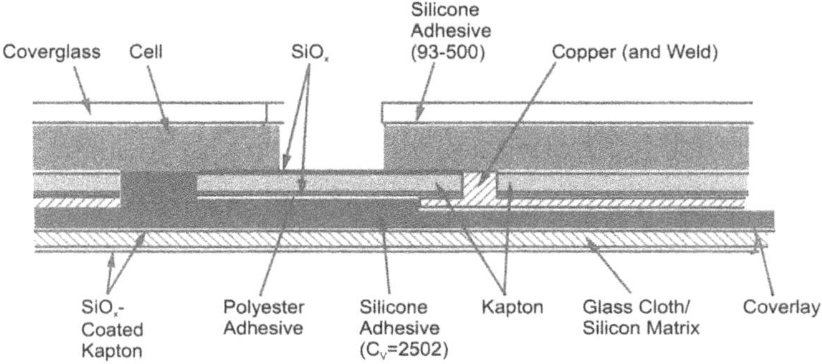

Fig. 5.7. Cross Section of a Silicon Cell aboard ISS

shows different options and reflects efforts to counter the inherent drawback of high cell prices by reducing the active semiconductor area. The necessary cell efficiency of approximately 30% has not yet been achieved to justify the higher specific mass of such a technology.

Figure 5.10 illustrates some typical properties of silicon solar cells. In the design process, an overall efficiency determined by the ratio of End-of-Life (EOL) to Be-

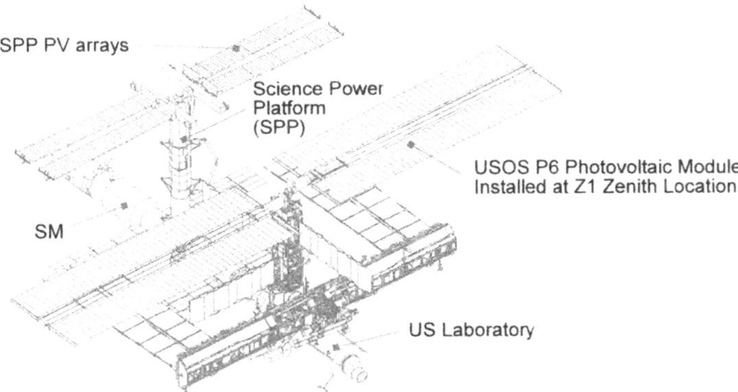

Fig. 5.8. ISS Configuration after Flight 11A in early 2001 with US and Russian Solar Collector

gin-of-Life (BOL) accounts for the degradation of cells due to aging. The degradation is higher in orbit planes with high inclinations than in the equatorial plane due to the South Atlantic Anomaly (SAA) and the higher level of radiation exposure present there (Chapter 3: "Orbital Environment"). In the initial phase, the International Space Station will rely upon proven silicon technology which might be replaced later with other semiconductors and innovative concepts.

Energy Storage for Photovoltaic Systems

In order to supply power during the eclipse period, large photovoltaic systems require large secondary energy storage units. The following are options to consider:

- Electrochemical storage systems, i.e. batteries
- Regenerative fuel cells
- Chemical storage units
- Flywheels
- Electrodynamic tethers

Chapter 10 ("Synergisms") will cover the advantages of regenerative fuel cells in conjunction with a powerful electrolyzer outfit. Flywheels have attractive storage efficiencies and energy densities; however, they require further technological development to handle large angular momentums which can cause perturbing influences on attitude control, and especially those of the converters. Figure 5.11 shows the comparison of different storage options. Past space stations used nickel-cadmium batteries, ISS will have nickel-hydrogen batteries. Important storage system parameters are energy density, e.g. Wh/kg, and storage efficiency. The energy densities computed for Fig. 5.11 are related to the system mass including solar arrays and thermal control; based on battery mass only, the energy density is (about 50–70%) higher. More energy must be input to a storage system that can be withdrawn from it. Typical electrical energy storage efficiencies are 65–80%. Thermal storage and flywheel storage systems can be as high as 90%.

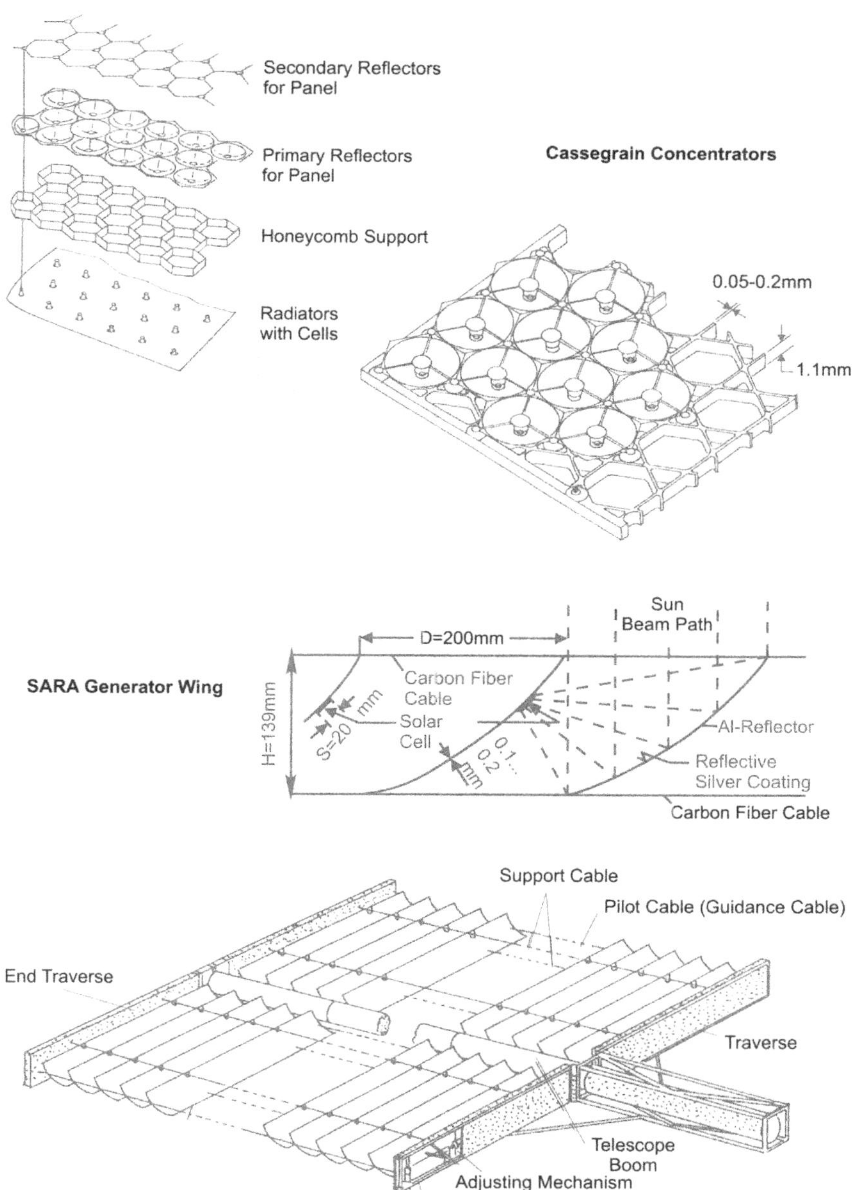

Fig. 5.9. Cassegrain Concentrators and Advanced Generator Arrays. SARA is a linear, "mini-blind"-like parabolic mirror with cooling fins on the back side, Si solar cells, and a concentration factor of about 10.

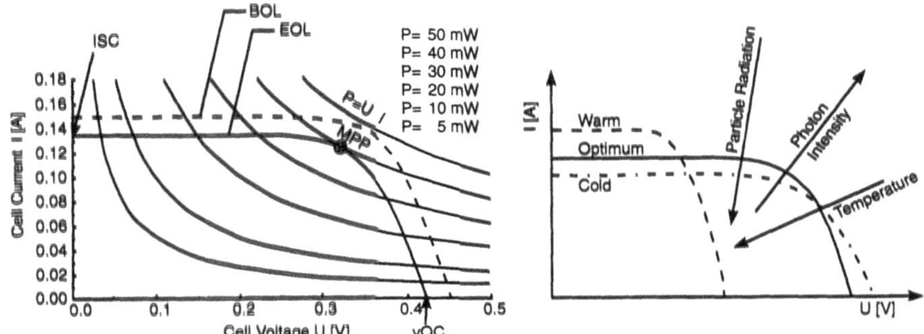

Fig. 5.10. Typical Characteristics for Silicon Cells. MPP is the Maximum Power Point. The short circuit current I_{sc} (U=0) increases with photon intensity and decreases with particle radiation. The open circuit voltage V_{oc} (i.e., I=0) increases logarithmically with photon intensity and decreases nearly linearly with temperature. Maximum power increases linearly with photo intensity and decreases rapidly with increasing temperature.

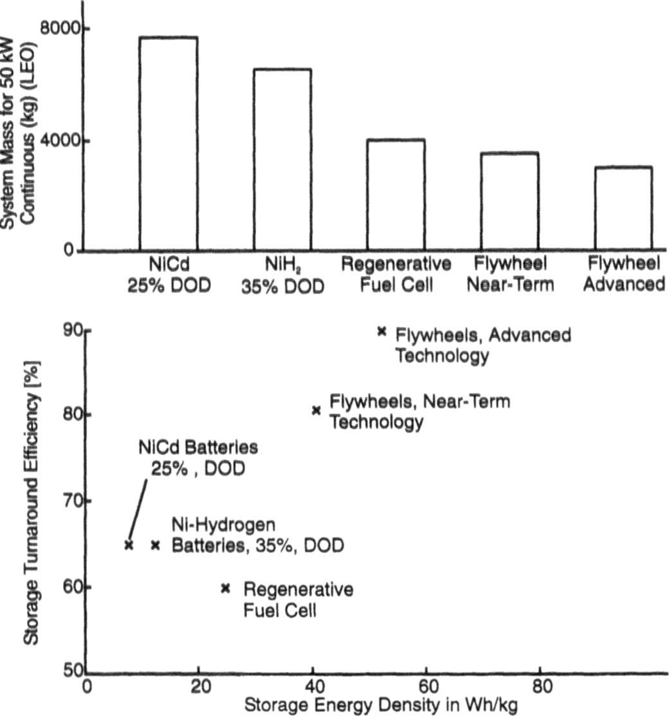

Fig. 5.11. Comparison of Several Secondary Energy Storage Systems [Woodcock 86]. The system mass includes solar arrays and thermal control. The values are computed for 92 min orbit period and 32 min eclipse.

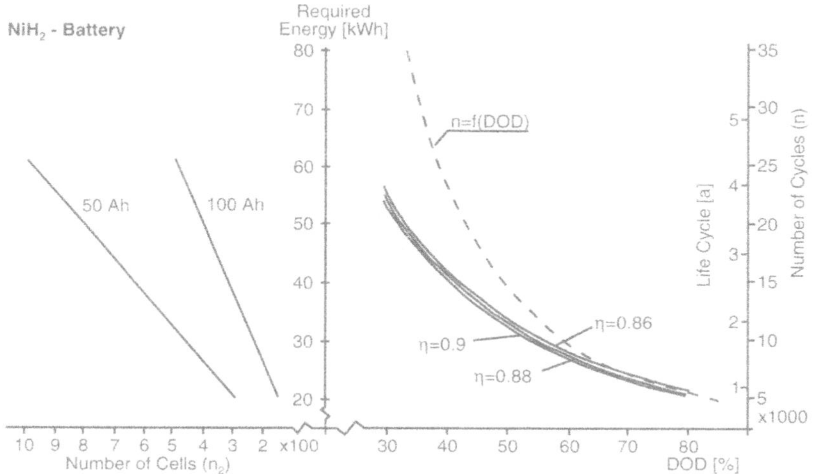

Fig. 5.12. Installed Energy and Cell Number as a Function of Lifetime, Discharge Efficiency, and Depth-of-Discharge to Provide 14.6 kW, i.e. an Assumed Station Power of 25 kW_e of Useable Electrical Power [SDR 87].

NiH_2 batteries are planned for all upcoming space station projects. These have an energy density of approximately 25 Wh/kg and a depth-of-discharge (DOD) of 35%, while conventional NiCd batteries supply 10–15 Wh/kg and 25% DOD. Regenerative fuel cells yield an even better mass. These fuel cells possess good integration capabilities with other subsystems and have ample overload capacities by a simple gaseous storage reservoir. However, their storage turnaround efficiency of 60% is significantly lower than that of batteries.

NiH_2 batteries are manufactured as Individual Pressure Vessel (IPV) cells. Nickel and hydrogen electrodes are placed in parallel and divided by a separator providing the electrolytic connection. Tie rods mechanically secure the electrodes into a compact stack and are installed in a pressure vessel. Currently typical IPV parameters are:

- Average cell discharge voltage: 1.25 V
- Cell capacity: 35–80 Ah
- Energy density: maximum 35–50 Wh/kg (40–60 Wh/l)

Further development of IPV cells projects a capacity growth to 300 Ah thereby increasing energy density to 70 Wh/kg. Figure 5.12 illustrates energy and cell number as a function of lifetime. The NiH_2 batteries of the USOS are shown in Figure 5.13.

Recently, tether systems (see Fig. 5.14) are being discussed as promising converter systems for transforming orbital energy into electrical energy and vice versa. A tether suspended from the space station aligns radially in the direction of or away from the Earth due to the gravity gradient force. Depending on tether length and tether direction relative to the magnetic field, the interaction with Earth's magnetic field

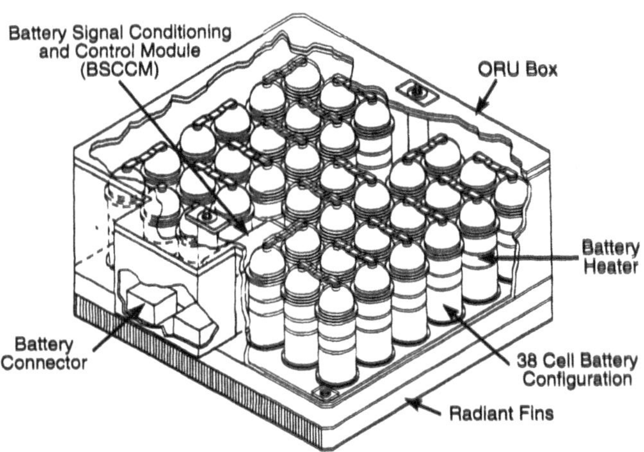

Fig. 5.13. NiH_2 Cells for the USOS Will be Exchanged as Orbital Replaceable Units (ORU). One such ORU contains at least 30 cells (35% DOD). Three ORUs will be used as one battery, while five batteries represent one unit delivering 81 Ah.

induces a voltage, which, in turn, produces a current when a load is applied. Unfortunately, little is known about the following: current passage through the atmosphere, technical potential of plasma contactors, or vibrational behavior of long tethers. The major drawbacks of power generation are forming electrodynamic forces that decelerate the tether and also the station, in addition to the usual aerodynamic forces colinear to the flight direction.

An orbital plane coinciding with the equatorial plane of the Earth's magnetic field (i.e. 11.3° orbital inclination) would be best for the current-generating tether operation (cross product is $|\vec{v} \times \vec{B}| \sim \cos[i - 11{,}3°]$). In this case the effective component of the horizontal magnetic induction B_h is at a maximum allowing a scalar representation. The induced voltage is $U_i = v \cdot B_h \cdot L$, where v is the orbit velocity and L is the effective length of the tether. The retarding force acting on the tether is $F_{ed} = I \cdot L \cdot B_h$ and the generated electrical power is

$$P = F_{ed} \cdot v = U_i \cdot I \sim (\text{Orbit Radius})^{-3.5} \tag{5.1}$$

The dependence of electrical power on orbit radius becomes clear only at high orbital altitude H or high orbital inclination i. Defining the most favorable case (100%) with the pair of variables 300 km/11.3°, the electrical power is reduced to 77% for 300 km/28.5° and to 47% for 300 km/51.6°. Compared with the SFF inclination of 28.5°, ISS's inclination of 51.6° is less favorable, but it is still perfectly suitable for applications of electrodynamic tethers. For example, they could be used as emergency power supplies or for regulating and damping tether oscillations with the help of which e.g. return capsules can be decelerated, see Fig. 12.6.

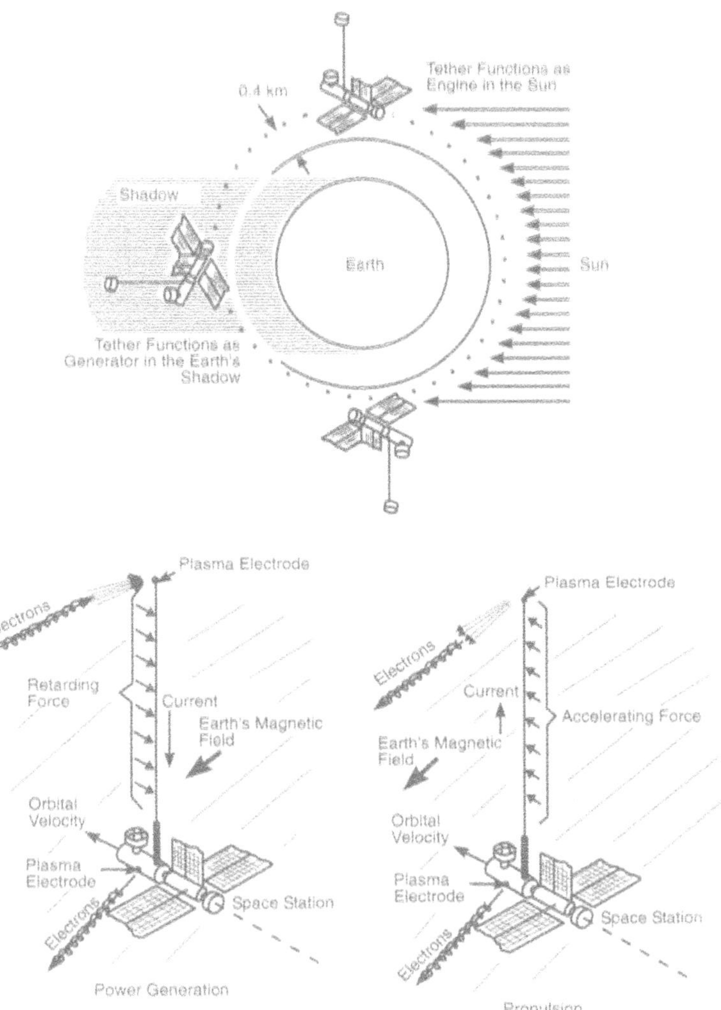

Fig. 5.14. Principle Operation of an Electrodynamic Tether Converting Orbital Energy into Electrical Energy and Vice Versa

5.2.2 Solar Dynamic Generators

Solar dynamic systems have not yet been tested in orbit but were investigated in the initial phase of the US space station studies. Since at that time the station targeted a power level of 300 kW at completion, only solar dynamic systems were able to keep the structure within acceptable dimensions (see Fig. 5.15 for the Power Tower configuration of Fig. 2.9).

In a solar dynamic system, radiation captured with the concentrators is used as thermal energy and is stored in a salt system for the shadow phase. A working me-

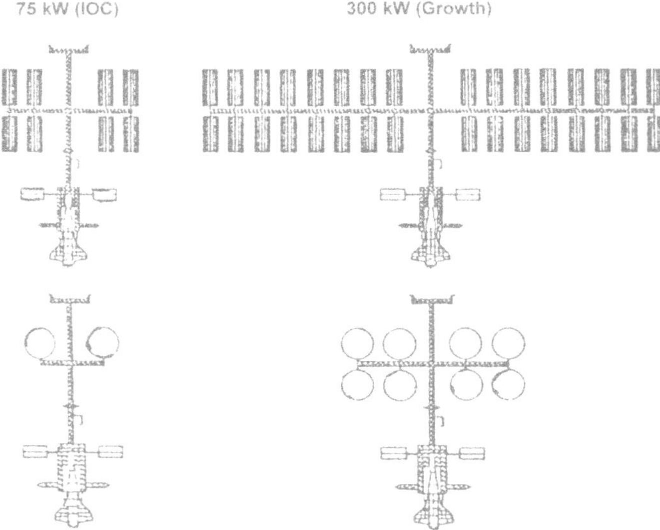

Fig. 5.15. Size Comparison of a PV (Top) and a SD (Bottom) Power Supply for Both the Initial Operational Capability (Left) and the Final Power Tower Configuration

dium drives a generator to convert heat into electrical power (Fig. 5.16). The three main generator options are the Rankine, the Brayton, and the Stirling processes, where the Brayton process has been favored for quite some time. It admittedly operates at a high temperature level and imposes specific requirements on the components but as a result, achieves a higher efficiency than do the other options. Development of solar dynamic systems is pushed due to their potential and relevance for terrestrial applications. At present all three processes are being pursued.

For the time being, a solar dynamic power plant will not be realized due to cuts in US plans. First, the power requirements were reduced to a size manageable with photovoltaic systems. Second, technological development risks, e.g., costs, are considered too high to benefit from a more efficient operating system. This is true even more so if the efficiency of solar cells can be significantly increased.

Figure 5.17 illustrates a typical configuration of a solar dynamic generator for a future orbital system. In addition to the precisely adjustable mirror, it shows both the receiver/turbine unit and the large thermal radiator necessary for this process. Along with the development of components for the thermodynamic process, transport and construction of the concentrator also may raise problems.

Currently, different countries are investigating solar dynamic systems for application on space stations:

- USA: A 25 kW system (Brayton and Stirling). For ISS a 2 kW system has been already ground-tested. 7.5 and 25 kW systems are considered to be of interest for terrestrial applications.
- Russia: Several SD approaches (Rankine, Brayton and Stirling) have been designed and tested for different power levels (5–2200 kW).

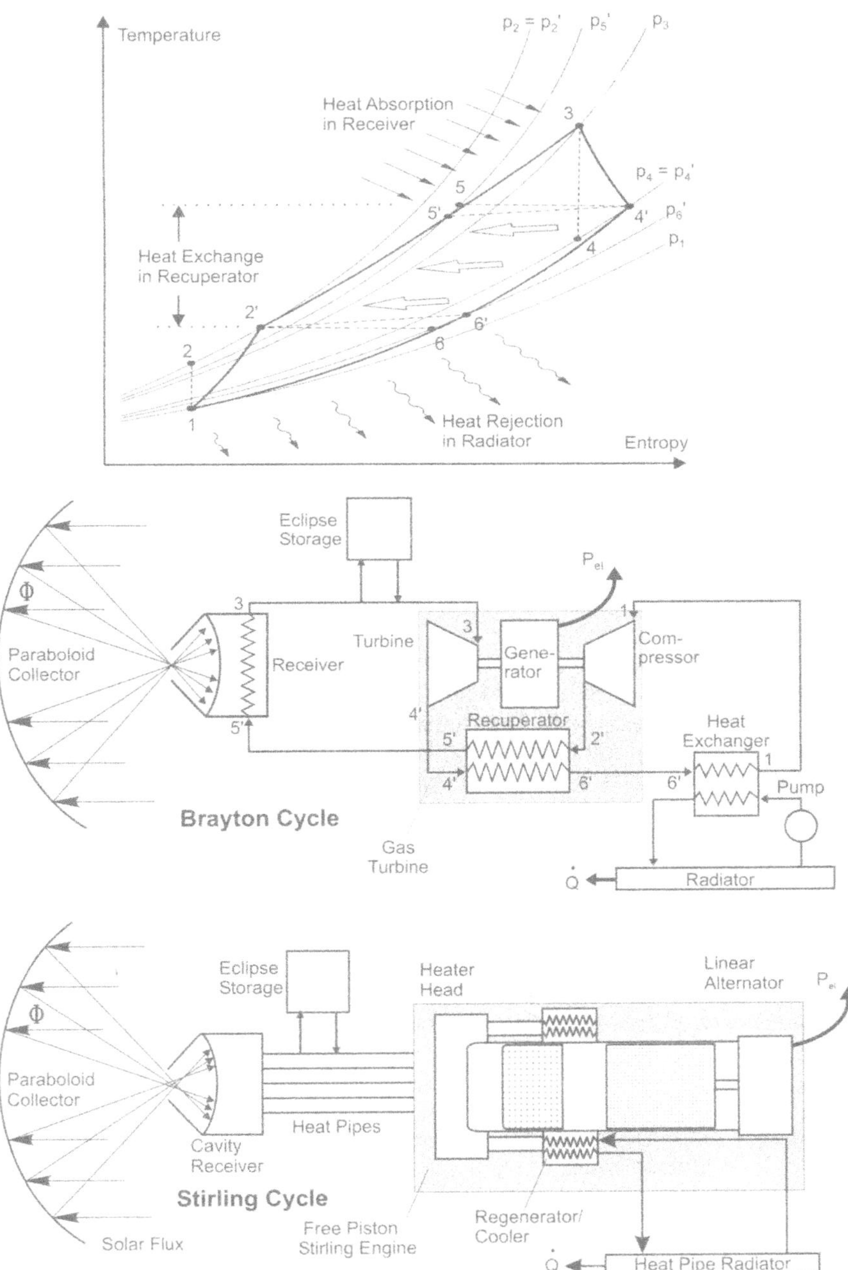

Fig. 5.16. Temperature-Entropy Diagram and Schematic of a Solar Dynamic System: Brayton Cycle [Oberle 93] and Stirling Cycle [Audy 99]

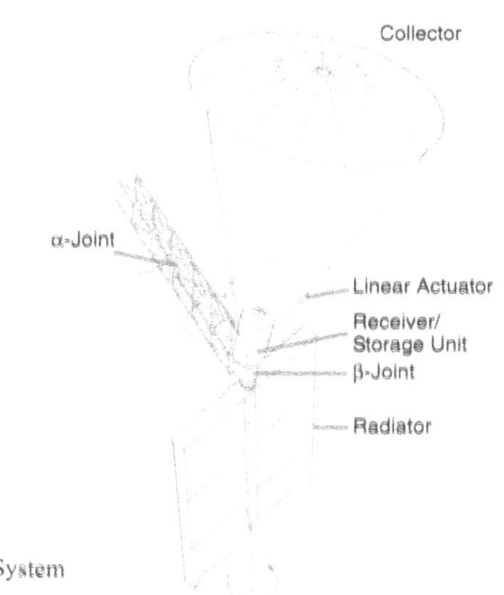

Fig. 5.17. Principal Configuration of a SD System for Space Missions [Oberle 93]

- USA/Russia: A joint flight experiment was conducted for a 2 kW system onboard Mir in 1997.
- Japan/Germany: Currently, a small Japanese Stirling engine having 300 W and DLR components are being tested for a joint flight experiment on JEM/ISS after the year 2002 [Sprengel 97]. The German Aerospace Center (DLR) analyzed in detail (theoretically and numerically) both the Brayton system [Oberle 93] and the Stirling system [Audy 99]. The construction and testing of critical components is still in an early phase.

5.2.3 Influence of Shadow Period on the Design of Solar Power Systems

The duration of the eclipse period depends on both the orbital parameters of the space station and the position of Earth with respect to the Sun. The eclipse and hence the shadow period for low orbits can be 40% of the total orbital period (see Table 5.5).

Table 5.5. Minimum Sun Periods and Maximum Shadow Periods as a Function of Orbit Elevation

Altitude [km]	200	300	400	500	600	700	GEO
Sun Period [%]	57.9	59.6	61.0	62.2	63.3	64.2	94.4
Shadow Period [%]	42.1	40.4	39.0	37.8	36.7	35.8	5.6

The eclipse period of a circular orbit is calculated according to

$$\frac{t_E}{t_O} = \frac{2 \cdot \alpha}{360°} \tag{5.2}$$

where the orbital period is

$$t_O = 2 \cdot \pi \cdot \sqrt{\frac{R_0}{g_0}} \cdot \left(\frac{R_0 + H}{R_0}\right)^{3/2} . \tag{5.3}$$

The shadow half angle α is determined according to

$$\alpha = \arcsin\left(\frac{\sqrt{(R_0/r)^2 - \sin^2\beta}}{\cos\beta}\right) \quad \text{for} \quad \beta < |\arcsin(R_0/r)| \tag{5.4}$$

In all other cases ($\beta > |\arcsin(R_0/r)|$) there is no shadow period. β is calculated according to:

$$\sin\beta = \cos\theta \cdot \sin\Omega \cdot \sin i - \sin\theta \cdot \cos i_E \cdot \cos\Omega \cdot \sin i + \sin\theta \cdot \sin i_E \cdot \cos i \tag{5.5}$$

where R_0 is Earth's radius, r is the radius of space station orbit, θ is the angle in the ecliptic between vernal equinox and the Sun, Ω is the right ascension of the ascending node, i is the inclination, and i_E=23.5° is the inclination of the ecliptic with respect to the equator (cf. Fig. 6.1).

Here, β is defined as the angle between the orbital plane and the direction of the incident solar radiation (see Fig. 5.18). For inclinations i<90° and considering the inclination of the ecliptic, a maximum of the eclipse period is obtained for $\beta = i - 23{,}5°$ and a minimum for $\beta = i + 23{,}5°$. For orbits of 500 km with $i = 28{,}5°$, the eclipse periods are 35.6 and 27.6 minutes (depending on the season), corresponding to the simulation results illustrated in Fig. 5.19. The influence of the atmosphere can be simulated by increasing the Earth's radius by 30 km.

System Design with Storage for Shadow Period. Solar systems have to be designed to convert enough solar energy per orbit to provide thermal or electrical energy for the orbital path which is in the planet's shadow and thus has no radiation energy available (Fig. 5.20). The collector area necessary for providing continuous supply is divided into two sub-sections. The sub-area A_1 of the collector providing energy for the shadow period is calculated according to:

$$P \cdot t_E = A_1 \cdot \Phi \cdot \eta_{PS} \cdot \eta_{Stor} \cdot (t_O - t_E) \tag{5.6}$$

- P electrical output power required (P=const.)
- t_E eclipse period (shadow period)
- t_O orbital period
- Φ solar radiation flux ($\Phi \approx 1371$ W/m^2, minimum value for maximum distance between Earth and Sun)
- η_{PS} power system efficiency (solar electrical efficiency)
- η_{Stor} storage efficiency (includes both charging and discharging)

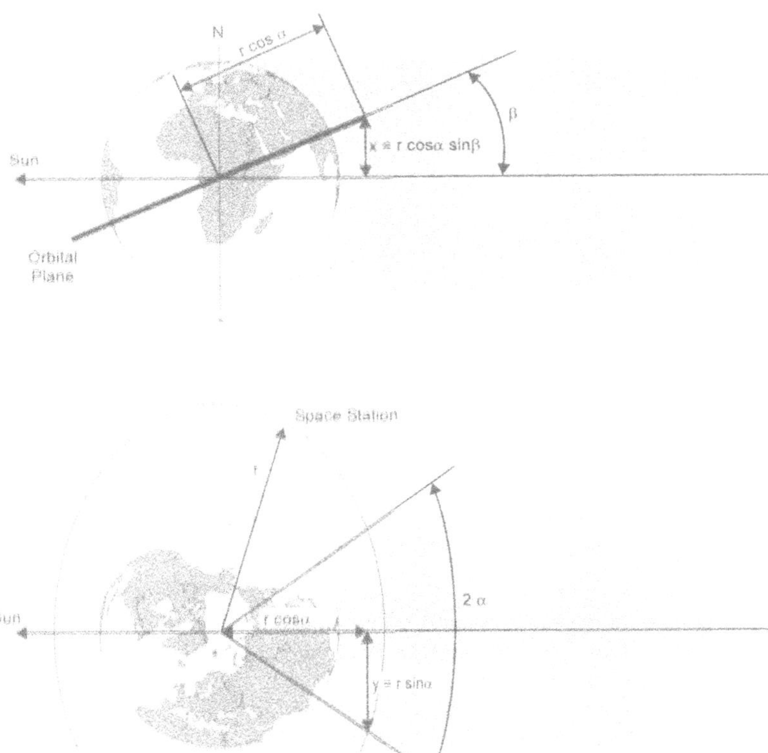

Fig. 5.18. Calculation of the Eclipse Duration

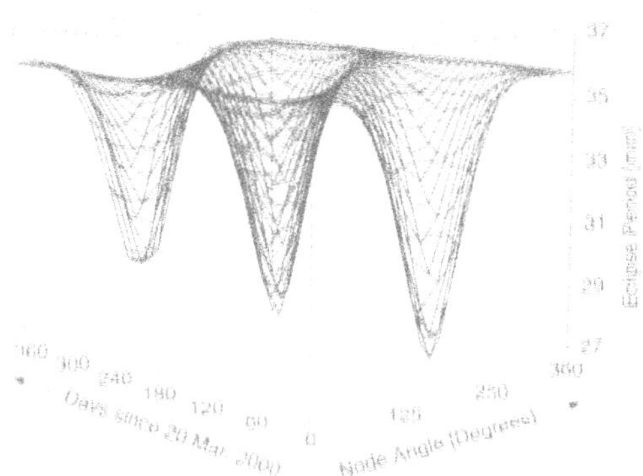

Fig. 5.19. Shadow Periods as a Function of Nodal Line and Season (Orbit Altitude: 500 km, Inclination: 28.5°, Θ=75.4° on March 20, 2002).

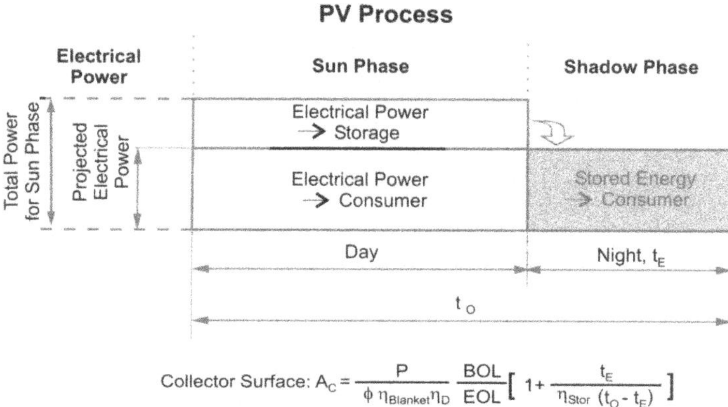

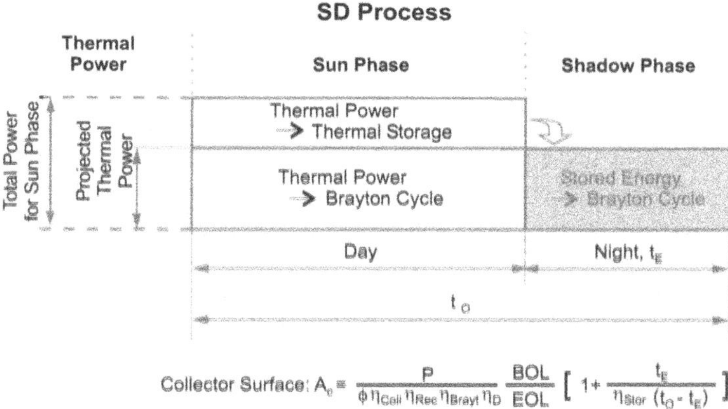

Comparison of Both Systems

	Photovoltaic	Solar Dynamic
Collector Surfaces	$A_C = 397\ m^2$	$A_C = 118\ m^2$
Surface Ratio	$\frac{A_{PV}}{A_{SD}} = 3{,}4$	
Efficiencies	$\eta_{total\ PV} = 4\%$	$\eta_{total\ SD} = 15\%$

Fig. 5.20. Performance vs. Orbital Phase and Comparison of Efficiencies for Photovoltaic and Solardynamic Process Units

$$A_1 = \frac{P}{\Phi \cdot \eta_{PS} \cdot \eta_{Stor}} \cdot \frac{t_E}{t_O - t_E} \tag{5.7}$$

The other sub-area A_2 represents the power supply during the Sun period:

$$A_2 = \frac{P}{\Phi \cdot \eta_{PS}} \tag{5.8}$$

Adding the areas of Eq. 5.7 and Eq. 5.8 results in the necessary total "energy collecting area" for the power generator:

$$A = \frac{P}{\Phi} \cdot \frac{1}{\eta_{PS}} \cdot \left(1 + \frac{1}{\eta_{Stor}} \cdot \frac{t_E}{t_O - t_E}\right) \cdot \frac{BOL}{EOL} \tag{5.9}$$

where the term *BOL/EOL* denotes a degradation factor which is available as an excess factor at the beginning of plant operation. The energy storage must be designed to deliver the required output power during the shadow period. Accordingly, the energy content (in [Ws]) of the storage is determined:

$$W_{Stor} = \frac{P \cdot t_E}{\eta_{Stor} \cdot DOD} \tag{5.10}$$

DOD stands for "depth-of-discharge" and is the part of the total energy storage content which is actually used both in the charging and discharging cycles. In particular, for electrical storage units, this parameter determines the number of cycles and thus the lifetime of the storage device (see Fig. 5.12). By knowing specific technological parameters such as the area-specific mass of the collector (kg/m^2) or the specific mass of the storage (kg/Wh), an estimate of the system mass for collector and storage can be made.

5.2.4 Comparison of Photovoltaic and Solar Dynamic Systems

The following example compares the efficiencies of systems presented in Sect. 5.2.1 and Sect. 5.2.2. An advanced silicon cell system will serve as a representative of photovoltaic systems, while a Brayton gas turbine power plant with latent heat storage will represent solar dynamic systems.

Example: Photovoltaic System

The secondary energy storage systems listed in Table 5.6 differ with regard to their structural mass, structural volume, durability, depth-of-discharge, etc. The total efficiency of photovoltaic power generators is determined by the definitions introduced in Table 5.6.

$$\eta_{PS,\,PV} = \eta_{PV} = \eta_{Blanket} \cdot \eta_D = 0.11 \cdot 0.9 \approx 0.1 \tag{5.11}$$

$$\eta_{Stor,\,PV} = 0.75 \tag{5.12}$$

Table 5.6. Efficiency Definitions of a Silicon Cell System

Efficiency		Comments
$\eta_{Lab\ Cell}$	up to 30%	• Subject to the clean processing and unwanted impurities and only achieved in a few selected single cells • No reproducible manufacturing conditions and no series bonding of cells
η_{Cell}	≈ 17–20%	• Reproducible efficiency • Manufacturing conditions with bonding
η_{Panel}	≈ 15–20%	• Efficiency including network losses with panel sizes up to 0.25 m^2
$\eta_{Blanket}$	≈ 10–12%	• Entire wing (up to several m^2) connected and wired including wiring losses (high ohmic losses) • Additional consideration of structural areas not covered with cells
$\eta_{Collection\ efficiency}$	≈ 80–90%	• Consideration of ratio of cell area and structure area • Already included in $\eta_{Blanket}$
η_D	≈ 90%	• Power processing and distribution efficiency • DC/AC conversion losses as well as losses occuring during transformation to signal voltage
$\eta_{Stor,PV}$	≈ 75–80%	• NiH_2 Battery Systems
	≈ 70–80%	• NiCd Battery Systems
	≈ 85%	• NaS Battery Systems
	≈ 55–60%	• RFC (Regenerative Fuel Cell) Systems*)

*) System Combining Electrolysis and Fuel Cells. The fuel cell is not operated in reverse, but instead a separate electrolysis apparatus is used.

Using electrical ouput power of 25 kW, an orbit altitude of 500 km, and a typical degradation factor of BOL/EOL=1.2 (10-year service life) in Eq. 5.9, the total collector area for a photovoltaic system is:

$$A_{PV} = \frac{25\ \text{kW}}{1371\ \text{W/m}^2} \cdot \frac{1}{0.1} \cdot \left(1 + \frac{1}{0.75} \cdot \frac{36\ \text{min}}{95\ \text{min} - 36\ \text{min}}\right) \cdot 1.2 \qquad (5.13)$$
$$A_{PV} \approx 397\ \text{m}^2$$

Example: Solar Dynamic System

In the case of solar dynamic systems, the efficiencies multiply according to (cf. Table 5.7):

$$\eta_{PS,\,SD} = \eta_{SD} = \eta_{Collector}\eta_{Receiver}\eta_{Brayton}\eta_D = 0.9 \cdot 0.9 \cdot 0.4 \cdot 0.94 \qquad (5.14)$$
$$\eta_{SD} \approx 0.305$$

$$\eta_{Stor,\,SD} = 0.95 \qquad (5.15)$$

Table 5.7. Efficiency Definitions of Solar Dynamic Power Plant with Brayton/Gas Turbine and Latent Heat Storage

Efficiency		Comments
$\eta_{Brayton}$	≈ 40%	Process efficiency depends on both heat source and heat sink temperatures, performance, etc.
$\eta_{Collector}$	≈ 90%	Considers reflectivity of mirror and possibly of structure, etc.
$\eta_{Receiver}$	≈ 90%	Receiver losses consist of reflection losses and thermal radiation losses through aperture surface.
η_D	≈ 94%	Power processing and distribution efficiency. Transformation losses are avoided since gas turbine produces AC in the generator.
$\eta_{Stor,SD}$	≈ 95%	Poor isolation and temperature differences during charging and discharging induces losses in latent heat storage (primary storage).

The collector area necessary for a solar dynamic system with output power of 25 kW and orbital parameters which are the same as those for the photovoltaic example results in:

$$A_{SD} = \frac{25\ \text{kW}}{1371\ \text{W/m}^2} \cdot \frac{1}{0.305} \cdot \left(1 + \frac{1}{0.95} \cdot \frac{36\ \text{min}}{95\ \text{min} - 36\ \text{min}}\right) \cdot 1.2 \tag{5.16}$$

$$A_{SD} \approx 118\ \text{m}^2$$

Similarly, a degradation factor BOL/EOL of 1.2 is assumed since in a solar dynamic system, both the collector surface and the radiator surface are subject to degradation. This causes a lower reflection coefficient and worse radiating conditions, leading to an increase in the heat sink temperature, and thus yielding lower efficiency of the Brayton process.

System Comparison. The ratio of the collector areas calculated above results in:

$$\frac{A_{PV}}{A_{SD}} \approx 3.4 \tag{5.17}$$

Considering a secondary storage unit similar to the PV system, i.e. batteries, the collector area of the solar dynamic system results in 137 m^2 and thus a collector area ratio of 2.9. This suggests a degradation in the SD system; however, the heat engine idles during the shadow period since the secondary system supplies the electrical power. Consequently, the thermal radiation losses of the receiver are lower and a gain is obtained due to improved cooling of the radiator during the shadow period.

The effects mentioned above cause improved results for the solar dynamic system. However, the favorable trends are hindered by durability problems, cycle endurance and material problems in the receiver, the design of the gas turbine (gap leakage), etc., leading to a deterioration of the positive effects.

Resulting from the above analysis, the main advantage of the solar dynamic system is the thermal storage of energy for the shadow period during the primary energy conversion.

5.2.5 Energy Distribution and Processing

The power provided by the generator and storage system must be delivered to the different end consumers. In the past, most spacecraft used a 28 or 50 VDC network system (direct-current) mainly due to the lightweight DC circuit components. Power levels of a space station go beyond the limits of conventional concepts, such that a level of 75 kW of power at 28 VDC would produce a current of 2700 A. In addition, transmission distances on a space station add up quickly to 100–200 m, thus increasing line leakage.

These factors cause two adjustments for design:

- The power supply is distributed in parallel through several bus networks.
- The main bus voltage is increased.

Unfortunately, the voltage cannot be raised to the usual terrestrial levels since it is limited by

- losses through Earth's plasma in LEO (approx. 1% loss at 400 VDC, risk of arcing),
- qualification of components (U_{max} = 400 VDC, thus $U_{operation}$ = 160 V), and
- protection against electrical shocks (U_{max} < 400 VDC).

Until now, the highest bus voltages were planned at 160 VDC for SSF and at 120 VDC for Columbus elements.

Individual consumers take the required power off the bus. Because of a variety of subsystems, a whole range of different voltages is required to supply energy. The transformation would be largely simplified if the network used alternating current. A 20 kHz network system was actually analyzed for SSF, particularly because it promised lower losses; however, the International Space Station will run on a DC system due to cost concerns.

Depending on concept design and favorable total mass, voltage management is achieved either with the individual consumer or centrally with the bus. In any case, it must be possible to isolate the individual bus terminals in case of failure such as a short circuit. Expert systems used in the USOS perform the operational management and error diagnostics of the bus terminals.

In most cases, main modules on large space stations are designed as a modular system and have separate, regulated network systems. For example, the JEM (which underwent the least changes during the transition from SSF to ISS) is illustrated in Fig. 5.21: an intermediate stage transforms the main voltage and can incorporate its own separate energy storage. Shown in Fig. 5.22, the secondary network systems operate with parallel bus terminals to increase redundancy and operating safety rather than reducing the current.

Originally, the Columbus project was designed for a main bus voltage of 150 VDC; in view of the availability of reliable components, however, the voltage was lowered to 120 VDC in phase B2. The Multiple Power Bus Management (MPBM) was introduced for power control including solar generators (Fig. 5.23: STAGE 00 up to STAGE 16), a battery control unit (BCU) with battery charge relays (BCR), and battery discharge relays (BDR). This is in contrast to EURECA

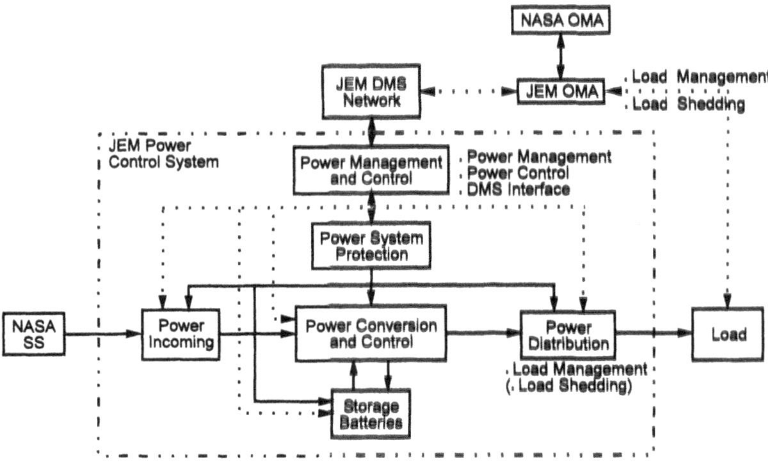

Fig. 5.21. Block Diagram of the JEM Power Control [Kawamura 89]

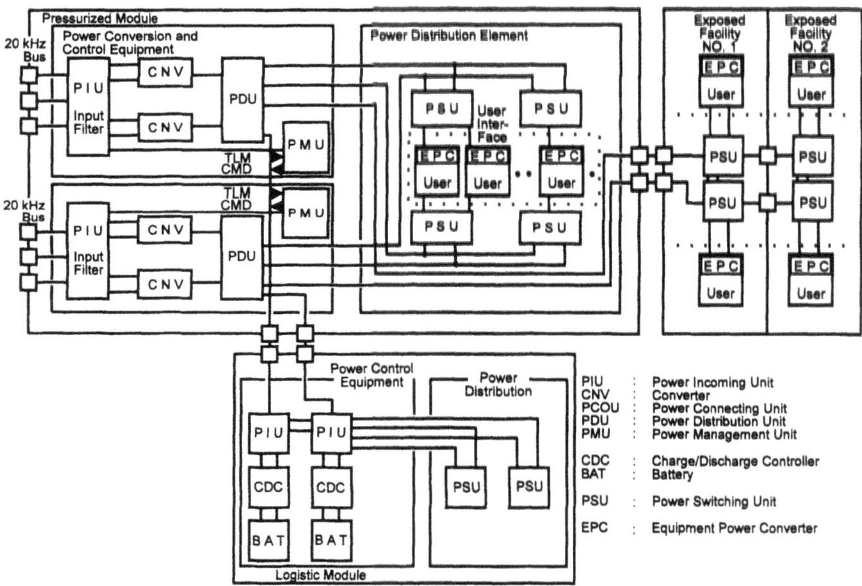

Fig. 5.22. JEM Power System

where a separate solar cell area is available for charging the batteries. According to Fig. 5.23, each component is assigned to a domain with MPBM. Individual sections can connect to the bus depending on the power need of consumers. Peak loads can be taken from the bus on short notice when the battery charging is interrupted by opening of the charge relays (BCR), and, additionally, the battery current can be used by closing the discharge relays (BDR) of the batteries.

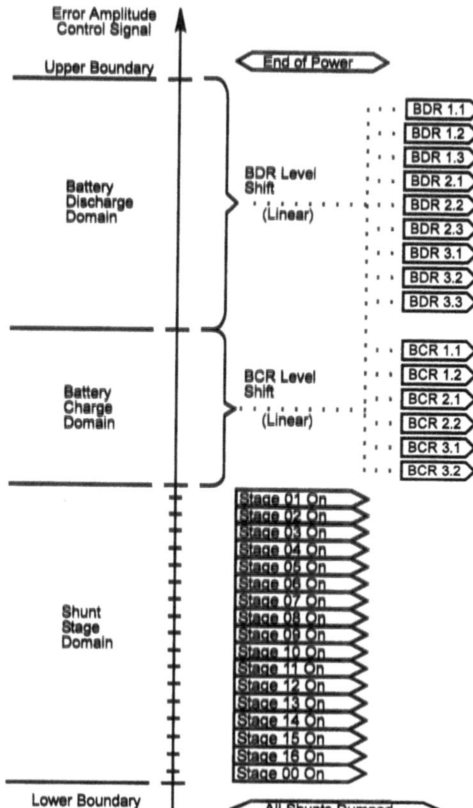

Fig. 5.23. Columbus MPBM Concept, Arrangement of Control Sectors

The following example will demonstrate that the design of a power distribution system forces development of new components as well. Switches are necessary to cut off individual circuits when short circuits or overloads occur. The conventional mechanical switches could not be considered for CFFL. First, their switching time of 10 ms is too long. Secondly, relay networks for a power volume of 150 kW would be too heavy, and switching would cause noticeable perturbing accelerations in the laboratory.

Thus, electronic switches such as the 120 V/10 A Solid State Power Controller (SSPC), shown in Fig. 5.24, were developed for Columbus. This unit is based on MOS transistors and has a switching time of less than 1 ms. However, thermal protection circuits must be incorporated to prevent excessively high temperatures on the transistors and thus the destruction of the switch. Likewise, the rate-of-change of current is limited in order to prevent extremely high voltage peaks in the bus resulting from short switching times.

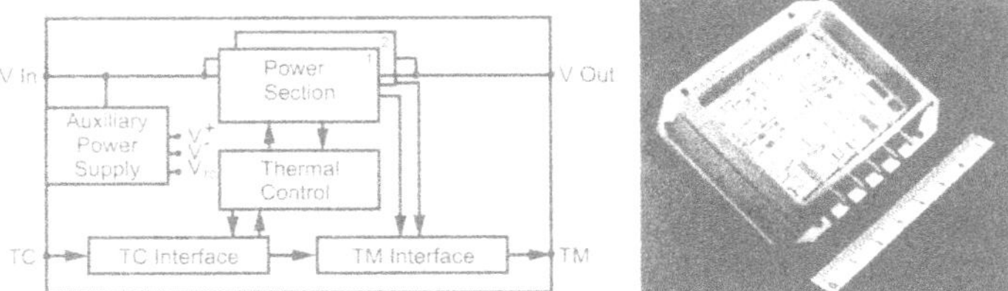

Fig. 5.24. Columbus Solid State Power Controller

5.3 Examples of Overall Systems

Due to the second OWS solar array severed during launch, *Skylab* only had a generator area of 162 m^2 and an effective power of 5.7 kW. The power supply systems of both the OWS and the ATM were completely separated. The battery systems consisted of NiCd cells with 264 Ah (3.8 kW for OWS) and with 360 Ah (3.7 kW for ATM), both at a voltage of $U = 25–30.5$ VDC. Thus, two separate power supplies in parallel provided redundancy for both electrical power generation and the battery system. There were two battery sets on the OWS, each connected to one of two bus terminals which distributed the energy to the main modules. To achieve a symmetrical load distribution, the bus terminals were interconnected. The ATM system was structured similarly to the OWS with the exception that all batteries were wired in parallel to the bus terminals. Power was transferred in both directions across a junction between the ATM and OWS systems. Electrical power transfer was possible between the main bus terminals in case of failures.

CFFL was designed for a 21 kW power level (BOL, solar panel size 25.5 x 3.9 m) and 19 kW (EOL after 8 years), of which 6 kW were available for the payload. The power supply proceeded through two 120 VDC channels regulated by the MPBM concept discussed above. After reserving the power necessary for the resource module, the pressurized module was operated via three single redundant bus terminals, each supplied by both main power channels (Fig. 5.25). Six NiH_2 cells serving as batteries were planned, each providing 50 Ah at 120 VDC. A separate micro-processing unit monitored the batteries in order to keep the cells at operating temperature during charging and discharging.

The planning for *SSF* envisioned four photovoltaic generators each supplying 18.75 kW. Two generator units were to be available during the assembly phase, while the remaining two were supposed to go on-line shortly before the beginning of the permanently crewed capability. Each generator unit consists of a solar cell wing, switching and control units, batteries, and thermal control devices. These assembly units are installed in the truss as ORUs. Each unit consists of five NiH_2 batteries providing 81 Ah, while each battery contains 90 cells of which 30 constitute one ORU (Fig. 5.26). The total (energy) volume of all the batteries is 1620 Ah operated at a DOD of 35% with a design life of 6.5 years. The energy is supplied to the main bus of the station at 160 VDC imposed by the generator system.

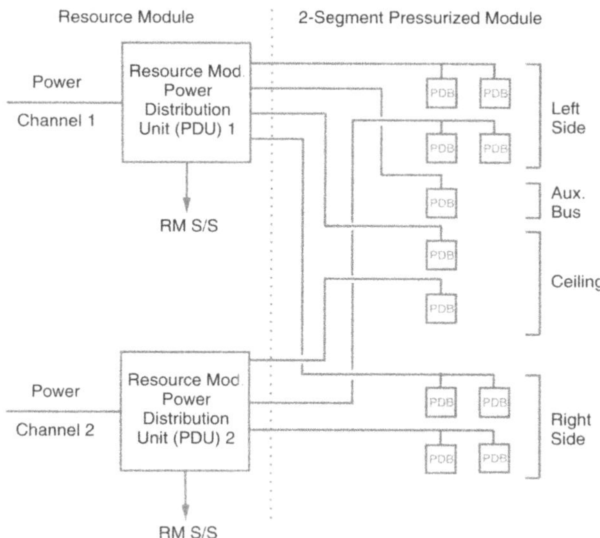

Fig. 5.25. Concept of CFFL Energy Distribution

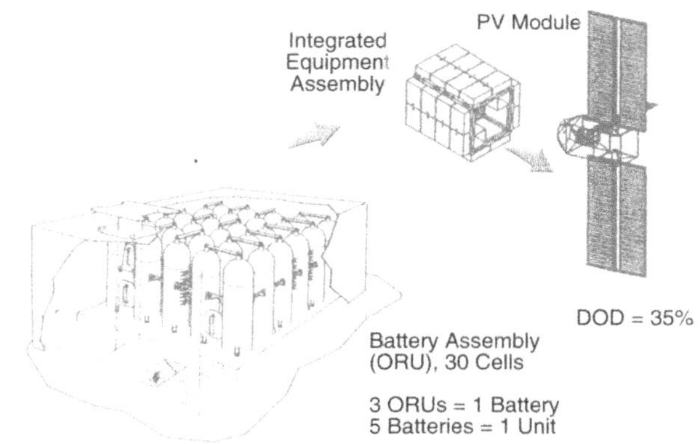

Fig. 5.26. Battery Configuration of ISS and ORU Concept

The transition from the regulated bus to the secondary network in the modules and payloads transforms the primary voltage to 120 VDC.

Figure 5.27 illustrates the power supply of the International Space Station, which was adapted nearly unchanged from the SSF, for two out of eight US solar arrays. The power originating at the generators is regulated via a Sequential Shunt Unit (SSU) at 53.3 kW. A power level of 22.9 kW is delivered to the Battery Charge/Discharge Unit (BCDU) via the direct current switching unit (DCSU) while the batteries receive a power level of 20.2 kW. The Main Bus Switching Unit (MB-SU) obtains 27.4 kW during the Sun phase via the alpha joint and 23.5 kW during

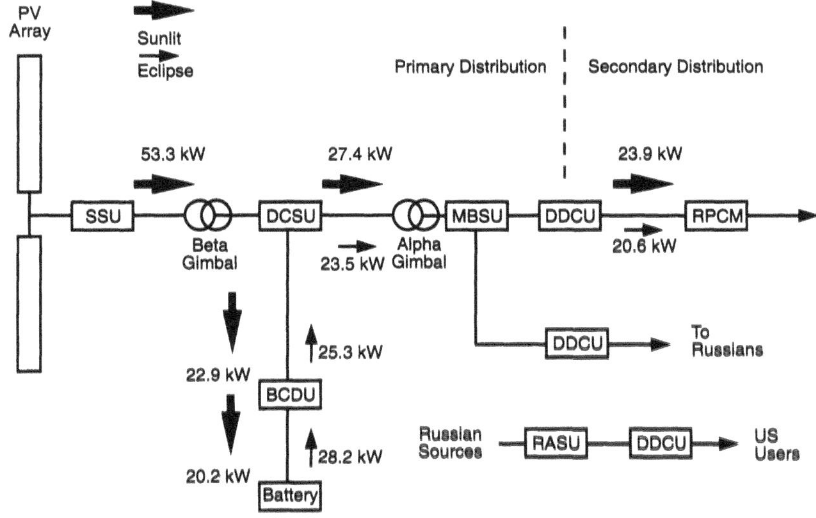

Fig. 5.27. Electrical Energy Distribution in the USOS

Table 5.8. Performance Data of US Solar Generators

Pointing Accuracy Accuracy; PV Module (Sun Vector and Rotary Joint Target Determination)	 3°, 3 Sigma RMS	**Power for Users** Voltage Annual Average Minimum Continuous	 120 VDC 30 kW 26 kW
Accuracy; Integrated Truss (S3 and P3 Rotary Joint Orientation) Accuracy; Solar Power Module (S4, P4 and S6 Rotary Joint Orientation)	2.22°, 3 Sigma RMS 2.00°, 3 Sigma RMS	**Power Distribution** US Lab US Hab Node 2 Integrated Truss Assembly	 37.5 kW 12.5 kW 50 kW 25 kW
Pointing Stability Solar Tracking Rate	0.50°, 3 Sigma RMS -0.0675 to +0.0675°/s per Axis	**US/Russian Power Transfer** Stage 1A-7R Stage 3A-10A Stage 17A and beyond	 1 kW Russian to US 13 kW Russian to US 19 kW US to Russia
Total Power: #arrays × #blankets × #panels × #solar cells × power/cell = 8 × 2 × 82 × 200 × 0.94 W = 246 kW Size: 8 × 32.9 m × 11.58 m = 3050 m^2, where 2415 m^2 are usable			

the shadow phase, distributing the power into primary and secondary circuits through the DC-to-DC Converter Units (DDCUs). The secondary system is electrically isolated from the primary system and regulated at 6.25 kW (two connections can also be established in parallel to obtain 12.5 kW). As indicated in Fig. 5.27, Russian power supplies can feed electrical power into the DDCU level via the Russian American Switching Unit (RASU). Table 5.8 lists some of the performance data for US solar generators. The Russian ISS solar cell generator is installed on a separate pylon oriented perpendicular to the US pylon. The Russian part of the power supply is shown in Fig. 5.28.

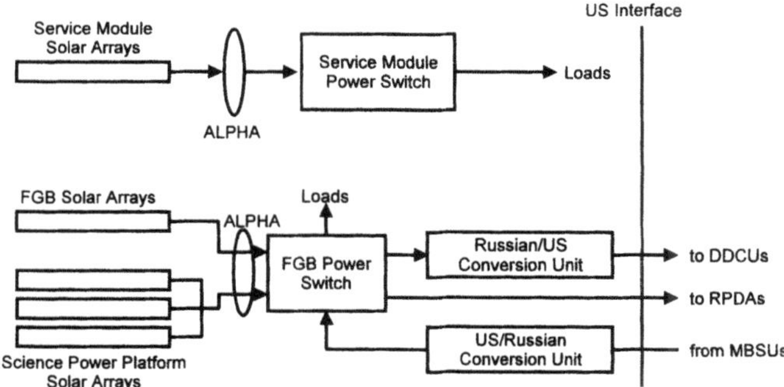

Fig. 5.28. Russian Part of the ISS Power Supply

5.4 The Tasks of the Thermal Control System

During all phases of a space flight mission, the Thermal Control System (TCS) is responsible for providing and maintaining a thermal environment as is required by the crew, equipment, subsystems and payloads. The TCS performs this task by collecting, distributing, transmitting, removing and/or providing heat. In this context, the most important parameters of influence and control for a TCS are the following:

- The natural thermal environment varying during one orbit (solar and albedo radiation; IR radiation of the Earth; deep space environment)
- The induced environment (debris, docked space vehicles, configuration, body shading)
- The boundary conditions of the mission (mission duration, tasks, attitude and orbit control maneuvers)
- The power level of the station (housekeeping, payload dissipation)
- The temperature and climate requirements of crew and equipment

For example, due to the different illumination cases from the Sun, the incidenting radiation induces varying temperatures from -160°C up to +125°C at the vehicle surface, and these temperatures can lead to significant mechanical stress. As shown in Fig. 5.29, the heat dissipation of payloads today, in many cases, clearly exceeds 1 kW. About 80% of these payloads and related instruments function optimally at an operational temperature of 20°C ± 5°C; examples of temperature limits for other components and subsystems are indicated in Table 5.9.

Usually, metabolic heat loads represent only a small percentage of the total heat generated in crewed systems. For example, even with a metabolic rate of 250 W per crew member (cf. Table 5.10, "heavy activity"), the sensible heat load of a six-person crew (as will be nominal for ISS) would amount to a maximum of 1.5 kW. With a total heat turnover of e.g. 74 kW (in the case of ISS), this would result in no more than 2%. The latent heat rate (atmospheric vapor production) in consideration of respiration, hygiene, food preparation, and washing would add with 73 W per person, i.e. a total of 438 W, which corresponds to 0.6% of the total turnover. The heat loss

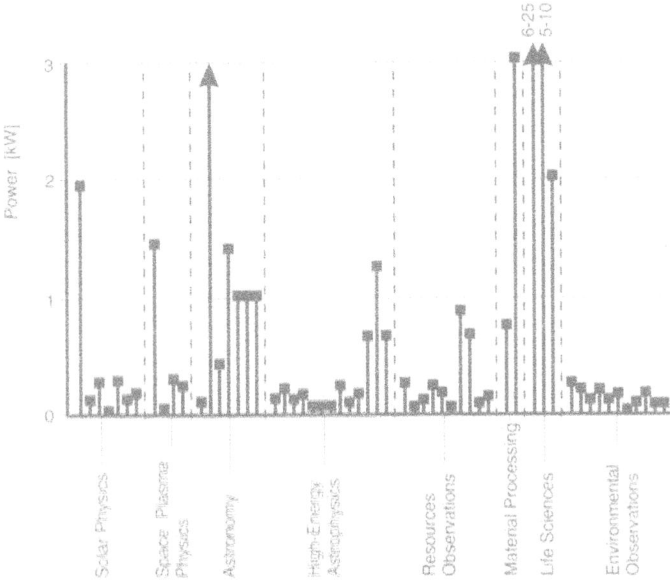

Fig. 5.29. Typical Internal Instrument Power Dissipation [Ollendorf 83]

Table 5.9. Temperature Limits of Spacecraft Components

Subsystem	Components	Temperature Limits [°C]	
		Operating	Non-Operating
Electrical Power System	Batteries Charge Control Units Solar Array (Si) Solar Array (GaAs)	-5 to 15 -15 to 45 -65 to 80 -105 to 150	-10 to 25 -30 to 60 -100 to 100 -150 to 200
Propulsion	Tanks, Vents, Pumps – Mono-Propellant – Bi-Propellant	5 to 40 0 to 40	
TC/TM	S-Band Transponder Decoder/Encoder	-15 to 45 -20 to 40	-30 to 60 -30 to 65
AOCS	Sun-Sensor IR-Sensor Gyros, CMGs Electronic Computer	-20 to 70 -15 to 40 -5 to 45 -5 to 45 -15 to 45	-80 to 80 -20 to 50 -15 to 55 -20 to 65 -30 to 60
Payload	Experiments	-10 to 30	-25 to 40

of the human body leads, under standard conditions, to the following distribution: 46% from thermal radiation; 33% from conduction and convection; 2% from warm, virtually saturated breathing air and 10% from water evaporation on the skin. When the temperature increases, the sensitive heat decreases by the same value with which the latent heat increases, so the total heat loss for constant activity at a room air tem-

Table 5.10. Sensible Heat Rates vs. Activity Level

Activity	Sleeping	Lying	Sitting	Light Activity	Moderate Activity	Heavy Activity
Activity Level	–	–	I	II	III	IV
Heat Rates [W]	60	80	100	≈ 150	≈ 200	≈ 250

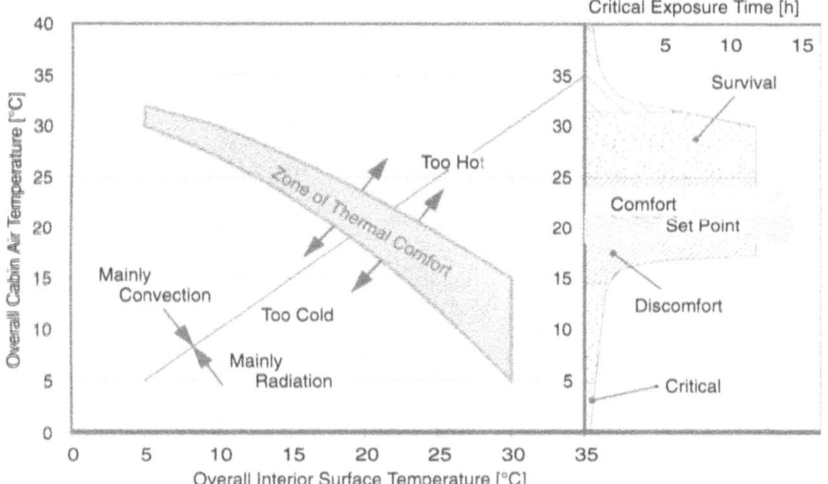

Fig. 5.30. Zone of Thermal Comfort

perature T_0 between 14°C and 34°C remains about constant. The control of the interior climate is necessary in order to prevent hygienic and/or health risks due to growth of mold and mildew, damage of electronic components or instruments due to the thawing of water, and to maintain the well-being and performance of the crew (cf. Chapter 4 "Environmental Control and Life Support System" and Chapter 11 "Human Factors"). Air humidity should be in the nominal range (set point) between 5 g and 8 g water/kg dry air (mean value of the relative humidity: 45%). Apart from humidity, air velocity and temperature, the temperature of the cabin wall T_W must be controlled as well.

Air temperatures T_0 above 25°C have proved to aggravate working processes; at temperatures above 30°C, a decrease in performance of 40% and more occurs. As a measurement of the comfort (see Table 5.11, Fig. 5.30) in enclosed rooms, the subjective temperature T_E is set. T_E is the arithmetic mean of the mean wall temperature T_W (= radiation temperature of the cabin) and of the air temperature T_0 (cf. Fig. 5.29). By efficient heat isolation of the modules, the temperature difference between air and module wall can be reduced, and thus, the radiation exchange between the crew and the module wall can be minimized. The subjective temperature, hence, nearly matches the measurable air temperature.

Table 5.11. Margins for Overall Habitat Air Temperature for Humans [ESA PSS-03-401]

	Effective Temperature [°C]		**Relative Humidity [%]**		**Critical Exposure Time [h]**
	Minimum	Maximum	Minimum	Maximum	
Set Point	18	21	40	50	∞
Comfort	18.5	24	28	65	∞
Discomfort	14.5	25	25	80	<12
Survival	5	31	20	90	<3

In order to avoid the formation of condensate, the temperature of the module wall T_w on the inner side must not reach T_{TP} (the dew point temperature of the air) at any location belonging to T_0. For that reason, the following results for the temperature T_{wa} at the outer cabin wall:

$$T_{wa} \geq T_{wi} - \frac{h_i}{k} \cdot (T_{wi} - T_{TP}) > T_{TP}(T_0) \tag{5.18}$$

with the inner heat transfer coefficient h_i:

$h_i = 3.5$ to $6.0\ \mathrm{W/(m^2{\cdot}K)}$	respite air (recess, behind paneling) (heat transfer coefficient)
$h_i = 7.0$ to $9.0\ \mathrm{W/(m^2{\cdot}K)}$	air in enclosures, small flow velocity (heat transfer coefficient)
$k \geq 70$ to $500\ \mathrm{W/(m^2{\cdot}K)}$	thermal conductance through wall of inhabitat (without thermal insulation)

Due to the high thermal conduction in the metallic module wall, there is only a small temperature difference between the inner and outer surfaces. The module wall thus must be brought to a certain temperature by suitable insulation and perhaps by additional heating.

5.4.1 Mechanisms of Heat Transfer

Heat can be transferred by means of heat conduction, convection and radiation. The fundamental equations for heat transfer are summarized in Table 5.12. By ventilation of large amounts of air throughout the pressurized modules, forced convection also occurs inside the modules at a non-negligible flow velocity. Apart from the atmosphere management system for climatization and regeneration of the cabin air, there are further air and liquid loops inside crewed modules that control the temperature inside the racks. The collected heat is usually transmitted to a heat exchanger and is there transferred into a second thermal control loop. Often, a small part of the heat is rejected at locally mounted radiator surfaces, and the major part is rejected into space by means of large, centrally installed radiator surfaces. At a comparable transmission capacity for heat transfer, air loops require considerably more energy than liquid loops. Since many of the even more effective two-phase working fluids have a toxic effect on human beings, mostly single-phase water is used for internal

Table 5.12. Fundamentals of Heat Transfer

Heat Conduction $\dot{Q} = \frac{\lambda}{d} \cdot A \cdot (T_j - T_k)$	λ: Thermal Conductivity of Material d: Distance from Point j to k A: Cross Section Area of Heat Transfer T_j, T_k: Temperature at Points j and k	[W/(m K)] [m] [m^2] [K]
Heat Convection $\dot{Q} = h \cdot A \cdot (T_F - T_w)$	h: Film Coefficient/Heat Transfer Coefficient A: Surface Area T_w, T_F: Wall Temperature, Bulb Temperature in Stream	[W/(m^2K)] [m^2] [K]
Heat Transfer $\dot{Q} = k \cdot A \cdot (T_1 - T_2)$	k: Thermal Conductance = $\left(\frac{1}{h_1} + \sum_i \frac{d}{\lambda_i} + \frac{1}{h_2}\right)^{-1}$ with λ_i, d: see above h_1: Heat Transfer Coefficient to Interior h_2: Heat Transfer Coefficient to Exterior A: Cross Section Area T_1, T_2: Temperature at Interior and Exterior	[W/(m^2K)] [W/(m^2K)] [W/(m^2K)] [m^2] [K]
Radiation $\dot{Q} = \varepsilon \cdot \sigma \cdot A_i \cdot (T_i^4 - T_0^4)$	ε: Emittance (Gray Body: 0 to 1) σ: Stefan-Boltzmann Constant: $5.67 \cdot 10^{-8}$ A_i: Body Surface i T_i, T_0: Temperature of Body i and Environment	[–] [W/(m^2K^4)] [m^2] [K]
Heat Transport $\dot{Q} = \dot{m}_F \cdot c_F \cdot (T_i - T_j)$	$\dot{m}_F$: Mass Flow c_F: Specific Heat of the Flowing Fluid T_i, T_j: Temperature at Two Points i, j in the Stream	[kg/s] [J/(kgK)] [K]

fluid systems. Mainly in Russian space stations, a mixture of glycol and water or silicon oils is used.

For the modules, the following equation is valid for the heat balance

$$\dot{Q}_{net}(t) = \dot{Q}_{env}(t) + \dot{Q}_{gen}(t) = m \cdot c \cdot \Delta T(t) \tag{5.19}$$

with the heat losses $\dot{Q}_{net}$, the heat gain $\dot{Q}_{env}$ from the environment and the heat Q_{gen} produced inside the module. The mass of the interior is designated by m and its thermal capacity by c. The temperature inside the module remains constant in steady-state cases.

If a heat budget is considered and Eq. 5.20 extended integrally for the entire space vehicle, the following equation is yielded:

$$\begin{aligned} & M_{Sun} \cdot \iint \alpha_S \cdot f_S(\Psi_S) \cdot dA_S + M_A \cdot \iint \alpha_S \cdot f_A(h, \Psi_E, \delta) \cdot dA_A \\ & + M_E \cdot \iint \varepsilon \cdot f_E(h, \Psi_E) \cdot dA_E + \dot{Q}_{gen} - \iint \varepsilon \cdot \sigma \cdot T^4 \cdot dA \\ & = \iiint c \cdot \frac{\partial T(\vec{x}, T)}{\partial t} \cdot dm \end{aligned} \tag{5.20}$$

The external parts of the load are caused by solar radiation, albedo radiation, and Earth IR radiation (cf. Fig. 5.31). Here, ψ_S and ψ_E are the angles between each in-

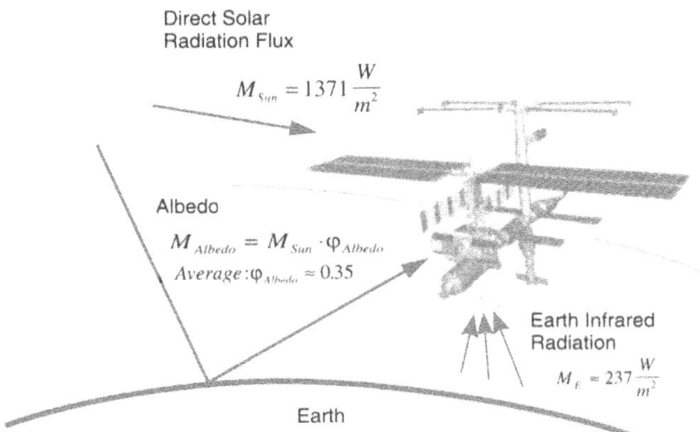

Fig. 5.31. External Heat Loads for Space Vehicles in LEO

dividual surface normal to the vehicle surface and the vector to the Sun and/or the Earth. The functions f_S, f_A and f_E are the view factors (dependent on altitude, position and geometry) for the solar/albedo (index "A") radiations and the Earth IR radiation. The surface areas dA_S, dA_A *and* dA_E designate those portions of the surface which are illuminated by the different radiation sources. Due to the complex geometry and the complicated radiation conditions in consideration of shadowing and multiple reflection, this equation is usually solved with the help of specific radiative heat transfer software packages (e.g. using Raytracing, Monte-Carlo method, etc.).

5.5 Thermal Control Systems

5.5.1 Passive Thermal Control Systems

Thermal control systems can be categorized into active and passive systems. The prerequisite for passive thermal control is that the temperature of a space vehicle or a subsystem can be controlled with the help of the geometric orientation of the parts themselves and their arrangement among one another as well as the use of surface materials by means of heat conduction and radiation. Passive systems are all those that fulfill their functions without mass flows, information flows or (electrical) energy flows and that do not need any control or regulation units. Often, even thermal control supported by heaters aboard space vehicles is also called "passive". Usually, passive systems are very reliable and cost-effective in applicable scenarios. Moreover, by using them, it is possible to keep the power consumption low and to avoid perturbations caused by pump vibrations and leakage. Most components that belong to the passive methods of thermal control are presented in Fig. 5.32. The main task of the passive, heater-supported TCS is to keep the temperature inside the vehicle within a tolerable range from -16.7°C to +45°C and to prevent condensation inside the vehicle. Although such systems can be used without the parallel installation of active components, passive methods alone cannot fully thermally control systems

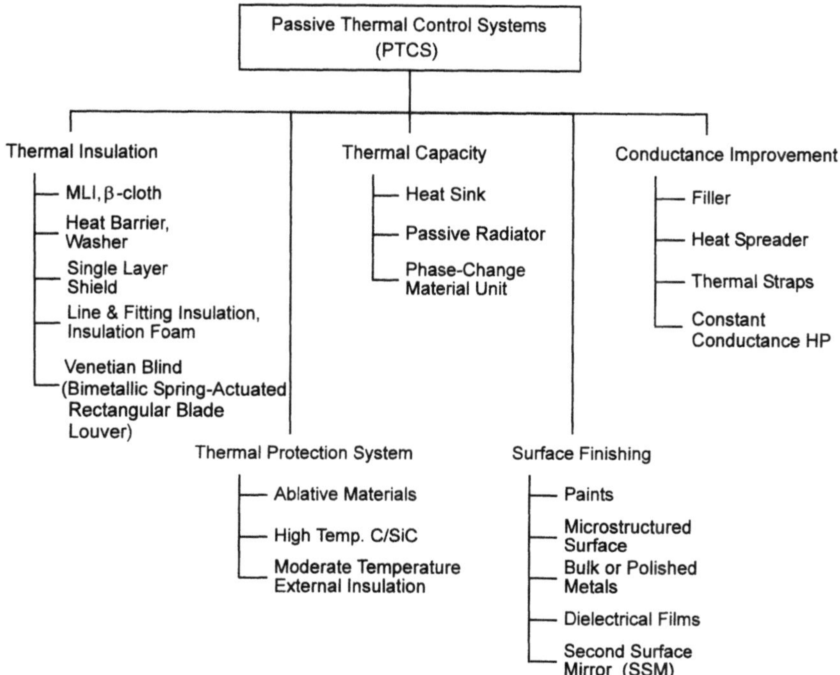

Fig. 5.32. Passive Thermal Control Systems of Space Vehicles

with a lot of waste heat and/or low heat capacity (≤ 850 J/kgK) such as batteries, electronic components and the like, at least this cannot be accomplished with acceptable efforts. Moreover, such systems are not able to react to changing loads in a flexible way, for example, external loads (irradiation on orbit, shadowing, multiple reflection) or internal loads (on-off cycles). Due to the high level of electrical power installed in the case of crewed systems (e.g. because of the ECLSS), there are high cooling demands. The presence of the crew additionally requires a considerably more precise and more flexible temperature control.

5.5.2 Active Thermal Control Systems

The power requirements of space stations lie far above 5 kW (cf. Skylab); for the International Space Station, for example, they are more than 100 kW. In Fig. 5.33, the qualitative distribution of the accumulated waste heat of a crewed system is shown, based on typical electrical power supply systems without taking into account the charge/discharge heat and dissipation from the voltage adjustment of batteries. As a value for the amount of heat to be rejected, approximately 90% of the installed electrical power can be estimated from a thermodynamic point of view. With transmission paths of 10 to 30 m in length, this heat can only be transmitted to the large rejection areas with the help of active systems. In the case of decentralized systems, the wasted heat, at the place of its occurrence, will be transferred to locally mounted

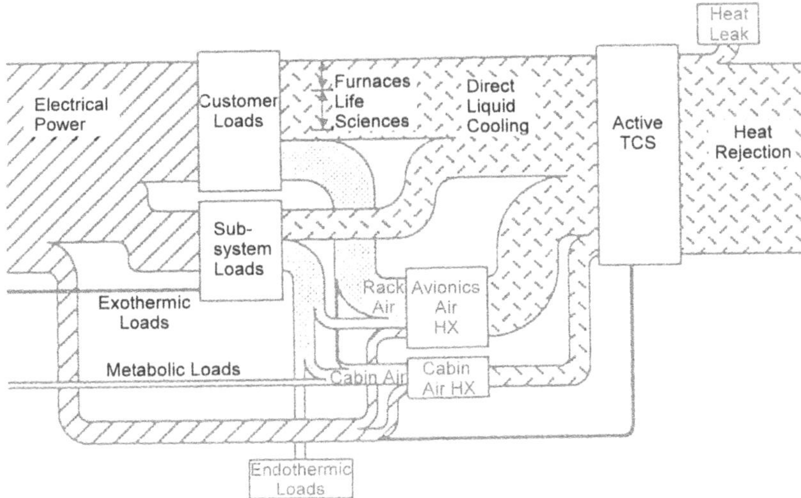

Fig. 5.33. Typical Energy Flow Throughout Space Station Modules [ESA PSS-03-406]

radiators. Centralized heat transport systems, on the other hand, usually consist of an internal and an external partial system with one thermal bus each. The internal systems collect the heat and transport it to a heat exchanger where it is transferred to the external loop and transmitted to the radiators. Within the module, the tasks of the internal TCS are often assumed by air loops (cabin air loop, avionics air loop) and/or single-phase fluid loops. Single-phase fluid loops mostly work with water at temperatures between 2°C and 50°C.

In the case of cooling demands of more than 5 kW and despite their more complex structure, two-phase systems are more advantageous than single-phase systems. Nevertheless, single-phase systems often using ammonia as working fluid are currently preferred for external loops, for example for ISS, due to their reliability and simple structure. Two-phase systems can react to changes in the heat load in a very flexible way and offer nearly isothermal conditions for the condenser and the evaporator. With comparable cooling demands, the mass flows are smaller and hence the cross sections of pipes and pump systems as well. For example: for the removal of 10 kW of heat by means of a water loop at a set-point temperature of 20°C, a mass flow of 0.12 kg/s is necessary, whereas with a phase change, the same heat flux can be transmitted with 0.0045 kg/s. Additionally, the pumping power can be considerably reduced (by a factor 10 to 60). However, up to the present time, no two-phase systems have been qualified for operational life of 10 to 15 years.

The operating temperature of the external loop is influenced by the temperature of the internal TCS (range from 0°C and 18°C) and is usually between -40°C and +50°C, therefore, in phases of low dissipation it can drop below 0°C. For that reason, mainly ammonia, which has thermal properties similar to those of water, but a freezing point of -77.7°C, is used as coolant.

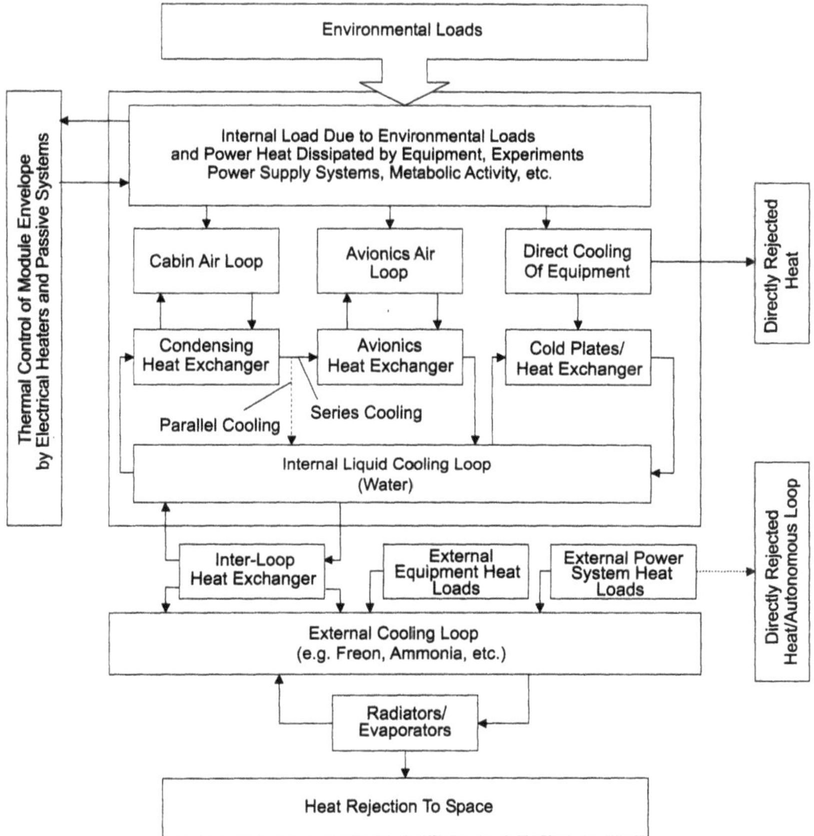

Fig. 5.34. Typical Active Thermal Control Systems in Space Stations

The components of the electrical power supply system are thermally controlled on the external platforms by the main thermal bus, by their own radiator surfaces, or by means of an autonomous TCS. In the case of high dissipated heat rates, the latter alternative has the advantage that the loop temperature can be selected independently from any central, external TCS. If, at a constantly transferred thermal energy, the working temperature in the cycle is increased, the radiator surfaces, as a consequence, can be considerably reduced (cf. the section on radiators). The elements of a space station thermal control system as described above are shown in Fig. 5.34.

There are many interfaces between the overall system and the TCS. The thermal control system has a topology similar to that of the power system, i.e. the heat is collected and/or transferred into distributed systems. Sensors and heaters are spread over the entire surface of the modules and inside of the racks. The degradation of the surfaces determines the station's temperature level and, together with external radiation exchange, considerably determines the performance of the radiators. Space debris and particle radiation may damage the superinsulation or radiator components, and that, in turn, leads to a degradation in performance. In addition to that,

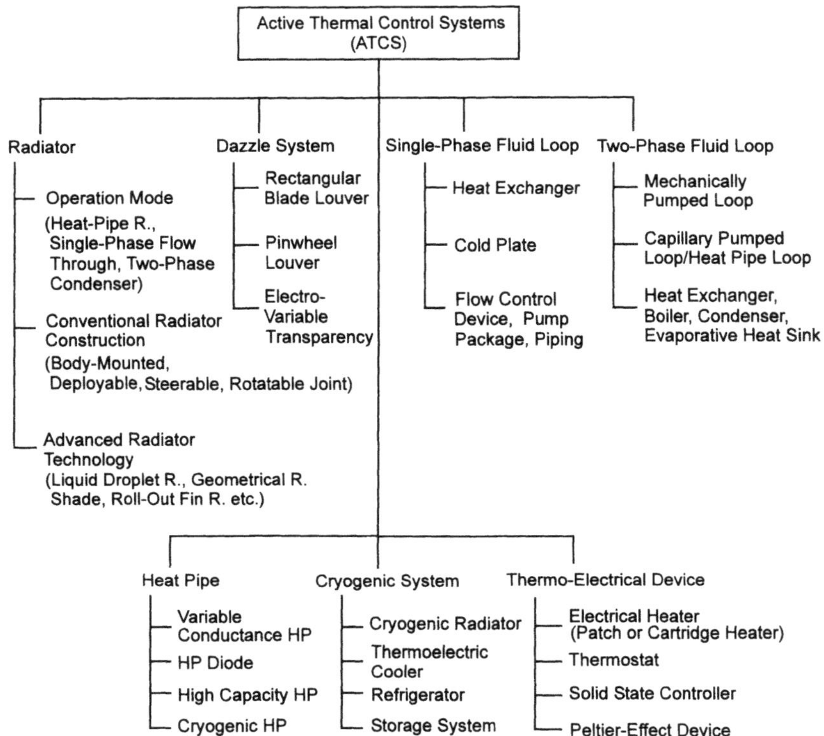

Fig. 5.35. Hardware of Active Thermal Control Systems

the large radiator surfaces can cause a significant part of the aerodynamic drag; that would make necessary the orientation of the station or corresponding tracking devices in order to fulfill pointing requirements.

The design of active thermal control systems is mainly determined by the amount of heat to be rejected, the reliability of the systems used, and the necessary isothermal quality of the controlled components. Additionally, there are requirements such as low power consumption; high system flexibility, in the case of changing heat load profiles (due to varying boundary conditions of the mission and payload requirements); and high reliability even for operation over many years. There exist a high number of components and subsystems that can be categorized into electrical, fluid dynamic and mechanical systems, as is shown in Fig. 5.35.

Electrical systems are very often used, since they can control the temperature in a very accurate and flexible way. However, they need power, and because of the high number of systems needed, they form a large part of the TCS overall system mass during space flight projects. A description of the individual systems can be found, for example, in [Gilmore 94] or [Krüger 97]. Because of their dominant role, the radiators will be described in more detail in one of the following sections.

5.5.3 Performance and Technological Data of TCS Hardware

In the case of crewed missions, the thermal control system will be very expensive with its costs amounting to 5% to 7% of the total costs of the flight hardware, and it also covers a share of 7% to 12% of the overall system mass. For this reason it is helpful for the design and selection of TCS components to have a good knowledge of the performance and system mass of the hardware. The so-called "figures of merit" (FOM) for passive and active systems are presented in Table 5.13 to Table 5.15.

Table 5.13. Typical Technology Data of Passive Thermal Control Hardware

Thermal System	Figures of Merit (FOM)	Typical Values
MLI, b-cloth	ε_{eff}, Specific Mass m	MLI: ε_{eff}<<0.1, m<1.0 kg/m^2 β-cloth: α_S<0.2; ε=0.9
Coatings	α_s, ε, α_s/ε, $\alpha_{EOL}/\alpha_{BOL}$ Specific Mass m	α_s/ε<1 (Non-Metal), α_s/ε>1 (Metal) $\alpha_{EOL}/\alpha_{BOL}$>3 m=0.05–0.5 kg/m^2
Second Surface Mirror	α_S, ε, α_s/ε, $\Delta\alpha_{BOL/EOL}$	α=0.08–0.14 α=0.08–0.14
Filler	Contact Heat Transfer Coefficient k, Temp. Gradient	$k<2\ \frac{kW}{m^2K}$, $\Delta T \leq 10$ K
Heat Barrier, Washer	Heat Flux Resistance R	R>100 K/W
Line and Fitting Insulation	Heat Flux Resistance R	R>10^3 K/W
Passive Radiator	Specific Mass, Effectiveness	>5 kg/m^2, h≤0.6
Heat Sinks	Specific Heat c and c · ρ	Beryllium: 190 J/kgK 3.65 J/m^{3}K
Phase Change Material	Heat of Fusion, Melting Point	variable
Ablative Material	Specific Mass, max. Temperature Load	800 kg/m^3, T_{max}<2000°C
Flexible Insulation	Specific Mass, max. Temperature Load	T_{max}<900°C, 200 kg/m^3
High Temperature Ceramic (C-SiC)	Specific Mass, max. Temperature, High Emissivity	2000 kg/m^3, T_{max}<2600°C, e>0.8
Thermal Straps	Mass, Thermal Conductivity λ	ε_{eff}, Specific Mass m

Table 5.14. Typical Technology Data of Internal Fluid Loops

Fluid Loop	HERP [kg kW^{-1}]	MERP [kg kW^{-1}]
Avionics Air Loop Cooling	225	280
Avionics Water Cooling	200	200
Water Loop Cooling	175	230
Freon Loop Cooling	155	210

HERP = High Efficiency Radiator Panel, 25 W kg^{-1} at Most Unfavorable Attitude (MUA)
MERP = Medium Efficiency Radiator Panel, 14 W kg^{-1} at MUA

Table 5.15. Typical Technology Data of Active Thermal Control Hardware

Thermal System	Figures of Merit (FOM)	Typical Values
Radiator	Heat Rejection (W/m^2), Thermal Efficiency Specific Mass	50–350 W/m^2 5–30 kg/kW 1.0–8.5 kg/m^2 (Lightweight: 1.0–3.5 kg/m^2) (Simple Tube-fin: 3.5–8.5 kg/m^2)
Heat Exchanger	Mass Penalty	<5 kW: [kg] = 1.0*[kW]+6.6 >5 kW: [kg] = 0.25*[kW]+12
Single Phase Cold Plate (Two Phase Cold Plate)	Heat Transfer Coefficient k, Specific Mass	$k \leq 2$–4 kW/m^{2}K 50–70 kg/m^2 ($k \leq 4$–6 kW/m^{2}K)
Constant Conductance Heat Pipes Variable Conductance Heat Pipes High Capacity Heat Pipes	Heat Transport Capacity, 0g FOM ($\sigma\rho\lambda/\mu$), typ. Mass Penalty [kg/W]	0.1–0.8 kWm; $2 \cdot 10^{-3}$–0.01 kg/W 0.5–3.0 kWm; $5 \cdot 10^{-3}$–0.03 kg/W 1–18 kWm
Single Phase Liquid Loop[1)]	Specific Mass, Heat Transport Capacity	0.12–0.35 kg/W 10–100 kWm
Two Phase Loop[1)]	Specific Mass Heat Transport Capacity	0.1–0.19 kg/W 50–1000 kWm
Capillary Pumped Loop	Specific Mass, Heat Transport Capacity	$5 \cdot 10^{-3}$–0.17 kg/W 5–80 kWm at 150 kW/m^2
Heaters (incl. 1m Harness)	Specific Mass, Maximum Heat Load	0.23–1.2 kg/kW (>10 W) 0.2–0.8 kg/kW ($\leq$10 W) 0.25–4 W/cm^2
Thermoelectric Devices	Heat Flux, Specific Mass	0.1–3.2 kWm^{-2}
Louvers	Mass, ε_{eff}	10–30 kg/m^2, $\varepsilon_{eff\,open} < 0.7$; $\varepsilon_{eff\,closed} < 0.15$

5.5.4 Boundary Conditions for the Design of Thermal Control Systems

The design of thermal control systems for crewed systems is determined by the mission-specific boundary conditions. A summary is presented in Table 5.16 of those boundary conditions that must be examined at the beginning of the design phase. In order to determine the radiator surfaces' size, it is necessary to estimate the cooling capability, select the radiator technology, the thermo-optical properties of the radiator, the operating temperature, and the worst-case boundary conditions of the mission. Requirements for thermal insulation and the heater power to be installed result from the worst-case scenario for boundary conditions of the orbit, the mission duration, and the temperature limits of subsystems and components.

The preliminary result of an analysis of the mission-specific boundary conditions for a space station similar to ISS is shown in Table 5.17. Here, an integrated thermal control system with central loops and large radiator panels for heat rejection were selected. In the specific case of the International Space Station, even an additional autonomous TCS was selected for the solar arrays. At a high level of installed electrical power, the demands for the thermal control system grow simultaneously and automatically since the largest portion of the waste heat is caused by electrical

Table 5.16. Mission-Dependent Boundary Conditions and Requirements for the TCS Design

Boundary Conditions	**Options**
• Integration Depth	⇒ Integrated System with Central Heat Transportation Loop and Central Heat Rejection ⇒ Autonomous System ⇒ Hybrids (Combination of Integrated System and Autonomous Components)
Example: Central Thermal Control System with Joint Fluid Loop and Radiator Surfaces vs. Local Heat Acquisition Equipment and Rejection by Body-Mounted Radiators	
• Transport Length and Dimensions of the Orbital System	⇒ Expanded Systems ⇒ Compact Systems
Example: Expanded Truss Structure (ISS) vs. Compact Mir Station	
• Thermal Environment Definition Range	⇒ General Validity Range of Thermal Environment Definition ⇒ Individual and Locally Varying Thermal Environment ⇒ Peak Loads and Orbit/Time/Annual Average Loads
Example: Space Station with Homogeneous Utilization vs. Station with Different Areas of Utilization e.g. Material Sciences, Habitation, Maintenance and Repair Docks for Satellites	
• Main Tasks of TCS	⇒ Collection, Transport and Heat Rejection from Internal Systems ⇒ Demands for Temperature Uniformity ⇒ High Level of External Heat Loads
Example: Heat Loads from Material Science Furnaces vs. Sun-Synchronous Orbits for Earth Observation	

Table 5.17. Mission-Dependent Boundary Conditions for an ISS-like Space Station

Integrated System with: • Main Thermal Bus (One or More) • Central Thermal Rejection (perhaps Additional Local Mounted Radiators) • Long Transport Distance Capability • Higher Heat Transport Capacity (Needs Higher Pump Power, always Needs a Working Fluid) • Overall Thermal Bus Environmental Definitions	**Typical Candidate Technology for Heat Transport System:** • Pumped-Liquid Loop • Capillary-Pumped Two-Phase Loop • Capillary-Pumped Heat Pipe • Two-Phase Pumped Loop **Typical Candidate Technology for Heat Rejection System:** • Heat-Pipe Radiator • Fluid-Loop Radiator • High-Capacity Heat Pipe Radiator
Additional Mission Requirements: • Mission Duration • Degradation (BOL/EOL) • Probability of Meteoroid Penetration • Contamination Sensitivity • Redundancy • Pointing Requirements for Attitude Control or Utilization • Non-Blockage Conditions • Electrostatic Charging/Discharging • Micro-g Requirements • Power Level	**Cost Factors:** • Transportation Cost Factor • Maintenance Cost Factor • Integration Cost Factor • Programmatic Cost Factor
Analysis Objectives: • Determination of Launch Weight and Volume • Heat Transfer Surface Area and Radiator Area Requirement • External Power Requirements Imposed by the Acquisition System for Each Module • Pumping Power Requirements • Thermal Bus Connection Requirements • Mass of Consumables, Working Fluids • Materials and Operating Temperatures	**Essential Analysis Parameters:** • Technology Parameter (Quality, Effectiveness, Pressure Loss, etc.) • Radiator Surface Temperature • Emittance of Radiator Surface • Internal and External Heat Loads

equipment, computers, voltage converters, etc. The TCS must be designed such that, in case of doubt, the total capacity of the electrical power system can be thermally removed. In the case of Skylab, for example, large amounts of the electrical power were carried away during the inactive phases by means of shunts (resistors). Candidate technologies for central heat transportation systems are mainly single-phase fluid (circulated) loops and capillar or mechanically pumped two-phase loops.

Due to the central heat removal, the radiator must be also highly effective and controllable. A simple criterion for the assessment of the radiator's efficiency is its isothermal quality which can be considerably improved by heat pipes. Also when taking into account aspects of safety and reliability in the case of usually long mission durations and a debris risk, it is favorable to use heat-pipe radiators since even in the case of damage, a reduction in the performance may occur, but not a complete failure.

5.5.5 Radiators

Radiators play a very important role when dealing with crewed orbiting systems since they are the most effective and often also the only possibility for removal of the waste heat generated in the space vehicle. If there are high cooling demands, usual degradation rates and radiator temperatures of 0° to 50°, the masses of the radiators can amount to 50% of the overall TCS mass. Due to the large radiation areas of common radiator panels, they also contribute considerably to the aerodynamic drag, as in the case of the International Space Station. Radiators used at present can be categorized according to the type and position of their installation on the space vehicle (see Fig. 5.36):

- Body-mounted radiators (installed directly on the modules)
- Steerable/deployable radiators that are transported into space in a stowed configuration

Another possible method of categorization is to group the radiators according to their structure:

- Passive radiator surfaces
- Liquid or vapor flow-through radiators
- Heat-pipe radiators

Moreover, future options currently being examined use even more advanced and hence highly dissipating space station TC systems. What all radiators have in common is the principle of operation: they exchange the heat between their radiative surface and a heat sink – in the ideal case, with the space environment (≥ 4 K).

Most radiators have surfaces that are coated with high IR emittance materials (typical: $\varepsilon \geq 0{,}8$) and solar absorptance ($\alpha_{s,EOL} < 0.25$). That allows maximization of the IR emittance and minimization of e.g. solar irradiation. The performance of radiators during long-term missions is limited mainly by the degradation of the thermo-optical surfaces, causing an increase of the solar absorptance. For this reason, radiators at the beginning of their operation (BOL) are often considerably oversized. Typical relations of $\alpha_{EOL}/\alpha_{BOL}$ today are at 0.25/0.08 (EOL to BOL). The famous Stefan-Boltzmann Law of Gray Bodies is valid for the rejection of heat as follows:

$$\dot{Q} = \varepsilon \cdot \sigma \cdot (T^4 - T_S^4) \cdot A \tag{5.21}$$

with the mean radiator temperature T, the emissivity ε of the surface A, the Stefan-Boltzmann constant σ, and the (effective ambient) sink temperature T_S. Due to the T^4 relation, the rejected heat is very sensitive to changes of the radiator temperature. For example, at 323 K it is twice as large as at 273 K, and it is even 300 times larger than that of a cryogenic radiator at a mean temperature of 70 K. An average value of 100–320 W/m^2 for the heat rejection (referring to the effective radiator surface area) of a radiator is considered good for a low Earth orbit at radiator temperatures of 20°C to 40°C (the upper value for radiators running at 40°C, modest heat back-loads or none at all). For that reason, the satisfactory use of radiators at or below a temperature of 70 K requires much more effort than those well above 70 K. The size of the radiator surface must be determined with the help of detailed analyses and verified by tests. Typical masses for radiators range from 1 to 12.5 kg/m^2 (support structure included). Apart from the radiator temperature, the isothermal quality of the radiator surface is also important for the radiator's efficiency. As a consequence, Eq. 5.21, at a linear temperature profile along a liquid flow-through radiator, must be multiplied by the following correction factor:

$$K = \frac{3 \cdot \Delta T}{T} \cdot \frac{(1 - \Delta T/T)^3}{1 - (1 - \Delta T/T)^3} \tag{5.22}$$

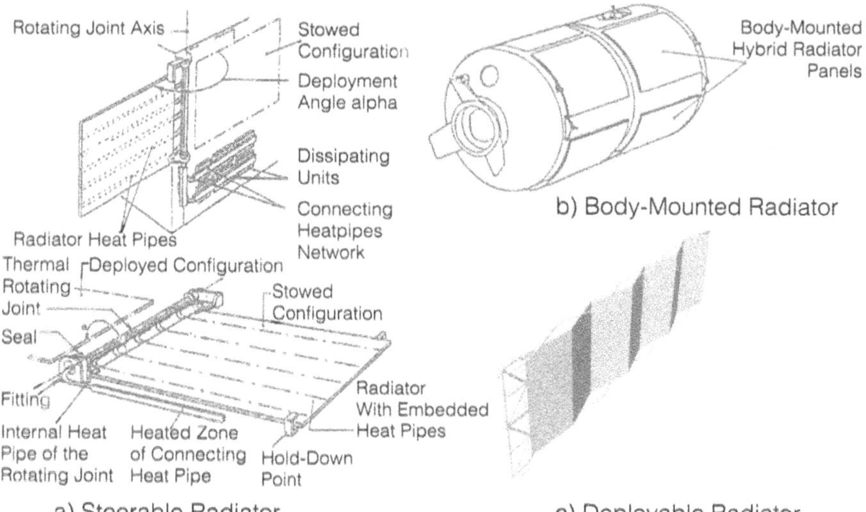

Fig. 5.36. Typical Radiator Design Options

A temperature difference ΔT between flow inlet and outlet of, e.g. 5°C, results, at a mean radiator temperature of 20°C, in a reduction of performance of about 4%; in the case of 20°C difference, this reduction amounts to about 15%. Often, a fin efficiency is also introduced into Eq. 5.21, in order to illustrate the influence of the sink temperature and the quality of the inner fin structure. This results in the following:

$$\dot{Q} = A \cdot \varepsilon \cdot \eta_e \cdot \sigma \cdot T^4 \tag{5.23}$$

The fin effectiveness h_e shows the relation between the net rejected heat of the radiator and the heat rejection of an isothermal fin structure at a reference temperature to an ideal radiative sink temperature as follows:

$$\eta_e = \eta_e^* \cdot \left(1 - \left(\frac{T_S}{T}\right)^4\right) \tag{5.24}$$

$$0.41 \leq \eta_e^* \leq 0.99$$

$$\text{usually: } 0.78 \leq \eta_e^* \leq 0.95$$

T is for heat-pipe radiators, for example, the heat-pipe temperature while T_S is the sink temperature and η_e^* is the *fin efficiency* for the finned structure. The *fin effectiveness* is hence a measure of the isothermal quality, the geometry, and the surface properties (emittance) of the radiator, and it takes the environmental conditions into account, as well.

Passive Radiators (Structure Panel) and Body-mounted Radiators (BMR)

Low dissipating components (<5 W) that do not have to satisfy isothermal requirements are mounted onto an already existing structural element on one of the side walls of the vehicle for the purpose of direct rejection. Usually, the structural mass of this surface is determined by structural requirements and thus, when calculating the system mass budget, it is also assigned to the structure subsystem. In the case of unfavorable material properties, the dissipating equipment is connected to simple radiator surfaces by means of so-called "thermal straps" in order to increase the heat conduction. At loads between 3 W and 10 W, often Constant Conductance Heat Pipes (CCHPs) are used. At heat loads that vary heavily and at higher isothermal requirements, fluid loops or Variable Conductance Heat Pipes (VCHPs), and heaters for temperature control are necessary. A special form of such radiators which is directly mounted on the vehicle is the so-called "body-mounted radiator". Since these often assume the additional function of a debris shield as well, they are generally designed as heat-pipe radiators. In the case of micrometeoroid impact, the functionality is completely maintained, except for a certain reduction in performance. For that reason, heat-pipe radiators (mainly in the case of long-duration missions requiring high reliability) feature less mass at equal performance levels (up to about 6 kW·m) as compared with flow-through radiators. In most of the cases, less than 75% of the module surfaces can be used for installing BMRs and therefore, the rejected power remains limited to less than 10 kW.

Steerable and Deployable Radiators

Steerable radiators offer some advantages. During launch, they are folded to the vehicle's outer structure, and in space they can be oriented to function in a favorable cold environment. Compared to body-mounted radiators, the heat rejection can be increased by a factor of 1.8 to 2 (nearly doubled, due to front side and reverse side effective radiation areas!). In addition to that, the surface degradation can be reduced. Since only one fluid joint is used, they are less complex than deployable radiators. During the low-load phases, the radiator can be rotated so external heat loads prevent freezing of the radiator. Since with crewed systems, the electrical power requirements increase considerably, the amount of waste heat also increases. Apart from the orbital vehicle and/or module surfaces, the maximum dimensions of transport volume within the launch vehicle, fairing, or cargo bay are also limited. For that reason, deployable radiator units that are mounted onto a truss-like structure far from the main vehicle are used. In view of the dimensions, it is necessary to assemble the radiator from individual elements which can be exchanged in orbit. Typical rejection capabilities for those systems are from 0.4 to 4 kW $m^{-2}K^{-1}$.

5.6 System Examples

Whereas *Skylab* used a methanol/water loop and body-mounted radiators with a total cooling capability of 4.7 kW, the *Space Shuttle*'s radiator surfaces are installed in the doors of the Shuttle cargo bay (about 110 m^2). When the doors are closed, the heat of the freon loop is transferred to a water-evaporator assembly and, below an altitude of 30 km, to an ammonia-evaporator assembly as a heat sink. *Spacelab* has a passive TCS that relies mainly on the use of Multi Layer Insulations (MLIs) for thermal insulation of external surfaces. The MLI consist of 19 layers of Kapton® foil which is gold-film metallized on both sides of the sheet, with Dacron® netting for separation of the layers, and a Teflon®-coated β-cloth as external protection layer. A water loop transmits the heat load generated by the payload to a freon heat exchanger in the cargo bay. The heat resulting from air cooling and climatization is transferred to a condenser and then to the water loop. The freon loop at the external pallet transfers its heat load first to an additional heat exchanger and then to the water loop. The Space Shuttle/Spacelab system can continuously reject up to 5.8 kW heat (see also Sect. 2.4.1 and 4.2.2).

The *Space Station Freedom* was to have a thermal control system consisting of a passive and an active TCS. Most of the TCS items foreseen for SSF were taken over for the ISS design. The passive TCS is used for the thermal insulation of the modules and for the prevention of the formation of condensate or ice inside the modules. The active TCS provides cooling capability for the ECLSS, for housekeeping and for the payload. The internal active thermal control systems (IATCS) transmit the heat load to the so-called "Interloop Heat Exchanger" which is the interface to the main Heat Rejection Transport System (HRTS). The Subsystem ITCS Loops of the US Lab and the US Hab feature redundancies in the form of a second loop each. The inlet temperature in both loops is 4°C; the return temperature is 21.1°C at a transport capacity of 25 kW each. An identical heat load is to be transferred by means of the Resource Node Support Loop as well. This loop provides the heat acquisition from Resource Nodes 1 and 2, the Pressurized Logistics Carriers, and the Airlock System. The second loop for Resource Nodes 3 and 4 is to be used only during the assembly sequence of the space station. About 90% of the housekeeping load will be directly collected by means of the coldplate-mounted equipment, and the rest by the air conditioning system. The Resource Nodes feature a further freon loop for heat transport to the body-mounted radiators installed on the external surface, and this loop is to be used until the HRTS is ready for use. The Pressurized Logistics Carrier, during docking, is provided with 10 kW cooling capability via the Node. In the external TCS, ammonia is used as the two-phase working fluid. Two of the three autonomous loops will be operated at temperatures of 1.7°C and one at 21.1°C. At this point in time, a maximum cooling capability of 22 kW is available for the Attached Laboratory, of which 10 kW will be directly transferred as water-cooled payloads and 12 kW will be indirectly transferred via the Cabin and Avionics Air Loop. Both cooling loops have a heat exchanger that connects them with both HRTS loops and are operated with an inlet temperature of 5°C (cold loop) and 18°C (warm loop). Several disadvantages, for example missing control valves for maintenance and exchange of individual payloads (US Rack approach), lead to the re-design of the IATCS.

5.7 The Thermal Control System of the International Space Station

The re-design of the Space Station Freedom (1985) lead initially to the design of the International Space Station Alpha (ISSA) and finally, in 1993, to the current International Space Station (ISS). Essential changes as to the TCS were carried out concerning mainly the external and the internal active TCS.

The originally planned ammonia two-phase loop was replaced by a single-phase system. Since the station was expanded by the Russian modules, some functions could be transferred to them, and the available thermal capability for some areas could even be improved. The thermal control systems for the Russian section, to a large extent, work autonomously. The overall TCS was re-subdivided, according to its functions, into the following:

- Passive Thermal Control System (PTCS)
- Internal Active Thermal Control System (IATCS)
- Photovoltaic Thermal Control System (PVTCS)
- External Thermal Control System (ETCS)
- Russian Segment Thermal Control System (RSTC)

In these subcategorized systems, components and equipment that have to fulfill similar functions were integrated. The segments of the Thermal Control System are presented in Fig. 5.37.

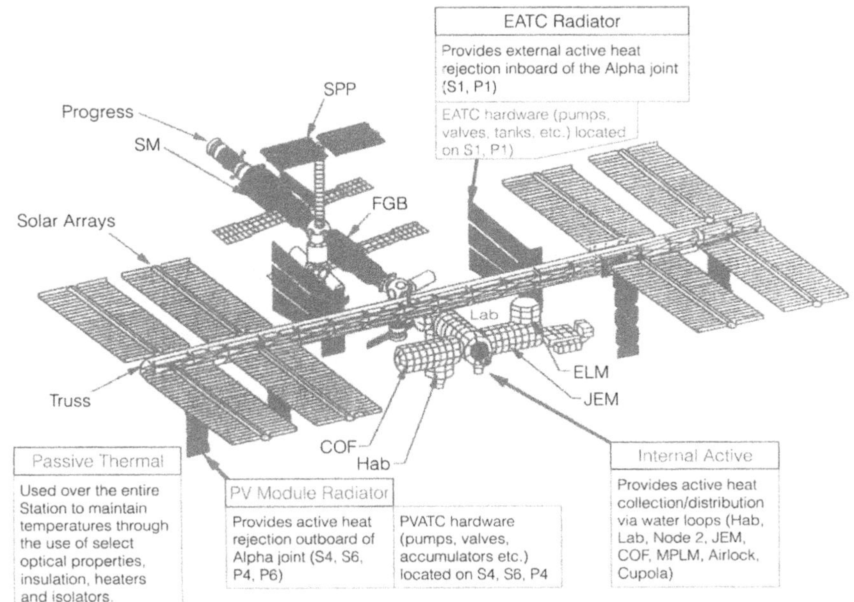

Fig. 5.37. The ISS Thermal Control System

5.7.1 Passive Thermal Control

A tolerable temperature range from 16.7°C (minimum) to 45.0°C (maximum) was defined for the wall temperature of ISS [Peter 92], and the air temperature should be within a range from 18°C to 27°C. The passive thermal control system consists of MLIs, surface coatings, and thermal insulations supported by active heater elements. The MLIs are used to thermally shield the crewed modules from space. Coatings are located on the outer surfaces, especially on the micrometeoroid and debris shields that are installed above the MLI. Apart from the heaters distributed on the module surfaces, further heaters are installed at thermal insulations of fluid lines and other components in order to prevent local condensation or freezing. Typical MLI blankets are composed of 20 layers of double aluminized Mylar® (DAM) with integrated Dacron® nets for the separation of the individual layers. An external surface material made from single-aluminized, Teflon®-coated β-cloth serves as a protection against mechanical damage and degradation caused by atomic oxygen. On the inside, this function is assumed by a double aluminized and reinforced Kapton® foil. The Mylar layers and the Kapton foil are connected by ultrasonic welding, whereas the β-cloth is sewn together with the DAM and trimmed at the edges. In Fig. 5.38, the layer structure of the MLI is shown. The entire blanket has an average thickness of about 5 cm at a specific weight of about 1.2 kg/m^2 and an average effective emittance ε of 0.035 (one single DAM layer $\varepsilon = 0.03$). In order to facilitate, for ex-

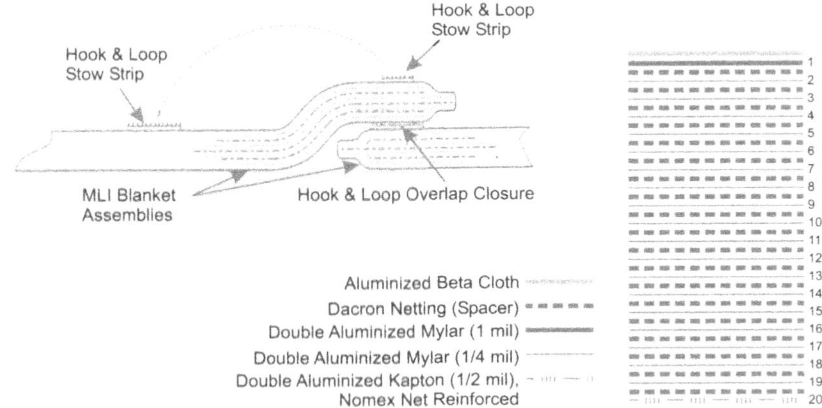

Fig. 5.38. Structure of MLI Blankets [Peter 92]

ample, the exchange of the MLI blankets for repair purposes, surrounding structural brackets were welded on each structure at which the blankets are fixed (partially overlapping) with the help of the so-called "Velcro fasteners".

Thermo-optical coatings are used in order to control the thermal rejection and the solar irradiation. The coatings must have high resistance against UV radiation, atomic oxygen, particle radiation, plasma influences, and thermal cycling degradation. A chromic-acid anodization ($\alpha_{S,BOL} = 0.3$, $\alpha_{S,EOL} = 0.5$ and $\varepsilon = 0.4$) of the micrometeoroid shields made from aluminum makes an effective thermal decoupling with regard to the environment possible. The micrometeoroid shields, whose average temperature varies between -120°C and +120°C, are fixed at the structure with the help of thermal insulators by use of titanium brackets. Pumps, valves and other active thermal control equipment are thermally padded with insulation foam and Kapton®-coated glass fiber materials (0.032 g/m^3, 0.029 W/(m · K)), when there is the risk that they will encounter temperatures below the dew point and to provide a touch temperature within safe limits for crew protection.

5.7.2 Active Thermal Control

The electrical power delivered by ISS's solar arrays amounts to a total of 102.5 kW (see Table 5.18). If losses in the power distribution and storage system and metabolic heat loads are added, a further 45.1 kW will be obtained (cf. Table 5.19 and

Table 5.18. Installed Electrical Power of ISS [Chambliss 94]

Subsystem	Installed Electrical Power [kW]
4 US Photovoltaic A.	75
RSA SPP Mast PV A.	19
RSA Elements PV	4.5
RSA FGB Tug	4.0

Table 5.19. Heat Generation Capability of the ISS Power System [Chambliss 94]

Equipment	Dissipation Power [kW]
RSA SPP Batteries	7.0
US PV Batteries	28.0
DDCUs and MBSUs (Electrical Transformation)	8.1
Passive Heat Gain	0.8

Table 5.20. Proposed ISS Heat Rejection Resource Allocations [Chambliss 94]

Equipment	Housekeeping Thermal [W] (Annual Average)
Airlock	344
Hab	6609
Lab	6059
Node 1	893
Node 2	2083
P1	1271
S0	3355
S1	1271
JEM	4431
ESA APM	2947
Total	29262
15% Reserve	5988
Payloads	30000
Payload Support	7950
Metabolic	1200
Total Necessary Radiator Capability	74400

Table 5.20). In the worst case, the active external thermal control system and the additional passive components must be in a position to reject the total power of 147.6 kW in order to keep the station within the thermal equilibrium. The peak heat rejection amounts to 140.5 kW (according to Table 5.21), of which 28 kW is available for the autonomous PVTCS. According to Table 5.20, the active TCS for the non-Russian section must transmit a total of 74.4 kW heat load to the radiator surfaces for housekeeping and the payload.

The *internal active thermal control system* (IATCS) is used inside the pressurized modules and transmits the heat to the interloop heat exchangers, where it is transferred to the EATCS. That mainly limits the maximum transferable heat (cf.

Table 5.21. Heat Rejection Capabilities of ISS TC [Chambliss 94]

US Segment	Rejectable Power [kW]	Russian Segment	Rejectable Power [kW]
Passively Cooled Equipment	4.0	FGB BMR	4.5
EATCS Radiators	70.0	SM BMR	4.0
PVTCS Radiators	28.0	SPP Radiator	30.0
Total	102.0	Total	38.5

Table 5.22. Margins of the Internal Active TCS Cooling Capability of ISS

Subsystem	Cooling Capability at the LT loop [kW]	Cooling Capability at the MT loop [kW]
US Lab	14	25 (Max. Design Load)
ESA APM	7	25
JEM	9	25
Node 2	2.48	2.48

Table 5.22). The IATCS consists of two single-phase water loops: the Low Temperature Loop (LTL) and the Moderate Temperature Loop (MTL). The Avionics Air Loop that was planned for the station Freedom was canceled in order to facilitate the detection of fires inside the racks and to reduce the risk of contamination in case of fire. If necessary, the payloads must transfer their heat load via a local heat exchanger to the water loop. The low temperature loop delivers cooling water at temperatures between 0.6°C and 6°C for the rejection of heat from the heat exchangers of the life support system's cabin air loop. The moderate temperature loop operates at a temperature from 13°C to 18°C and serves mainly the purpose of cooling electronic equipment that is mounted on different cold plates (cooling capability of cold plates >620 W at 35°C surface temperature). In the European APM there is only one water loop that flows in series through LT and also MT heat exchangers. The loop's temperature is adjusted with the help of by-pass valves. The cooling capability of the IATCS for the individual modules (cf. Table 5.22) is limited to a maximum of 28.7 kW due to the interface inside the heat exchanger.

The important components are the pump package assembly (separate water pumps providing MT and LT coolant to racks and cold-plated hardware), the system flow assembly (fluid flow control), the rack flow control assembly (to access MT and LT loops and to provide additional temperature control at the racks), the three–way mixing valves, and the loop crossover assembly (for cross strapping and failure redundancy). Heaters are distributed over the entire module surface area; they control the temperature at the cabin wall. In the case of the APM, this network is composed of 144 heating elements (72 of which are primary and 72 redundant) forming groups of 6 primary and 6 redundant loops with a total power of 940 W (120–179 W per element) at an expected heat leak of 790 W for the APM. For each of the circuits, there are three temperature sensors which define the heater hysteresis at a temperature between 20°C and 23°C. The heating elements for the PMA 1 and 2 have a lower heater power of 50 W to 68 W (50% over-sized).

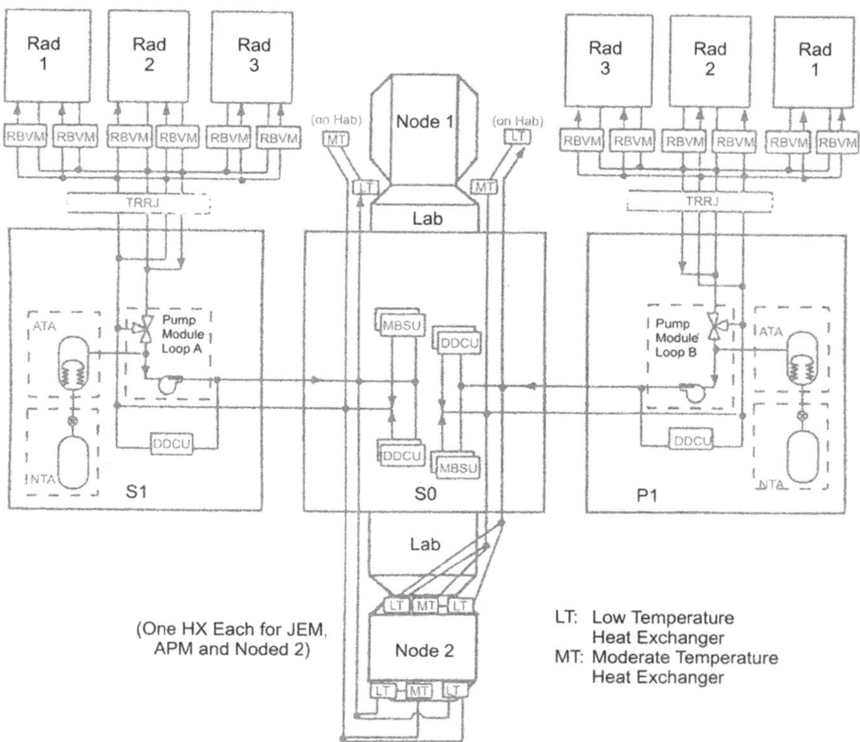

Fig. 5.39. External Active TCS [Cannon 95]

The *external active thermal control system* (EATCS) (see Fig. 5.39) collects heat from the individual modules by means of the heat exchangers described above, transmits the heat to the radiators, and rejects it into space via radiation. A dual loop single-phase ammonia system is used for this process whose design was based on the freon loop of the Space Shuttle. Compared to other cooling agents such as heptane, toluol or freon, anhydrous ammonia requires a pumping power that is lower by a factor 6 to 16, at the same time that the specific heat is four to seven times higher. Due to technical risks, this two-phase concept was abandoned. From the original three EATCS loops, two remain since part of the equipment is stowed in the Russian segment and is cooled there by the local TCS.

Major truss-mounted EATCS components are the single-phase pump assembly, ammonia and nitrogen tanks, control units/instrumentation (ORU packaged) and plumbing, cold plates (for the DDCUs and the MBSUs), the thermal radiator rotary joint (TRRJ) (quick disconnect couplings), and the rotatable radiator wings. Each ammonia loop makes its way through the heat exchangers (in counter flow) coupled with the LT loops of the IATCS (called LT HX), and then through the heat exchangers of the MT loops (MT HX). Thus, a minimum heat transport capability is available even if one of the EATCS loops fails. Entrance temperatures of 0.5°C to 3.9°C for the LT HX (accurately controlled to prevent freezing on the water side) and 13°C to 18.3°C for the MT HX are specified. One of both loops transmits the heat to the so-called "Starboard TCS"-radiator wing, the other transmits the heat to the so-called "Port TCS"-radiator wing. Each radiator wing contains three radiators designed as Orbital Replaceable Units (ORUs). The radiators can be rotated about an axis parallel to the flight direction to minimize incident solar flux (or maximize system heat rejection). The aerodynamic drag of the radiators is minimized because they are mounted perpendicular to the truss. Each of these ORUs is composed of eight radiator segments connected in series with two redundant independent flow passages each. The individual panels are connected by a crossbar control mechanism by means of which the radiator is deployed.

An electronic valve control assembly at the truss controls the fluid flow of the segments with the help of valves. In the case of damage, the individual segments can be closed by a valve unit (RBVM). The arterial heat pipe radiators chosen during the SSF phases, because of their high tolerance with regard to micrometeoroids and freezing, were replaced during the re-design of the station by simple radiators through which ammonia flows. Each of the two loops can transmit up to 35 kW of heat and reject it via the radiators; this value, however, is currently limited by the radiators.

The reconfiguration of the SSF which yielded the International Space Station had very little influence on the design of the four photovoltaic modules (PVM) that comprise the PV arrays, batteries and their charge/discharge units (BCDU), the *Photovoltaic Thermal Control System*, and other electronic equipment that is used for transforming and providing electrical power. The PVTCS (see Fig. 5.40) is used to

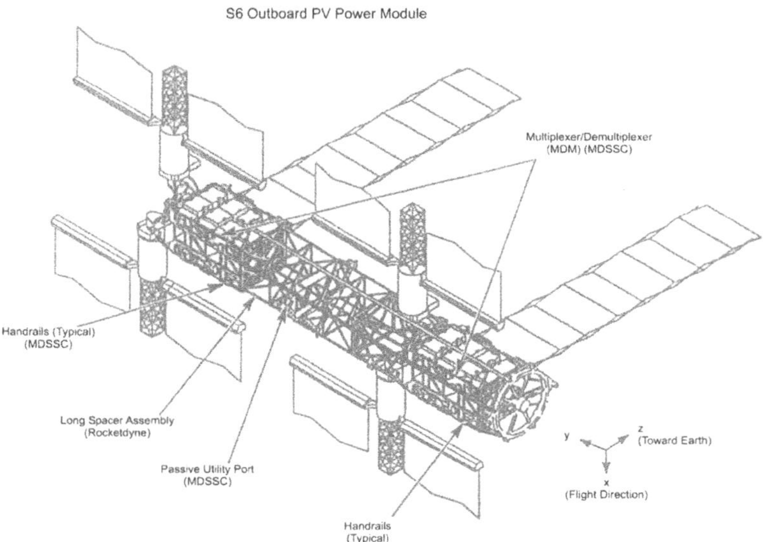

Fig. 5.40. PV Thermal Control System [Hajela 92]

Table 5.23. Temperature Margins of Photovoltaic Module Hardware

Equipment	Temperature [°C]
Batteries	
– Nominal Operation	0 to +10
– Mean Value	-5 to +10
– Maximum Discharge	0 to +20
Charge/Discharge Control Units (BCDU)	-24 to +40
DC-to-DC Converter Unit (DDCU)	-24 to +54

remove heat from the PVM and to reject it into space. It consists of two independently operating, mechanically pumped single-phase loops with ammonia as working fluid. Each of the radiator wings is composed of seven detachable radiator segments that feature two independent flow passages.

In the place where the radiator is mounted on the truss, there is a thermo-hydraulic joint called "thermal radiator rotary joint" (TRRJ). This joint allows the orientation of the whole radiator to space in such an optimal way that the external heat loads acting upon the radiator are minimized. The components to be cooled are mounted onto cold plates through which ammonia flows. As can be seen from Table 5.23, the isothermal requirements for the batteries are relatively high. At BOL +5, each PVTCS of the individual modules can transmit a maximum of 16 kW of heat. In this context, maximum heat loads occur for the PVTCS during the shadow phases since the batteries, mainly during their discharge, produce large amounts of heat.

The thermal control system of the *Russian Segment* (RSTCS) assumes the tasks of internal and external thermal control. The air temperature inside must be kept between 18°C and 28°C. Single-phase water/glycol loops with a return temperature of 15 ± 1.5°C are used for heat removal inside the cabins. These glycol mixtures, despite their somewhat sweet taste, are highly toxic and in the case of several leaks aboard the space station Mir in 1997, they caused considerable contamination aboard the station. During the free-flying phases, the internal loops can transfer their heat to the body-mounted radiators that are installed on the module surface. About 75% of the module surface is covered with BMRs ($\alpha_S \approx 0{,}45$, $\varepsilon \approx 0{,}8$) that are operated with benzene; the rest is covered with an MLI (20-layer aluminized Mylar) as insulating material and several layers of Kevlar® foil for protection against mechanical damage. During later assembly phases, the Russian segment will use a central ammonia two-phase loop and a second single-phase fluid loop. The heat is transported from the cold plates via seven heat exchangers to the two-phase loop and is transferred to the heat pipe radiators at a set-point temperature of about 10°C. Temporarily, the US Lab is being provided with cooling capability by the use of a heat exchanger mounted on the US Lab module. This heat exchanger allows the transmission of heat (nominal 3 to 8 kW, maximum 15 kW) to the Russian TPS (cf. Fig. 5.41). The radiators mounted at the SPP truss have an operational area from 8.4 m to 9 m and can be rotated about the same axis as the radiator surfaces mounted on the ISS main truss. The heat pipes used have a length of about 2 m at a transport capacity of 120 W per heat pipe. The two-phase loop has a transmission capacity of

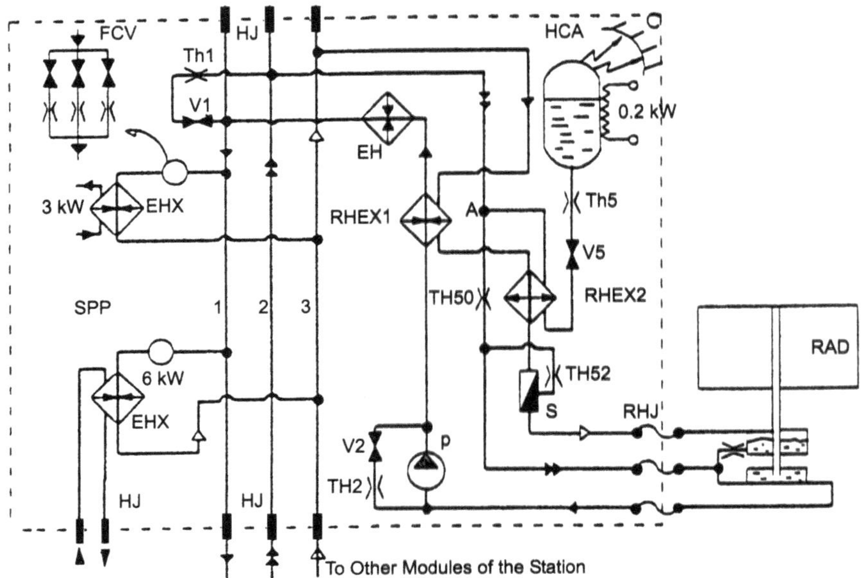

SPP: Scientific Power Platform; RAD: Radiator; HCA: Heat-Controlled Accumulator; P: Pump; RHJ: Rotary Hydraulic Joint; RHEX: Regenerative Heat Exchanger; S: Separator; EH: Electrical Heater; V: Valve; TH: Throttle; FCV: Flow Control Valve; HJ: Hydraulic Joints; EHX: Evaporative Heat Exchanger

Fig. 5.41. Principal Schematic of ISS Russian Segment Two-Phase TCS [Grigorov 97]

about 30 kW; the rejectable heat is limited to 20 kW due to the size of the available radiator surface. In later phases of assembly, the heat rejection is to be increased to 91 kW by installation of additional radiator elements.

One particular problem posed by the air and humidity transported by natural air exchange is that the control and monitoring area for the modules of the US, European and Russian parts of the station are not identical and hence, the amounts of these heat and humidity flows are difficult to quantify.

6 Attitude and Orbit Control System

The next subsystem typical of a space station, the Attitude and Orbit Control System (AOCS), will be discussed in this chapter. On one hand, the orbit control and orientation of a large orbiting structure is a particular challenge for engineers, but on the other hand, special concepts must be developed for attitude and orbit control because of utilization aspects, logistical requirements and relatively low orbit altitudes.

After a short description of the problems related to attitude and orbit control, this chapter will analyze the most important perturbations influencing it. Additionally, different attitude and orbit control strategies will be presented, and technologies of propulsion systems will be addressed, in particular, that of the International Space Station (ISS).

6.1 The Attitude and Orbit Control Problem

For a better understanding of this chapter, this section will briefly describe the general problems of attitude and orbit control existing for all types of space vehicles (satellites, transfer vehicles, space stations) and also present some of the terms and definitions related to this subject.

Orbit control is the control of shape and orientation of an orbit, which is determined by the six Kepler elements (cf. Fig. 6.1). The orbit's shape is typically described by two parameters, the *semi-major axis* a and the *eccentricity* $e = 1 - r_p / a$, with r_p indicating the perigee radius. The *true anomaly* Θ, an angle measured from the perigee in the direction of the revolution, describes the position of the spacecraft on the orbit. The angle between orbital plane and equatorial plane is called the *inclination* i. The orientation of the orbital plane in space, called Right Ascension of Ascending Node (RAAN) Ω, is defined with respect to a reference direction, the direction of the vernal equinox γ. The orientation of the orbit ellipse in the orbit plane is defined with the help of the angle between ascending node and perigee, denominated *argument of perigee* ω. A typical space station orbit is nearly circular ($e \approx 0$), has an altitude of 200–500 km and an inclination of anything between 0° (equator) and 65° (beyond which the radiation may become too high). Typically the inclination is close to the latitude of the launch sites, i.e. 28.5° or 51.6° (cf. Sect. 12.1.2). The orbit (and thus single orbit parameters) is influenced by perturbing forces (cf. Sect. 6.2).

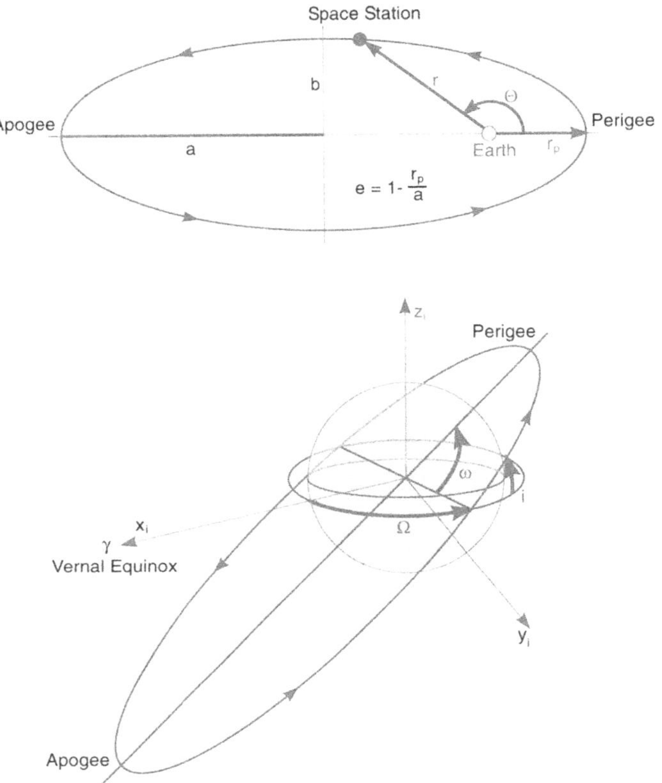

Fig. 6.1. Definition of an Orbit by the Six Kepler Elements (Semi-Major Axis a, Eccentricity e, True Anomaly Θ, Inclination i, Right Ascension of Ascending Node Ω, Argument of Perigee ω)

What is understood by attitude control, is the control of a body-fixed coordinate system's orientation of a vehicle with respect to a reference system. Common reference systems are e.g. Earth-centered and inertial systems, where

- the x_i-axis lies in the direction of the vernal equinox,
- the z_i-axis lies in the direction of the Earth's rotation axis, and
- the y_i-axis lies in the equatorial plane (cf. Fig. 6.1).

Moreover, a local orbit system serves as a reference system. Here,

- the z-axis points to the Earth (to the Nadir),
- the x-axis lies perpendicular to the z-axis in the orbital plane with its positive axis in the direction of the velocity vector, and
- the y-axis is perpendicular to orbital plane, completing the right-handed orthogonal system.

Further modes of orientation for the description of a flight attitude (also referred to as flight mode) are, for example, the local vertical (Nadir) as well as the local horizontal, together called "Local Vertical Local Horizontal" (LVLH). If the orbit is nearly circular, the x-axis and y-axis of the local orbit system will be located in the local horizontal plane, and thus the local vertical and the z-axis of the orbit system will be identical.

The orientation of a body-fixed system with respect to one of the aforementioned reference systems can take place by the following different forms of "parameterization":

- An attitude transformation matrix
- A rotation sequence about three axes (Euler or Cardan angles)
- The indication of a rotation axis and a rotation angle (typically, this axis does not correspond to any axes of the body-fixed or reference system)

The latter parameterization simultaneously illustrates the theorem that every attitude transformation can be generally described by indicating a rotation vector and the rotation angle belonging to it. If these are expressed in complex notation, the quaternion parameterization is obtained, which is advantageous especially for numerical simulations and the formulation of kinematic equations.

Within the framework of this chapter, the orbit of a space station will be described by the classical Kepler elements and the station's attitude with the help of attitude matrices and Euler and Cardan angle sequences, respectively.

There are two preferred flight attitudes for space stations: Earth-oriented and inertial. Unlike orbits, the attitude of a space vehicle is not influenced by perturbing forces but by perturbing torques instead.

A fundamental description of general problems of attitude and orbit control can, for instance, be found in [Wertz 78].

6.2 Perturbations

The task of the Attitude and Orbit Control System of a space station is to keep the station at an orbital altitude and orientation according to the mission requirements. A sub-component of this system including the control system is a propulsion system in the classical sense, e.g. a system of thrusters.

The AOCS must compensate all perturbing forces and torques that influence the station. In a Low Earth Orbit (LEO), these are, for example, the following:

- Aerodynamic deceleration due to the residual atmosphere
- Torques caused by asymmetrical incidence of aerodynamic forces
- The Earth's gravity gradient
- Terrestrial perturbations such as Earth oblateness, electromagnetic forces
- Stellar perturbations, e.g. solar radiation pressure
- Perturbations originating from subsystems, like torques caused by gyro drift
- Mission-dependent influences such as docking maneuvers, cyclic motion of the solar arrays, etc.

Table 6.1. Perturbations Influencing Attitude and Orbit

Type of Perturbation	Orbit	Attitude
Residual Atmosphere	Δa, Δe, (Δi)	M_{Drag}
Gravitation: Earth's Mass Distribution, Gravity Gradient, 3-Body Influences	$\Delta\Omega$, $\Delta\omega$, $(\Delta e, \Delta i)$	M_{GG}
Magnetic Influences	(Tether)	M_{mag}
Solar Radiation Pressure	Δe, Δi	M_{sol}

This list contains perturbations resulting from interactions between the space station and its orbital environment as well as perturbations induced inside the station during its operation. Table 6.1 offers an overview of the environmental influences as well as their effect on the orbit and attitude of a spacecraft. Chapter 3, on the subject "Orbital Environment", comprises all environmental influences. Only the most important factors influencing the attitude and orbit control system are listed here; in Table 6.1, they are set off by shading. Although for simulations, all other external perturbing influences as well as internal perturbations must, in each single case, be taken into account, they generally do not constitute a design parameter.

According to this table, the most effective perturbation of a space station's orbit is the aerodynamic drag, whereas satellite systems in geosynchronous orbits, for instance, are mostly influenced by gravitational and solar radiation forces. The most important perturbation for the attitude of a space station is the asymmetrical incidence of the aerodynamic forces as well as the gravity gradient.

6.2.1 Aerodynamic Drag

In low Earth orbits, the effect of the aerodynamic drag occurs especially in the case of large incidence areas such as pressurized modules, solar arrays or thermal radiators. The decelerating force $\vec{F}_D$ can be described by the dynamic pressure p_{dyn} with:

$$\vec{F}_D = -p_{dyn} \cdot C_D \cdot A_p \cdot \hat{v} = -\frac{1}{2} \cdot \rho \cdot v^2 \cdot C_D \cdot A_p \cdot \hat{v} \tag{6.1}$$

In this context, the drag coefficient C_D can be set to 2.2–2.5 for estimations. To be exact, it can be determined by integrating the exchange of impulses between the gas particles of the residual atmosphere and the space station's outer surface. The orbital velocity v_c for a circular orbit is:

$$v_c = \sqrt{\gamma \cdot M/r} = \sqrt{\mu/r} \qquad \text{with} \qquad \mu_{Earth} = 3.989 \cdot 10^{14}\ m^3/s^2 \tag{6.2}$$

The effective incidence area A_p (perpendicular to the flight direction) can vary due to moving space station components such as solar arrays or due to attitude changes of the station itself.

The local density ρ of the residual atmosphere is influenced by several factors, among them the solar activity, seasonal effects, the day/night cycle, orbit geometry and the Earth's magnetic activity. Table 6.2 shows the different factors with their relative influence on the density of a circular orbit at 400 km altitude.

Table 6.2. Density Variations Indicated for a Circular Orbit at 400 km Altitude [Marcos 94]

Flux (Solar Cycle)	Flux (Daily)	Geomagnetic Activity	Local Time	Semi-Annual Variations	Latitude	Longitude
1165%	5%	60%	115%	50%	60%	5%

In order to predict the atmosphere density, there is a series of models [Marcos 94], usually available in the form of computer programs which allow the projection of already measured data. The models referred to as "Reference Atmosphere" are in a position to cover different environmental conditions or points in time. The models Jacchia or CIRA (COSPAR International Reference Atmosphere) and MSIS (Mass Spectrometer and Incoherent Scatter) are widely used. They differ in the number of effects taken into account and the measuring methods applied. The three-layer model GRAM-90 (Global Reference Atmosphere Model) will be used for the design of the International Space Station [SSP 30425]. This model contains density values up to an altitude of 25 km in the form of four-dimensional tables (longitude, latitude, altitude and time). The Groves model covering altitudes between 30–115 km, follows immediately afterwards. The most important atmosphere model for the design of ISS is the so-called "Marshall Engineering Thermosphere Model" (MET) which covers altitudes between 90–2500 km. All these are empirical models whose coefficients have been obtained by drag measurements of satellites. Less realistic are the so-called "Standard Atmospheres". They are standardized density tables and therefore cannot take dynamic atmosphere effects into account (see also Sect. 3.6.1).

The power spectral density of the solar radiation at 10.7 cm wavelength $F_{10.7}$ with the unit 10^{-22} W/(m^2 Hz) can be used as a measure for solar activity. On the one hand it can easily be observed on Earth, on the other hand it is closely related to the solar radiation in the high UV range, which is responsible for the heating of the atmosphere and thus for the density variations. $F_{10.7}$ fluctuates between values of 70 and 260 over a period of about 11 years. For design problems, the worst case should be considered, i.e. a high radiation flux $F_{10.7}$. Figure 6.2 shows the predicted course of the spectral solar radiation flux for the present solar cycle. According to this figure, the solar activity had its minimum in mid-1997, whereas in the year 2002 the next maximum solar activity is expected.

The day/night variations of the atmospheric drag and the seasonal effects can be explained by the variable solar radiation and the density variations in the atmosphere related to it. The amplitude of the day/night variation decreases with the altitude, but still amounts to ± 115% at $h = 400$ km (cf. Fig. 3.22 and Table 6.2). The daily density maximum is at about 14:00 h local Sun angle for orbits above 300 km altitude. The seasonal maxima are reached in October and April. Also the orbit geometry has an influence on the local atmosphere density: on orbits with high inclination, regions with a larger solar incidence angle (i.e. less heating and correspondingly less atmosphere density) are crossed. During geomagnetic storms, large amounts of charged particles are "pumped" into in the atmosphere near the poles, where they thermally excite the gaseous particles in the atmosphere due to collisional interactions. This also results in an increased atmospheric density.

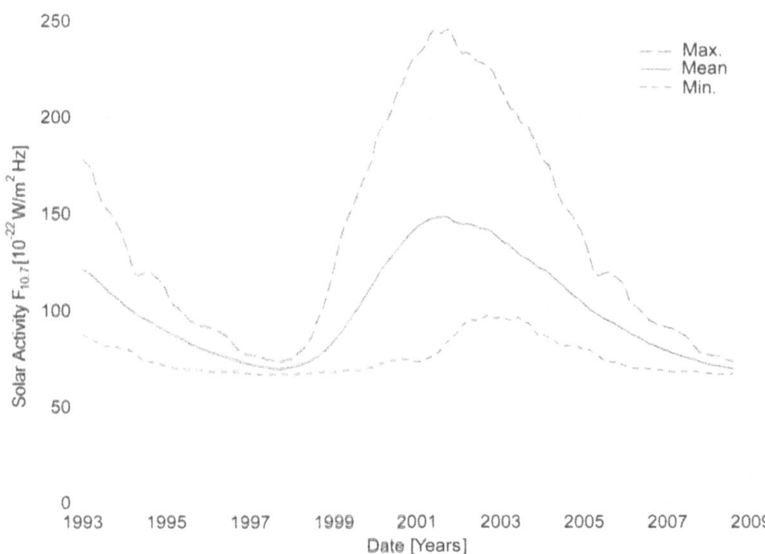

Fig. 6.2. Prediction of the Spectral Radiation Flux $F_{10.7}$[SSP 30425]

The estimate of the residual atmosphere must be carried out carefully since the atmosphere density has a linear influence on the dynamic pressure, and thus on the thrust necessary for drag compensation. Considering the effects described, the atmospheric drag must be formulated in a more general way as follows:

$$\vec{F}_D(t) = -\frac{1}{2} \cdot \rho(h, t) \cdot v^2(t) \cdot C_D \cdot A_p(t) \cdot \hat{v} \tag{6.3}$$

This leads to a differential equation which cannot be solved analytically because of the unknown function for ρ. There is, however, a permissible simplification for design computations that allows all quantities that depend on the altitude for one or several revolutions to be kept constant and then the adaptation of v, h and ρ to the impulse loss accumulated up to that point.

All in all, for a space station without permanent orbit control, the atmospheric drag varies in its absolute value with a short period of $A(t)$ and $\rho(t)$ and with a long period due to the solar cycle. With decreasing orbital altitude, the aerodynamic drag becomes increasingly stronger until at an altitude of about 100 km, the orbit turns into a reentry trajectory.

6.2.2 Aerodynamic Torque

In the case of a space station, the incidence line of the aerodynamic drag will not always run through its center of mass. Torques, whose values follow the variations of the drag force, result from the distance between center of pressure (incidence point of the atmospheric drag) and center of mass (cf. Fig. 6.3).

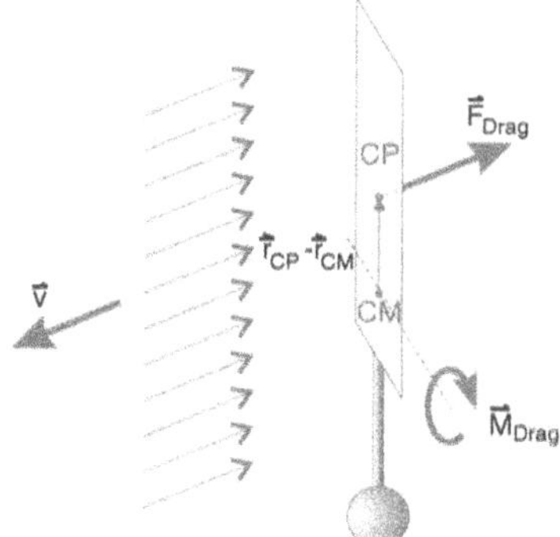

Fig. 6.3. Description of the Aerodynamic Torque

According to Fig. 6.3, the aerodynamic torque is obtained from:

$$\vec{M}_{Drag} = (\vec{r}_{CP} - \vec{r}_{CM}) \times \vec{F}_{Drag} \tag{6.4}$$

Indeed, the aerodynamic torque can lead to a considerable torque in the case of a space station (cf. the example of ISS in Sect. 6.3.1). Therefore, the spatial arrangement of aerodynamic incidence surfaces within a space station configuration needs to be carefully determined. This subject is covered in more detail in the Sect. 6.3.1 ("Strategies for Attitude Control") and also in Chapter 9 ("System Design").

6.2.3 Gravity Gradient

Classical orbit mechanics deal with point masses. Real space vehicles with finite dimensions, by contrast, are exposed to the spatially varying gravitational field of the Earth. As a result of the gravity force varying along the vehicle, the body will tend to orient itself in a preferential attitude. The torque due to the gravity gradient is given by the relation

$$\vec{M}_{GG} = \frac{3 \cdot \mu}{R^3} \cdot [\hat{R} \times (\underline{I} \cdot \hat{R})] \tag{6.5}$$

with μ as the gravitation parameter of the Earth, R as the radius vector (from the Earth's center to the vehicle's center of mass, cf. Fig. 6.4) and R as its direction. $\underline{I}$ stands for the inertia tensor, a matrix containing the products and moments of inertia (cf. Eq. 6.6). This relation is formally obtained by integrating the terms of torques which result from Newton's law of gravitation in the case of every mass element of the vehicle. Here, the gradient-related information of the gravitational field is included in the inertia tensor and is no longer explicitly visible. In the following, the effects of the gravity gradient will be discussed for both the Earth-oriented and the inertial flight modes.

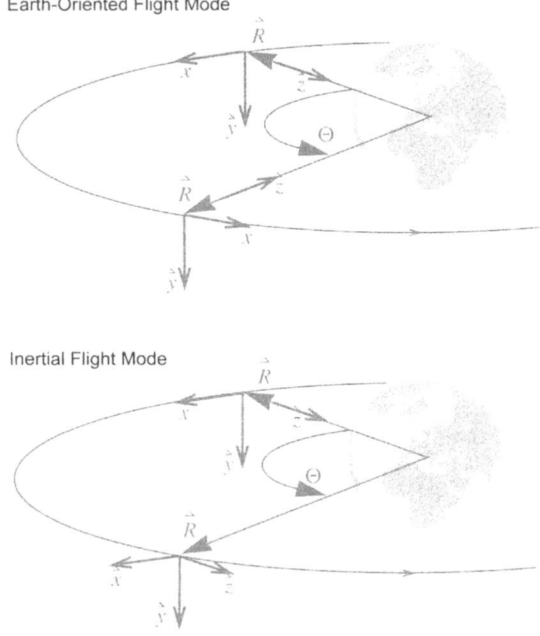

Fig. 6.4. Body-Fixed Coordinate System, Earth-Oriented and Inertial Flight Attitudes

Earth-Oriented Flight Mode. In the case of this attitude, $\hat{R}$ must always point in (-z)-direction in an orbit-reference system (cf. Fig. 6.4):

$$\hat{R} = \begin{bmatrix} 0 \\ 0 \\ -1 \end{bmatrix} \qquad \underset{-}{I} = \begin{bmatrix} I_{xx} & -I_{xy} & -I_{xz} \\ -I_{xy} & I_{yy} & -I_{yz} \\ -I_{xz} & -I_{yz} & I_{zz} \end{bmatrix} \tag{6.6}$$

Thus, Eq. 6.5 is simplified as follows:

$$\vec{M}_{GG} = \frac{3 \cdot \mu}{R^3} \cdot \begin{bmatrix} 0 \\ 0 \\ -1 \end{bmatrix} \times \begin{bmatrix} I_{xz} \\ I_{yz} \\ -I_{zz} \end{bmatrix} = \frac{3 \cdot \mu}{R^3} \cdot \begin{bmatrix} I_{yz} \\ -I_{xz} \\ 0 \end{bmatrix} \tag{6.7}$$

As a consequence, only roll torques (torques about the x-axis) and pitch torques (torques about the y-axis) result from the gravity gradient in an Earth-oriented flight mode. If the stations's principle axes of inertia are co-aligned with the Earth-oriented coordinate frame (I_{xy}, I_{xz}, I_{yz}=0), these perturbations would completely disappear.

Considering a small deviation, e.g. $\hat{R} = (\varepsilon\ 0\ -1)$ instead of $\hat{R} = (0\ 0\ -1)$, it can be demonstrated that only for the conditions $I_{xx} > I_{zz}$ and $I_{yy} > I_{zz}$, torques counter-

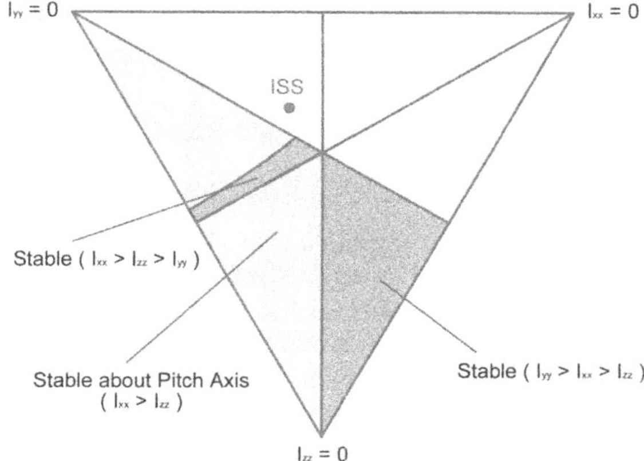

Fig. 6.5. Gravity Gradient Stabilization: Stability Areas in the "Form Triangle"

acting the deviation occur. Thus, a flight attitude is called "gravity gradient stabilized", if the following equation is satisfied:

$$I_{yy} \overset{!}{>} I_{xx} \overset{!}{>} I_{zz} \tag{6.8}$$

As to the technical realization, the condition $I_{yy} > I_{zz}$ (stability of yawing) is less relevant. Eq. 6.8 clearly illustrates that the axis with the smallest moment of inertia must be radially oriented for a gravity-gradient stabilized flight attitude. The stability areas for an Earth-oriented flight mode may also be represented in a "form triangle", as seen in Fig. 6.5. The triangle containing different shades of gray is that area in which there is stability of pitch ($I_{xx} > I_{zz}$). The two areas colored in dark gray are those in which stabilization of the gravity gradient is possible. In addition to the larger of both areas described by Eq. 6.8, there is another small area of stability in which $I_{xx} > I_{zz} > I_{yy}$. This area, however, is not interesting from the technical point of view, since it only contains a small stability margin.

Inertial Flight Mode. If a vehicle is kept in an inertial flight mode (cf. Fig. 6.4), the radius vector in the body-fixed coordinate system can, after introducing a revolution angle Θ corresponding to the true anomaly, be described as follows:

$$\hat{R} = \begin{bmatrix} \sin\Theta \\ 0 \\ -\cos\Theta \end{bmatrix} \tag{6.9}$$

After multiplication, the following is obtained from Eq. 6.5 for the gravity torque:

$$\vec{M}_{GG} = \frac{3 \cdot \mu}{R^3} \cdot \begin{bmatrix} -I_{xy} \cdot \sin\Theta \cdot \cos\Theta & + & I_{yz} \cdot \cos^2\Theta \\ (I_{zz} - I_{xx}) \cdot \sin\Theta \cdot \cos\Theta + I_{xz} \cdot (\sin^2\Theta - \cos^2\Theta) \\ I_{yz} \cdot \sin\Theta \cdot \cos\Theta & + & -I_{xy} \cdot \sin^2\Theta \end{bmatrix} \tag{6.10}$$

It is interesting to note that this expression, on one side, consists of terms with $\sin\Theta \cdot \cos\Theta = 1/2 \cdot \sin(2 \cdot \Theta)$, i.e. perturbations which compensate themselves after integration over $\Theta = 0...180°$, and, on the other side, of quadratic terms which result in secular torques. Cyclic perturbations can be compensated with adequate dimensioned momentum storage devices, whereas the accumulated momentum has to be desaturated, e.g. by means of thrusters. If the station is in the inertial flight attitude with one principle axis of inertia being perpendicular to the orbit plane, only cyclic perturbations will remain according to the above equation.

In Sect. 6.3.1, perturbation torques due to residual atmosphere and the gravity gradient will be estimated, the space station ISS being taken as an example.

6.2.4 Operational Influences

Moving station components may also cause considerable perturbations of the attitude and orbit of a space station. This applies especially to the Earth-oriented attitude, when large collector surfaces have to be tracked facing the Sun. In the case of the International Space Station, mainly the solar double panels of the US segment can cause cyclic drag forces (Fig. 6.6).

The rotation of the solar arrays causes a cyclic variation of the incidence area according to $A_{Solar} = A_{Solar,max} \cdot |\sin\alpha|$. This superposes the rotation of the radiators which stand perpendicular to the solar arrays. As a consequence, the drag force will show the same sine-shaped history. If this variation is not compensated by the orbit control system, the result is an increasing eccentricity of the orbit and hence a perturbed altitude profile.

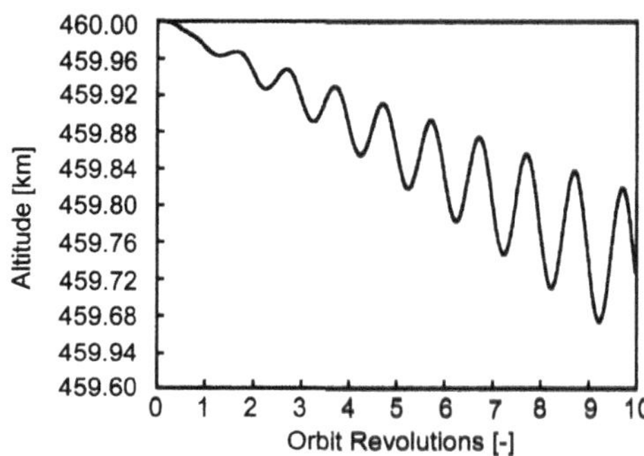

Fig. 6.6. Oscillations of the Altitude Profile of ISS due to Solar Array Tracking

Further operational perturbations influencing the system are e.g. the docking maneuvers of vehicles such as the Space Shuttle or mission-dependent maneuvers, such as space debris collision avoidance.

6.3 Flight Strategies

The aforementioned perturbations can be handled by means of different strategies. Main objective is, however, to minimize propellant consumption and, at the same time, to observe the operational boundary conditions such as maintaining a certain degree of alignment accuracy of the vehicle or of a given minimum orbital altitude.

6.3.1 Strategies for Attitude Control

The most important strategy for attitude control is the avoidance, or at least minimization, of perturbation torques by means of careful selection of a flight attitude in combination with an appropriate space station architecture. This section will therefore treat those aspects important to the selection of the flight altitude. Design guidelines that arise from attitude stability are dealt with in Chapter 9 ("System Design").

Figure 6.7 illustrates terminology used for describing the flight attitude of a space station as follows:

- POP: *Perpendicular to Orbit Plane*
- IOP: *In Orbit Plane*
- ASL: *Aligned with Sun Line*
- PSL: *Perpendicular to Sun Line*
- LH: *Local Horizontal*
- LV: *Local Vertical*

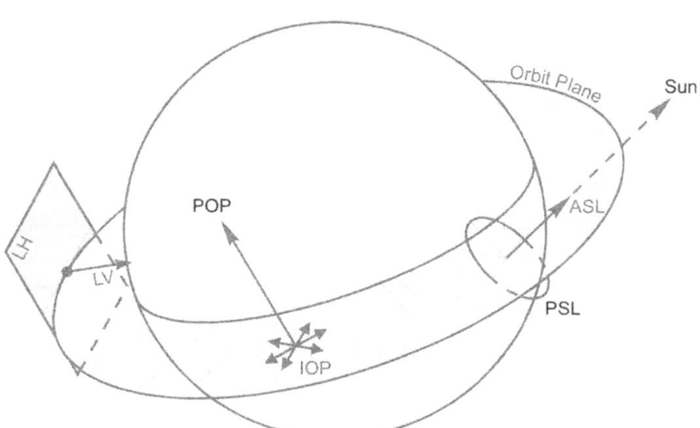

Fig. 6.7. Terminology for the Description of Flight Modes of a Space Station

With the help of these terms, the orientation of single space station axes can be defined relative to reference directions such as Sun line, orbit plane or local horizontal plane.

Selection of Flight Attitude. Important factors that must be taken into consideration when comparing and selecting a flight mode are e.g. the following:

- *Mission Requirements*. Depending on the intended utilization of a space station, different flight attitudes are suitable. For example, an Earth observation mission requires an Earth-oriented flight mode, whereas for astronomical experiments, the inertial flight attitude is more favorable. Earth-oriented flight modes are also advantageous for microgravity experiments since in this case no tidal perturbations will occur along the direction of the velocity vector. This feature will be discussed in more detail in Chapter 8 ("Microgravity").
- *Minimization of Atmospheric Drag*. Seen from the point of view of atmospheric drag, the solar arrays with their surface normals should be aligned perpendicular to orbit plane (POP). Of course, this contradicts the requirement that they should also be permanently pointing towards the Sun (i.e. their surface normals hence ASL). The same applies to thermal radiators: their surface normals should always be aligned perpendicular to Sun line (PSL) to achieve the maximum efficiency. In both cases, an appropriate tracking strategy that constitutes a compromise between both requirements must be worked out. For example, the solar arrays could, during shadow phases, be turned "out of the wind" and, during the Sun phases, be Sun-tracked, despite the atmospheric drag. In addition to that, the pressurized station parts must expose the least possible incidence area to the residual atmosphere.
- *Effect of the Gravity Gradient*. In an Earth-oriented flight attitude, the gravity gradient can be used as a stabilizing factor. In order to achieve this, the body axis with the smallest moment of inertia must be oriented towards the Earth. In an inertial flight attitude, the gravity gradient will always be a perturbing factor, especially if the station's inertia tensor has products of inertia. In such a case, the gravity gradient torque must be kept as small as possible by symmetrical mass distribution (principal axis attitude).
- *Mechanical Effort for Solar Array Tracking*. When the solar arrays of a vehicle flying in Earth-oriented flight attitude are tracked, two movements must be compensated: Firstly, the vehicle (viewed from the Sun's position) rotates once on its axes during one orbital revolution – this can be compensated by the so-called "α-tracking". Secondly, the Sun direction relative to the equator changes during each year by about ± 23.5°. This variation, together with the orbit inclination, can be compensated by means of the so-called "β-tracking". In the ideal case, tracking mechanisms are unnecessary in an inertial flight attitude, since the vehicle does not rotate relative to a inertial coordinate system (thus also relative to the direction of the Sun). Figure 6.8 shows the principle of α- and β-tracking in the picture on the left. The picture on the right illustrates how complete tracking can be achieved in a gravity-gradient stabilized flight mode with only one joint. Here, the entire station constantly rotates on the yaw axis such that the tracking joints of the solar arrays can align them to perpendicular solar incidence. In order to

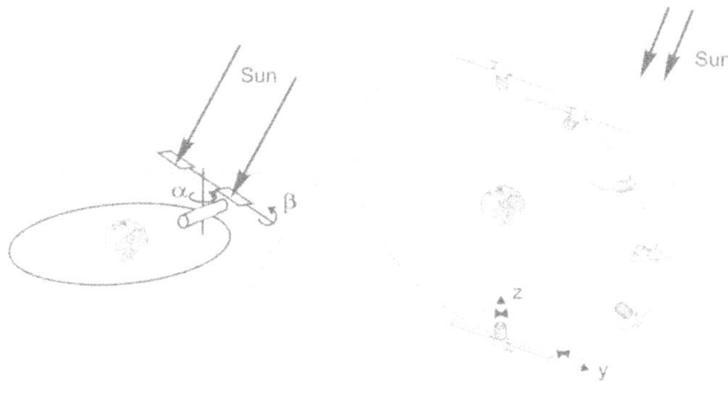

Fig. 6.8. Tracking of the Solar Arrays for Two Earth-Oriented Flight Attitudes

achieve that, the y-axis (= the axis on which the solar arrays can be rotated) must be aligned perpendicular to the projected Sun line (y-PPSL); i.e. the Sun line being projected into the local horizontal (xy-) plane. This attitude control strategy (also referred to as "yaw steering") will, for the first time, be applied on the mobile communication satellites Globalstar (start of service: 1999).

- *Minimizing Micrometeoroid Impact.* The maximum debris flux occurs in the local horizontal plane (debris plane) within an angle range of 45°–75° left and right of the flight direction. Especially the pressurized parts of the station should be oriented in such way that they offer the least possible incidence area for the debris flux or that they shield one another. These aspects will be described in detail in Sect. 9.3.2 ("Characterization of System Elements").
- *Rendezvous and Docking Maneuvers.* For reasons of safety and orbit mechanics, rendezvous and docking maneuvers will take place either along the radius line or along the direction of orbital velocity. For that reason it is appropriate that in a stable attitude, the docking adapters are correspondingly aligned, and hence no additional attitude maneuver of the space station is necessary.
- *Psychological Aspects.* The Earth-oriented flight mode is advantageous for Extravehicular Activities (EVA) or docking maneuvers because the astronauts have a fixed visual reference. By contrast, the light conditions will be stable in an inertial flight attitude.

The International Space Station will fly in the so-called "Local Vertical Local Horizontal" (LVLH) attitude as can be seen in Fig. 6.9. This means that the reference attitude is determined by the local horizontal plane in orbit and the local vertical direction (Nadir). In this context, the truss with the large solar generators is perpendicular to the orbit plane (POP), and the longitudinal axis of the US laboratory module points in the direction of the velocity vector. However, due to the station's mass distribution this reference attitude and the stable attitude of the station are not identical. The following example will show the reason:

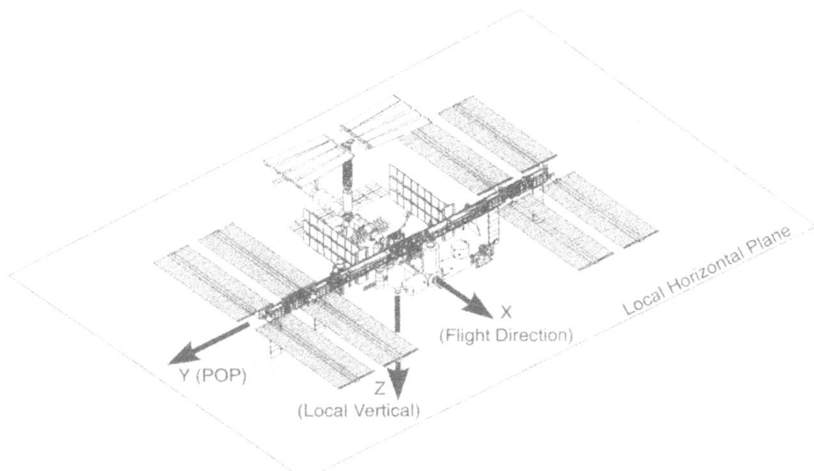

Fig. 6.9. Reference Attitude of the International Space Station

Example ISS. When completely assembled, the International Space Station has the following tensor of inertia:

$$\underline{I} = \begin{bmatrix} 127.29 & 2.23 & 8.15 \\ \div & 90.35 & 0.92 \\ \div & \div & 182.52 \end{bmatrix} \cdot 10^6\ kg \cdot m^2 \tag{6.11}$$

The first fact to notice is that the products of inertia I_{xy}, I_{xz} and I_{yz} are unequal to zero. If ISS were kept exactly in the reference attitude, the following perturbation torque (due to the gravity gradient) could be expected at an average orbit altitude of 400 km:

$$\vec{M}_{GG} = \frac{3 \cdot \mu}{R^3} \cdot \begin{bmatrix} I_{yz} \\ -I_{xz} \\ 0 \end{bmatrix} = \begin{bmatrix} -3.53 \\ 31.43 \\ 0 \end{bmatrix} Nm \tag{6.12}$$

Apart from a non-negligible roll torque (about the x-axis), an extreme pitch torque of 31.43 Nm occurs due to the product of inertia I_{xz}. Products of inertia result from asymmetrical mass distribution, in the case of I_{xz}, from asymmetries in the xz-plane($I = \int_m xz\ dm$). This is caused by the imbalance between the Russian and the US/Japanese/European orbit segments.

An eigenvalue analysis allows the calculation of the principal moments of inertia as well as the orientation of the principal axes. The principal inertia tensor has the form

$$\underline{I}_{PA} = \begin{bmatrix} 126.23 & 0 & 0 \\ \div & 90.21 & 0 \\ \div & \div & 183.71 \end{bmatrix} \cdot 10^6 \; kg \cdot m^2 \qquad (6.13)$$

and the LVLH coordinate system can be transformed into the principal-axes system of ISS by means of a Cardan rotation sequence (first rotation angle on x-axis, second rotation angle on y-axis, etc.) of (-0.27°, 8.28°, 3.34°). In order to make the above mentioned pitch torque of the gravity gradient vanish, the roll axis (x-axis) of the International Space Station must be tilted towards the Earth by 8.28°.

In addition to that, it can be seen from the tensor of inertia Eq. 6.11 and also from the principal-axes tensor Eq. 6.13 that the principal moments of inertia in flight direction and perpendicular to orbit plane (I_{xx} and I_{yy}) are markedly smaller than the radial main moment of inertia (I_{zz}). The relations between the individual moments of inertia for ISS are marked with a black point in Fig. 6.5. ISS is not located in one of the stability areas, thus the condition from Eq. 6.8 is not fulfilled, and the torques resulting from the gravity gradient have no stabilizing effect on the flight attitude. Hence, when ISS's principal axes are aligned to the LVLH reference system, this is only an unstable equilibrium attitude, and the attitude control system has to maintain this torque-free flight attitude by active control.

In order to estimate the aerodynamic perturbation torque, first of all, the area projected in velocity direction is described by the relation:

$$A_{total} = 739m^2 + 3070m^2 \cdot \sin\alpha + 181m^2 \cdot \cos\alpha \qquad (6.14)$$

The following values apply to the International Space Station's center of mass and its center of pressure:

$$\vec{r}_{CM} = \begin{bmatrix} -7.03 \\ -0.46 \\ 3.93 \end{bmatrix} m \qquad \vec{r}_{CP} = \begin{bmatrix} -7.03 \\ -0.82 \\ 3.12 \end{bmatrix} \qquad (6.15)$$

From a circular orbit velocity of 7671.3 m/s, a density of 10^{-11} kg/m^3, and a drag coefficient of 2.2, a drag force of 0.596 N is obtained. As a consequence, the aerodynamic torque resulting from Eq. 6.4 is:

$$\vec{M}_{D,min} = \begin{bmatrix} 0 \\ 0.485 \\ -0.215 \end{bmatrix} Nm \qquad (6.16)$$

In the present case, the solar arrays as presented in Fig. 6.9 are located in the local horizontal plane, and thus they are not perpendicular to the aerodynamic flow. The worst case aerodynamic torque is obtained when the solar arrays are perpendicular to the velocity direction. With unchanged center of mass, the center of pressure in this case "moves" to the coordinates

$$\vec{r}_{CP} = \begin{bmatrix} -1.61 \\ -0.82 \\ 0.37 \end{bmatrix} m \tag{6.17}$$

and the aerodynamic torque will have the following values caused by a total drag force which is now at 2.47 N:

$$\vec{M}_{D,max} = \begin{bmatrix} 0 \\ 8.8 \\ -0.89 \end{bmatrix} Nm \tag{6.18}$$

If both perturbation torques are compared, it can easily be seen, especially in the case of the pitch axis, from the identical signs that the individual components will not compensate one another. For example, for the solar arrays perpendicular to the velocity direction, the following is obtained:

$$\vec{M}_{pert} = \vec{M}_{D,max} + \vec{M}_{GG} = \begin{bmatrix} -3.53 \\ 40.23 \\ -0.89 \end{bmatrix} Nm \tag{6.19}$$

During daily operation, however, the space station could not keep the reference attitude for long periods of time. This becomes clear by looking at the propellant consumption that would be necessary for such a correction:

The thrusters for attitude control are, among other places, located in the two propulsion units called "Autonomous Thruster Facilities" (ATF, cf. Fig. 6.10) which are installed at the truss (running in z-direction) of the Science Power Platform in the Russian segment. The distance of the ATF modules from the center of mass of the station in z-direction is about 21 m and constitutes the lever for the generation of pitch control torques. If the pitch perturbation torque were compensated by permanent reversal of thrust from the ATF thrusters, the propellant requirements would result in the following:

$$\dot{m} = \frac{M_{pert}}{l \cdot I_{sp} \cdot g_0} \tag{6.20}$$

In the case of a specific impulse for a hydrazine propulsion system of 280 s, the numerical value for the mass flow is $6.96 \cdot 10^{-4}$ kg/s, corresponding to 60.1 kg per day. For that reason, it is easy to understand that most of the time, a flight mode other than Local Vertical Local Horizontal must be applied (cf. Section 9.4.4). Indeed, the International Space Station is intended to fly in an LVLH attitude for only about 10% of its period of operation [SSP 41000D], namely every time the Space Shuttle is docked to ISS.

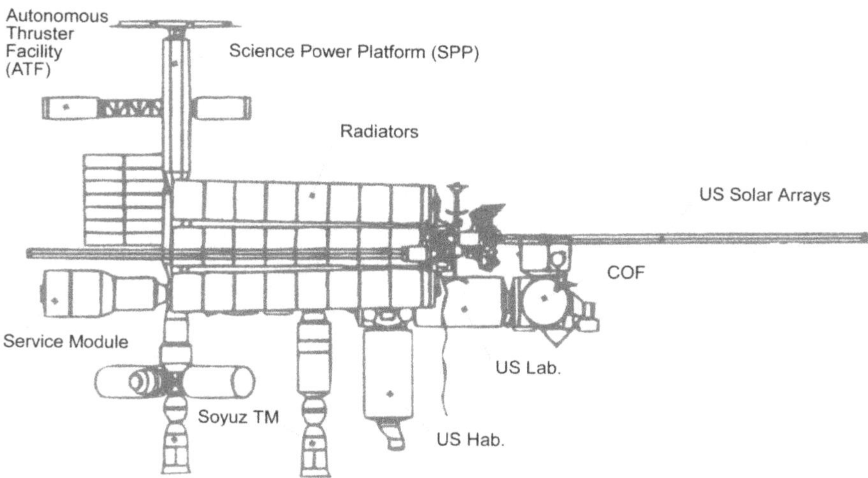

Fig. 6.10. The Autonomous Thruster Facility (ATF) of ISS

Torque Equilibrium Attitude (TEA). When the perturbation torques (especially those caused by aerodynamic drag and gravity gradient) compensate one another in a given orientation, this orientation is called torque equilibrium attitude. Each deviation from TEA has to be maintained by control torques, and that, in the long run, leads to a considerable level of propellant consumption, as it was illustrated in the previous section.

In practice, the determination of TEA is no trivial problem since e.g. in the case of ISS, the solar arrays are in constant motion. Consequently, the position of the center of pressure as well as the inertia tensor, and thus the gravity gradient torque are permanently subject to cyclic variations. A TEA that makes sense from the operational point of view must have a mean value at which (calculated over one orbital revolution) no angular momentum accumulates in the momentum storage devices. This requires constant examination and/or re-determination of the nominal value of the TEA in order to take all the different influencing factors into account, from space station geometry, orbit altitude and solar activity.

6.3.2 Orbit Control Strategies

The perturbations discussed before, mainly the aerodynamic drag have led to a permanently increasing deviation from the nominal orbit, so correctional maneuvers have to be carried out. Different strategies can be devised. Firstly, the orbital perturbations can be permanently compensated by means of thrusters. The common method for space stations is to correct the orbit in regular time intervals and during the time between these intervals, the stations coast. Combined propulsion systems offer considerable advantages with respect to propellant economization, operational safety and flexibility under the condition that appropriate synergism potentials with other subsystems are utilized. The following sections will introduce permanent compensation as well as impulsive orbit control; synergistic propulsion concepts will be illustrated with the help of examples.

Constant Reboost. The problem with constant reboost is that very small forces (up to only a few N) have to be steadily applied. Additionally, the thrust level must be variable in order to adapt the compensation to the instationary course of the perturbing forces. For this purpose, electric propulsions such as resistojets or arcjets are well suited, since their thrust can be controlled either over the electrical power input or over the mass flow rate. Compared to cyclic orbit reboost, constant reboost has the important advantage of perturbations being actively compensated, and this, in principle, improves the station's microgravity conditions ("drag free" concept). Moreover, no special reboost phases must be integrated into the operational cycle, during which utilization (e.g. of the laboratories) would be impaired.

Cyclic Orbit Reboost. In the case of cyclic orbit reboost, a conventional Hohmann transfer is carried out after every coast phase. Due to the higher thrust level, conventional and tested thruster technologies and operational procedures can be applied.

The permanent and impulsive reboost strategies do not make a difference in the level of propellant consumption when it comes to corrections that are small and thus appropriate from the operational point of view. On one hand this can be shown analytically, but on the other, it can simply be understood when we think of the fact that the density gradient $\rho(h)$ is not very large in the considered region.

In practice, however, a series of operational and safety-related aspects have to be taken into account for orbit control. This is made clear by the following example of the orbit control strategy of the International Space Station:

In principle, the orbital altitude should be kept as low as possible in order to ensure maximum payload capacity for launch vehicles in the course of the International Space Station's assembly and operation. The most important boundary condition for the orbit control of ISS, however, relates to the minimum orbital altitude. During nominal operation, the International Space Station is intended to be able to coast at any time for about 180 days without entering denser atmosphere layers which would lead to a reentry trajectory. The period of 180 days normally includes two supply flights; so even if one supply flight is cancelled, propellants for reboost will still be available. According to the variation of the atmospheric density with the solar activity (cf. Sect. 6.2.1), the minimum altitude must be adapted to the solar cycle to ensure the margin of orbital decay. Figure 6.11 clearly shows that the altitude profile of ISS in its long-period oscillation follows the predicted solar activity ($F_{10.7}$) and the flight attitude is the highest at the point of maximum solar activity (for example in the year 2011).

The long-period cycle of the mean orbital altitude is superimposed by altitude variations with a period of 92 days. These variations come about by coast phases with subsequent reboost maneuvers. Figure 6.12 schematically shows this short-period cycle. After rendezvous with the supply vehicle at minimum operational altitude, the supplies are transferred into ISS and perhaps, new crew members go on duty while others go off duty. The subsequent reboost maneuver lifts ISS to maximum operational altitude. Two periods of 30 days duration for microgravity exper-

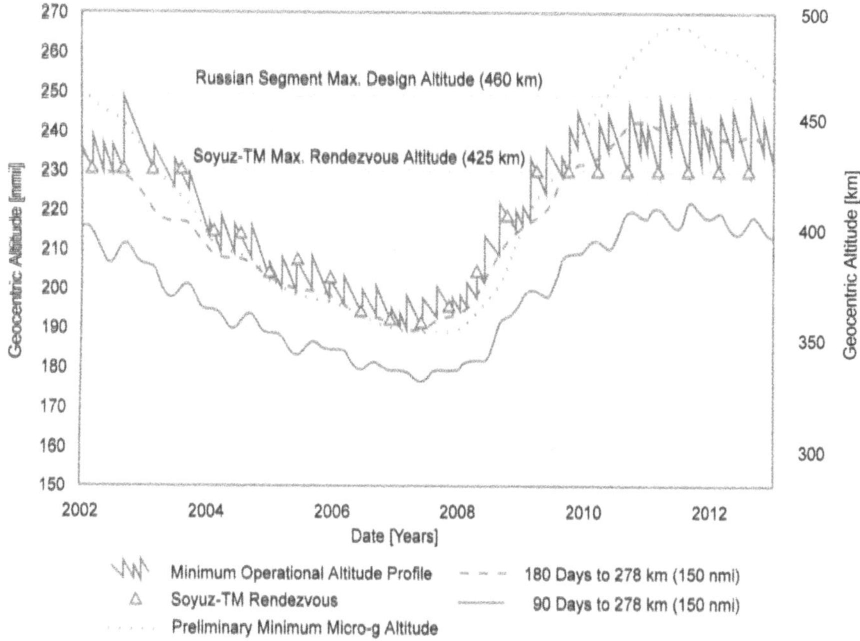

Fig. 6.11. Altitude Profile of the International Space Station [ESA 95]

iments are available during the subsequent coast phase of the International Space Station. Due to the already mentioned safety-related aspects, the maximum and minimum operational altitudes depend on solar activity. In addition to that, they are limited by operational boundary conditions: on one hand, the Russian modules have been designed for a maximum of 460 km altitude; thus, the upper limit for reboost maneuvers is fixed. On the other hand, the maximum rendezvous altitude for Soyuz is 425 km, hence limiting the maximum altitude at the end of the coast phase. Both factors together have the effect that, at times of maximum solar activity, the decay margin is reduced from 180 to 120 days. The altitude profile in Fig. 6.11 also shows that the station will not be flown with the maximum operational altitude until autumn 2002 despite the maximum of the solar radiation flux $F_{10.7}$. The reason is that, during the last phase of complete assembly, the maximum rendezvous altitude of Soyuz-TM should be reached as often as possible.

For conventional reboost, a thrust level of about 600 N and a specific impulse I_{sp} of about 280 s are typical values. Thrusters for reboost are located at the Service Module (SM) of the Russian Orbit Section (ROS) as well as at the docking transfer vehicle Progress-M or, later on, the European Automated Transfer Vehicle (ATV). An overview of ISS's reboost capacities is presented in Table 6.3.

The propellant used for the attitude and orbit control system is the propellant combination of NTO/UDMH. The tanks are pressurized with nitrogen and helium. Since the propellant tanks are connected with one another, propellants are available for both attitude and orbit control tasks.

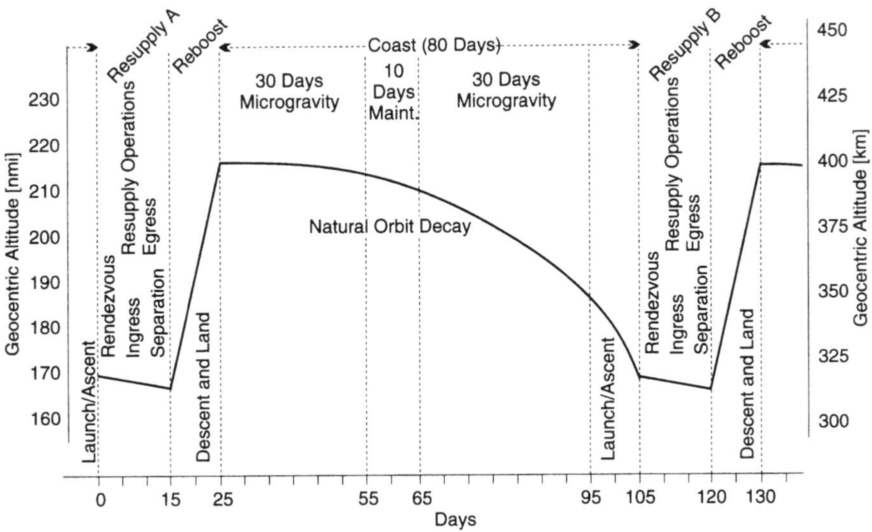

Fig. 6.12. Schematic of the ISS Altitude Short Period Cycles [SSP 41000D]

Table 6.3. Orbit Correction Capacities of ISS [DARA 95]

Reboost Module	Thrust [N]	Propellant Mass [kg]	Δv Capacity [m/s]
Service Module	2×600	390 (+5700 in FGB)	40
Progress-M (M2)	600	≈900±tbd.	≈6
ATV	2000	4000	25

The Δv-requirement, depending on the reboost maneuver, is between 5 and 19.6 m/s for a reboost from 425 km to either 434 km or to 460 km, respectively. This corresponds to a propellant consumption of 762 to 2994 kg per maneuver.

Synergistic Orbit Control. Synergistic propulsion concepts try to partially merge the propulsion system with other subsystems. From their interplay, additional advantages are meant to be obtained for the overall system level (for details cf. Chapter 10 "Synergisms"), such as improved operational flexibility, redundancy or reduced resupply needs. In this context, the life support system offers suitable starting points since large amounts of air and water are circulated throughout it. Often, waste from the ECLSS is simply dumped "overboard" or is disposed of by means of controlled re-entry in logistics capsules, since in principle, the available methods of processing (cf. Chapter 4 "Environmental Control and Life Support System") are only reluctantly applied due to high development costs and risks. With electric propulsion systems, these waste products could, for example, be used for orbit control. In the following, this shall be presented by taking the International Space Station as an example.

Also with ISS, significant mass flows will be vented "overboard". For instance, in the United States On-Orbit Segment (USOS), no processing of the carbon dioxide collected in molecular sieves will take place in the near future. Only the Russian Orbit Section (ROS) has a Sabatier reactor in the so-called "Life Science Module"; the reactor is planned to be temporarily operated. Normally, however, all CO_2 will be vented overboard. This means that during the phase of operation of ISS's complete assembly, a daily amount of 6 kg CO_2 will be available for propulsion purposes; that corresponds to a mass flow of 0.0694 g/s.

Another possible option for thrust generation is the use of water. According to the International Space Station's state of operation, water will be available e.g. in the US laboratory module that normally is dumped "overboard". The water results from either excess water reserves from the Space Shuttle's fuel cells or from excess condensate that could not be accommodated by condensate storage. Moreover, not yet purified solutions such as urine or hygiene water can be used for the generation of thrust. As a consequence, energy-consuming processing would be avoided and expendables as well as energy could be saved.

With the example of the International Space Station, Fig. 6.13 shows how synergistic propulsion concepts can be implemented under consideration of operational boundary conditions [Laible 95, Bertrand 96]. As an example, a point in time was chosen with maximum solar activity (maximum operational altitude) for the year 2011. In order to accelerate the waste substances, arcjet thrusters were suggested. A possible strategy is to delay the decay of ISS by permanent operation of the electric thruster. Figure 6.13 shows this strategy for CO_2-propellant. Such an arcjet motor would need about 6.2 kW of electrical power. As an alternative, the orbit altitude could be kept stable for a certain time (in our example, 56 days) by permanent reboost before ISS would have a coast decay to the operationally required rendezvous

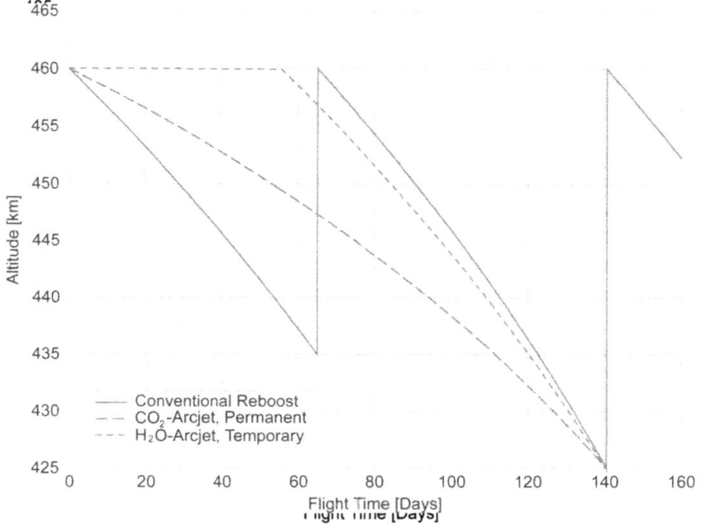

Fig. 6.13. Conventional and Synergistic Orbit Control Scenarios of ISS in the Year 2011

Table 6.4. Equivalent System Mass of a Reboost Cycle for the Orbit Control Strategies from Fig. 6.13

	Reboost	CO_2-Arcjet	H_2O-Arcjet
Consumed Propellant Mass	2110 kg NTO/UDMH[a]	840.0 kg CO_2 +84.0 kg H_2	895.0 kg H_2O +89.5 kg H_2
Required Electrical Power	–	6.20 kW (Permanent)	Ø 2.5 kW
Effective Resupply Mass	2127 kg NTO/UDMH[a]	–	765.0 kg H_2O +1.3 kg H_2
Fixed System Mass	–	1265 kg	385 kg
Equivalent System Mass	2127 kg	1265 kg	1152.3 kg

[a] for Acronyms See Glossary

altitude of the Soyuz system. In our case, steam was chosen as propellant for a 5 kW arcjet thruster. Compared to conventional orbit control (also seen in Fig. 6.13), both arcjet variants allow the avoidance of one reboost maneuver.

Table 6.4 shows the equivalent system masses of a logistical cycle for all three reboost concepts. The conventional strategy needs 2110 kg propellants for a reboost maneuver after each period of 62 days. This will be completely avoided when applying the synergistic option, but in turn, corresponding amounts of carbon dioxide or water have to be taken into account. Additionally, the arcjet thrusters need small amounts of hydrogen as an inert gas for the anode, which is also planned to be taken out of the life support system. Carbon dioxide as a waste substance will not occur in the logistical balance, whereas with the steam variant, 895 kg of water will be required. This will e.g. be covered by the daily amount of urine of about 11.8 kg if the urine collected during coast phases of previous cycles is also used. The average electrical power of approximately 270 W necessary for processing the urine is also available for thrust generation. Since, with the urine processing methods used in the Russian life support module and the US Habitation Module, the water recovery levels are at 80% and 91%, respectively, the effective resupply mass of water is reduced (due to the arcjets) to 765 kg.

Compared to the CO_2 option, significantly smaller, fixed system masses occur with the H_2O thrusters, since the water steam arcjet requires markedly less electrical power. When comparing both mass balances, the water steam arcjet comes out on top under the conditions given in this case. Typically, a break-even point can be determined in the case of such estimates. The scenario discussed here was limited to the operational constraints of ISS mentioned before. Nevertheless, already with the second synergistic reboost maneuver, the fixed system masses of the arcjet concept would be cost-effective for themselves, so the breakeven-point would already have been exceeded.

Except for the mass advantages, there is series of advantages to be found on the overall system level, such as:

- *Operational Flexibility and Safety*. The arcjet system available in addition to the hydrazine system provides a parallel redundancy. Moreover, an obvious safety advantage results from the possibility of using water from the ECLSS for propulsion purposes: If, due to problems with the launcher systems, propellant resupply was interrupted, hygiene water, e.g., could be used for the vital reboost of the space station.
- *Improved Utilization Conditions*. The presented orbit control strategies featuring a delayed decay and temporary maintenance of altitude promise an improvement of the quasi-stationary microgravity level since the atmospheric drag will be partially or entirely compensated by the arcjet thrusters. In both cases, the duration of microgravity periods will be prolonged by about 20% [Bertrand 96] since the period of time required for the additional reboost maneuver in the case of the conventional strategy can now be used for experiments.
- *Simplification of Logistics and Storage*. Apart from the reduction of resupply mass flows, resupply of water requires less effort than transport and storage of the toxic hydrazine. Moreover, as already mentioned, resupply masses (e.g. multifiltration beds or filters) can be reduced in the water processing unit by using waste water for propulsion purposes; in this case, "new" water must be introduced into the water cycle.

This example shows how the coupling of orbit control and life support systems can result in notable advantages on the overall system level. Further advantages can be achieved by integrating additional subsystem functions, i.e. the introduction of further subsystem couplings, for instance with the power system. Details on the subject will described in Chapter 10 ("Synergisms").

Experimental and theoretical investigations on the mixing of reacting liquids in a thermal arc-heated plasma generator have shown recently that a water steam plasma generator can be run efficiently [Caesar 98, Glocker 98]. A small plant for the destruction of halogenated hydrocarbons using this steam plasma technology is currently tested in industry showing that the injected liquids or gases (mainly CFC's) can be 100% disposed. Figure 6.14 gives an indication of this promising thruster technology.

6.4 Propulsion System Technology

On the space station side, there are various possibilities in the form of actuators for compensation of the individual attitude and orbit perturbations. Perturbation torques resulting from gravity gradients and drag, for example, are partially cyclic and partially accumulating. In order to compensate the cyclic part, the momentum storage devices known from satellite technology will be used. Aboard space stations, they have been designed in the form of the so-called "Control Moment Gyros" (CMGs). Secular momentums are, firstly, also stored by these systems, but in the end, they have to be removed by cyclic desaturation via the thrusters. The forces necessary for

Fig. 6.14. Water Steam Plasma Thruster in Operation at Stuttgart University (P = 8 kW, $\dot{m}$ (H_2O) = 0.2 g/s, $\dot{m}$ (Ar) = 0.1 g/s, estimated I_{sp} = 450 s)

orbit control usually must be produced by thrusters. This section will also include a short description of unconventional propulsion components such as tethers.

6.4.1 Thrusters

According to their application, two classes of thrusters can be found aboard a space station, which mainly differ in their thrust levels as follows:

- High thrust, $F \approx 100 \ldots 1000\ N$ (impulsive orbit transfers, collision avoidance maneuvers)
- Low thrust, $F < 10\ N$ (orientation, balancing structure movements, orbit correction with constant thrust)

A choice of possible propellant combinations for different thrusters is listed in Table 6.5. Basically, propellants with a high specific impulse are preferred since they create the same thrust but need a lower propellant mass flow. When selecting a propellant, also transport, storage, distribution and processing as well as technological maturity and availability of the systems for thrust generation and handling of propellants must be taken into account. In the case of electric thrusters, additionally, the availability of electrical power as well as reliability and the attainable life time must be taken into consideration.

In this list, hydrazine thrusters (N_2H_4, MMH, UDMH) are representatives of the traditional technology; they can be operated as cold-gas thrusters or catalytically. Initially, a hydrazine system was also chosen for the Space Station Freedom (SSF) because the necessary technologies were available, but also because satellite servicing was planned to be carried out.

In the course of system integration, minimization of the use of the Space Shuttle, and the cancellation of servicing tasks within the spectrum of a mission, the concept was converted to a system of GH_2/GO_2 thrusters and resistojets. Contrary to the hydrazine system, the use of hydrogen and oxygen (obtained from water by electrolysis) became possible: the necessary water should either be taken from the ECLSS,

Table 6.5. Propellant Combination for Space Station Thrusters [Woodock 86, Laible 95]

System Type	Propellant	I_{sp} [s]
Cold Gas	H_2	228
	N_2	60
	CO_2	49
Monopropellant	N_2H_4	230
Resistojet	N_2H_4	280
	H_2	800
	NH_3	280
	CO_2	120
Arcjet	N_2H_4	450
	H_2	1200
	NH_3	450
	H_2O (90%), H_2 (10%)	450
Bipropellant	N_2O_4/MMH[a]	285
	N_2O_4/UDMH[a]	286
	H_2/O_2	280
	LH_2/LOX[a]	400

[a] for Acronyms See Glossary

or it should be produced by the Space Shuttle's fuel cells whenever the Shuttle docked at SSF.

Resistojets can be operated using various waste gases from the station. During the development of SSF, an engineering model of a resistojet was tested with H_2, CH_4, CO_2, NH_3 and steam: at heating temperatures of 500°C to 1400°C, specific impulses of 100–320 s were achieved (thrust at 0.05°N to 0.08 N).

In the course of the SSF studies [Jones 87], this combined system of electrolysis and resistojets was also identified as the most cost-effective (life-cycle costs of $214 million (US) compared to $1078 million (US) for the classical hydrazine system). Nevertheless, in the course of the redefinition of the station in 1989, again a hydrazine system was opted for in order to avoid development risks and to reduce development costs. This reduction, however, was carried out at the expense of operational costs, which, as a consequence, increased. For the same reason it is merely a conventional system that will be operated aboard the International Space Station [Foley 96].

Looking at the options listed in Table 6.5, cold-gas thrusters can also be well integrated into the overall system, whereas other bipropellant systems in the form of H_2/O_2 markedly increase system complexity. Ion and Stationary Plasma Thrusters (SPT) are less interesting for a space station because of their high power require-

Table 6.6. Thrusters in the Russian Orbit Segment of ISS [ISS 95]

	Service Module		**SPP ATF**[a]	**Progress-M**		**Progress-M2**	
Engine Design Specification	Reboost Engine	Attitude Control Thrusters		Reboost Engine	Attitude Control Thrusters	Reboost Engine	Attitude Control Thrusters
Thrust [N]	300 ± tbd	13.3 ± tbd	10 ± tbd	300 ± tbd	13.3 ± 3	300 ± tbd	13.3 ± tbd
Quantity	2	32	18	1	28	1	36
Steady State Specific Impulse [s]	300 ± tbd	250 ± tbd	> 290	300 ± tbd	250 ± tbd	300 ± tbd	280–285
Max. Number of Firings (Life)	200	450000	150000	30	400000	30	400000
Total On-Time Limit (Life) [s]	25000 ± tbd	45000	up to 20000	880	5000	1600	200000
Total Propellant Throughput (Life) [kg]	25000 ± tbd	2500	700 ± tbd	880	350	1600	600–800
Engine On-Times [s]	10–400	0.03–600	0.03–1500	0.5–300	0.03–600	0.5–600	0.03–2000
Gimble Angle Range [Deg]	± 5 in 2 Axes			± 5 in 2 Axes		± 5 in 2 Axes	

[a] for Acronyms See Glossary

ments and comparatively low thrust. Arcjets, on the other hand, offer a comparatively high specific impulse with acceptable power requirements. In the field of satellite technology, arcjets also have a solid reputation for tasks such as station keeping, etc. [Anselmo 96, Messerschmid 96]. Their application aboard space stations may well be thought of for the compensation of atmospheric drag (as it was presented in the example of synergistic orbit control in Sect. 6.3.2), since an arcjet thruster allows a certain variation in the thrust level.

In Table 6.6, the thrusters of the Russian Orbit Segment of ISS are listed. Apart from the reboost thrusters with 300 N thrust, thrusters of the 10 N class are used for attitude control.

Electrodynamic Tethers. One still unconventional method of propulsion is the use of electrodynamic tethers. An electrodynamic tether is a thin, electrically conducting "cable" connecting two orbital masses with each other. Since the whole system of the two connected masses moves according to the orbit of the joint center of mass, in the lower mass the gravitational force is dominant, whereas in the upper mass, the centrifugal force, respectively. As a consequence, the tether will be stabilized towards the Earth in radial orientation (cf. Fig. 5.14).

To a first approximation, the thus oriented tether on its orbit moves perpendicular to the Earth's magnetic field, i.e. the tether lying in the orbit plane is perpendicular to the local direction of the magnetic field vector $\vec{B}$. According to Faraday's law, an electric field is induced in the tether element of a length l, causing a difference in the electric potential and thus a voltage of:

$$U_i = (\vec{v} \times \vec{B}) \cdot \vec{l} \tag{6.21}$$

If the conductor is insulated, and if the ends of the tether are brought to the electrical potential of the surrounding plasma in LEO using appropriate contactors, an electron flow driven by U_i will be able to proceed through the tether. Electrons from the ionosphere are transported from the tether's upper end (anode) to the lower end (cathode). According to the definition of the Lorentz law, for the current I the following force is applied to the tethered system:

$$\vec{F} = (I \cdot \vec{l} \times \vec{B}) \tag{6.22}$$

This force, to a first approximation, is opposite to the orbital velocity vector. As a consequence, the vehicle (as one of the end masses) decays at the expense of the electrical energy produced by the tether.

If a larger opposite voltage U_G can be put up against the voltage U_i (e.g. by means of a power source aboard the vehicle), the electrical current flow in the tether is reversed: at this point, the cathode is located at the upper end of the tether and the anode at the lower end. Correspondingly, the Lorentz force also changes its direction and now acts in flight direction. Thus, by generating electrical power inside the vehicle, thrust can be created, for example, in order to compensate the drag force (cf. Fig. 5.14) [Pohlemann 92].

In line with this idea, it would also be possible to deliver more thrust on the Sun side of the orbit than is necessary for drag compensation. During eclipse, in turn, the additionally obtained altitude could be re-transformed into electrical energy by using the tether as a generator. However, one fundamental problem of such an energy-storing mission might be a steadily increasing eccentricity due to the resonant excitation of the orbital altitude.

From the technological point of view, some questions with respect to the use of tethers remain unanswered that occurred during the Shuttle-supported test missions TSS-1 (in August 1992) and TSS-1R (in February 1996). Apart from mechanical problems with the deploy mechanism, mainly the knowledge of interactions between the ionosphere and the tether, or, to be exact, between the ionosphere and the plasma contacts, is insufficient. For example, the induced current measured during the mission TSS-1R exceeded the predicted values by a factor of three and the tether tore due to an arc discharge shortly before the deployment was completed.

As for operational utilization aboard space stations, it must be kept in mind that the station's center of gravity can be significantly shifted in the case that the tether is deployed on one side only. If, for instance, the same type of tether used during the TSS experiments (length 22 km, end mass about 518 kg) was installed on the International Space Station, the station's center of gravity would be radially shifted by about 24 m. As a consequence, the quasi-stationary microgravity level of ISS would be increased by about 10^{-5} g. How to control the occurring tether oscillations and how to protect the tether system against collisions with debris has to be still answered.

Apart from a tether's electrodynamic application in thrust generation, the purely mechanical application is also of importance. There is a possibility of achieving gravity gradient stabilization for a space station in an Earth-oriented flight mode by means of two radial tethers. Consequently, the conventional attitude control system would receive some support. Such a "double tether" could, moreover, be used to control the station's center of mass or to adapt it to variations in the station configurations. As illustrated in Fig. 12.6, the de-orbiting of payloads could also be effected by means of such tethers [Burkhardt 96]. A tether of the same length as the TSS tether, deployed in a controlled way and cut while swinging back through the local vertical, could, for example, lower the perigee of a return capsule from 400 km to about 100 km, i.e. to a favorable altitude for the reentry corridor, without propellant consumption. Increasing the tether length would provide a steeper reentry.

6.4.2 Generation of Control Torques

Control Moment Gyros. The most important actuator of attitude control of a space station is the "Control Moment Gyro" (CMG), a gyro mounted on gimbal. The space laboratory Skylab was equipped with three CMGs for attitude control, whose dimensions and arrangement in the three spatial axes are shown in Fig. 6.15.

When applying a torque about one of the cardanic CMG-axes with a servo motor, a precession movement perpendicular to the motor torque is initiated. This movement aims at positioning the rotor's momentum axis in parallel to the axis of the applied torque. As a consequence, it is possible to "induce" a torque for the compensation of perturbations with the CMG either until the precession movement reaches the CMG's constructural limits, or until the momentum axes of several CMGs are oriented parallel to one another. For desaturation, i.e. repositioning of the momentum axis, opposite torques must be generated. This can be achieved by using other torque generating actuators, for example thrusters or magnetic torquers. Adequate flight strategies, however, can also have the effect that after an orbital revolution, no precession deflection will remain. For example, during orbital eclipse, the flight attitude of the US Skylab was changed in such a fashion that during the Sun period of the orbits, the accumulated angular momentum could be brought back to

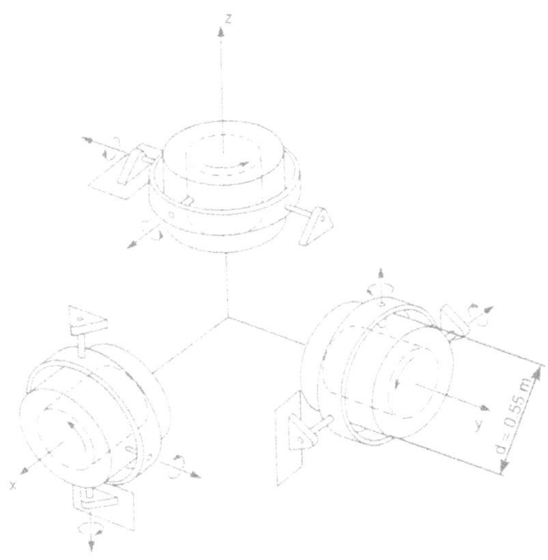

Fig. 6.15. Skylab Control Moment Gyro System [Belew 73]

zero. Which one of these methods is applied depends, in particular, on the specific mission. For example, the Skylab method, however, may need low amounts of propellants, but over a long period of time, it requires a change of attitude compatible to the mission.

Contrary to a momentum wheel, whose momentum axis is fixed relative to the vehicle, generating control moments by the changes in the angular frequency, a CMG has the advantage that torques can be produced about two axes. When mounting three CMG units, it is possible to produce a torque about each of the three spatial axes alternatively by two different CMGs. Consequently, a system redundancy is achieved.

Example: If a Skylab-CMG were used to compensate a perturbation torque about the pitch axis of ISS, a control torque of M_{pert}=40.23 Nm would have to be produced, cf. Eq. 6.19. If the axes were positioned perpendicular to one another, the gyro moment of the CMG-rotor would be given by its moment of inertia Θ, the angular frequency ω and the precession movement ω_p:

$$M_G = \Theta \cdot \omega_p \cdot \omega \tag{6.23}$$

The Skylab-CMG has a mass of 181 kg rotating at an angular frequency of 9000/ min. A cylindrical rotor of 0.55 m diameter would result in a moment of inertia with:

$$\Theta = \frac{1}{2} \cdot m \cdot r^2 = 6.85\ kg \cdot m^2 \tag{6.24}$$

With the perturbation torque of 40.23 Nm, the precession rate would be:

$$\omega_p = \frac{M_{pert}}{\Theta \cdot \omega} = 6.23 \cdot 10^{-3}\ \frac{rad}{s} \tag{6.25}$$

Assuming constructural limits of $\varphi_{max} = 30°$, the CMG has reached its maximum allowed deflection after $t_{max} = \varphi_{max}/\omega_p = 84\ s \approx 1.4\,\text{min}$. Considered a “black box”, the CMG would reach a storage capacity of $\Delta D = M_{y\text{effective}} \cdot t_{max} = 3380 Nm \cdot s$.

Also Russia makes use of CMGs aboard the Mir station (where they are called “Gyrodyne”). They are partially located inside the pressurized part of the station and partially outside the Kvant 2 module. They were gradually installed as single components, corresponding to the station’s control moment requirements.

ISS will include Gyrodynes at the Science Power Platform 1 in the Russian Orbit Section as well as four CMGs at the Z1 truss segment of the United States Section (cf. Fig. 6.16). Each of the CMGs in the US segment has a momentum storage capacity of 4742 Nms at a constant rotation rate of 6600/ min.

Magnetic Torquers. This is an elegant method for the generation of torques for attitude control: it makes use of the interaction of a magnetic torquer with the Earth’s magnetic field. A coil with the magnetic moment $\vec{m}_{mag}$ in a magnetic field $\vec{B}$ generates the torque:

$$M = \vec{m}_{mag} \times \vec{B} \tag{6.26}$$

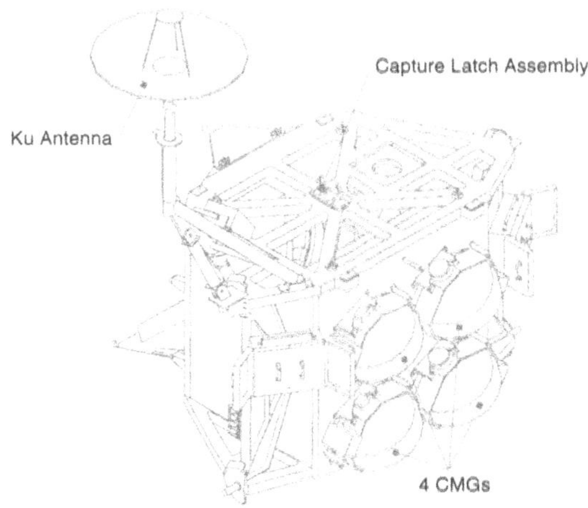

Fig. 6.16. Arrangement of CMGs in the Z1 Truss Segment of the International Space Station

The magnetic moment $\vec{m}_{mag}$ of a coil with the circular cross section is obtained according to the equation:

$$\vec{m}_{mag} = \frac{\pi}{4} \cdot n \cdot I \cdot d_c^2 \tag{6.27}$$

Here, n is the number of wire turns, I stands for the current flowing through the coil and d_c for the coil diameter. If the coil has an iron core, the magnetic moment additionally multiplies with the magnetic permeability

$$\vec{m}^*_{mag} = \mu_r \vec{m}_{mag} = \mu_r \cdot \frac{\pi}{4} \cdot n \cdot I \cdot d_c^2 \tag{6.28}$$

which, for example in the case of cast iron, can reach values of up to about $\mu_r = 254$. The first candidate for the material for the coil wire is aluminum, since the relation between the electrical conductivity (essential for the resistive losses) and the density (decisive factor for the system mass) is very good.

Until the present time, magnetic torquers have been used for desaturation of the momentum wheels onboard satellites: usually, coils with an iron core are applied which, due to their rod-shaped appearance, can easily be integrated into the satellite structure. For example, Table 6.7 shows the characteristic data of magnetic torquers which have actually been constructed:

Table 6.7. Characteristic Data of Magnetic Torquers [Hitachi 92]

Magnetic Momentum [Am^2]	Diameter × Length [mm]	Mass [kg]	Electrical Power [W]
30	20 × 500	0.9	0.5
60	25 × 640	1.4	0.6
300	50 × 900	4.9	6.6

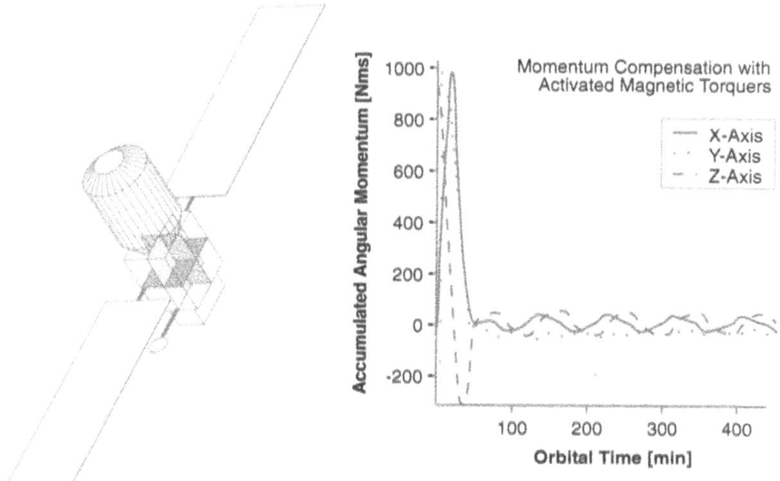

Fig. 6.17. Integration of Magnetic Torquers into a Model Space Station (CFFL)

Of course, the achievable torque depends on the value and direction of the local vector of the Earth's magnetic field (cf. Chapter 3 "Orbital Environment"). The magnetic flux density for an orbital altitude of 400 km, for example, varies between 25631 nT at the magnetic equator and 46925 nT near the magnetic poles. Consequently, the maximum possible torque of a magnetic torquer (with a magnetic moment of 60 Am^2) in this altitude range varies between 1.41 and $2.82 \cdot 10^{-3}$ Nm.

If control torques are to be generated about a determined vehicle axis, usually the magnetic torquers must be installed in all three spatial directions. Additionally, the station's computer must have a model of the Earth's magnetic field so it can, depending on the position of the vehicle in the Earth's dipole field and its orientation in space, control the correct coils.

Magnetic torquers have not been used aboard space stations, yet. Nevertheless, this method seems to be promising: especially for desaturation of momentum storage devices, with the large dimensions of e.g. a pressurized module or supply module, large cross sections for the realization of coils without iron core are available. A study of a model space station similar to the Columbus Free Flying Laboratory (CFFL) [Hans 95] showed the following: At equivalent electrical power requirements and an equivalent magnetic moment, the mass of a coil that is "wrapped" around a supply module is about 9% smaller than the mass of a coil with an iron core. At a maximum magnetic moment of 12104 Am^2, a total of three coils with 1 kW electrical power and a mass of 15 kg each is sufficient to reliably and completely reduce the accumulated angular momentum of the Control Moment Gyros within one orbital revolution. Figure 6.17 illustrates the positions where the coils were to be installed in the supply module and gives an example of the reduction of the angular momentum with activated magnetic torquers.

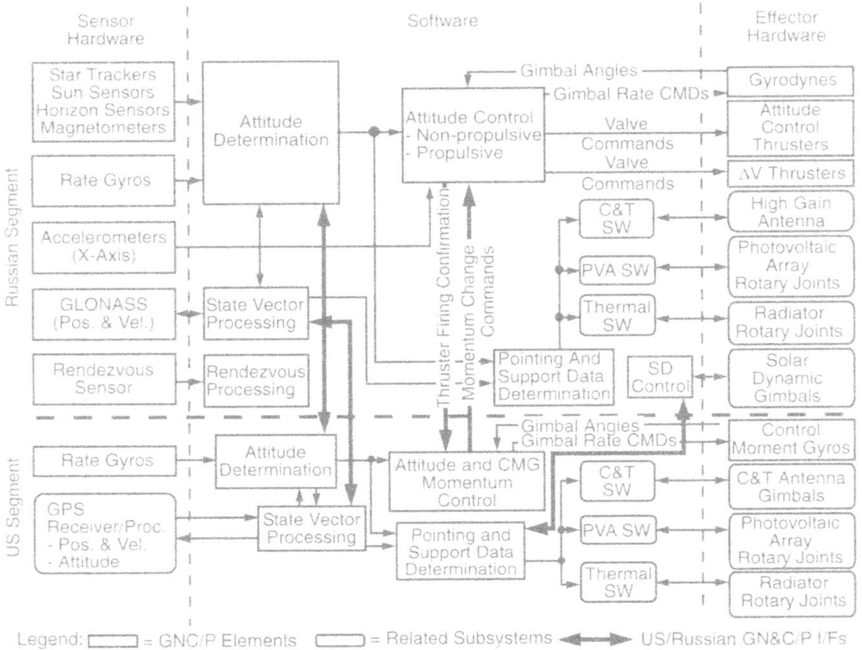

Fig. 6.18. Block Diagram of the Attitude and Orbit Control System of the ISS [ISS 95]

Compared to a conventional hydrazine propulsion system, the break-even point according to the criterion of the equivalent system mass for the chosen CFFL scenario would already be reached after 2.33 years.

6.4.3 Sensors

All components presented for thrust and control torque generation are, as actuators, the active elements of the attitude and orbit control system. The complete control loop, however, also contains, apart from the necessary control units, different sensors that provide information on the state of the system, in this case, on orbit and orientation in space. Generally, the technology used for space stations does not differ considerably from that applied in satellite systems with one significant exception: the performance data with respect to e.g. the orientation accuracy obtained for a space station seems moderate compared to those required for a telecommunications satellite. For example, the specific orientation accuracy in the Torque Equilibrium Attitude is maximum ±3.5° (±5° in the LVLH flight mode), compared to ±0.1° for a typical telecommunications satellite. For that reason, payloads with high requirements for their orientation must be provided with special steering platforms.

Figure 6.18 schematically shows the sensors, control units and actuators of the attitude and orbit control system of the International Space Station. On the sensor side, there are conventional gyro platforms, Earth and Sun sensors and star trackers, but also modern navigation systems such as GPS and GLONASS. The corresponding software is contained in the GNC Data Management System shown in Fig. 12.8. Within the framework of this chapter, the subject of sensor technology will not be treated in greater detail.

6.5 Overall System

When integrating the attitude and orbit control system into the overall system "space station", the systems engineer is faced with some demanding problems, namely:

- *Configuration Management.* Unlike satellite systems, an orbiting structure like the International Space Station has several configurations during its assembly phase. At any time during the assembly sequence, it must be assured that orbit and orientation can reliably be controlled within the specified limits. Apart from the logistics necessary for propellants and spare parts, this means that the configuration of every assembly phase must meet requirements such as mass distribution (gravity gradient) and minimization of external perturbations (atmospheric pressure incidence areas) to make efficient control possible. Hence, this requirement is valid for the design of both: assembly and operation of ISS. Moreover, a sufficient number of sensors and actuators such as thrusters, CMGs, coils, etc. must be available from the very beginning.
- *Safety and Reliability.* The safety and reliability of the components of the attitude and orbit control system must be guaranteed in any case. Especially on a crewed platform where failures that may lead to safety risks such as explosions or contamination of the pressurized modules due to emerging propellants must be excluded. Also in the case of a malfunction or failure of individual components, vital functions like keeping a minimum altitude or the orientation of the solar arrays for emergency power supply must be guaranteed at least in the sense of a "graceful degradation". Figure 6.19 shows how safety requirements influenced the layout of the attitude control system of Skylab's cold-gas system:
  - The propellant is stored in several small tanks outside the pressurized volume.
  - The propulsion system itself features two entirely independent thruster modules (Thruster Module 1 and 2).
  - Each thruster, in turn, is connected to the tank system by valves connected in series/parallel.

  Hence, valve malfunction in both open and closed switching conditions can be accepted with a single redundancy.
- *Impairment of Utilization.* Thrusters may contaminate payloads and subsystems outside the modules such as sensors, optical instruments or material samples, or impair their functioning, cf. Sect. 3.9. Also, utilization aspects have to be taken into account while operating the station. For example, the attitude and orbit control system must not create overly large perturbation accelerations during periods of microgravity. Consequently, also the influence of the chosen reboost strategy must be taken into account, and as it was already indicated in Sect. 6.3, a constant reboost does have operational advantages in this context.

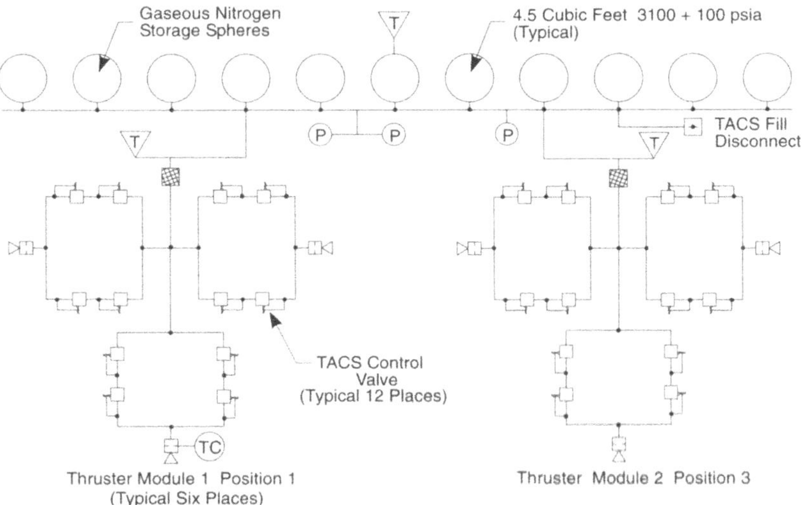

Fig. 6.19. Skylab Thruster Attitude Control System (TACS) [Belew 73]

- *Positioning of Thrusters*. When integrating the thrusters for torque generation into a large structure such as ISS, it is advisable to install them with the largest possible distance from the center of mass either on the main truss or even onto beams standing perpendicular to it. This was the case with the aforementioned Autonomous Thruster Facility (ATF), which, as an integrated unit of thrusters, tanks and control valves, is installed at the Science Power Platform (cf. Fig. 6.10). Orbit control thrusters, on the other hand, must be installed such that their line of force lies in the velocity direction and also through the station's center of mass.
- *Structure Oscillations*. Due to their size, space stations, especially in low frequency ranges, are susceptible to structure oscillations. This has to be considered when mounting the actuators and also when defining the mission procedures. Careful examination must be carried out of whether or not the attitude and orbit control maneuvers or the thrusters' exhaust gas streams have effects on filigreed structures such as solar arrays. This not only applies to the propulsion system of the station itself, but also to supply vehicles that will dock to the station.
- *Repair and Maintenance*. In the final count, we must not forget that the station's operational life will markedly exceed that of its actuators. For that reason, special precautions must be taken for servicing and exchange of sensors, actuators and control elements. An expensive, but nevertheless very elegant, solution is an exchange module composed of the above mentioned integrated units with several thrusters and tanks.

All points discussed in this chapter make clear that a system-wide point of view is necessary, as it was pointed out to some extent in the examples on synergistic propulsion concepts. A detailed view of factors influencing design and operation will be offered in Chapter 9 ("System Design") and Chapter 10 ("Synergisms").

7 Utilization

Space stations are multi-purpose facilities in an Earth orbit that, similar to laboratories for nuclear research or elementary particle research, are large and complex experiment facilities. Assembly and operation take place mostly through international cooperation. The time available for experiments and other resources is usually divided among the numerous users according to criteria such as scientific merit and special boundary conditions which are either technological, user-oriented or political in nature. Another typical characteristic is the fact that there are operators or operational organizations of such a large experiment facility on one hand, and its users on the other. The latter usually have no (or if so, a very limited) direct access to experiment operation.

A space station offers access to the special environmental conditions in near-Earth space to the multiple user disciplines, that space being characterized by terms like weightlessness, vacuum, space radiation and specific observation opportunities. The International Space Station is the current climax of previous development and will combine interdisciplinary and international cooperation in an extremely interesting multicultural and extraterrestrial environment, as it has never been the case before.

This chapter will introduce the development of space laboratory utilization, the environmental conditions, and the different disciplines. Further sections will address the individual utilization possibilities with the help of examples. The structure, contents and objectives of the individual user disciplines were taken from [ESA 95, NASA NRP 98]. The experiments described originate mainly from the detailed Space Shuttle/Spacelab/Mir investigations and documentation. General questions concerning access to and the operation of a space station will be analyzed in Chapter 12 "Logistics, Communications and Operations"; and Chapter 13 "The International Space Station" will outline examples of experiment facilities and the existing opportunities for user support of ISS.

7.1 Environmental Conditions and User Disciplines

The history of the utilization of near-Earth space already spans several decades, but it is only in this decade or even the next that a broadening of the individual research areas will begin. A space station, supported by humans, makes the following activities simultaneously possible:

- Scientific research and practical application in the field of physics, chemistry, astronomy, biology, geo sciences, etc.

- Investigation and testing of new technologies in the field of engineering sciences, of devices and operational procedures for use in future space systems
- It serves as a starting point for further space exploration and use beyond near-Earth orbits.

A space station as a scientifically operated research and development "institute" offers laboratory conditions and numerous observational opportunities both in pressurized modules and at externally mounted experiment and observation platforms. The following conditions are either not present on Earth at all, or they are not available at the same level of quality over a long period of time:

- Weightlessness
- Vacuum
- Space radiation (particles, electromagnetic and/or thermal radiation), Earth's magnetic field
- Observation of space undisturbed by the atmosphere
- Permanent view of the Earth and its atmosphere

7.1.1 Weightlessness and Microgravity

Thanks to nearly perfect weightlessness (microgravity), a space station offers an opportunity to create laboratory conditions in space as they cannot be achieved on Earth.

Research performed under microgravity conditions is of interest to a number of disciplines that are usually grouped under the term "microgravity research or µg research". In fact, microgravity is not a discipline in its own right; it is merely a kind of tool used by several disciplines in order to investigate certain phenomena and processes. Among the disciplines concerned, there is not necessarily a connection. Often, the only common denominator is that they have an interest in using the same laboratory conditions.

Utilization of the microgravity environment is of interest especially with respect to the following disciplines: life sciences, physical sciences, space engineering and technology research as well as – to a certain degree – space sciences. The scientists' interest in these disciplines stems from the following motivation:

- To unveil and discover phenomena, relations and processes that are unknown or on which there is insufficient knowledge
- To try not only to separate the various factors which are supposed to be of influence, but also to observe the effect of each individual factor in an isolated sense in the laboratory
- To fully understand a given process, and beyond the knowledge of a single, selected process with determined, fixed parameters, where a full, theoretical knowledge covering a wide parameter range of all relations influencing a process is necessary.

Gravity, the weakest of the four physical fundamental forces, has a relatively small effect, unless the forces of the strong, weak or electromagnetic interactions that are many times larger (in relation to the unit mass, e.g., of a proton) are not effective or,

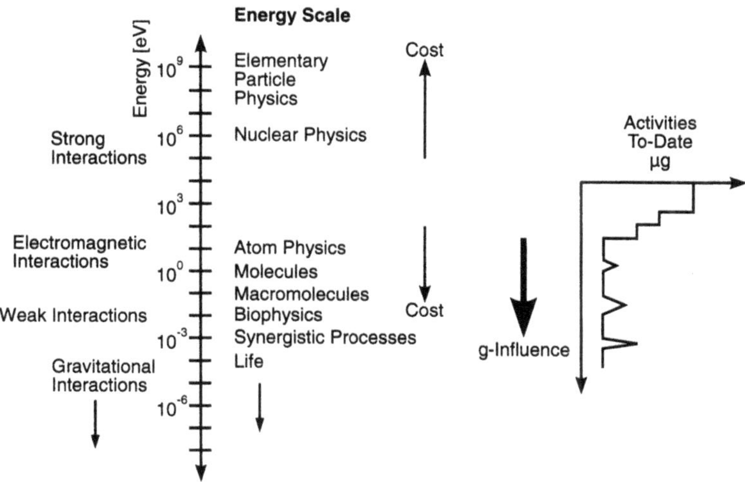

Fig. 7.1. Energy Scale of Interactions from the strong forces to the electromagnetic and weak to the weakest force, i.e. gravity. Some processes can only be examined in greater detail when gravity is not present.

by neutralization, only slightly effective. Despite its visible effect on celestial mechanics or orbit mechanics, gravity hardly influences the physics of the elementary particles, nuclei, atoms or molecules. At much lower interaction energies far below 1 eV (electron volt), however, its influence on physical, chemical and biological phenomena, relations and processes becomes visible, as it is indicated in Fig. 7.1.

Despite all this, on Earth, gravity has such a strong effect that it overshadows other effects or at least prevents them from being perceived. The main effects caused by gravity on Earth are convection (gravity-driven flow), sedimentation, and the force created by a mass in the gravitational field (i.e. weight, cf. Fig. 7.2). The elimination of these effects in a space laboratory allows the experimental investigation of second order physical effects, which are normally hidden due to gravity's dominance on Earth. The knowledge of the influence of these effects allows an improvement of the present theoretical and numerical models that describe a physical, chemical or biological process as a whole – i.e. not only under conditions of weightlessness in space, but also under the conditions present on Earth.

Additionally, the microgravity environment aboard a space station allows not only the passive observation of the course of biological, chemical or physical processes, but also the active influence of these processes in a way which is impossible on Earth. That opens up new prospects for the production of alloys and composites, large crystals, complex proteins and other products that, on Earth, are difficult to obtain or can only be obtained at a lower quality.

The disciplines interested in the microgravity environment benefit mainly from the numerous research facilities aboard a station and from large resources of energy. The latter is especially necessary for maintaining pre-defined stable thermal conditions for test items and for providing electrical power to furnaces and other experiment facilities. The disciplines also profit from the presence of a crew whose help

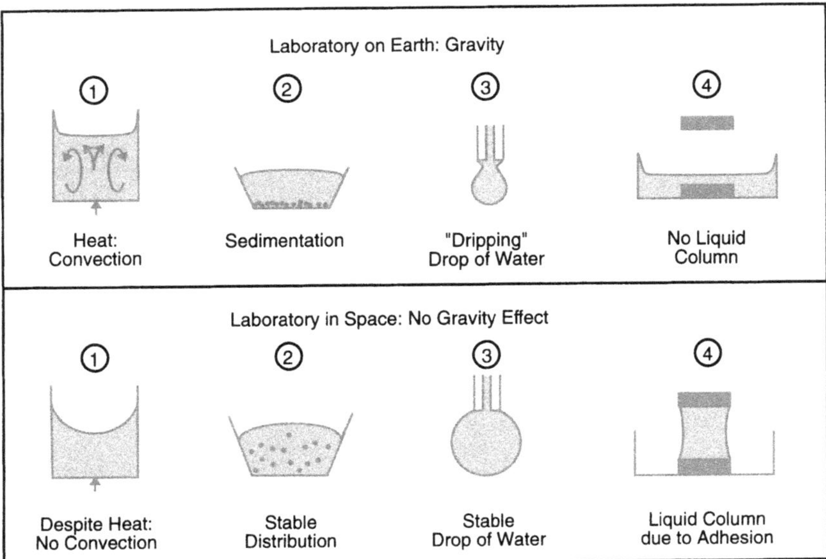

Fig. 7.2. The Effects of the Gravity Force on Liquids

makes it possible to reconfigure experiments and to react directly to expected or unexpected intermediate results. Additionally, new equipment and samples can be brought to the station, processed aboard, and end products can be returned to Earth rapidly and safely.

Finally, a space station is a well-suited base for intensive studies of how the space environment influences biological systems and human physiology.

Investigations in space on the influence of gravity on all kinds of lifeforms – from cells and molecules to whole organisms – have already yielded some surprises and brought some common theories on basic biological mechanisms and processes into question. Thanks to the onboard crew, small plants and animals can be bred, observed and tested. Additionally, the astronauts themselves are test subjects for studies in the field of medicine and human physiology. In this context, the investigation and better understanding of the basics of human physiology are of importance, but also the determination of the influence on and limitations for the astronauts due to their stay in space.

For thousands of years, physical phenomena have been observed, and theoretical ideas on the influencing causes have been developed. For the past 400 years, scientific facts have been collected and throughout all of known history, the influence of gravity has been omnipresent. Only during the last 40 years, have we been able to observe these phenomena for longer than a few seconds without the effects of gravity, and it was only a few years ago that, for the first time, a coordinated series of scientific experiments was conducted in space.

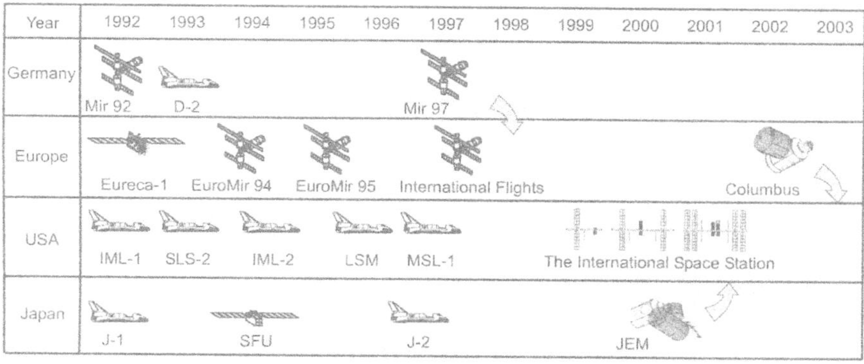

Fig. 7.3. Past and Future Flight Opportunities in an International Environment

Research in weightlessness is still in its infancy. This becomes clear when taking into account that over half of all microgravity experiments were not conducted until the past four years. As a consequence, it is only now that reliable statements concerning potential and future prospects of this field of experimentation are possible. Whereas the first flight opportunities carried discipline-specific experiments and offered only a limited means of utilization capacity, the large systems, especially of the past decade (Spacelab, EURECA and the Mir space station), were complex and multidisciplinary experiment carriers (cf. Fig. 7.3). In the field of materials science and physics, as well as in life sciences, the opportunities resulting from this development are being used by thousands of research groups all over the world. Thus, at German universities and other research laboratories alone, about 100 research groups are continuously preparing or conducting microgravity experiments [DARA 95].

7.1.2 Vacuum

The natural vacuum of space offers nearly unlimited experimental possibilities: high pump speed and capacity, no back flow. Additionally, this environment offers outstanding opportunities for observation of the behavior of gaseous, especially ionized matter and electromagnetic radiation as well as their interaction with one another, but also with solids or with the electromagnetic or gravity fields present. Especially the disciplines of plasma physics, astrophysics, radiation physics, and magnetospheric physics benefit from it.

The technical disciplines, however, can also make good use of the vacuum. First, the vacuum conditions around a space station can be used for the *in situ* testing of components and processes for their use in other space systems under similar or equivalent conditions. Examples are: investigations concerning electromagnetic compatibility, the outgassing of solids and liquids, heat and mass transfer processes, etc. Secondly, pump systems for the laboratories can be saved, if the vacuum in the vicinity of the station is of sufficient quality. Evacuation of furnaces, for example,

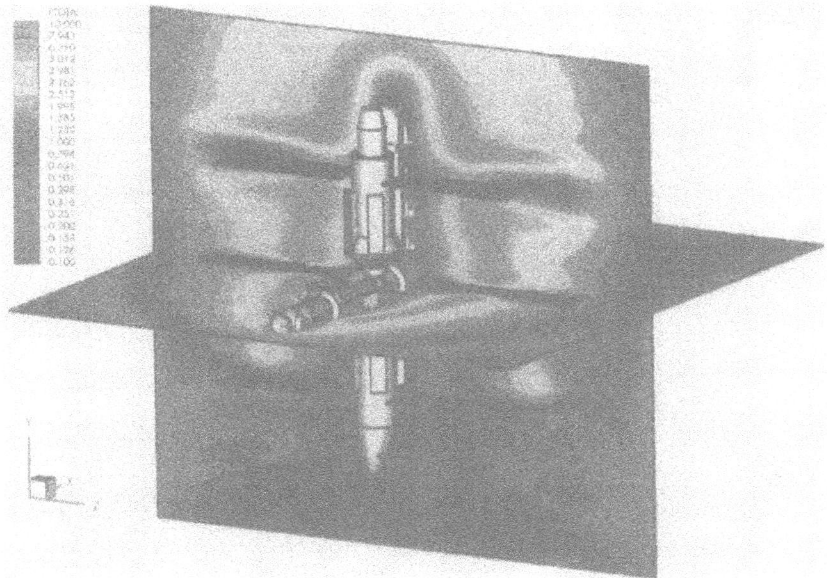

Fig. 7.4. Increase in pressure due to an attitude control thruster in the vicinity of the space station Mir (F3U thruster of the approaching space shuttle at a distance of 10 m in x-direction) [Haas 96]

is effected by simply laying a vacuum line from the furnace out into space. According to the station's altitude and other influences, a pressure on the order of magnitude of 1 Pa can be achieved. The outside pressure can, of course, vary considerably, according to whether

- the vacuum is "tapped" in the flight direction or in the opposite direction,
- there is heavy outgassing of the pressurized modules or surfaces, or
- attitude control thrusters or ECLSS vent lines are being operated nearby (cf. Fig. 7.4).

Accordingly, the pressure can be higher or lower by several orders of magnitude. By using a special device known as "Wake Shield Facility", NASA researchers produced a vacuum near the Space Shuttle better than anything achieved on Earth. With the help of this device, semiconductor thin films were grown.

7.1.3 Space Radiation

Space research in the field of biology and physiology is not only to examine the influence microgravity has on life, but also to examine the radiation in space and the radiobiological effects resulting from it. Due to its long operational phase, ISS offers a unique opportunity to measure the radiation fields both inside and outside a station over a long period of time.

Space radiation is also of interest to investigations in the field of engineering sciences and for the qualification of space system components. For example, radiation problems occurring with satellite-based telecommunication and navigation systems can be analyzed. The origins of space radiation and its effects on astronauts and experiments and the space station itself are described in detail in Chapter 3 ("Orbital Environment").

7.1.4 Overview of User Disciplines

Thanks to its physical, technical and operational advantages, a space station offers the opportunity to perform a wide range of different scientific, technological and industrial experiments and to continue tasks, on a broader basis, which have already been carried out in the past by the existing space systems.

In order to gain a better impression of the entire user potential, it might be helpful, within the framework of this chapter, first to summarize the experiences acquired up to the present point, and then to have a look at the objectives of and priorities for future tasks. Finally, activities as they could be performed aboard the International Space Station will be presented. The latter will include an outlook on possible industrial and commercial applications. For the sake of clarity, the disciplines in question are categorized as seen in Fig. 7.5.

Physics and Materials Science

Fundamental physics, thermodynamics, fluid dynamics, combustion processes and materials science belong to those disciplines that are of relevance for the space station experiment scenario.

In physical sciences, gravity is only one parameter among many others whose effects can be examined in order to deepen our scientific understanding. The most im-

Disciplines / Qualities	Microgravity: Physics and Materials Science	Microgravity: Life Sciences	Space Sciences	Earth Observation	Engineering Sciences, Development of Technologies, Commercialization
Weightlessness	●	●	○		●
Vacuum			●	●	●
Space Radiation (Particles, Thermal, Electromagnetic), Earth's Magnetic Field	○	●	●	○	●
Space Observation Undisturbed by the Atmosphere			●	○	●
Permanent View of Atmosphere and Earth				●	●

● Strong Connection ○ Weak Connection

Fig. 7.5. Laboratory Conditions and Observation Possibilities of a Space Station and their Resulting Research Categories

portant effects gravity has on these fields of research is the absence of gravity-driven convection, sedimentation and hydrostatic pressure in fluids. The possibility of containerless processing in space is, for these disciplines, an advantage of the microgravity environment as well.

Life Sciences

Gravity is a fundamental force that has a marked influence on all life on Earth. Therefore life scientists and biomedical researchers exploit the space environment, and in particular the near weightlessness, in order to answer fundamental questions in basic biology relating to humans, plants and animals. Various investigations on the response of microgravity on the lungs, brain, nervous system, bones and muscles have already been performed. This research is not only of interest to the health and survival of astronauts but is also relevant to the understanding of balance disorders, osteoporosis, and cardiovascular disease. This research is just at the very beginning of clarifying the role of gravity at molecular level. Present basic studies address the reaction-diffusion feedback pattern forming processes which occur in self-organizing systems.

For cell-cell relations or for the respective cell functions within a differentiating tissue, the origin of a change is likely to involve subtle modifications of their mechanical and biochemical micro-environment. Experimental investigation of such minute changes and their eventual application would benefit from the increased stability of fluid systems in microgravity and from reduced mechanical loads.

Fluid dynamical models describing macroscopic systems are not valid in the submicron range. Experiments in microgravity are needed to observe and to determine the influence of gravity on the processing and amplification of signals involved in the gravity sensing and response of cells, cell aggregates and tissue or whole organisms [ESA DMSM 98]. The investigation of space radiation and its effects on plants and live organisms as well as the effect which other environmental factors have on life-forms in space (i.e. biorhythm) may now be done on a large scale.

Space Sciences

The so-called “space sciences” comprise astronomy, astrophysics, cosmic ray physics and magnetospheric physics, as well as the study of the Sun and the solar system. The study of the effects of weightlessness and space radiation along with investigation and observation from outside the Earth’s atmosphere are of interest to this subject matter, since the Earth’s atmosphere has an optically distorting effect and absorbs electromagnetic radiation.

All these disciplines benefit mainly from the numerous external accommodation sites on the truss structure of a space station and on other external platforms. They also profit from the available power and from the opportunity to correct errors in the design or manufacture of experiment hardware directly in orbit, with the help of astronauts. Especially the latter aspect must not be underestimated in view of the high complexity and extreme costliness of instruments for space sciences.

Observation of the Earth and its Atmosphere

The International Space Station is especially interesting as a platform for observation of the Earth and its environment due to the station's orbital parameters and the available electrical power, data processing and accommodation capacities. The station's Earth observation opportunities are of interest to the disciplines of meteorology/climatology, ecology, geodesy, geology and agriculture.

Thanks to the permanent presence of a crew and the frequent opportunities for access, ISS is advantageous for both space sciences and Earth observation, for testing and optimizing prototypes of sensors and other expensive instruments before they are finally put to use aboard uncrewed space vehicles.

Engineering Sciences and Technology Development in Space

Apart from scientific research, studying the space environment is also useful for rather application-oriented areas such as space engineering and the testing of new technology. Space has become the "workplace" for many commercial satellites and other automated space systems. The production, launch and operation of such systems alone amounts to a value of several tens of billions of US dollars per year. Any malfunction of a satellite may entail severe financial losses.

A space station offers new opportunities to test, adapt and optimize less proven materials, technologies and instruments in their real environments. The operational environment of such systems is not only characterized by weightlessness, but also by vacuum, space radiation and the extreme temperatures of space. Utilization of the space environment as a testbed for new technology reduces the risk of development when using new technical and operational solutions for space systems.

As far as the influence of microgravity is concerned, there will be laboratories and laboratory equipment available for all disciplines that have an interest in the "testbed function" of ISS. In order to study other characteristics of the space environment, the station offers sufficient external accommodation sites which are fully linked to the station's power resources and data processing resources.

The special attractiveness of space station utilization of this kind lies in a station's proximity and guaranteed availability over a long period of time and the frequent opportunities for access and return. Equally interesting is the opportunity to rely on both crew and robotic capabilities for assembly, reconfiguration, and observation of experiments.

7.2 Physics and Materials Science

7.2.1 Results Obtained and Areas for Future Research

When solving a certain problem, well-thought-out μg experiments can be an integral part of a more general approach. Usually, these experiments are embedded in extensive programs of theoretical and experimental research taking place in laboratories on Earth.

Despite the fact that in the past the number of flight opportunities for microgravity experiments was limited, the results obtained had considerable influence on the scientific development of the disciplines in question. This is proved by a high number of publications on experiments conducted under weightlessness [Walter 87, NASA 88, MSFC 94, NRC 95, ESA 95, D1 85, D2 93, D2 95, NASA NRP 98] as well as in numerous discipline-specific papers and books.

Fundamental Physics

In the field of fundamental physics, a new class of experiments has been formed, using the microgravity environment for investigations. Apart from tests of the theory of General Relativity, different projects are planned such as the development of ultra-precise atomic clocks, studies on the subject of fractal materials, aggregation phenomena and thermal radiation forces. It can thus be expected that in the future, more fundamental physicists will discover microgravity as a tool of research.

The development of ultra-precise atomic clocks is not only of academic interest, since today's satellite-based navigation systems need a very exact reference time in order to calculate the exact position of an object. Any improvement on the accuracy of atomic clocks will have direct influence on the precision of satellite-based navigation systems.

Example: A time comparison between two atomic clocks that are perfectly synchronized before a space flight mission had the following result: compared to the clock on Earth, the clock in space has a longer oscillation period at orbital altitudes smaller than half of the Earth radius, $H < R_0/2 = 3189$ km. At higher altitudes, this period will be shorter. This effect of the theory of General Relativity was predicted by Albert Einstein and was proved, more precisely than ever before, by a two-way clock comparison during the D1 Spacelab mission.

Since at low orbital altitudes the relativistic effect of velocity is stronger than the relativistic effect of gravity, one second in space is longer than one second on Earth. As a consequence, astronauts in space age slightly more slowly than on Earth. During the D1 mission, this effect for an orbital altitude of 324 km was exactly 26 μs per day, see Fig. 7.6.

The real objective of this experiment was to learn to master the technology of atomic clocks and radio transmissions for navigation systems like GPS (Global Positioning System) [D2 93].

Example: Atomic Clock Ensemble in Space (ACES). This is a program for testing the performance of a new type of atomic clock which exploits and also depends on microgravity conditions. The project was recently approved to fly on the International Space Station as an external payload, starting in 2002 for about 18 months. ACES consists of the following elements (contributions): a laser-cooled atomic clock "PHARAO" (France), a hydrogen maser as a reference clock (Switzerland), a laser link for optical transfer of time and frequency (France), and a microwave link for transfer of time and frequency (ESA) [ESA DMSM 98].

The principle of an atomic clock such as "PHARAO" is to lock an oscillator to the atomic resonance frequency ν_0, see Fig. 7.7. Heisenberg's uncertainty principle

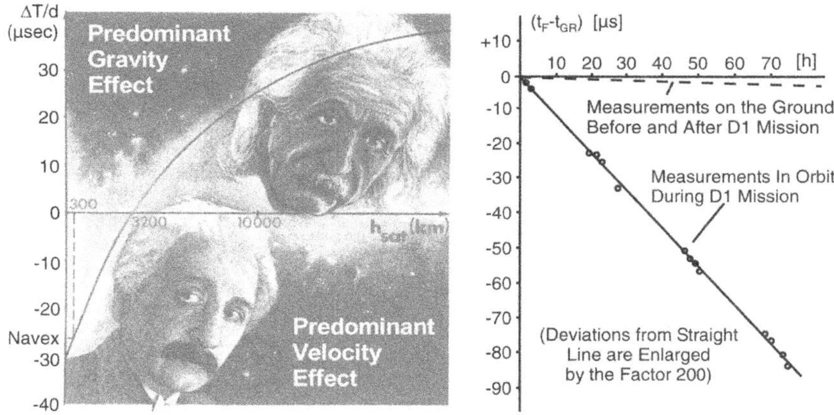

Fig. 7.6. Aging in Space – Results of the D1 Mission

shows the following: the longer the interaction time of the atoms with the radiation emitted by the oscillator, the narrower the resonance. Two key points determine the ultimate performance, namely a narrow resonance and a high signal-to-noise ratio. PHARAO uses laser cooling in order to reduce the thermal velocity of atoms to a few cm/s, which corresponds to a temperature of about 1 μK. Under microgravity conditions, the atoms remain at these low velocities while on Earth, they would increase their speed rapidly due to gravitational acceleration when the lasers are switched off for signal interrogation. Laser cooling allows an interaction time which is 100 to 1000 times longer than for a conventional caesium clock. Since 1995, an atomic fountain clock has been in operation at the "Observatoire de Paris" (France) with a relative stability of $1.3 \cdot 10^{-13}\ \tau^{-1/2}$ (where τ is the measurement time in seconds) [ESA DMSM 98].

With the help of another **example** from the D1 Spacelab mission concerning the measurement of adhesion forces, some typical characteristics of experiments in fundamental physics shall be shown. The measurement of van-der-Waals forces at the interface between solid and liquid matter is performed in a fluid physics module (see Fig. 7.8). The astronauts fill silicone oil in through a drill hole in a circular disc until it touches a second disc opposite of the first one and distributes itself between both discs in the form of a cylindrical liquid bridge. The result of changing either the distance between both discs or the volume of the liquid precisely determines the shapes of such zones covered by a fluid: large amounts of liquid yield spherical zones; when reducing the distance between the two discs, the liquid forms a nodoid with a significant convex bulge. Smaller amounts of liquid result in the zone shapes shown in Fig. 7.8 with a concave surface.

The shape in the case of which on each surface point, the bulge has the same curvature radius as the liquid column in this place, is particularly interesting. This is called a catenoid with which the pressure along the surface becomes zero. The slightest perturbation or shock will lead to visible deviations of this shape and that

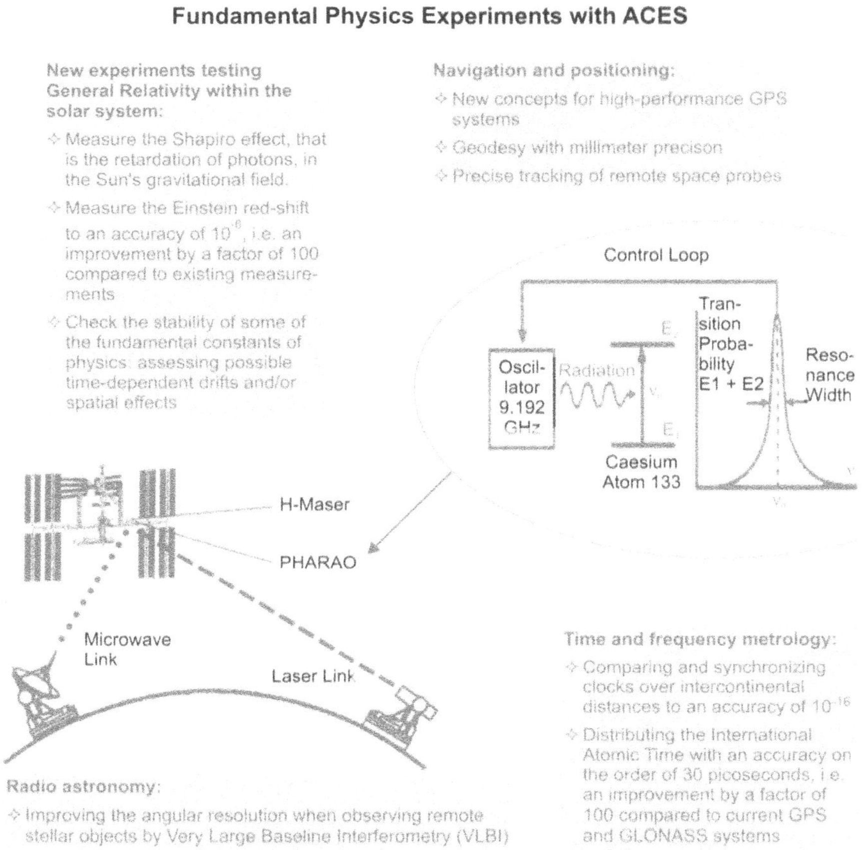

Fig. 7.7. Principle of the Atomic Clock PHARAO [ESA DMSM 98]

is the reason why such a zone covered by the fluid can be used as a very sensitive instrument for the measurement of pressures and forces.

At the interface between the larger disc (which takes the form of a truncated cone) made of titanium and silicone oil as a well-wetting liquid, adhesion forces occur which make a relatively thick silicone film appear over the entire disc. The stronger the penetration effect of the adhesion forces, the thicker the film becomes, the less silicone oil remains in the "neck", i.e. the smallest diameter of the silicone column. The degree of deviation from the catenoid form is an indicator for the entire adhesion force; with the help of the law of distance, the van-der-Waals forces between the interface and the solid body can be determined. Compared to the experiment on Earth, an increase in sensitivity by a factor of 100 000 was obtained during the experiment in space.

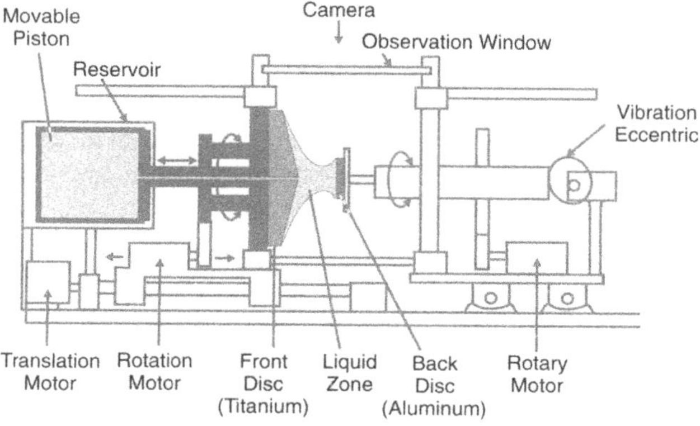

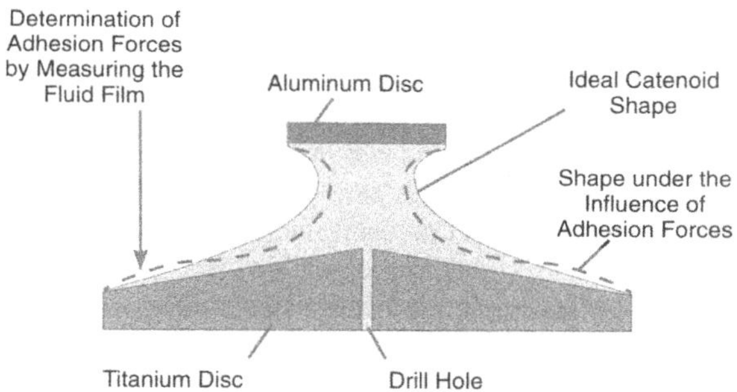

Fig. 7.8. Cross-Section of the Fluid Experiment Module. This module allows the measurements of the adhesion forces near the stability limit as well as the fluid shape typical under the influence of these forces.

The advisory committees from ESA and NASA recommend the following research areas for future experimentation on the subject of fundamental physics:

- Laser cooling of atoms with the aim of developing ultra-precise atomic clocks in space, gravitational and relativistic physics
- Low temperature and condensed matter physics
- Plasma crystal – a new state of matter, aggregation of protoplanetary dust as contribution to the better understanding of the origins of our solar system
- Search for antimatter. In this context, the Nobel prize laureate Sam Ting (USA) suggested an experiment with the aim of detecting, aboard ISS, antimatter particles from remote galaxies. This is done by the Alpha Magnetic Spectrometer (AMS) which serves as a particle detector and contains a large, permanent magnet. The scientific objective is to search for cosmic sources of antimatter and dark matter. Current plans call for operating the AMS aboard ISS as an attached payload for a three-year period. The AMS will benefit from ISS's power and communications resources which will enable it to operate for a much longer period than would be otherwise possible.

Investigations in the field of fundamental physics aboard ISS will also support a number of concurrent technology development efforts. Examples of such efforts include the development of superconducting magnetometers for more efficient resource mining and non-invasive medical diagnostics, the development and operation of high-temperature superconducting components for satellite-based transponders as well as the management of extremely low-temperature fluids for life support systems and manufacturing purposes.

Thermodynamics

In the field of thermodynamics, considerable progress was made on the way to understanding transport processes in fluids (liquids and gases) as well as of relaxation processes near the critical point.

Density gradients in liquids and gases are found in places where there are temperature or concentration gradients as well. Under the influence of gravity on Earth, density gradients lead to convection flows that are caused by buoyancy. The effect of convection flows can dominate heat and mass transfers as well as prevent the performance of certain experiments in order to confirm theoretical models. Under μg conditions, by contrast, there is virtually no convection caused by gravity. As a consequence, experimental research of a totally new kind is possible in space.

The μg research led to the first exact measurement (based on temperature gradients) of an atomic diffusion in mixtures, the so-called "Soret effect", including the observation of isotropic separation. Another important finding was the observation of the fact that the critical heat flux during boiling does not depend on gravity and hence contradicts present understanding.

Experiments in μg make it possible to prove a new type of thermalization phenomenon near the critical point, the so-called "Piston effect". The experiments delivered the quantitative confirmation of a thermophysical divergence in the immediate vicinity of the critical point.

Example: The question of how liquids boil in the reduced-gravity space environment (cf. Fig. 7.9) has puzzled scientists for decades. To the surprise of many investigators, Space Shuttle experiments have found that, compared to normal gravity under certain conditions, steady boiling is not only feasible in microgravity, but it can also result in enhanced heat transfer. If steady microgravity boiling can be attained on a predictable basis, it has the potential to generate major breakthroughs in heat transfer equipment for space. This research is also providing greater fundamental understanding of the boiling process itself and may result in more efficient designs of steam generators which are used to produce most of the world's electrical power [NASA NRP 98].

Example: During the Spacelab missions SL1, D1 and D2, the selfdiffusion in metallic melts was investigated with the isotopes Sn112 and Sn124 (cf. Fig. 7.10). Compared to measurements on Earth, the determination of the diffusion constant in space was 50 times more exact: without the convection disturbances present on Earth it was possible to precisely determine the diffusion constant and to find out that it was markedly smaller than expected.

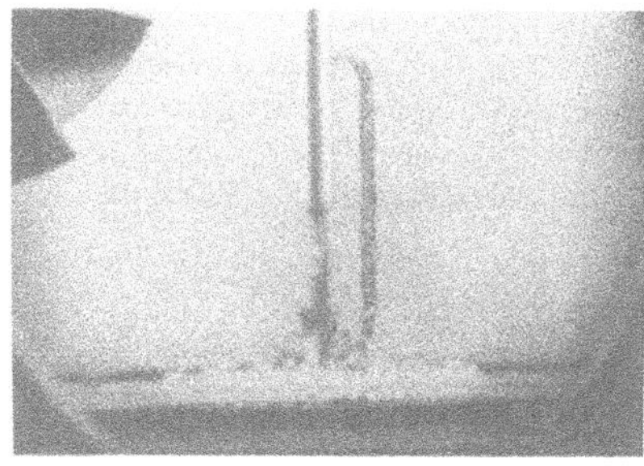

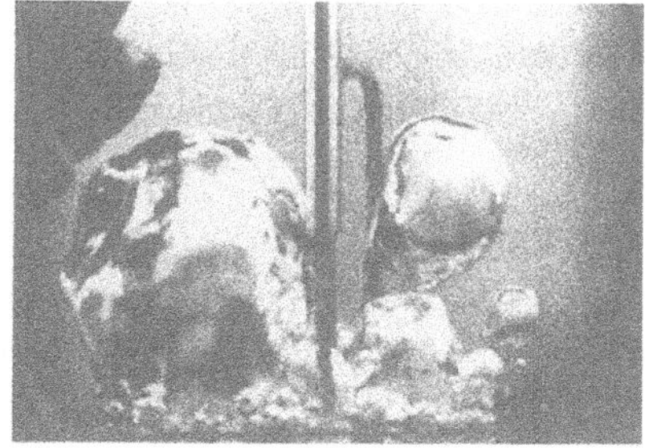

Fig. 7.9. Boiling Liquid in the 1g Environment of the Earth (left) and in Microgravity (right)

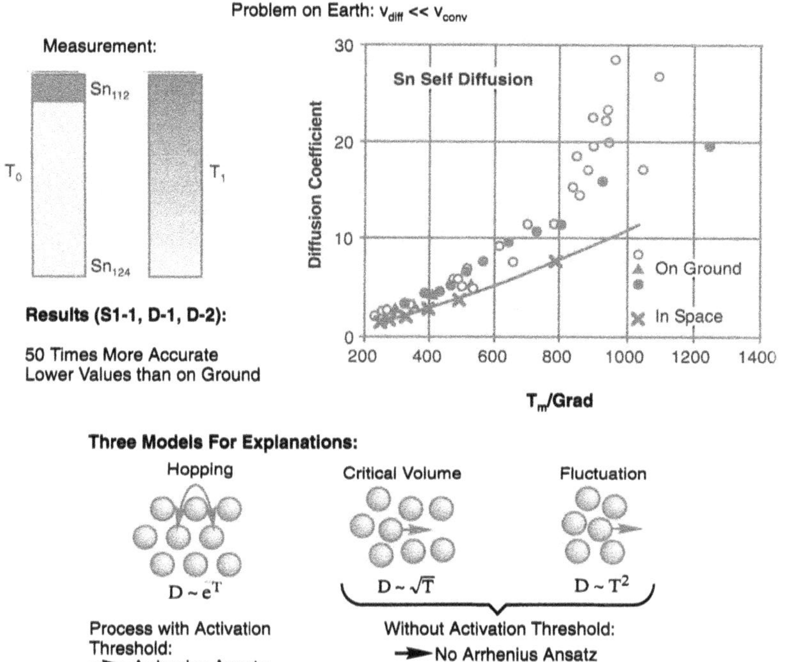

Fig. 7.10. Diffusion in Metallic Melts

The NASA and ESA Advisory Committees recommend concentrating on the following subjects for future experiments:

- Time and length scales for phenomena near the critical point (Fig. 7.14 shows the asymptotic behavior of transport coefficients and the time constant at the critical point.)
- Thermophysical properties of liquids

- Mechanism of boiling
- Interface tension and adsorption
- Metastable phases

Example: Enhancement of oil recovery. Oil is one of the resources on Earth which greatly affects our daily life and the global economy. Therefore, all oil companies are constantly working in order to enhance recovery methods and to discover new reservoirs. Modern geophysical and geological exploration methods allow detection of hydrocarbon reservoirs at depths of up to 7 km.

Understanding fluid physics in crude oil reservoirs is a major challenge in optimizing exploitation. Present modeling methods are based on pressure-temperature equilibrium diagrams and on gravity fractionation. However, rising exploitation costs are forcing oil companies to develop improved models which more accurately predict reservoir yield. Advanced models account for constituent concentrations, and, based on the chemical analysis of a sample taken from a reservoir, thus allow the determination of the reservoir's volume, its vertical extension and the quality of its oil. However, the development of such models requires precise diffusion coefficients, which to date are not available, since convective motions and buoyancy affect their measurement on Earth, resulting in inaccurate forecasts.

The availability of accurate diffusion coefficients allows for a reliable prediction of oil prospect and the validation by numerical codes. Such codes simulate laws which describe the kinetics of pressure/temperature-dependent physicochemical processes that generate hydrocarbon mixtures from organic matter. The concentration distribution of constituents in these hydrocarbon mixtures is mainly driven by phase separation and diffusion caused by concentration differences and temperature gradients.

The role of thermo-diffusion (Soret effect) in petroleum reservoirs is not yet fully understood. Numerical modeling neglecting thermodiffusion results in the predicted vertical extension of a reservoir being inaccurate on an order of 100 m, and thus indicates the need to take thermodiffusion into account.

Since in normal gravity, buoyancy and convection affect the distribution of constituents in fluids, accurate coefficients of pure diffusion can be measured only in microgravity. Several days or even weeks are required for such measurements. Consequently, ESA is sponsoring two projects on the precise measurement of diffusion coefficients in crude oil mixtures. The projects' relevance is underlined by the active participation of the European petroleum industry.

The first experiment was carried out in March 1998 aboard the Space Shuttle (STS-91), where diffusion coefficients of various crude oil and hydrocarbon mixtures were accurately measured by means of a specific experimental set-up in a Get-Away Special (GAS) canister (see Sect. 8.2.5). A second experiment, scheduled for a GAS experiment in autumn 1998, is aimed at precisely measuring Soret diffusion coefficients of crude oil mixtures with temperature gradients and pressure conditions as they are found in oil reservoirs [ESA DMSM 98].

Combustion Processes

Combustion processes provide energy sources for surface and air transportation, home heating, electrical energy production, waste incineration, and material synthesis. However, combustion can also cause undesired effects, like unwanted explosions, fire hazard, and toxic exhausts. The challenging goal of combustion research is to enhance the advantageous aspects and to diminish undesired effects.

Gravity has intense influence on many combustion processes. Reducing its effects can therefore help to focus on underlying mechanisms which are normally overshadowed by convection, buoyancy or sedimentation. Thus, microgravity offers unique possibilities of improving the physical understanding of combustion processes, of precisely measuring data relevant to combustion, and of developing or validating numerical models.

Droplet cloud combustion and spray combustion are examples of processes where microgravity may contribute to optimization. These processes are employed in piston and turbine engines and also in industrial burners. In order to maximize efficiency and minimize fuel consumption, soot production (and thus pollution), a detailed understanding of the processes is necessary.

Microgravity facilitates the investigation of droplet interactions in clouds, since an environment without convection and buoyancy allows for homogeneous droplet distribution and for more precise measurements of evaporation, ignition and flame propagation.

Currently, four major topics are being discussed with representatives from industry:

- High pressure droplet cloud combustion
- Particle cloud combustion
- Soot formation in diffusion flames
- Stability and structure of lean premixed flames

Combustion processes often occur very rapidly, so that detailed studies on droplet vaporization and spray ignition can be effectively performed during short microgravity periods like those provided in drop towers (see Sect. 8.2.1). However, mapping of the parameter μg requires long-duration microgravity experiments. To that end, several experiment modules are presently available for use in precursor sounding rocket experiments. These activities are expected to expand with the advent of ISS [ESA DMSM 98].

Example: The burning candle illustrated in Fig. 7.11 shows the influence of gravity-driven convection on Earth and its absence in weightlessness. Figure 7.11 (left) shows the conical form of a candle flame burning such as it occurs on Earth due to the hot gases and combustion products rising within the flame. The hot gases will cool down as they move further up and will then be replaced by denser air that gushes out from underneath. Figure 7.11 (right) shows the spherical form of the weakly glowing candle flame in weightlessness [NASA 95].

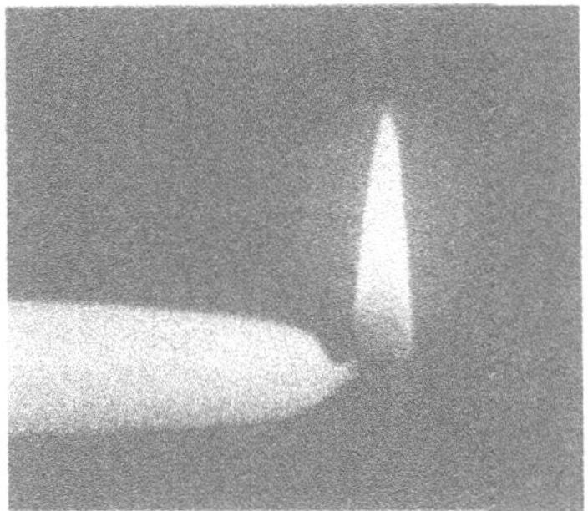
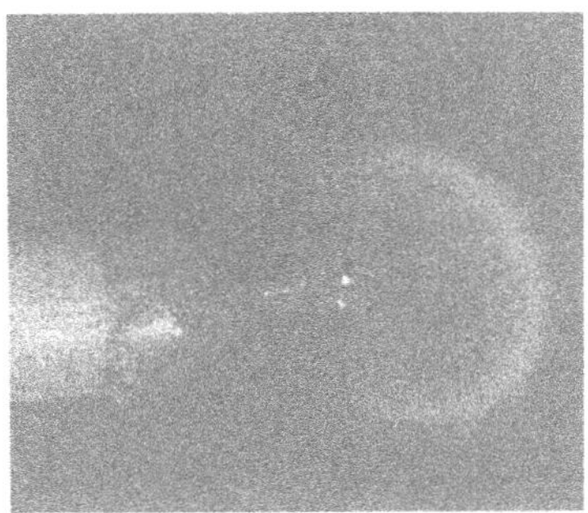

Fig. 7.11. Burning Candle – Left: on Earth, Right: under Weightlessness [NASA 95]

Fluid Physics

The absence of hydrostatic pressure and gravity-driven convection in liquids under weightlessness is the basis for optimal conditions for the investigation of capillarity phenomena, phase changes, and phenomena in liquids near the critical point (cf. also "Thermodynamics"). The results of such studies yield valuable conclusions on the further development and verification of theories that describe laminar, oscillatory and turbulent flows generated by some other different driving forces.

Figure 7.12, e.g., shows how fluids flow differently in microgravity. On Earth, the fluids flow through the bottom of the pipe while the air remains trapped in the top half of the pipe (Fig. 7.12 below). In space, however, the fluid flows more symmetrically, clinging to the sides of the pipe all along its length. A column of air is left down the center of the pipe (Fig. 7.12 above)

Thus, phenomena that cannot be observed on Earth, because they are obscured by gravity-driven convective flow or hydrostatic pressure, can be observed in space in a very detailed way. The absence of gravity in space also allows the observation

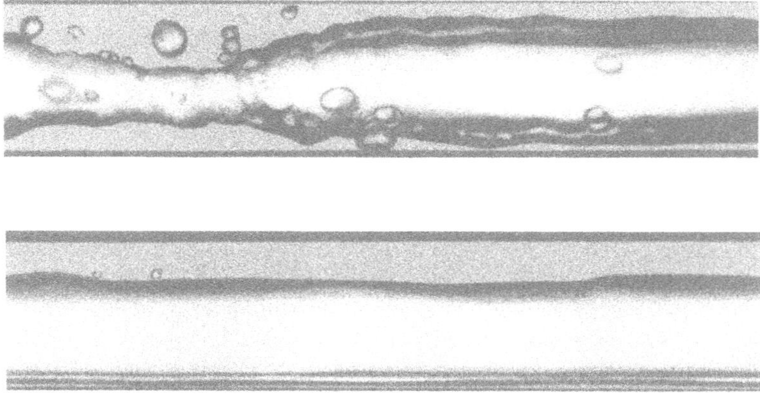

Fig. 7.12. Fluid Flow – Above: in Space, Below: on Earth [NASA NRP 98]

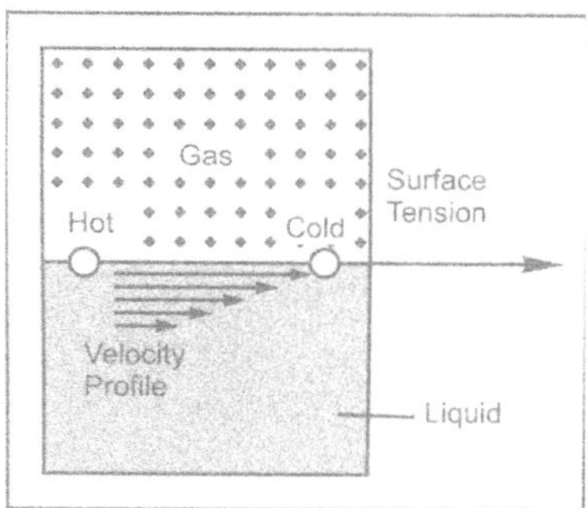

Fig. 7.13. The Marangoni Convection: a) Principle
b) Visualization by Stroboscopic Illumination of Tracer Particles

and quantitative measurement of instability or bifurcation phenomena in pure Marangoni flow and leads to the quantitative understanding of the static and dynamic stability limits of large liquid interfaces.

Example: Investigation of the Marangoni convection at the interface between a liquid and a gaseous medium. If one side is warmer than the other (e.g. as shown in Fig. 7.13: warm on the left and cold on the right, i.e. the temperature gradient is parallel to the boundary), forces predominantly near the boundary pull the molecules from the warm zone into the cold zone. In rare cases, this can also be observed on Earth, for example, when the remains of a candle's wick fall into the liquid wax of the candle and move, very quickly on the surface, radially away from the wick, hence from "warm" to "cold". The small particles drifting in such a liquid, when filmed while being illuminated with a stroboscope, show the velocity profile and stream lines of a forming Marangoni convection (see Fig. 7.13b). The prerequisite for the formation of such a convection is a free surface of the liquid or an interface between different, immiscible liquids as well as a temperature-dependent surface tension and a temperature gradient, i.e. for example $\sigma(T_1) > \sigma(T_2)$ and $T_2 > T_1$. In Fig. 7.13b, there is a higher temperature on the left (350°C) and a lower one (320°C) on the right; this excites the convection roll.

Critical Point Phenomena

When approaching the critical point, the compressibility of liquids is increased by several orders of magnitude. For that reason, on Earth, such liquids are compressed by their own weight. The liquids form density-dependent layers, so critical properties only occur in an infinitesimally thin plane. Homogeneous samples of finite dimension that are necessary for the quantitative examination of phenomena near the

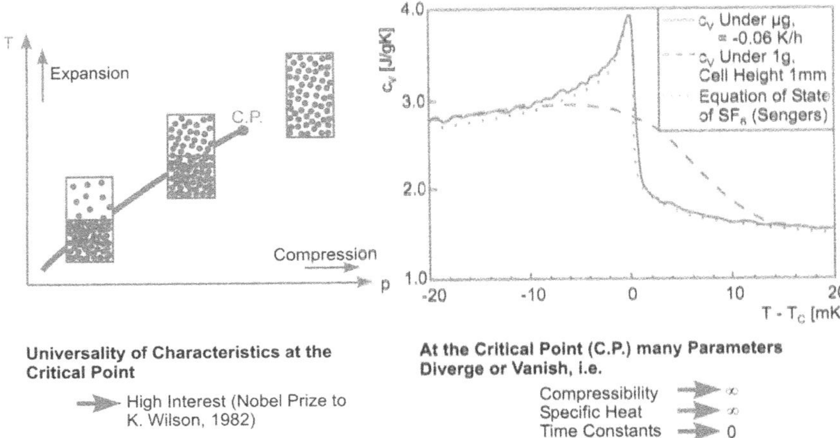

Fig. 7.14. Phenomena at the critical point and the influence of gravity. Up to the present, the singularity of the isochor specific heat capacity c_V could only be observed in space.

critical point are only available in a µg environment (cf. Fig. 7.14). Important future subjects of research are the following:

- Flows and instabilities, induced by surface or thermal radiation forces
- Doubly diffusive instabilities
- Convection and morphological stability

In summary, research conducted in microgravity is increasing our understanding of fluid physics to provide a basis for predicting, controlling and improving a vast range of technological processes. The behavior of fluids is crucial for many phenomena in materials science, biotechnology and combustion science. The performance of a power plant depends on the flow characteristics of vapor-liquid mixtures. Oil recovery from partially depleted reservoirs depends on how fluids flow through porous rocks. The safe engineering of buildings in earthquake-prone areas requires an understanding of fluid-like behavior of soils under stress. Advances in materials engineering require a better knowledge of how fluid behavior determines the structure of solid materials during solidification.

Many products that require high-precision manufacturing processes could directly benefit from greater control over fluid flow. Surface-tension-driven flows, for example, affect some techniques of semiconductor crystal growth, welding and the spread of flames on liquids. The behavior of liquid drops is an important aspect of chemical process technologies and meteorology [NASA NRP 98].

Material Sciences

In the field of material science, the behavior of gaseous, liquid and solid mixtures can be investigated without sedimentation. The latter normally occurs on Earth and leads to the separation of the mixtures' components according to their density. Unlike on Earth, unusual composites such as highly porous alloys or particle-and-fiber dispersions can be produced in space.

Research under microgravity has resulted in new insights into physical phenomena at the growth interface of metallic materials and a better understanding of phase separation and ripening processes in immiscible alloys. In the meantime, the relevant results are used on Earth in the production of materials e.g. for sliding bearings.

In the absence of gravity, melts can be positioned and manipulated without using a container. Containerless processing is interesting insofar as crucibles can be the source of contamination, heterogeneous nucleation, and mechanical stress. Containerless processes as described in Sect. 8.4 offer the opportunity to study undercooling, nucleation and the formation of metastable phases.

Example: Directional solidification means the orderly arrangement of atoms at the solidification front. It plays a decisive role for solid materials as they form from a melt during cast processes of metallic and also non-metallic materials. The direction of solidification is determined by the direction of the heat flux which is given by the furnace. The form of the solidification front and the crystal structure of the solidified body are determined by the material properties and the experiment parameters. Decisive influencing parameters are the temperature gradient at the solidification front, the concentration of the components, and the geometry. In this context, the growth front can be either planar, cellular, dendritic or globulitic (cf. Fig. 7.15). The mechanical properties, the corrosion resistance, and other material properties are heavi-

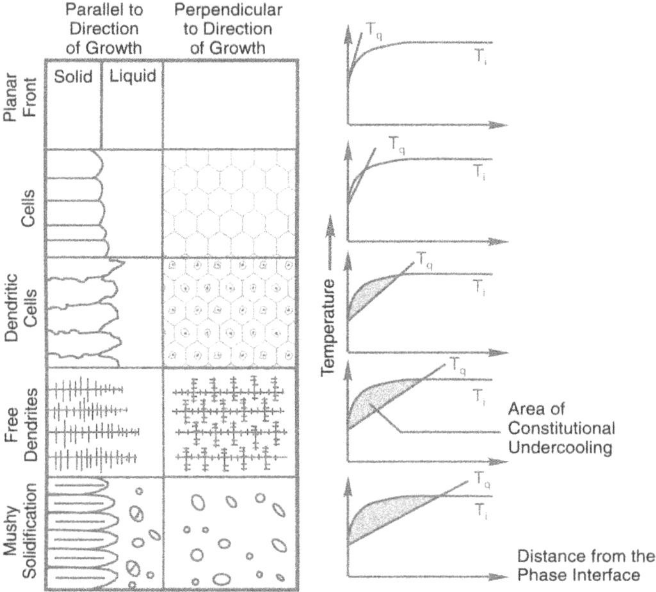

Fig. 7.15. Morphology of the Solidification Front (on the Left) at Different Values of Constitutional Undercooling (on the Right) [Access 90]

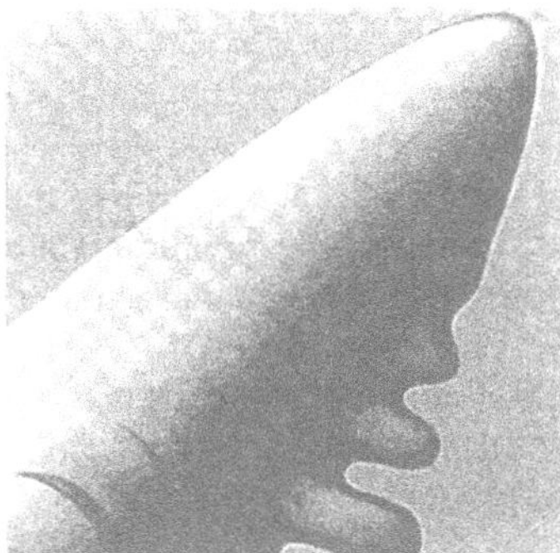
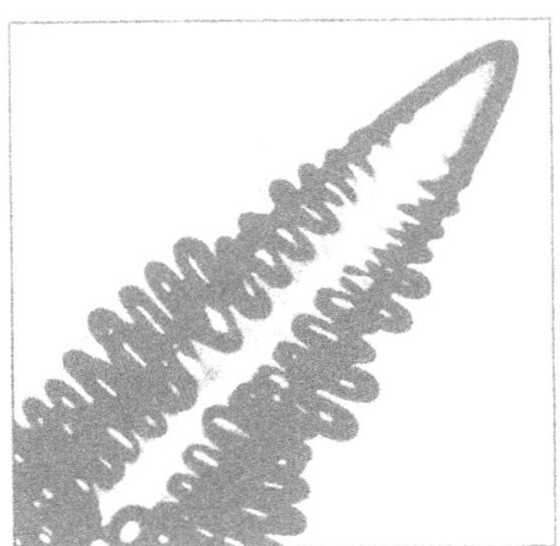

Fig. 7.16. The dendritic crystal enlarged by microscope (on the left) is marked by the deforming influence of gravity during its process of development. The dendrite on the right developed under weightlessness.

ly influenced by structures forming during the solidification. Figure 7.16 shows the homogeneous growth of a free dendrite produced in space.

Example: Planar directional solidification with a planar solidification front allows the production of homogeneous, well-defined materials and, for that reason, is also used in the production of semiconductor crystals. The stability of such planar solidification fronts can be described well with the help of theoretical models. Experiments and calculations show that the planar, moving solidification front of alloys can become unstable mainly in two different ways:

- Constitutional undercooling: The movement of the solidification front at constant velocity creates an increase in the concentration resulting in a local undercooling in front of the phase boundary (hatched area in Fig. 7.17, lower left). This leads to instability of the planar solidification front corresponding to the illustration in Fig. 7.15 (lower right).
- Convective instability: With the directional solidification, this can emerge due to an increase in the concentration of a less dense component in front of the solidification front. Consequently, concentration and temperature fields will determine the solidification morphology. The area in which a planar solidification front grows in a stable way is illustrated qualitatively in Fig. 7.17 for a constant temperature gradient.

Example: Manufacture of a composite consisting of a nickel-aluminum matrix that is used for mechanically and thermally heavily loaded structural elements such as turbine blades. The regular embedding of molybdenum fibers in weightlessness can markedly improve these properties (cf. Fig. 7.18). The lifespan of heavily loaded bearing alloys can be improved in a similar way by the embedding of a finely distributed, soft material in a matrix (cf. Fig. 7.19).

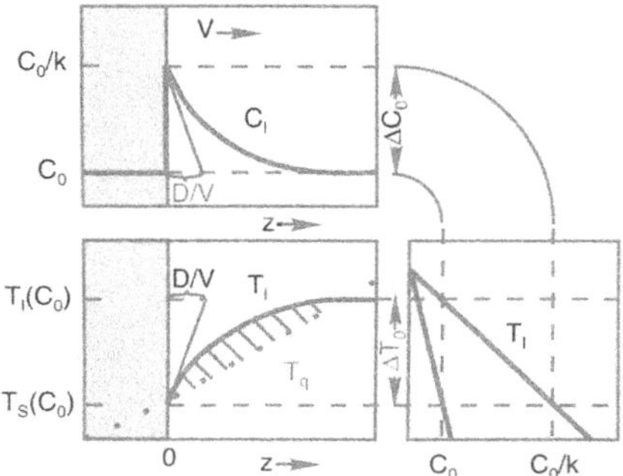

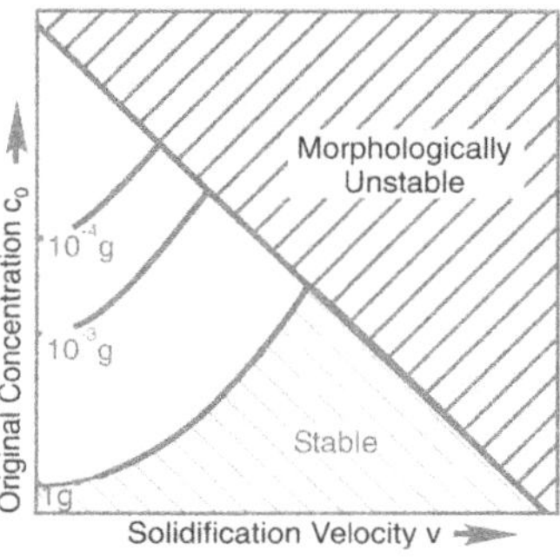

Fig. 7.17. The formation of the constitutional undercooling by increase in the concentration in front of an advancing solidification front (on the left) and a stability diagram for the vertical, planar and directional solidifications at a constant temperature gradient and different levels of gravity (on the right) [ACCESS 90]

Fig. 7.18. The transverse section of a sample of a directional solidified eutectic alloy, where the reinforcing molybdenum fibers were uncovered by selective etching. The seemingly perfect composite, important for the mechanical strength at high temperatures, displays flaws on Earth (see below). Under weightlessness, a homogeneous fiber separation and reduction of flaw incidence are obtained (see above) [D1 85]

A number of attempts to obtain a homogeneous distribution in various alloys and model materials in microgravity unexpectedly yielded separated phases as well. This is attributed to a droplet migration mechanism called Marangoni migration which is driven by gradients of interfacial tension along the surface of droplets. These gradients of interfacial tension relate directly to the local temperature gradients in the matrix upon cooling and solidification and are thereby independent of

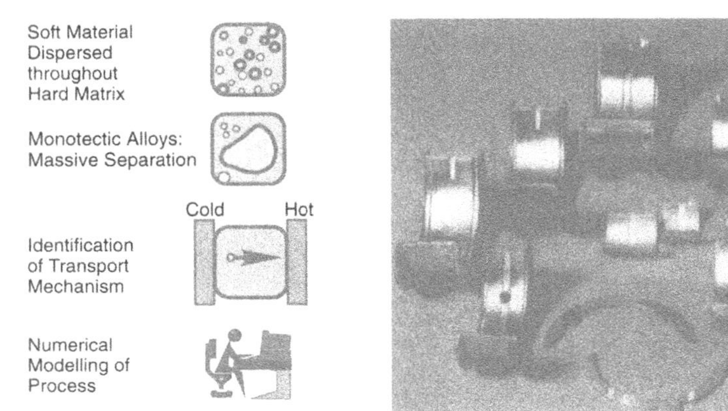

Fig. 7.19. Manufacture of durable bearing half-shells. A new casting process could be devised for the fabrication of bearing alloys consisting of soft particle dispersions in a strong matrix obtained from immiscible alloys [ESA DMSM 98]

gravity. This mechanism was studied on a theoretical basis and models were developed. Dedicated microgravity experiments permitted the validation of numerical models, particularly as far as multiple droplet migration is concerned. The possibility of controlling the direction of migration and the velocity of droplets through tailoring the heat flow in molten alloys has been demonstrated. The next natural step is to transfer this new knowledge to production processes.

The ideal way would be to minimize the sedimentation caused by gravity by balancing it with a Marangoni migration effect. If the droplets reach a state of dynamic equilibrium, a stable disperson in a solidifying matrix will become attainable. This concept has been tested and newly produced materials (consisting e.g. of an aluminum alloy matrix with evenly distributed inclusions of lead or bismuth) are currently being tested as candidates for advanced sliding bearings in combustion engines (see Fig. 7.19).

In these materials, the hard matrix sustains the dynamic loads from the combustion process, and the soft inclusions provide self-lubricating behavior. The transfer of this new manufacturing method into an economically viable casting process would permit the development of engines consuming significantly less fuel – a promising prospect regarding environmental pollution.

Future research will address the optimization of a strip casting process and its application to different materials. Further quantitative measurements and model validation will be needed for which microgravity will be a tool of critical importance [ESA DMSM 98].

Crystal Growth

Another important area of research which falls under microgravity deals with the growth and solidification of crystals. Experiments in space have very much promoted the general understanding of transport processes in the field of crystal growth.

During crystallization, the control of heat and mass transfer is a sizable problem, which is additionally complicated by the effects of convective flows on Earth. For that reason, microgravity is basically of general interest to crystal growth. Throughout practically all phases of crystal growth, e.g. from a melt to a solution or to a vapor phase, temperature and concentration gradients occur that generate a driving force for convection on Earth. Crystallization under the conditions of diffusion-controlled heat and mass transfer in weightlessness makes the optimization of these processes possible and therefore leads to a better mastering of production.

In space, not only chemically homogeneous semiconductors, but also crystals, with controlled doping grown from vapor phases and solutions, have been produced. Both products had a lower rejection rate due to less dense inhomogeneities. The identification of the relative influence that the different transport processes have on the perfection of a crystal also leads to improvements of production methods on Earth, such as growth in a magnetic field.

Example: Crystal growth of cadmium-telluride (CdTe). Materials for electronics, such as CdTe and related compounds are used in highly sensitive detectors and photorefractive devices. Due to the difficulties of growing crystals of sufficient quality, to date these applications are limited and expensive (Fig. 7.20).

CdTe-based X-ray and gamma-ray detectors have high potential in real-time dental imaging or mammography including three-dimensional (3D) tomography. This promises faster and more reliable medical diagnoses while exposing the patient to lower radiation doses. CdTe IR detectors enable high-resolution thermal imaging and high data-rate optical telecommunication. Additionally, the photorefractive properties can be employed for high-performance devices in optical ultrasonic nondestructive testing.

Fig. 7.20. Earth-grown crystals of the alloy cadmium-zinc-telluride (left) exhibit many more imperfections than space-grown crystals (right). By reducing the imperfections in the crystal, the overall performance of devices using the alloy, such as IR detectors and X-Ray and medical imaging instruments, would be improved [NASA NRP 98]

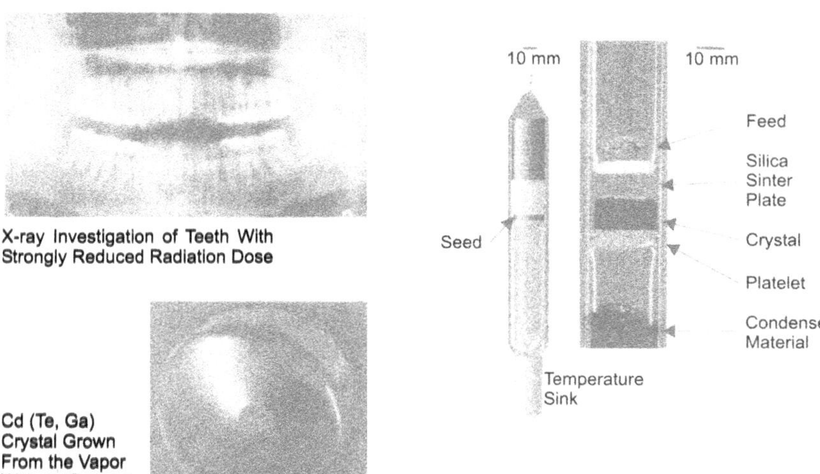

Fig. 7.21. The application potential of CdTe (left above) has not been exploited due to the difficulty in growing sufficiently large crystals with the required quality. Recent results of microgravity experiments (right) demonstrated that new techniques can be employed to grow high-quality crystals (left below) [ESA DMSM 98]

At present, the commercialization of such advanced detection systems is impeded by the difficulty in growing large CdTe single crystals with the required quality. The melt growth method usually employed for the production of CdTe crystals is the vertical Bridgman technique. Crucible contact contributes to increasing the impurity content and generates extensive twinning. In addition to that, the stress induced by the crucible in a material with pure mechanical properties results in very high dislocation densities so, eventually, the yield in terms of the usable fraction of the crystals does not exceed 5%.

Recent results of microgravity experiments have demonstrated that new techniques with a substantially improved yield can be envisaged (see Fig. 7.21).

One technique to be used in microgravity is to provide for the progressive detachment of the melt with a crucible during directional solidification. When the material solidifies from a detached melt, its impurity content is lower and no mechanical stress is induced by the differential contraction of both crucible and crystal during cooling. As a result, the density of twins and dislocations in the crystal is decreased by several orders of magnitude. Laminar convective flows can be imposed in the melt by applying a rotating magnetic field in order to minimize fluctuations at the solid-liquid interface and thereby to enhance the homogeneity of the crystal.

Another technique is growth from vapor. It takes place at significantly lower temperatures and, in the case of semi-closed configurations, without contact with the walls of the growth ampoule. By means of this technique, the above mentioned crystal quality can be achieved as well. Nevertheless, there is compelling evidence that gravity-driven convective flow in the vapor phase has salient effects on the compositional homogeneity of the crystals, particularly those with large dimensions. A thorough understanding of the convective flow and its coupling with the growth process, however, will permit major progress in optimizing CdTe crystal production.

Biotechnology researchers also use microgravity in order to produce, for pharmaceutical research, protein crystals superior to those that can be grown on Earth. Results from experiments, as they were first obtained in 1983 during the SL1 Spacelab mission, were met with great interest from pharmaceutical industry – both scientifically and commercially. Although there are about one million different proteins, until now, the molecular structures of only a few thousand proteins have been determined. This is due to the difficulty of generating protein crystals that are large and perfect enough to be able to analyze them by X-ray scattering. The growth of suitable protein crystals on Earth is hindered by the effects of gravity. Today, there is experimental evidence that crystallization of proteins in space is enhanced, since, compared to production on Earth, new experiments have brought about a number of protein crystals of superior quality. By using high-energy X-ray beams for the study of high-quality crystals, scientists are in a position to unravel their 3D structure. Due to their findings on space-grown crystals, researchers and pharmaceutical companies have already increased our knowledge about diseases such as AIDS, emphysema, influenza and diabetes. New drugs are now being tested for future markets.

Even if we could fly six Space Shuttle flights per year, each successfully producing crystals to reveal the structure of 1000 different proteins per flight, it would still take 35 years to learn the structure of all human proteins. Based on Mir experience, access to ISS promises to increase the rate of advancement in this field by a factor of ten. Space station facilities will enable investigators to analyze crystals on-orbit, decreasing the cost and increasing the quality of research. ISS could become one of the world's premier sources for critical data on the protein structure needed for this new method of drug development. In addition, ISS will be used to study and understand the physics involved in protein crystal growth, helping overcome the difficulties that currently limit much of this research on Earth [NASA NRP 98].

Example: During the Spacelab-1 mission, for the first time, protein crystals were produced whose volume was significantly larger than that of crystals produced on ground [Littke 92]. Currently, the only possible way of determining protein structures is through the X-ray scattering analysis of sufficiently large crystals: X-rays are sent through a crystal and from the diffraction pattern of the rays caused by the arrangement of the atoms, the spatial structure of the protein molecule can be deduced. This approach requires rather perfect grown and sufficiently large monocrystals (cf. Fig. 7.22 and Fig. 7.23).

Thus, researchers have found that microgravity provides new tools in order to address two fundamental aspects of biotechnology: the growth of high-quality crystals for the study of proteins, and the growth of 3D tissue samples in laboratory cultures. On Earth, gravity distorts the shape of crystalline structures while tissue cultures fail to take on their full 3D structure (see Fig. 7.24). NASA's bioreactor, developed to simulate low gravity, proved to be outstandingly successful as an advanced cell-culturing technology. This lead to a successful cooperation with the US National Institute of Health (NIH) and numerous commercial partners. Work with bioreactors has already produced advanced cultures of lymph tissue for studying the infectivity of

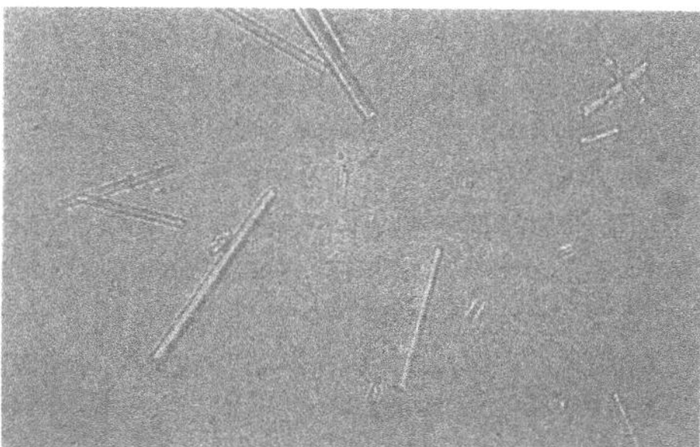

Fig. 7.22. Crystals of a few millimeters in length of the enzyme "Beta-Galactosidase" after undisturbed growth in weightlessness (on the left) [Littke 92] and a computer reconstruction of a protein structure found by structural analysis (on the right)

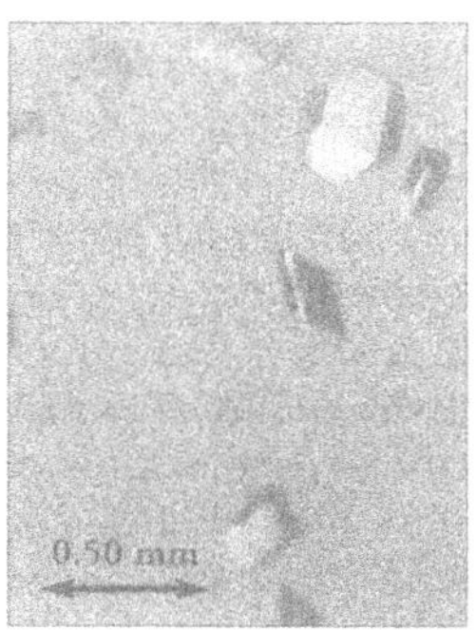

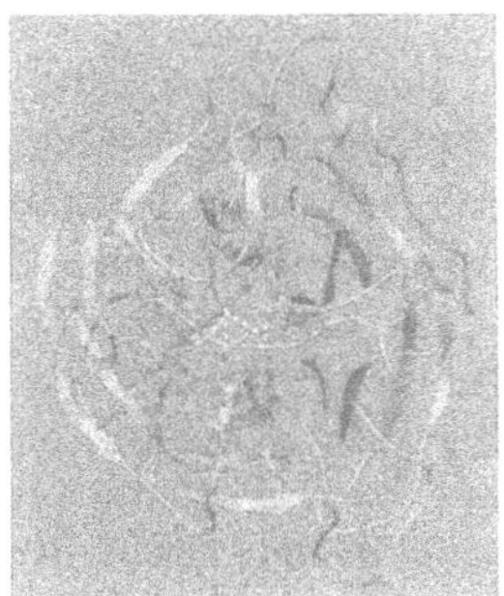

Fig. 7.23. New results on crystal growth of biological macromolecules [NASA NRP 98]. *Left and center:* Diabetes is the leading cause of death in the United States. The disease also causes complications such as blindness and the need for appendage amputation, for which treatment costs the US more than US$ 100 billion a year. Space-grown insulin crystals (left) are much larger and more highly defined than their Earth-grown counterparts (center). In collaboration with NASA's Commercial Space Centers and Eli Lilly and Company, the Hauptman-Woodward Research Institute, Inc., of Buffalo, New York, is using data from space-grown crystals of human insulin to design a drug that will bind insulin, improving the treatment of diabetic patients.
Right: The virus that causes influenza – commonly known as the flu – is shown here superimposed with an inhibitor made possible through space research. Inhibitors block, or "inhibit", the functioning of certain enzymes in the body. The inhibitors here target the neuraminidase enzyme, which plays an essential role in the infection cycle of the flu. A new class of influenza neuraminidase inhibitors started in 1997 clinical trials. The design of these inhibitors was based, in part, on the structure of influenza neuraminidase obtained by NASA's industrial partners from crystals grown on Space Shuttle flights.

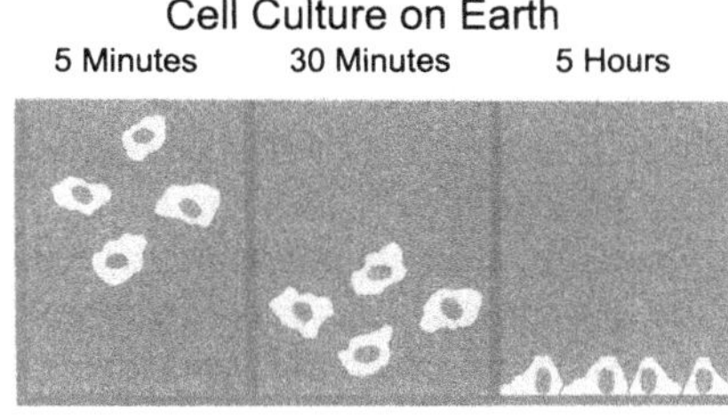

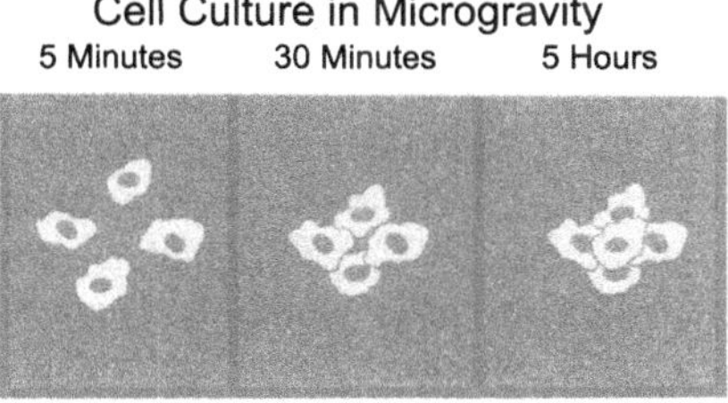

Fig. 7.24. On Earth, the attempt to culture tissue in the laboratory results in a single layer of cells (left) instead of the 3D shape actually taken in the human body. In microgravity, the cells group in the form of 3D structures that more closely resemble the natural state of the tissue (right).

the human immunodeficiency virus (HIV), the virus which causes AIDS. Other areas of success include cultures of cancerous tumors and the growth of cartilage [NASA NRP 98].

The following areas of research shall have top priority in the future:

- Investigations of crystallization mechanisms of biological macromolecules
- Growth mechanisms and perfection of the crystal structure during growth from solutions and vapor phases

The applications that resulted from space experiments conducted up to the present supported different areas and lead to the following achievements:

- Optimization of the cast structure of technical aluminum alloys for Airbus and automobile body parts
- Introduction of new terrestrial casting methods for sliding bearing metals (automobile engine) with improved properties
- Improvement of terrestrial techniques for crystal growth
- Further development of the application of capillary forces (e.g. surface tension tanks)
- Increase of knowledge of both diseases and design of new drugs

7.2.2 Summary of Prospects for the International Space Station

ISS is well-suited for investigations in the fields of fundamental and fluid physics as well as thermodynamics, since it offers the necessary laboratory environment for the hardware as well as scientists for the carrying out of experiments.

The planned activities comprise investigation of critical phenomena in unsteady states and in equilibrium, thermophysical measurements of condensed matter and evaluation of the equivalence principle for the theory of General Relativity.

The microgravity environment aboard the International Space Station is a new experimental tool for investigation of certain fundamental physical processes. For example, the aggregation of dust particles under microgravity will be simulated in a space laboratory in order to replicate the formation of our solar system on a small scale.

ISS also offers an environment for the realization of ultra-exact atomic clocks. This will extend the limits of fundamental research in time-dependent phenomena and moreover will improve the accuracy of time standards.

In the framework of the planned research programs, the fundamental properties of liquids and gases under microgravity will be examined. Fluid experiments will help us to acquire a better understanding of fundamental processes, and these, in turn, will be used in processing methods for a wide range of materials.

There are many practical ideas on how to improve, with the help of ISS, the knowledge of gravity's influence on combustion processes. These processes have a pivotal role in the case of energy conversion, air pollution, traffic, propulsion of aircrafts and space vehicles, global heating of the atmosphere, material processing, and the disposal of dangerous substances by means of combustion. Corresponding experiments dealing with combustion problems allow scientists to study fundamental processes in space which are not accessible on Earth. Such experiments aim at improving the efficiency of combustion on Earth and hence reducing the global consumption of fossil sources of energy. These experiments will also contribute to improvements in fire safety. Finally, the study of combustion processes is also important for the development of future space transportation systems.

The present plans for the equipment aboard ISS with regard to materials science include several furnaces, in which metals, alloys, electronic and optical materials, glass, ceramics and polymers are melted and solidified at different temperatures, thermal gradients, and heating rates. Each furnace optimizes the advantageously reduced fluid convection at critical points in time for every single material. Main advantage: Materials result which can be used for various purposes such as high temperature-resistant material, computers, electro-optical instruments, or protheses. Finally, ISS will also be used for crystal growth, especially for the growth of protein crystals.

There is also a plan to directly expose test items of different materials to space by means of the external accommodation sites of the International Space Station. This will improve our knowledge of their behavior under the simultaneous influences of vacuum and space radiation as well as extreme temperatures in space. This aspect will also be addressed in Sect. 7.6 on "Space Engineering and Technology".

7.3 Life Sciences and Biotechnology

7.3.1 Results Obtained and Areas for Future Research

In the field of life sciences, which includes biology and human physiology, research results obtained in space during the last decade have had a remarkable impact on the current thinking of the role of gravity in all areas of life. Experiments in space have disproved quite a number of previously existing theories on the mechanics of dynamic behavior in certain plants and animals [ESA 95, Moore 96].

Biological experiments under microgravity have shown that some aspects of cellular activity depend on the vertical, directionally stable gravity vector on Earth, and are thus disturbed under conditions of reduced gravity. This surprising scientific observation gives rise to numerous further questions.

One of the most important life sciences results obtained in the early years of space flight was the remarkable tolerance higher organisms have with regard to microgravity, even over long periods of time. However – it is clear that considerable changes take place throughout the human body as soon as the influence of gravity is missing. Certain effects occur within minutes, such as the loss of orientation or rapid shift of body fluids from lower parts of the body into the thorax and head. When being exposed to microgravity for a long period of time, reduction of muscle mass and muscular strength together with loss of bone minerals and deterioration of bone structure have been observed (cf. Sect. 11.2).

Although the results obtained in life sciences experiments under microgravity have brought about new insights, new questions were still raised as well. As will be explained in more detail below, it is expected that scientific research aboard ISS will provide answers to these questions. Any advances in the understanding of the fundamental aspects are important for the future of medical science.

Molecular Biology and Cell Biology

In the case of biological cells, the effect of gravity is very small in the cell compared to the forces between atoms and molecules. Despite this, in both fields of research there is obvious proof to support the fact that microgravity has a strong influence on cellular functions. Microgravity is a new, noninvasive tool to investigate these cellular functions. It will provide new insight into how cells perceive signals and react to them. This is essential for the better understanding of biological and physiological processes with potential applications in molecular medicine, such as wound and tissue repair.

In this context, presumably the most interesting observation was delivered by *in vitro* experiments concerning the strong influence of microgravity on the signal transfer in lymphatic cells, epidermic cells and bone-building cells.

The repeatingly occurring signs of immunological disturbances and problems with the healing of wounds, in astronauts and animals alike during a mission, as well as the occurrence of demineralization of the body, have their origins probably rather on the cellular than on the organ level. Up to the present point, there is no key to the understanding of the phenomena observed. This is the evidence for the existence of a huge number of potential discoveries of unknown biological mechanisms that are active in cells.

Especially in plants and single-cell organisms, gravity influences cell growth, the increased occurrence of damage to chromosomes carrying genetic information, and it also influences the generation of reproductive cells in the different organisms (see Fig. 7.25).

The interaction between the cytoskeleton and the organelles as well between the skeleted substance and the plasma membrane seems to play an important role for gravisensitivity in plants. This new finding is of perhaps general importance for the understanding of the influence gravity has on the cellular level. The very important

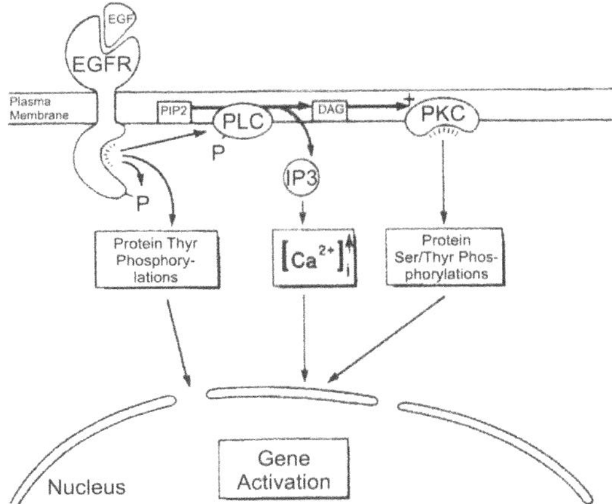

Fig. 7.25. Regulation of Cell Growth and Differentiation. Cascade of early cellular responses to growth factor binding to a membrane receptor; a process that may amplify microgravity sensitivity.

cytoskeleton is probably of decisive importance for the perfect course of numerous cell processes and probably reacts very sensitively to differences in gravity.

All new findings invalidate earlier views that assumed the influence of gravity on cells to be insignificant. Modern opinions with regard to cells and their interaction with the extracellular environment go together remarkably well with this conclusion. Utilization of ISS's potential for repeatable experimental research with re-adjusted parameters will contribute to the clarification of mechanisms that, due to their basic character, are of considerable importance to medicine, pharmacology and biotechnology.

Example: The ability of plant roots to perceive gravity (gravisensitivity) is important for the growth of plants on Earth. If, for example, a watercress seed is planted into the soil, its root will always grow downwards. When trying to prevent the plant from growing into this direction, e.g. by a change in orientation of the seed (cf. Fig. 7.26 a), the effect of gravity will, after only a few minutes, be such that the root tip will again grow downwards. It has been known for a long time that at the end of the root tip there are sensitive, gravitation-perceptive cells called the statocytes.

In Fig. 7.26 b, a cross section of such a cell is shown: in the gravitational field of the Earth (TM 1g) or in the artificially simulated gravitational field of a centrifuge in space (FM 1g), massive cell components, the amyloplasts (designated "a" in the illustration) deposit themselves in the direction of the acceleration vector and apply a pressure on the pressure-sensitive cellular substance of the organelles (designated "er"). The cell receives the information "downwards" through the location of the occurring pressure and, with the help of the organelles, transfers it to those cells that are able to react to the given information by growth control. During the D1 mission, an important finding was obtained: a cell grown in weightlessness (FM 0g), due to

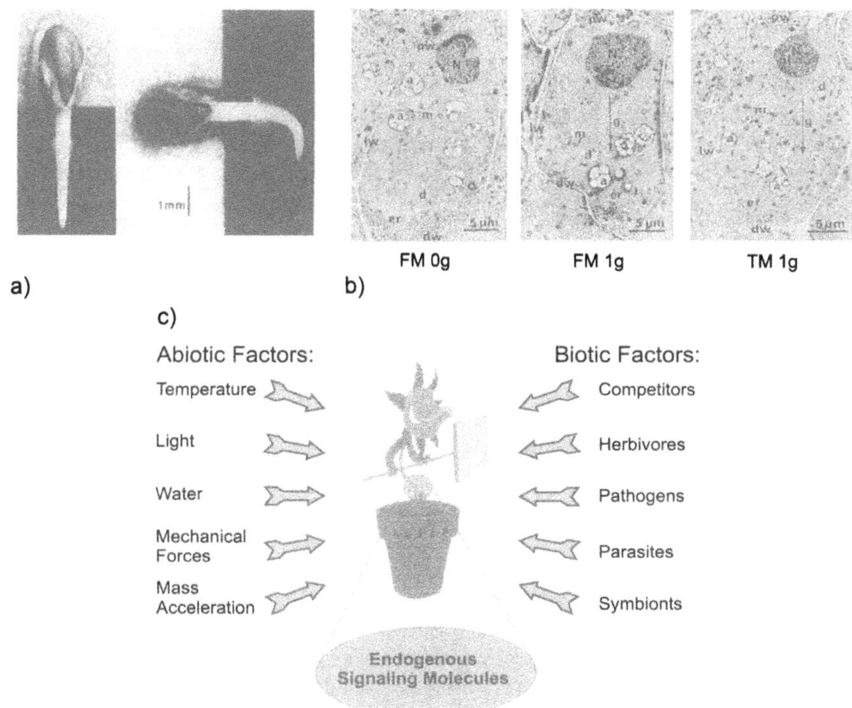

Fig. 7.26. a) Growth Behavior of a Watercress Root on Earth
b) The Root's Gravity-perceiving Cells on Earth (TM-1g) and in Weightlessness (0g and 1g by Onboard Centrifuge; FM-0g and FM-1g)
c) All Biotic and Abiotic Factors Influencing Plants, the Most Important Representative of the Latter Being Weightlessness

the existing genetic information, has the same basic "sensory" components as a terrestrial cell. As a consequence, the cell in space as well would be in a position to perceive very quickly, acceleration and hence gravity.

The actual growth under permanent weightlessness appears to be directed, although the orientation mechanism described does not work without any acceleration. The orientation of the individual plants appears, however, to be arbitrary. The experiment unit NIZEMI, a centrifugal microscope which resulted from the experiences acquired during D1, was used during the IML-2 mission. With the help of NIZEMI, the threshold value was determined for different organisms indicating the point in time when each individual organism begins to react to gravity. The determination of a threshold value is important in order to clarify the mechanism of gravity perception and processing. In Fig. 7.26 c, the different biotic and abiotic influences to which plants must react in order to ensure their survival are shown. Some of these influences are either strengthened or weakened by either gravity or its absence and, for that reason, are the subject of experiments in space.

Developmental Biology and System Biology

In the fields of developmental biology, botany and system biology, certain plant motions that occur on Earth such as tropism, nutation and stochastic motion have also been observed under microgravity. This observation caused common theories (according to which these motions are only possible under gravity) to be reconsidered.

A low-gravity environment causes disturbances in the early embryonic state; however, malformations seem to be balanced by the time the plants, insects or amphibians have completed their development. When studying those internal mechanisms which may correct the initial disturbances occurring in the development phase of embryos, there is the basic question of biological redundancy and its control: an area that is developing in the field of biology at the moment. Of course, the question of DNA repair processes being influenced in space is closely related to it.

One of the advantages of the long-term availability of ISS is the opportunity to observe changes in the development of biological systems through the long string of subsequent generations that would grow up in space.

Human Physiology

While we know that humans can live and work in space, we have found that the microgravity environment induces a number of changes in the body itself (cf. Sect. 11.2). Spaceflight affects almost every aspect of the human body, including the heart, lungs, muscles, bones, immune systems, and nerves. More specifically, astronauts experience a significant loss of both bone and muscle mass as a result of spaceflight. Their hand-eye coordination is also visibly affected. By means of the results available today, several important observations can be identified in this field of research [Moore 96, NASA NRP 98].

For example, apart from the thermoconvection indicated in common theories as the excitation mechanism for rapid eye movement at thermal stimulation (caloric nystagmus), there must be additional causes. A precise diagnosis is only possible when the actual mechanisms are identified; this would be of value in finding suitable therapeutic measures for different diseases related to mechanisms of the brain, inner ear or eye reflex. And it becomes even more important considering the fact that every day throughout the world, methods based on the principle of caloric nystagmus are used to examine the function of the inner ear. A new understanding can therefore lead to better diagnoses of the diseases of the inner ear.

Example: By the thermal stimulation of the inner ear (vestibular apparatus), the kinetic and proprioreceptive stimulus (nape muscle, tactile sensors), the central nervous system and hence also the control skill of the eye movement is influenced (cf. Fig. 7.27). As a consequence of accelerating the sled facility during the Spacelab mission D1, the vestibular apparati of the astronauts were stimulated.

The caloric nystagmus was surprisingly also observed in this case (as it was before during the SL-1 mission), even though in weightlessness, thermoconvection in the inner ear is virtually impossible.

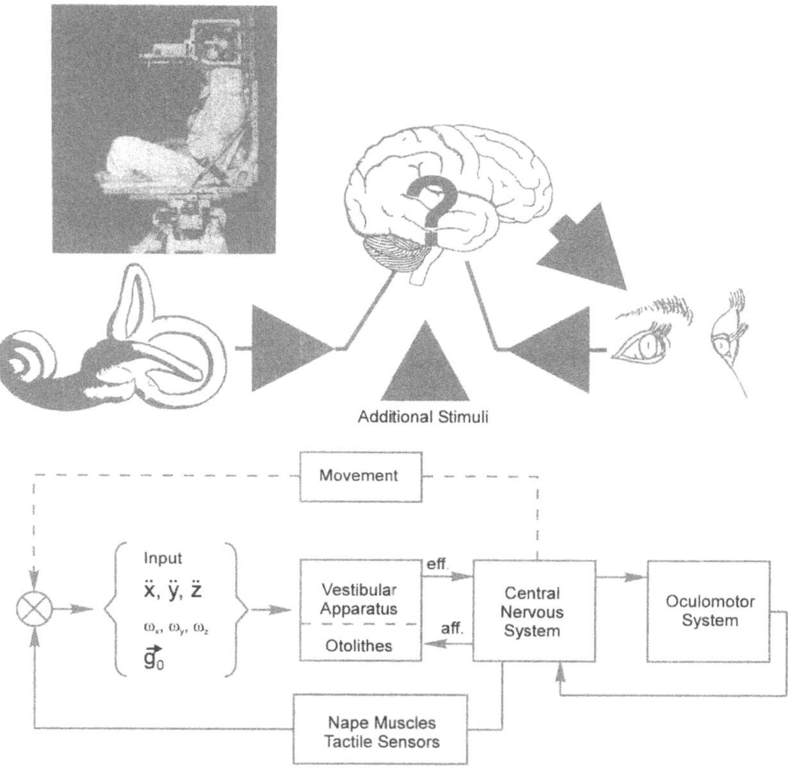

Fig. 7.27. Investigation of the Vestibular System with the Help of the "Sled Facility". The astronaut's head is put into a helmet that is equipped with sensors for recording eye movement. The control of the vestibular apparatus is markedly disturbed due to the influence weightlessness has on the otolithic system in the inner ear. This often leads to the "Space Adaptation Syndrome" (SAS), commonly known as "Space Sickness".

The ventilation-perfusion ratio for human lungs turned out to be almost gravity-dependent. This fact contradicts common theories as well. It is an important diagnostic finding, since numerous diagnostic and therapeutic measures rely on models that are based on the normal gravity conditions on Earth and have now turned out to be insufficient with regard to the overall functioning of human lungs.

Also the current knowledge on regulation and distribution of body fluids must be re-examined with regard to the results obtained during experiments in microgravity. The adaptation of the intravascular volume under microgravity progresses differently than on Earth; its mechanism is of great importance for the understanding of the regulation of blood pressure and blood volume. Moreover, interstitial and extracellular fluids were confirmed as the main reservoir for rapid fluid transfer throughout the human body – a fact that is of life-saving importance for patients experiencing severe shock or trauma, after surgical operations or with systemic diseases.

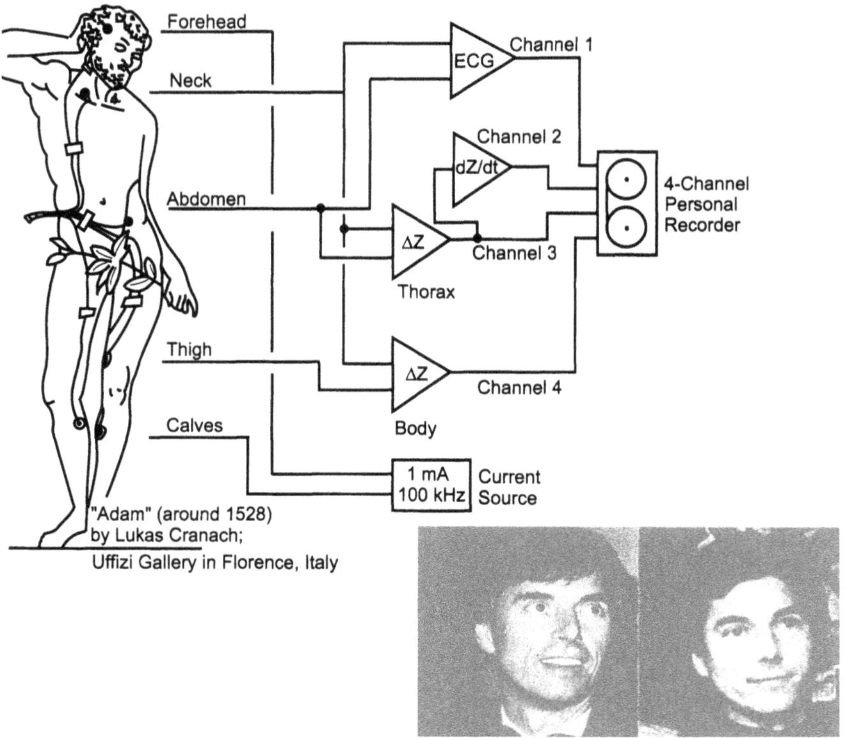

Fig. 7.28. Measurement and effect of the shift of body fluids into the upper half of the body: an astronaut's face immediately before flight (left) and only a few hours after arrival in weightlessness (right). The measurement of this shift in body fluids is carried out by recording the electric current and determining the corresponding electric resistance between the different body electrodes.

Examples: Under the influence of weightlessness, up to 2 liters of blood and tissue fluids are shifted from the legs into the abdominal cavity, ribcage and head. That not only changes the cardiac output but also the site-specific blood pressure, mainly in the veins. During the D1 mission, the astronauts' venous pressure near the heart as well as their intraocular pressure were measured daily (cf. Fig. 7.28 and Fig. 7.29) in order to find out how the changes in pressure due to fluid shifts are regulated throughout the body and the effect fluid shifts have on peripheral fluid systems. Astonishingly, the changes in pressure during the transition into weightlessness were not as high as expected; moreover, blood pressure and local blood volume, independently of one another, level out within a few hours or days at nearly terrestrial values. The shift of peripheral fluids into the limbs takes the longest time as was found out with the help of an impedance measurement between individual body electrodes.

An interesting symptom of fluid shift is actually visible in the astronauts' faces (cf. Fig. 7.28). The additional body fluid in the upper part of the body leads to a

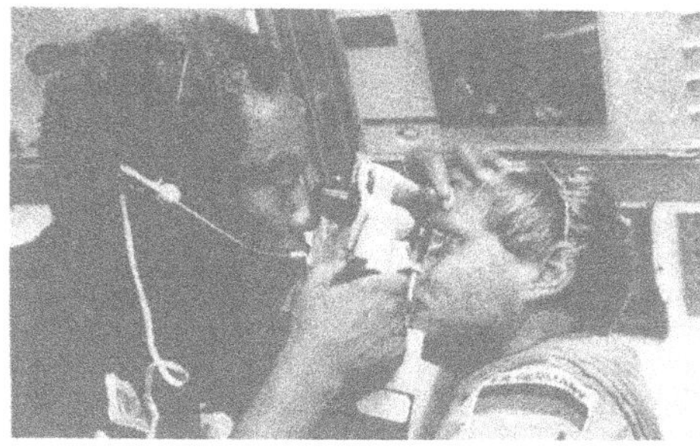

Fig. 7.29. Measuring the intraocular pressure of astronauts by means of a tonometer (on the left: D1 mission) and, on the right, a self-measuring instrument that was developed for the D2 and the Mir-92 missions, based on suggestions of the D1 crew.

puffy face with smoothed skin, and that makes the astronaut look younger by several years (after the mission and back on Earth, however, the astronaut looks at least as old as he or she did before the flight).

The self-tonometer illustrated in Fig. 7.29 offers various possibilities. In general, early diagnosis prevents the patients from suffering serious consequences. This also applies to the eyes, which, among other risks, are endangered by glaucoma, the leading cause of blindness in the western world. As a result of developments made for the D2 mission, a self-tonometer is now available to patients suffering from glaucoma, which will then be able to measure intraocular pressure themselves and thus spare frequent visits to the eye specialist. The practical self-tonometer will make possible the efficient early detection of complications and the corresponding preventive measures [D2 95].

Experiments in space showed that the chronological course associated with and the net loss of muscle mass and muscular strength change by several orders of magnitude. This breaks with the current concept of atrophy due to lack of use as it is observed in bedridden patients or with illnesses that are connected to a muscle weakness.

The loss of calcium and trabecular structure in bones cannot only be attributed to hormonal regulation of the bone homeostasis which will lead to osteoporosis if the endocrine system does not provide any support (cf. Fig. 7.30). Given these new results it would make more sense to take into consideration rather mechanistic, i.e. gravity-dependent, processes which would probably lead to new concepts for the diagnosis and therapy of health problems which are spread throughout the world.

Many of the physiological changes in astronauts actually resemble changes in the human body normally associated with aging on Earth. For instance, in addition to losing mass in the microgravity environment, bones and muscles do not appear to heal normally in space. For astronauts, time spent in microgravity seems to result in dissociation between their physical and chronological ages. By studying the changes in astronaut's bodies, a model describing the consequences of getting older could be developed.

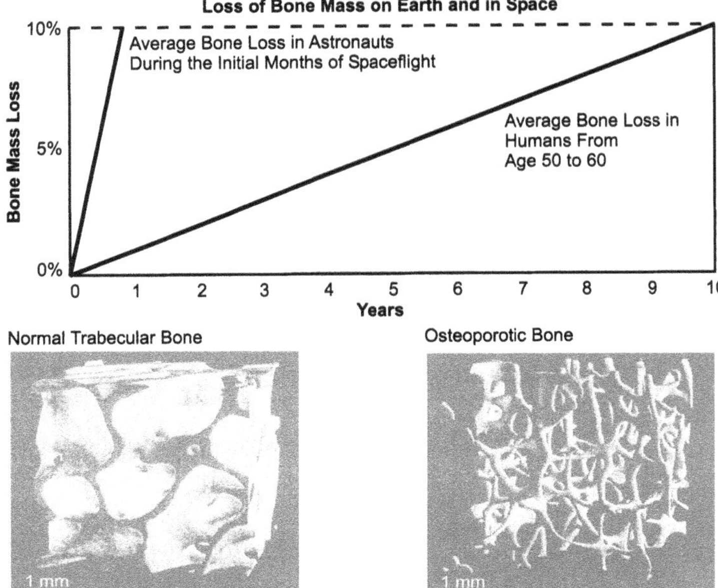

Fig. 7.30. Loss of bone mass on Earth and in space. Astronauts who spend over eight months in space lose about the same amount of bone mass as human beings on Earth do, on average, from age 50 to 60. Examinations in space offer the opportunity to find, over a very short period of time, strategies for the treatment and prevention of osteoporosis. Osteoporosis is a silent disease, often named the "women's" disease since about 80% of cases are related to women. In Europe, at present, there are over one million osteoporotic fractures per year and their occurance is expected to increase with aging of the population by 65% in the year 2025. Current treatment costs in Europe: $27 million per day. [NASA 95, ESA DMSM 98]

Radiation Biology

It must be reiterated that biologists and human physiologists are not only interested in space stations because of the opportunity for studying the effects of low gravity. They also want to examine the characteristics of an environment with space radiation and its effects on plants and live organisms as well as the effect which other space environmental factors have on lifeforms in space.

In the field of radiation biology, a series of experiments was conducted in space which proved in a highly precise and unambiguous way that individual heavy ions of cosmic radiation can cause severe damage in nearly all test organisms. The consequences were mutations, abnormalities, and even cell death. The results obtained in space showed qualitative agreement with the results obtained with heavy-ion colliders on Earth; quantitatively, however, the effects in space are greater than those on Earth. Thus there is the question of how the reaction to radiation is changed by massive ions and gravity simultaneously. The 1g centrifuge in the Biorack of the Spacelab yielded two important results that suggest that a low-gravity environment intensifies the effect of radiation. Before being able to give proven answers, these results call for further treatment and supplementary experiments.

Although large amounts of dosimetric information have been collected in space, further measurements of particle fluxes and spectra of all radiation components are necessary in order to carry out exact dosage calculations and to be able to make precise dosage predictions for future missions. The data obtained will serve as basic information in order to estimate the radiation risk for humans in space flight missions and also to establish radiation standards for humans in space.

Exobiology

In the field of exobiology, studies of microorganisms in space support the view that dormant forms of living organisms such as seeds and spores can survive a very long time under the influence of the space environment. This can be achieved by shielding them from the ultraviolet radiation of the Sun, e.g. by embedding them in stone. This, however, has the consequence that life is not necessarily limited to our Earth and that the possibility of traces of pre-existing life being found on other celestial bodies in the solar system exists as well.

As a summary of life sciences, the following list indicates those subject matters in which important new findings were obtained:

- Vestibular research with regard to the mechanism of the caloric nystagmus under refutation of the Barany hypothesis (area of application: diagnosis and therapy of equilibrium disturbances for human beings)
- Regulation of fluid distribution throughout the human body (area of application: patients with shock or trauma and patients suffering from edema)
- Muscle and bone physiology (area of application: patients with muscular atrophy and osteoporosis)
- Role of the peptide-like hormone Urodilatin in the regulation of water and salt excretion (area of application: maintenance of kidney function, e.g. post-surgical)
- Clarification of stimulus transmission mechanisms as well as effects of gravity on the cellular level (area of application: knowledge of the mechanism of the immune system and of bone demineralization as well as of the formation and treatment of diseases on the cellular level in general)
- Measurement of space radiation and the clarification of its effect mechanism on organisms (area of application: radiation protection in space and on Earth)
- Electrophoresis and electro-cell-fusion as well as crystallization of biological macromolecules, processes that are principally more effective under microgravity (area of application: from biotechnology to the specific design of medicines).

7.3.2 Emphasis on Further Research in the Field of Life Sciences

Within the life sciences, the various advisory committees recommended concentration of research activities on the main subjects which are shown in the following lists [ESA 95, Moore 96, NASA NRP 98]. In the field of biology they are:

- Regulation mechanisms of proliferation and differentiation on the cellular level
- Events during early development
- Neurobiology of development
- Organization of nuclei
- Perception and signal transfer with tropism and taxis
- Programmed cell death (apoptosis)
- Mechanisms of radiation damage in cells and tissues
- Repair of cell and tissue damage
- Long-term physiological and genetic stability including radiation effects
- Survival of microorganisms in space

In human physiology they are:

- Musculoskeletal system
- Cardiovascular function
- Fluid equilibrium and kidney function
- Respiratory function
- Sensory-motor function
- Hormones and metabolism

For those subjects listed under "biology", a space station is not absolutely necessary in each case since certain research activities can also be conducted without human intervention in space. If, on the other hand, astronauts are already present anyway, and all other resources of the International Space Station are available as well, most research activities can benefit from their presence.

In the case of research activities recommended in the field of human physiology, however, it is inherently understood that crew members must be present to serve as test subjects in space.

In its present state, the research program of biology for the International Space Station was based on the recommendations of leading scientists. The objective of this program is to achieve a better understanding of the role gravity plays in biological processes. Special emphasis is placed on research in the field of cell biology, molecular biology, developmental biology, botany and system biology.

The laboratory facilities for research in weightlessness and the planned centrifuge aboard ISS will be used for these research activities in order to examine the influence of gravity at different degrees of intensity and to compare the results obtained.

The research objective will be to find answers to the following questions: How is gravity information transferred? How do cells, plants and animals react to short-term and long-term changes in gravity with regard to their growth, development, reproduction, genetic wholeness, lifespan, aging and consequences for subsequent generations?

Although this is definitely a program for fundamental research, we must not forget that knowledge of the functioning of living systems in space is of decisive importance for long-duration stays in space.

The research program in human physiology for the International Space Station comprises tasks aimed at describing and understanding physiological changes caused by weightlessness. Moreover, within the framework of this program, therapeutic countermeasures and life support techniques will be developed, allowing the astronauts to live and work under microgravity and to reduce the risks after return to the terrestrial gravity field to a minimum. The main objective is the optimization of safety, well-being and the work performance of the crew. New methods for non-invasive measuring methods during flight are to be developed. This research program is also expected to entail improvements for health care and quality of life for humans on Earth. Concrete applications can be defined especially in the newly-established field of telemedicine. This term includes, on one hand, the measurement of critical data of a patient under emergency treatment on the way into the hospital and the real-time transfer of these data to the emergency room physicians, and, additionally, the transfer of a patient's data from a remote place to a special medical care center.

Moreover, this research program pays special attention to the problem of radiation and its influences on an astronaut's health. The International Space Station offers the unique opportunity to measure the radiation fields inside and outside the station and to measure the biological effects of space radiation comprehensively. In this context, special emphasis is placed on creating a solid base of knowledge to support the present and future exploration and utilization of space. Examinations which have top priority address the carcinogenic mechanisms of radiation and the reliability models for species-independent extrapolation of biological effects of radiation with special regard to their application for humans.

Research in the field of human physiology aboard ISS will also concentrate on Human Factors and Man-Machine Interfaces in order to develop new processes and procedures for the improvement of human performance in space and the improvement of the design of complex automated systems.

Special attention will be paid to the psychological and social problems astronauts have to face during their isolation in space. In the case of the Human Factors (see Chapter 11) program, it is expected that the main emphasis will gradually shift from the collection of knowledge to the application thereof.

The relatively new area of **biotechnology** is expected to play an important economic role in the 21st century. Biotechnology means the investigation and handling of molecules, tissues and live organisms; it covers a large number of disciplines: cell engineering; growth of protein crystals; polymer science; cell biology; biochemical separation; micro-carrier and micro-capsule preparation; cell culture; bio-molecule production. Biotechnology is becoming increasingly more important in the fields of health care, agriculture, production and process engineering, and environmental protection.

Biotechnological research aboard the International Space Station has the following objectives:

- To increase the understanding of gravity-influenced biotechnological processes
- To gain insight into the sequence of biotechnical processes with the help of experiments under low gravity
- To contribute to Earth-bound processes relevant in biotechnology
- To develop new technologies that especially promote biotechnology in space and on Earth

This will be performed through international cooperation among researchers from the fields of medicine, biotechnology and engineering.

Example: Millions of people suffer organ or tissue loss from diseases and accidents every year. Yet, transplantation of tissues and organs is severely limited by the availability of donors. Growing tissue samples outside the body is one of the major goals of current medical research, and the microgravity environment has great potential for advancing this research.

Experiments in bioreactors (cf. Fig. 7.31) are performed to study the ways in which cells multiply and interact to form skin, bone and organs; and these cell- and tissue-culturing techniques also aid in the study of cancer cells and tumor formation.

Knowledge gained in microgravity on the regulation of cell growth and differentiation will also help improve the cultivation of sensitive and highly differentiated cell strains like those needed to produce artificial organs. Studies of cells' ability to migrate in reduced gravity may produce new insights into the factors that allow can-

Fig. 7.31. ESA's Space Bioreactor for suspension culture of sensitive cells and multicellular systems. The benefits of low gravity for growing cell cultures can be attributed to the absence of convection and sedimentation, which for the three-dimensional cell aggregates, favors the creation of a tissue-like environment [ESA DMSM 98].

cer to spread. When combined with biomedical research on Earth, these investigations would contribute to the development of new ways to prevent and treat related diseases.

Example: DNA transfer between biological cells of different weight is nearly impossible on Earth, since, due to sedimentation and gravity-driven convection, the cells in the respective solution do not stay together long enough. In space, however, DNA transfer is possible using electro-fusion. Figure 7.32 shows:

a) Two different cells approach each other under the influence of the electric field; in the contact zone, protein-free areas occur in the membrane.
b) The electric discharge disturbs the membrane structure, the lipid molecules no longer adhere to their original membrane.

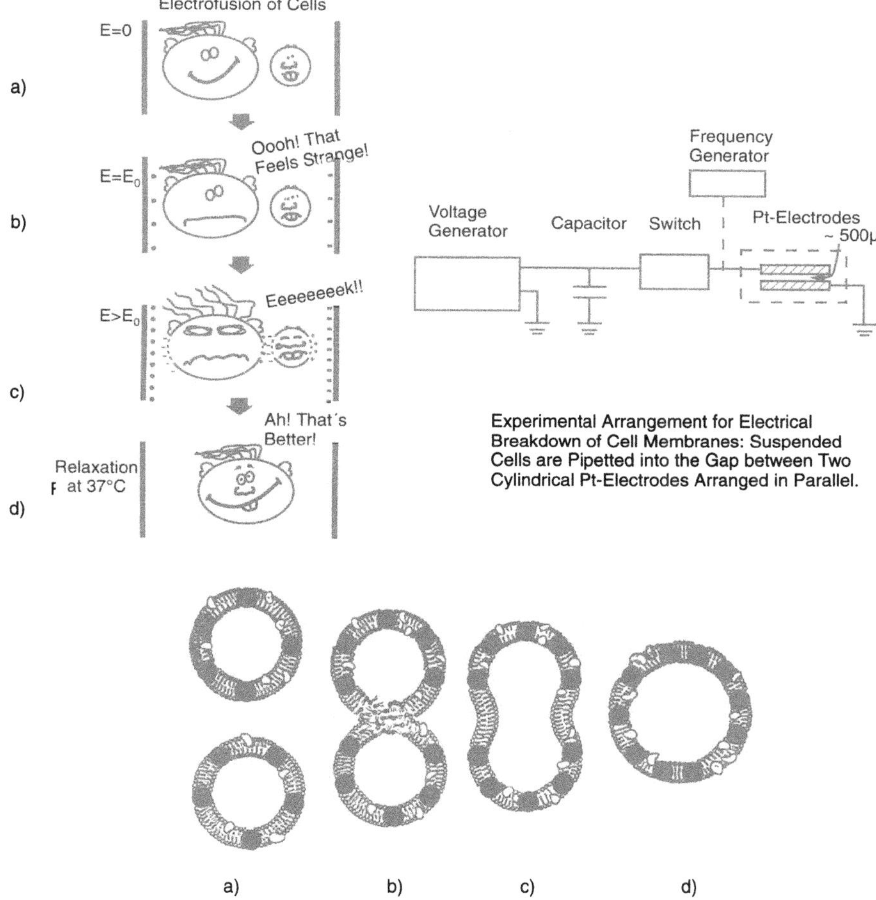

Fig. 7.32. Apparatus (above) and Model of the Mechanisms of Electro-Induced Cell Fusion (below)

c) If bridging occurs, the fusion product will be smoothed for energy reasons.
d) After the fusion of the membranes, the individual components of the membrane mix.

7.4 Space Sciences

Utilization of the International Space Station is of interest to the following space sciences disciplines: astrophysics, radiation physics, magnetospheric physics, and the various subdisciplines that observe the Sun and the solar system [ESA 95].

7.4.1 Typical Disciplines of Space Sciences: Astrophysics and Radiation Physics

For astrophysics, the observation of high-energy processes is an invaluable means of support in order to investigate unusual interstellar objects such as:

- Supernovae
- Black holes and neutron stars
- Structure and formation of galaxies
- Fundamental cosmological events such as the production of antimatter by the Big Bang, from which the universe emerged

This area of astrophysics covers the investigation of the origins, the spreading and the interaction of high-energy cosmic particles, and of electromagnetic radiation.

The study of these processes in space can yield new findings on naturally occurring, extremely high-energy processes that currently, in most cases, cannot be reproduced on Earth.

Examples: The Spacelab missions that have focused most on astrophysics are Spacelab 2 and Astro-1 with important contributions coming from missions such as ATLAS-1. Both Spacelab 2 and Astro-1 flew large telescopes, most of which were mounted on the Instrument Pointing System (IPS). The Spacelab 2 experience indicated that the environment of a large crewed spacecraft in Low Earth Orbit, together with pointing systems and the operation of thrusters and which can generate light-scattering particles, results in an induced emissions level that is not suitable for infrared observations. The major contributions to science have been in the vacuum ultraviolet and the X-ray ranges, due largely to the fact that Spacelab opportunities emerged just as the technology in these wavelength regions was becoming accessible. These contributions have been many and significant and have produced datasets that could not have been acquired through other missions.

Figure 7.33 shows (above) the detail possible in UV images acquired from outside the Earth's atmosphere. Observations from Spacelab 1 found lanes of new star formation in the bridge between the Large and Small Magellanic Clouds (Fig. 7.33 below). Some of the telescopes flown on Spacelab missions have had features not available in any of the other orbiting astrophysics facilities.

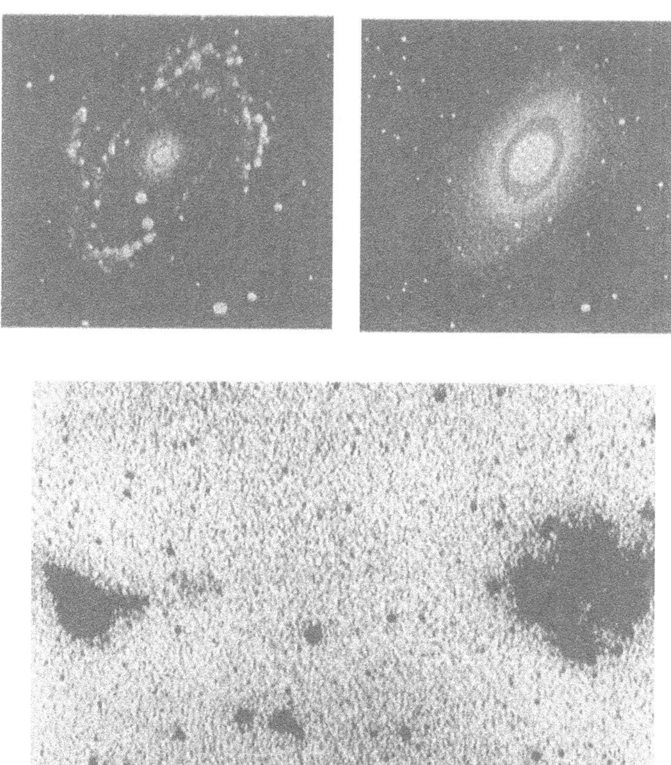

Fig. 7.33. *Above:* Two Images of the Spiral Galaxy M81 in the Constellation Ursa Major. The left panel was obtained with the Ultraviolett Imaging Telescope during the Astro-1 mission on December 9, 1990. The right panel is an image of the galaxy taken in red light using the 36-inch telescope at the Kitt Peak National Observatory.
Below: UV photograph obtained with the Wide Field Camera on Spacelab 1, showing a new star-forming cloud in the Large Magellanic Cloud–Small Magellanic Cloud Bridge [NASA 88].

Magnetospheric Physics

In the field of magnetospheric physics, the Earth's magnetosphere and the ionosphere and also near-Earth space have been investigated since 1958. Research on this subject yielded information on the Earth's magnetic field and its interaction with the solar wind, the radiation environment, and the dynamics of charged particles in the magnetosphere (cf. Sect. 3.2.1).

Solar Physics

Studying the Sun is important for our lives on Earth since the Sun is the most important source of thermal energy for our planet. A change in the solar radiation by only 1% could have lasting effects on our climate; for example, it could cause the beginning of a new ice age.

Studying the Sun includes measuring changes in its radiation intensity, i.e. its total energy dissipation. Such long-term measurements have great importance for long-term studies of the Earth's climate.

Moreover, a continuous observation of the Sun is useful insofar as solar flares may impair electronic instruments and telecommunication systems on Earth, and they are also a serious radiation risk for astronauts in space (cf. Sects. 3.4.2 – 3.4.5).

Study of the Solar System

Comets and asteroids are unique sources of information of the origins and formation of our solar system. Today's approach for getting hold of this information is to send interplanetary probes to direct encounters, for example, the probe for the mission to Halley's Comet.

Cosmic dust particles of comets and asteroids flying in the direction of Earth contain interesting information. As they rarely survive the thermal and mechanical strains upon entering Earth's atmosphere, they have to be collected prior to the entering. For that reason, lengthy collection times and large collection areas are necessary – two preconditions that cannot be fulfilled by the vehicles that are currently available and are used for research in the field of space sciences.

A comet or an asteroid has no problem penetrating the Earth atmosphere. However, the collision of a comet or an asteroid with the Earth could have devastating consequences for the environment and human civilization. Although events of this kind are extremely rare, they cannot be completely excluded as was shown by the collision between a comet and the planet Jupiter in 1994. As a consequence, Russian scientists suggested using the International Space Station as an outpost to search for objects in space that have a potential collision course with the Earth. ISS would be perfect for such a meteor watch function; its external truss offers sufficient space for the accommodation of instruments that are capable of locating meteors and other dangerous objects up to a distance of 1 AU away and analyzing their respective trajectories. Such a "warning system" would probably provide enough lead time to take suitable measures to protect the Earth from a global natural disaster.

Excerpts from the research agenda and the corresponding questions are:

- What physical processes take place in extreme environments such as black holes?
- How did the universe begin, and what is its ultimate fate?
- How and why does the Sun vary, and how do the Earth and other planets respond?
- How is the evolution of life linked to planetary evolution and to cosmic phenomena?
- How do galaxies, stars, and planetary systems form and evolve?

7.4.2 What does ISS Offer in the Way of Benefits for Space Sciences?

In the past and present, most space scientists have mainly been using their own mission-specific satellites and probes. A permanently crewed multi-purpose platform in a LEO would open up new possibilities to them.

Therefore, the Review Panel of the ESA-Directorate for science (which defined Europe's space science program "Horizon 2000 Plus") examined the potential utilization of the International Space Station for space science experiments.

The panel expects the opportunities offered by ISS to complement the uncrewed and task-specific platform missions (taken into account in the framework of "Horizon 2000 Plus") in a profitable way.

The external accommodation sites for payloads, the rapid access (and return alike) of experiments thanks to Express Pallets and Express Racks, and also the station's role as a testbed for prototypes of certain instruments are especially attractive. ISS also offers a special opportunity to assemble, adjust, operate and maintain large instruments directly in orbit. Transfer to coorbiting positions can be assumed by the transfer vehicles slated for logistical tasks such as the ATV.

The International Space Station is considered to be well-suited to accommodate instrumentation for high-energy astrophysics and radiation physics. Since the XMM and "Integral" missions from the "Horizon 2000 Plus" program, a group of leading representatives of high-energy astrophysicists exists which demands continuity for the development of the next generation of instruments. In order to use ISS for large experiments, the Review Panel recommends taking the possibility of a large facility for high-energy astrophysics onboard the station into consideration.

There is also the demand for long-duration observation of known astronomical sources in a wide spectral range. Such observations can be carried out with the help of a small multi-purpose telescope. In addition to that, such a telescope could also facilitate the observation of objects whose occurrence cannot be predicted.

Continuously taken measurements of the integral and especially the spectral radiation intensity of the Sun all the way from the ultraviolet to the infrared range could be interesting for research on the Sun and on the solar system. The International Astronomical Union explicitly encourages such an activity as could be performed by a group of instruments especially suited for measuring the various wavelength ranges.

Another subject of interest is the permanent observation of the immediate vicinity of ISS with the aim of examining the distribution of natural particles and artificial space debris. In addition to that, the use of a tether would enable active plasma experiments.

Example: The GAUSS camera used during the D2-Spacelab mission made it possible to study low-contrast structures in the ultraviolet range in the solar system and in the Milky Way (cf. Fig. 7.34). Aboard the International Space Station, there will be different cameras covering the optical range. The fact that the cameras are also able to be used in frequency ranges, where observation from the Earth is either impossible or very limited, is most important.

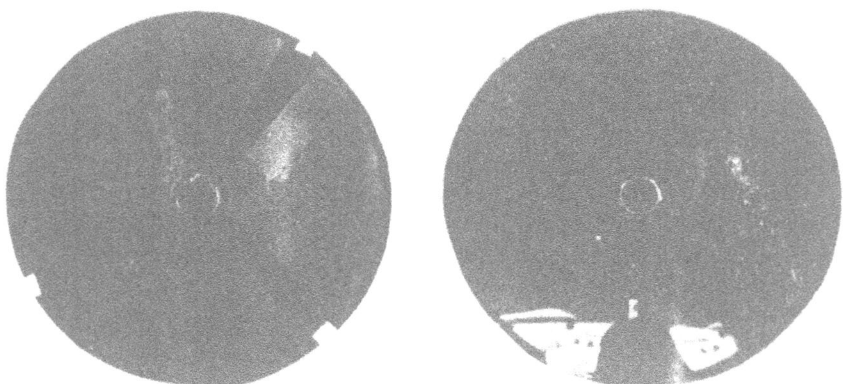

Fig. 7.34. Observation of coronal structures in the Milky Way at the hydrogen-Lyman-α-wavelength (102.4 nm). The picture on the left shows the so-called "geocoronal structures" of which only some traces are visible. The picture on the right is used for the identification of similar structures against the background of the stars of the Milky Way at a longer wavelength [D2 95]

One unique aspect of ISS for Earth and space science investigations is the ability to include human intervention in the operations loop. The ongoing operations of the Hubble Space Telescope have demonstrated the advantages such intervention can bring. Space Shuttle servicing missions have now upgraded Hubble's science equipment twice, and future Shuttle rendezvous missions with the telescope are planned. The dynamic nature of our planet and the Sun means that the capacity for operations run and managed by humans on a day-to-day basis will be a valuable resource. The astronauts will be able to take full advantage of human adaptability in order to observe trends, respond to the unexpected, and alter the research approach whenever necessary. They will respond quickly to observe, record, characterize, and assess the impact of natural events as they occur, thereby advancing our understanding of our planet as a system.

7.5 Earth Observation

What does a Space Station Offer in the Field of Earth Observation?

Apart from some research activities in cooperation with NASA aboard the Space Shuttle and some with Russia aboard the space station Mir, Europeans have exclusively used uncrewed satellites for Earth observation until the present time.

For that reason, research in the framework of crewed space flight is a new chance for Europe's Earth observers – and space scientists alike – which must be examined more carefully in order to see whether or not it is suitable for Earth observation purposes and fulfills the operational demands of the users.

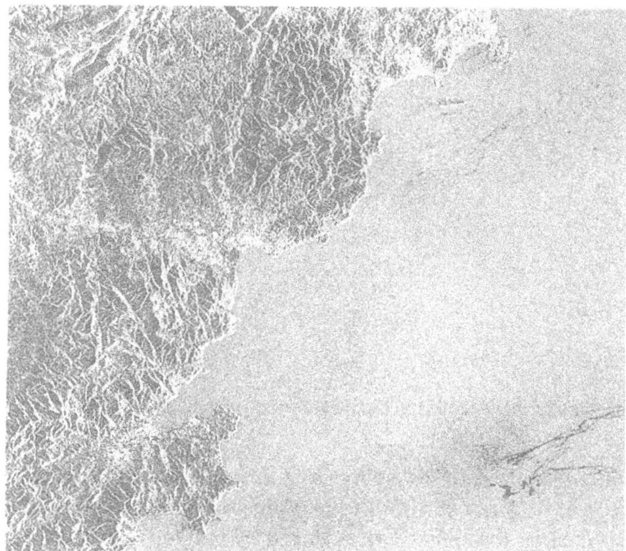

Fig. 7.35. The European Radar Satellite ERS-1 discovered the oil field in the right bottom corner of the picture in front of the French Riviera one day before it was discovered by an airplane.

Example: Infrared telescopes as they were demonstrated by the satellite platform CRISTA in 1994 and 1997, can register greenhouse-effect gases and can also register the smallest flows, circulations and turbulences in the stratosphere in a three-dimensional way. For example, on the basis of corresponding data obtained by platforms similar to CRISTA, global ozone maps can be drawn up.

Example: As it is shown in Fig. 7.35 with the example of an environmental survey satellite, ocean pollution by oil due to the cleaning of tanks aboard oil tankers can be detected throughout the Earth's hydrosphere with the help of radar. Early detection enables minimization, or in the long run, total avoidance of damage by supporting effective law enforcement and by putting offenders under the threat of being held responsible for environmental pollution and its repair.

Earth Observation from the Vantage Point of the International Space Station

The potential interest in ISS with regard to observation of the Earth and its environment results from its orbital parameters and the station's resources of electrical power, data processing and accommodation space available for payloads.

Although the orbital altitude deviations and attitude stability of the station are not ideal for the purpose of Earth observation, the station's inclination of 51.6° with respect to the Earth's equator allows the observation of 85% of the Earth's surface on which 95% of the Earth's population is living (cf. Fig. 7.36). These are favorable preconditions for testing new methods and accompanying technologies for the ob-

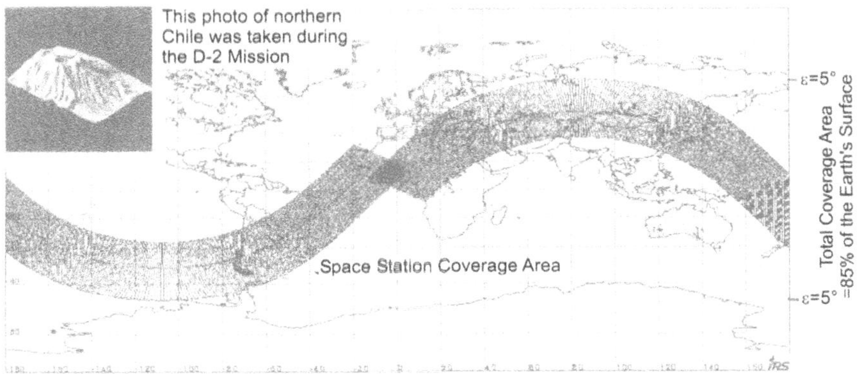

Fig. 7.36. Groundtrack/Swath of the International Space Station for One Orbit

servation of the atmosphere, hydrosphere and lithosphere of the Earth. The station's orbital altitude is at only about 50% of the typical altitude of Earth observation satellites. For that reason, the sensitivity required by active (radar) and passive signal detection methods is four times more favorable.

Example: In 1993 during the D2 Spacelab mission, a new dimension of Earth observation was introduced with the help of the stereo scanner MOMS because its data attained a level of accuracy and complexity never reached before. The scanner has a modular configuration and five different lenses of which three stand perpendicular to the Earth's surface and two are tilted. Due to the different angles of the lenses, three-dimensional "pictures" can be created. The central lens has a focal length of 600 mm and does ground surface imaging with a resolution of 4.4 m × 4.4 m possible.

Meteorology

In the field of meteorology, observations from space have long since provided the standard information for extensive weather forecasts. For example, image series of cloud patterns and the cloud movements derived from them are relatively easy to obtain and hence very common. Further parameters of great interest are wind fields and precipitation patterns. These are basic parameters for the global energy and water cycles, but they are still insufficient for the requirements of meteorologists and the creation of "climate models". Passive instruments such as cameras can only deliver limited information about these parameters. For that reason, active systems like radar or lidar, which can directly scan the structure of the atmosphere with wind and precipitation sequences, must be used. The problem is not only the size and cumbersome nature of such instruments but also their enormous energy requirements. From the technical point of view as well, they are very demanding with a correspondingly high risk level. The International Space Station with its large photovoltaic generators will provide a sufficient amount of power for the operation of sensor systems for the observation of wind and precipitation phenomena.

The resolution achievable with active sensors depends, among other factors, on the size of the antenna which receives the reflected signals. Thanks to its size, ISS can support high-resolution antennae as they are necessary for the thorough observation of many important characteristics of and processes in the atmosphere.

Ecology

In the field of ecology, the observation of Earth's atmosphere from space is of special interest. It includes the examination of different layers with regard to chemical composition, temperature, pressure, wind dynamics and interactions with the Earth's magnetic field. Important data which will deepen our understanding of ozone decomposition, air pollution, annually occurring changes in the atmosphere, and the general change of climate on Earth can be obtained from such studies.

The International Space Station could contribute to the observation of these important parameters and hence to the broadening of our knowledge. In addition to that, ISS can be used to observe phenomena that change over a period of a single day since, unlike Earth observation satellites, the station does not fly in a Sun-synchronous orbit.

Long-term lithospheric and hydrospheric observations by means of suitable instruments aboard the International Space Station could promote a better scientific understanding of the physical processes related to them. Possibly, such observations could also prepare the development of applications that stand in relation to renewable and finite resources, to natural risks, and to environmental pollution caused by human beings.

Example: One of the several interesting remote sensing instrument proposals was retained by the ESA peer group and selected for flight: FOCUS – a predecessor for a spaceborne high-temperature environment disaster recognition system. From ISS, the infrared sensors of FOCUS will allow the detection and analysis of such high-temperature events as vegetation fires and volcanic eruptions (see Fig. 7.37). Volcanic activity and large forest/savannah fires have severe global environmental consequences, such as the greenhouse effect, cloud generation and climate change. Measurements from ISS will classify the events and determine atmospheric composition. The data will be geocoded and transmitted to the worldwide user community.

Example: A primary Earth science payload already identified for ISS by NASA is the Stratospheric Aerosol and Gas Experiment (SAGE). SAGE is the third instrument of its kind to sample Earth's atmosphere using the Sun as an energy source. SAGE III will take global profiles of atmospheric aerosols, targeted molecular constituents, temperature, and pressure while assessing the spatial and temporal variability of these measures. This will lead to a better understanding of the role of certain compounds in climatic processes, biochemical cycles and atmospheric chemistry. SAGE III will fly on ISS as an attached payload for a minimum period of five years, overlapping in part a higher altitude and inclination flight of an identical instrument aboard a Russian spacecraft.

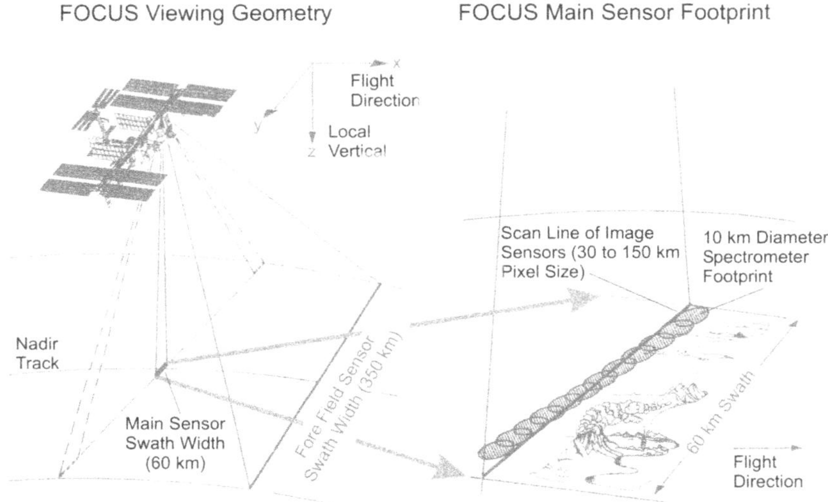

Fig. 7.37. Fire Detection IR-Sensor System (FOCUS) aboard ISS

7.6 Engineering Sciences and Development of Technology

New technological developments are decisive factors for progress in the utilization of space. The development of a new technology or of a newly engineered solution for application in space will not be finished until its successful demonstration *in situ*, i.e., in space. Even in areas where a demonstration is not absolutely necessary, the results of a test during a mission considerably increase the credibility of a new product.

Engineering tasks and demonstration of technology in space have the following objectives:

- Testing and optimizing the efficiency of space system hardware
- Demonstrating and qualifying space-supported instruments and satellite components *in situ*
- Validating and calibrating terrestrial test methods and numerical simulation methods
- Examining new hardware after its use and, if necessary, recalibrating it
- Reducing risks by subsequent flights at short intervals
- Shortening the development period

In order to guarantee cost efficiency, missions for the demonstration of new space technologies or procedures require availability of a space platform of sufficient size as well as power with sufficient supply and data transmission capability. European experiment platforms suitable for such missions are, at present, only available in a very limited selection: EURECA, the Ariane Structure for Auxiliary Payloads

(ASAP) platform and the microsatellites deployed by the Ariane launch vehicle. None of these experiment platforms offer the possibility of reconfiguring or exchanging experiments during flight or of optimizing the operational parameters directly in space.

The International Space Station is ideally suited for engineering and technological activities. It can provide extensive experiment resources for long periods of time; it allows the reconfiguration, inspection and adaptation of experiments in space, and it offers the possibility of returning the tested equipment to Earth for further analysis, making improvements, and perhaps even conducting a subsequent space flight.

The missions planned for the International Space Station in the field of Engineering Sciences and Technology Demonstration have the following three objectives:

- Validation of new technologies
- Testing and use of new materials directly in the space environment
- Collection of engineering data

7.6.1 Validation of New Technologies

Validation of new technologies aboard the International Space Station has the overall objective of improving the following:

- Space stations and platforms as well as synergisms from combining different subsystem functions, e.g. ECLSS, power system, propulsion system, attitude control system, operations, and logistics
- Database and models for the space station's interaction with the environment (structure, thermodynamic/electromagnetic radiation, particles, space debris, etc.) in order to optimize materials, structures, sensors and instruments
- Operational issues, e.g. by increasing automation and the use of robotic systems, improving Man-Machine-Interfaces and information systems (cf. Fig. 7.38)
- Communication, navigation, and other space-based systems and components as well as payloads, e.g. by testing higher frequency ranges up to optical transmission and adaptive antennae
- Manufacture of commercial products

7.6.2 Examples of the Development of Systems and Components

Power Systems. The space station is an ideal testbed for the verification of progressive technologies with regard to the generation and storage of electrical power. Photovoltaic systems are, and will be in the foreseeable future, the most important power source for space flight application. Photovoltaic generators, with a solar cell efficiency of about 25%, have yet to be tested and demonstrated during flight. Alternative technologies, as described in Chapter 5 "Power and Thermal Control System", are for example solardynamic generators combined with thermal energy storage. This is a very attractive solution for space flight application because, when compared to photovoltaic systems, its energy potential is larger and the drag cross-section is smaller, thus reducing atmospheric drag and resulting propellant needs. The plan to make ISS a platform which will lead the technology of solardynamic power to its operational readiness for routine application, is in place.

Facilities for the generation of electrical power are not only interesting for space flight systems, but also, in view of the limited reserves of fossil sources of energy on Earth and the pollution of the atmosphere by combustion gases, for building up solar power systems in space, and for transferring electrical energy back to Earth in the form of microwave radiation. Russian scientists suggested using the International Space Station as a testbed for the technology necessary for power generation and its transfer.

Life Support Systems. Crewed long-duration missions require life support systems that exceed those already in existence. Both the weight and volume of non-recyclable consumables for life support increase with mission duration and crew size. For that reason, new, increasingly more closed life support systems are necessary in which a higher fraction of consumables are reprocessed. This is possible by integration of physical, chemical and biological processes. It is assumed that this new technology will also become useful on Earth in the future.

Propulsion Systems. Numerous spaceflight propulsion systems emit exhaust trails that may have negative influences on sensitive space systems and external payloads. Tests and analyses on Earth can deliver data on some effects of these exhaust trails, but not on all of them. Experiments in space are therefore planned in order to measure characteristic parameters and effects of propulsion systems directly aboard ISS and to correlate these data with those obtained on Earth. The results from these experiments will yield support to the design of future propulsion systems for all kinds of space vehicles.

Automation and Robotic Systems. In space flight, these systems are at present limited to deployment and retrieval of satellites, but their tasks are to be broadened to include maintenance and repair. From the technological point of view, the robotic systems used here differ from Earth-bound industrial robotic systems insofar as they need to be light-weight, reliable and sturdy enough to be used in space. Moreover, their tasks often change, whereas industrial robotic systems on Earth always perform identical tasks. Scientists expect that the use of space robotic systems featuring increased reliability and the possibility of remotely controlling them will widen the possible operation of space vehicles and will contribute to cost reductions (cf. Fig. 7.38).

Communications and Navigation Systems. The International Space Station can serve as a demonstration platform for the operation of new communications systems in the radiofrequency range and in the optical range. This would have an influence on the development of new technologies for space-based components and also for the ground systems belonging to them. Not only would the manufacturers of space vehicles benefit from this development, but also future space flight and planetary missions, as well as commercial communications markets. Finally, progress in this field would also help to improve navigational equipment for vehicles which travel by land, water and air.

Fig. 7.38. ROTEX, a Highly Developed Robotic Arm with Six Joints. This arm was used for the first time during the D2 mission. ROTEX is typical of a modern, sensor-filled robotic system that will also find application with complex research and production tasks in the future. Despite a signal transmission time of two seconds, ROTEX was able to grip free-floating objects.

Figure 7.39 gives an **example** of an early ISS experiment, taking full advantage of the orbital features (low altitude, high inclination) for the synchronization of wrist watches and for the deactivation of electronic keys in stolen cars.

Influence of Space Environment on New Materials. Directly exposing new materials to space conditions is primarily done with new materials and coatings which are intended to be kept in space for a considerable length of time. This serves to monitor radiation, to test the protection against atomic oxygen, to examine coatings with variable emission coefficients and absorption properties for thermal control tasks, and is also done for obtaining a fundamental understanding of detectors for meteorite particles and space debris. The International Space Station will have an external facility for exposing material samples. This will be ideal for the investigation of materials since crew members and robotic systems will be permanently available for the performance and exchange of experiments. Moreover, there is the opportunity to bring new samples aboard the station and return those samples that have already been exposed by means of resupply flights.

Data Collection for Engineering in Space. The collection of technical and operational parameters for engineering purposes aims at the development of new space vehicle hardware and instruments. These data serve for the validation of mathematical models and the calibration of test methods on Earth. Thus, the development and qualification of new technologies for application in space will be accelerated. In certain cases, this approach could even lead to the development of simulation and test methods on Earth that are so precise and reliable that an additional demonstration in space of hardware intended for space stations and satellites will no longer be necessary.

GTS Technology

Frequencies: 1.428 GHz ("GPS Band") Using Spread Spectrum Technology, 400.1 MHz UHF Band With 50 KHz Bandwidth

Distribution: Rotating Beam With Angle-Dependent Modulation, CDMA for Encoding, Phase-Locked Coupling of All Frequencies and Signals for Increase of Resolution

Data Rates: 300-9600 Bit/s

Data Transmitted:
UTC and Local Time
Daylight Savings Zone Information
ISS Orbit Data
License Information
User Data

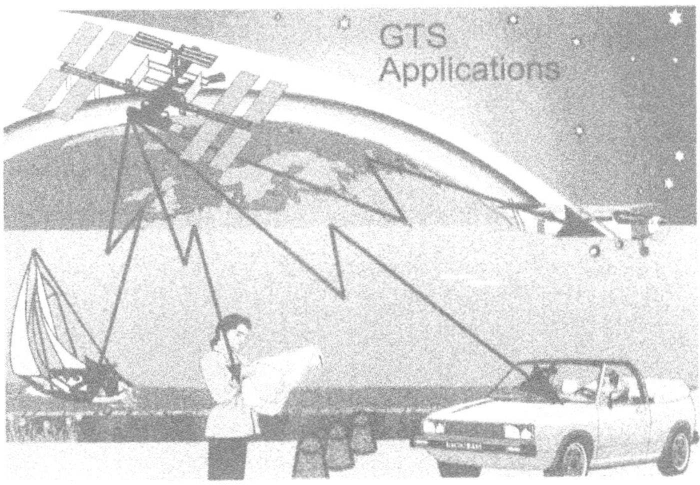

Fig. 7.39. ISS Experiment "Global Time System (GTS)". Synchronization of wrist watches and deactivation of electronic keys from stolen cars

7.7 Outlook for Industrial and Commercial Applications

The expectation of significant industrial innovations and commercial applications is an important motivation for international partners to participate in the International Space Station Program.

Consequently, the USA has reserved about 40% of its accommodations and resources for technological and commercial activities aboard ISS. In Japan, one quarter of all letters answering a first call for utilization proposals for the International Space Station came from industry.

In Europe as well, interest has been spurred for this new chance even with those industries who do not participate in space flight themselves. Apart from the advantages for fundamental research, ISS is ideal for the support of application-oriented activities. The corresponding interested parties demand mainly frequent and short-term access to space: maximum period of time from the first concept of an experiment to its performance in space of less than one year, but definitely not more than two years. Moreover, the interested parties need the opportunity for rapid subsequent flights as well as a guarantee of confidentiality and the protection of intellectual property rights (patents) regarding the experiments' results. All these conditions can be satisfied by specific access conditions for the user group in question (cf. Sect. 13.5.3).

Potential Areas of Application. It is generally assumed that the industrially-oriented potential of ISS addresses two areas, namely research and applications under microgravity and testing of progressive technologies for use in the space environment for the purpose of operation, maintenance or repair.

In earlier times, the utilization of microgravity conditions was mainly associated with the in-space manufacture of products with high added value. Nowadays, it is certain that the chance to gain deeper insight in space into processes determining the production of materials on Earth is highly advantageous for application-oriented research.

The results of experiments conducted during the past few years have unanimously shown that research in a space environment brings about additional knowledge that can be applied to an increase in the quality of industrial products. Recent and relevant examples are found in the areas of material development, biotechnology and medicine.

Mainly crystal growth and controlled casting processes benefit from the absence of gravity-driven convection; and from the opportunity for containerless processing and the absence of sedimentation which on Earth is caused by differences in density, multi-phase processes can be investigated in space without demixing of components. Generally, the absence of gravity allows the study of processes under fewer parameters, so validation of numerical models of technically relevant processes is easily feasible.

7.7.1 Fluid and Materials Sciences

A series of results obtained from fundamental research in fluid and materials sciences indicates that there is a potential for future commercial applications to be put to use aboard the International Space Station.

In the field of development of industrial processes on Earth using fluid particle movement induced by Marangoni convection (in order to counteract sedimentation), the most recent results obtained indicate that there is a broad potential for the study of multi-phase dispersed systems. Considerable progress was made with alloys for the bearings of combustion engines for automobiles. It was proved that new dispersion cast alloys could be generated by making use of the Marangoni convection in order to prevent sedimentation of dispersed particles during solidification. In such cases, mainly microgravity experiments are necessary to validate the results of numerical modeling for the optimization of Earth-based processes.

Another example is the introduction of the integrated cast technique during which aluminum cast alloys for automobile bodies and the load-bearing structure of the Airbus airplanes are optimized with the help of *controlled convection*. The findings obtained from investigations of the convection in microgravity experiments were applied to the production of a fine-grained, homogeneous micro-structure. The result is better mechanical strength and stiffness of the components processed in the casting process. This new method allows the replacement of the expensive traditional production process (during which up to now several parts have been separately mechanically processed and then assembled) by a more economical casting process.

In the field of fluid physics and dynamics, the knowledge acquired during the experiments conducted in space leads to a better understanding of the phenomena of phase changes and also of phenomena in the vicinity of the critical point. The examination of boiling phenomena and heat transfer can be used to improve heat pipes and heat exchangers and to increase the efficiency of heat exchangers and industrial furnaces for power plants.

The possibility of making use of the capillary forces in order to obtain a reliable propellant transport in the tanks of satellites was proven with the help of experiments performed in a drop tower. The use of microgravity for the verification of theoretical ideas lead to important modifications in the design of tanks which make use of the principle of surface tension for the bubble-free propellant supply for geostationary satellite thrusters.

Research in the field of aerosols and droplet combustion under exclusion of convection facilitates a considerably simplified study of the processes in question. The results obtained to date have shown that the ignition process significantly deviates from previous assumptions. They all led to several new projects which took place under international cooperation for investigating combustion processes with the aim of increasing the efficiency of industrial furnaces and power plants, and at the same time, of reducing environmental pollution.

7.7.2 Biotechnology and Medicine

The results obtained from the crystallization of large bio-molecules such as proteins, enzymes and viruses show, at least in a significant number of cases, an increase in the perfection of the arrangement of the inner structure of the crystals that allows the determination of the three-dimensional structure of these substances to be made. The reasons responsible for this structural improvement are not yet completely known, but their possible use in the creation of pharmaceutical products is apparent.

Up to the present day, most medicines were produced according to "trial and error". As a consequence, they not only have an effect on the targeted noxious proteins, but also on other proteins of the body, and that leads to adverse reactions. If, on the other hand, scientists know the three-dimensional structure of a protein involved in a specific illness, they can tailor a medicine that has a strong effect on the protein in question but does not influence other proteins. That reduces the level of undesirable side effects and makes the remedy in question more suitable for the treatment of the disease.

The influence of the absolute value of as well as the change in gravity – be it reduced or increased gravity – on the human organism is not only of importance to the health care of astronauts. Studies dealing with the perception of gravity by animal and plant cells and the influence gravity has on metabolism and cellular functions have been carried out successfully with the help of experiments by groups of scientists all over the world. In view of future medical and pharmaceutical applications on Earth the following results are important:

The investigation of the behavior of single cells through studying complex processes of cell growth and cellular differentiation under microgravity has yielded important results. The missing movement of immune cells, like lymphocytes, in weightlessness facilitates the study of their reaction with tumor cells. On the subcellular level, the signal transfer among and inside the cells in space may differ from observations on Earth. That offers the opportunity to test theoretical models that describe the relationship between the structure and function of basic proteins. The knowledge acquired is thus important for the development of new, active substances and treatments for diseases such as cancer, osteoporosis and viral infections.

New experiments showed that microgravity influences the cellular response to growth factors and transport factors. It is not only of scientific but also of practical interest to examine the role gravity plays in the growth and differentiation of mammalian cells. These results can actually be used to develop application-oriented models that would make the study of the healing of wounds, neural induction, and bone homeostasis possible.

In order to improve biotechnical processes, microgravity can be used to increase resolution and the yield of purification and separation processes, for example free-flow-electrophoresis and two-dimensional gel-electrophoresis. Microgravity is also used in order to achieve a better artificial transfer of genetic material: Electrofusion carried out under microgravity (cf. Fig. 7.32) makes possible longer contact times among cells with different densities and thus increases the yield of fused cells formed from individual cells with different densities.

7.7.3 Summary of Industrial Applications

In the following, an overview is given of those areas of research for which a utilization potential of microgravity as a means of applied industrial research was identified. The areas of research selected for microgravity experiments that might be conducted in the year 2000 and later can be divided into three main categories listed as follows:

- Areas in which significant results have already been obtained during earlier microgravity experiments, and that have already aroused an interest in the industry. This interest can be expressed in the form of direct commitment and/or direct technology transfer, but also in the form of products that were directly derived from experiments.
- Areas of research that are the subject of current examination on Earth and in space, and for which industry has outlined objectives and defined research instructions
- Examinations of future technologies that are assumed to be decisive for the competitiveness of the process engineering industry

Different areas have already roused interest in the industry, such as:

- Casting under controlled convection: application in the production of parts and components in the automobile and aerospace sector
- Immiscible and monotectic alloys: application in the field of self-lubricating bearings
- Production of electronic composites like CdTe and HgCdTe by crystal growth
- Growth of protein crystals

The industry itself defined objectives for further investigations in the following areas of research:

- Biomedical experiments in space for the improvement of quality of life on Earth: osteoporosis treatment, healing of wounds, other treatments
- Microgravity experiments on the subject of biotechnical processes such as electrophoresis done in order to increase their efficiency on Earth
- New methods of materials processing that would benefit from experiments under gravity: aerosols, chemical vapor deposition methods, examination of multiphase fluids
- Investigation of combustion processes in order to increase their efficiency and to reduce fuel consumption in industrial furnaces, power plants and compression ignition engines. This, at the same time, is expected to reduce environmental pollution considerably.

As far as activities for future technologies are concerned, it is generally taken for granted that the most important progress can be expected in the following areas:

- Information technology
- Biotechnology
- New materials and manufacturing methods

Research trends in these areas follow the idea of miniaturization in every respect: to achieve higher storage capacity and data transfer rates (information technology); to manipulate living systems on the molecular level (biotechnology); and to create micro-structures in the nanometer range with the help of which specific properties can be tailored (new materials and manufacturing methods).

8 Microgravity

The term "microgravity" (μg) describes the state of approximated weightlessness in which the observer does not perceive the acceleration g_0 constantly effective on Earth. This state is reached, for example, aboard a space station in an Earth orbit, but also during free fall whenever a body can freely move within the dynamic equilibrium of forces between gravitational force and force of inertia. Physically, gravity cannot be "switched off". In the case of a space station at an altitude of 400 km, for example, gravity still reaches 88% of the value it has on the Earth's surface. The centrifugal force resulting from the orbital motion affects the mutual cancellation of both forces. Superior to the equilibrium of forces, perturbing accelerations act in reality, caused by the orbital environment, the space station's structure or the crew. Therefore, it became a convention to characterize this state by the term of *microgravity* where "micro" means considerably smaller than 1g, but not always the factor 10^{-6}. In some cases, it would be more appropriate to talk of "milli-g", i.e. realistically of the range μg to mg.

8.1 Microgravity as a Locational Advantage

The use of the microgravity environment has become an important application area in space flight (cf. Chapter 7, "Utilization"). Phenomena which, under normal gravity conditions, cannot be observed at all or at least not separately, can here be analyzed and used for a definite objective. This is obvious by looking at the example of fluids. Figure 8.1 shows a summary of transport processes occurring in fluids.

While on Earth, the gravity-driven effects, such as convection by buoyancy or differences in temperatures, dominate other transport mechanisms like diffusion or surface tension driven convection, the latter are mostly accessible in an isolated way under microgravity conditions. Accordingly, there are a number of fields of research within fluid physics which use this microgravity environment for the purpose of studying convection, diffusion and surface phenomena (Fig. 8.2).

Fluid physics, however, is only one of the many overlapping and interrelated subdisciplines of μg-research (Fig. 8.3). The different areas of application can be divided into three major disciplines:

- Fluid physics (mentioned above)
- Materials science
- Biology and biomedicine, summarized under the term of "Life Sciences"

In the field of materials science, mainly questions relating to meltings or solutions and their functions concerning the manufacture of new materials are considered.

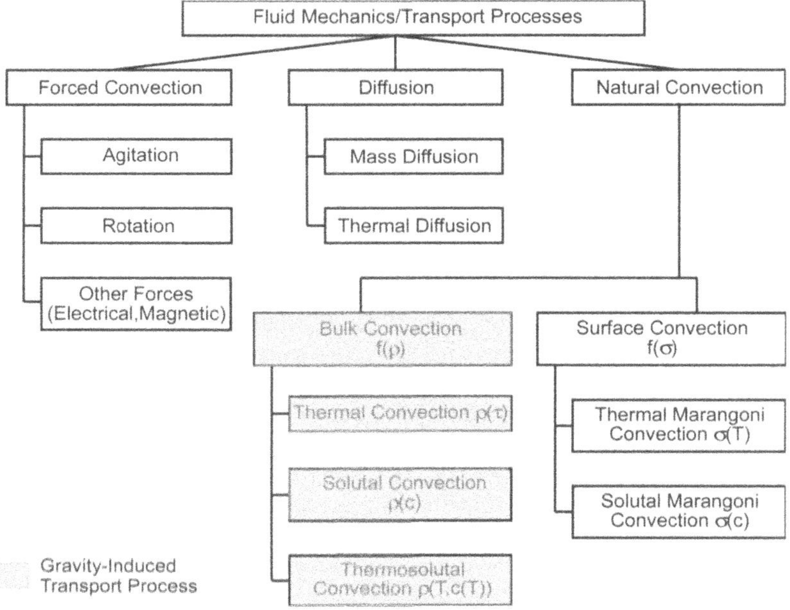

Fig. 8.1. Effects Observable in Fluids [Greger 87]

Particularly processes involving phase boundaries or phase changes are of great importance, for instance in the field of solidification of semiconductor crystals from the melting.

The area of biological and biomedical research is slightly different. On one hand, global changes in biological systems arising from the absence of Earth acceleration are examined, such as the absence of the information of growth direction for growth control. On the other hand, the detailed study of individual biological phenomena leads to a better understanding of complex physiological processes. In this context, calcium degradation in human bone structure under weightlessness (which may lead to explanations for the disease named osteoporosis) shall be mentioned as an example.

8.2 Ways to Obtain Microgravity

There are many ways of obtaining a microgravity environment, ranging from relatively simple drop experiments, aircrafts and sounding rockets to actual space flight experiments involving capsules, platforms or crewed missions such as the US Space Shuttle or space stations. In the following sections, typical systems for microgravity missions with their pros and cons will be briefly described.

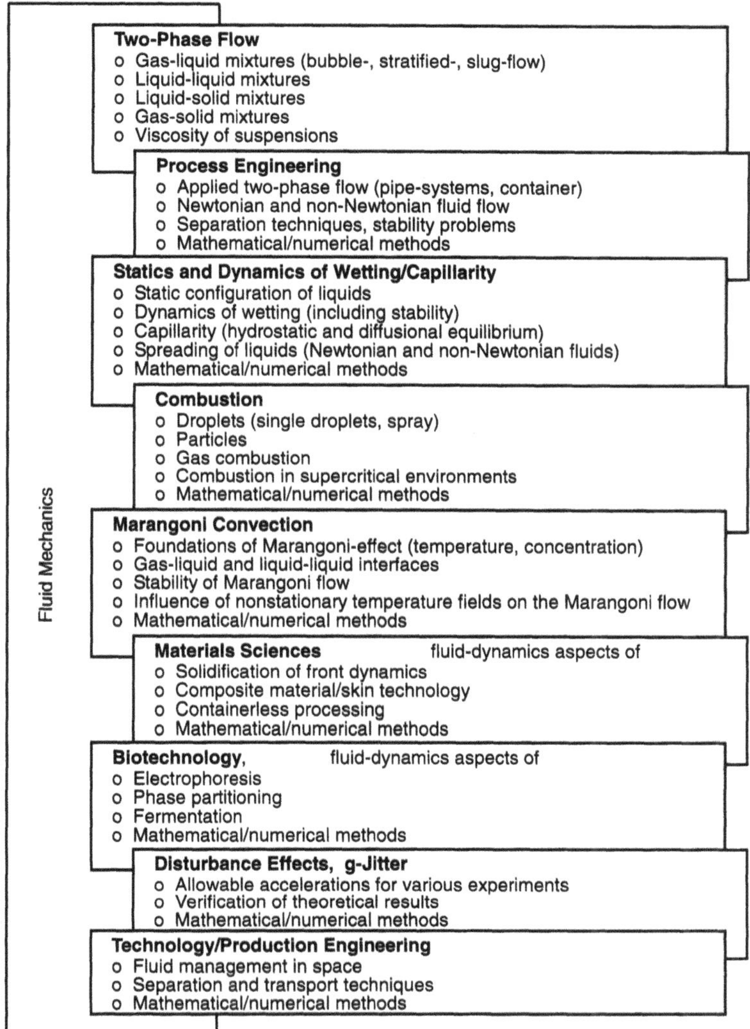

Fig. 8.2. Microgravity-relevant Research Areas in Fluid Mechanics [Greger 87]

8.2.1 Drop Tower

Since 1989, the Bremen Drop Tower [ZARM 96] (height 145 m, drop distance 110 m, see Fig. 8.4) has been used by the "Center for Applied Space Science and Microgravity" ZARM (German acronym) of the University of Bremen, Germany. At this point, it shall be presented as a typical representative.

Installed in TEXUS-compatible drop capsules (cf. Sect. 8.2.3), the experiments fall through a partially evacuated tube (for decrease in atmospheric drag) in

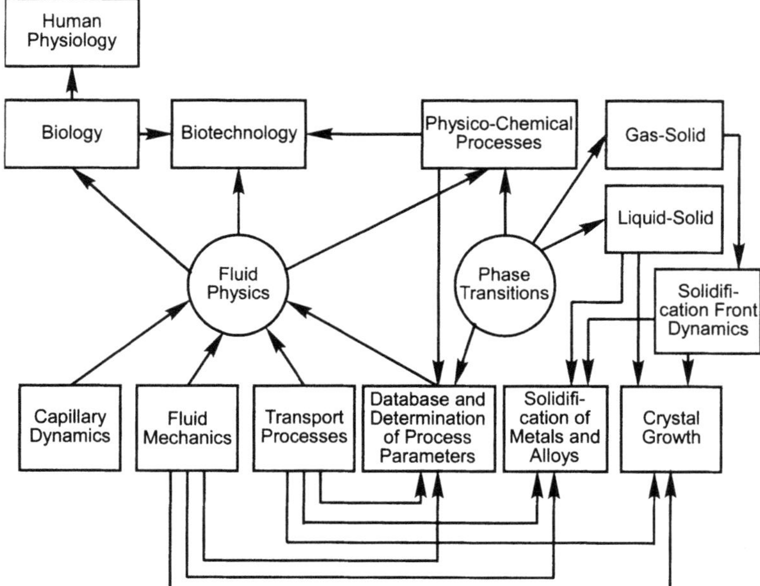

Fig. 8.3. Relationships between Fluid Physics and other Areas of Research

Fig. 8.4. The Bremen Drop Tower [ESA SP-1116]

4.7 seconds and are eventually slowed down by "plunging" into a deep bed of polystyrene granulate. In order to double experiment duration, a catapult to launch the capsule vertically from the floor is planned. The relatively high initial acceleration of about 25 g, however, is a disadvantage of this extension to double µg-duration.

Other drop towers are also run in the USA (NASA Lewis Research Center in Cleveland, Ohio: 2.2 s and 5.18 s; NASA Marshall Space Flight Center in Huntsville, Alabama: 4.2 s), in Grenoble, France (3.1 s) and in Japan (1.3 s and 4.5 s). The longest drop duration of 10 s for an Earth-bound drop system is the drop tube of the

JAMIC (Japan Microgravity Center, Kamisunagawa) on Hokkaido [Mori 93]. In an old and abandoned coal mine, a magnetically guided drop capsule falls for a distance of 710 m. The atmospheric drag is compensated by a cold gas thruster system.

Drop towers or drop tubes offer relatively uncomplicated and cost-efficient opportunities for the study of short-lived phenomena in microgravity at a comparatively good level of perturbing accelerations. In particular, phenomena of phase flow and capillarity, but also of combustion processes, are examined in drop experiments. Another application of these test units is the qualification of experiments or devices intended for a mission on a long duration μg platform.

8.2.2 Parabolic Flights

About 20–30 seconds of microgravity can be obtained during parabolic flights. Jet airplanes such as the KC135 (NASA), the Caravelle or the Airbus A300 (ESA), or the Ilyushin IL-76 MDK (Gagarin Cosmonaut Training Center, Moscow) are used. Their interiors are completely empty and padded with foam rubber.

The flight profile of a parabola is the following: After the introductory entry pull-up at increased acceleration (2g), the pilot reduces thrust and, with throttled or idle engines, the airplane follows the parabolic trajectory of a free-flying body (Fig. 8.5). As a consequence, after a short phase of transition, microgravity is obtained for about 20–30 seconds. After the recovery maneuver at increased acceleration (2g), the airplane flies horizontally to the Earth's surface for about 1 minute before introducing the next parabola. During one flight mission, between 20–40 parabolae are carried out in that manner.

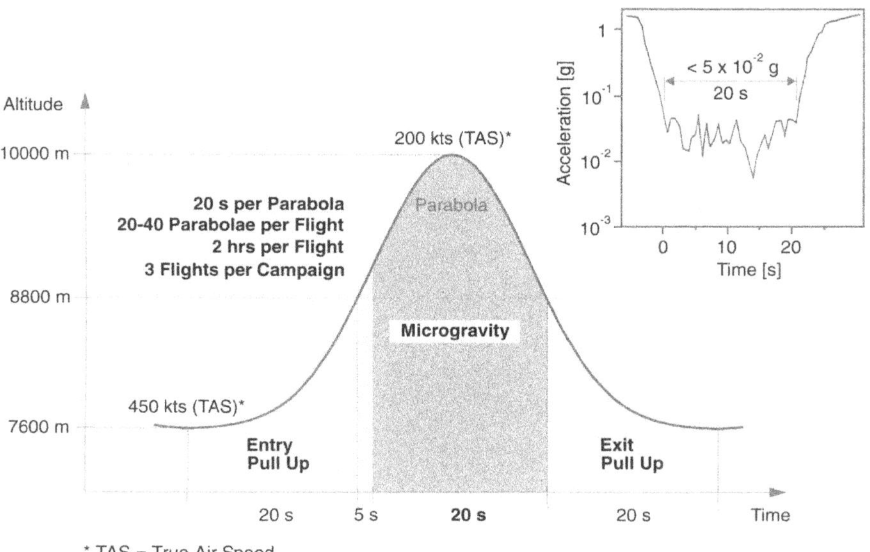

Fig. 8.5. Parabolic Flights: Flight Profile and Course of the Perturbing Accelerations

The utilization spectrum of parabolic flights ranges from the testing of technology and procedures to the qualification of experiments and subsystems to astronaut training. Due to flight perturbations and the presence of many crew members, however, there is a comparatively low microgravity level of only about 10^{-2} g. But, this disadvantage is counterbalanced by relatively uncomplicated access to this flight opportunity and low mission costs.

8.2.3 Sounding Rockets

Also sounding rockets are used for parabolic flights, but the parabolae's summit height and acceleration, however, are significantly above those obtainable with airplanes. In the German TEXUS program, the summit height amounts to about 250 km. The launch vehicle is a Skylark rocket launched at the test area Esrange in Kiruna, Sweden (cf. Fig. 8.6). After cut-off of the rocket engine at an altitude of about 100 km, about 6 minutes in microgravity remain before the payload reenters the atmosphere and performs a parachute landing. It is recovered by helicopter, so the experimenters have access to their instruments just one hour after launch.

The TEXUS-payload contains a pile of 5–6 cylindrical experiment modules of 0.3 m diameter (cf. Fig. 8.6). The modules are standardized and allow for short periods of development for hardware due to standardized interfaces. The entire payload is about 200 kg in mass.

The ESA successor program MAXUS was extended to a duration of 15 minutes of microgravity by using a reinforced Skylark-7 rocket. The inaugural flight of this system took place in autumn of 1992.

Fig. 8.6. TEXUS – Launch and Experiment Module [MBB]

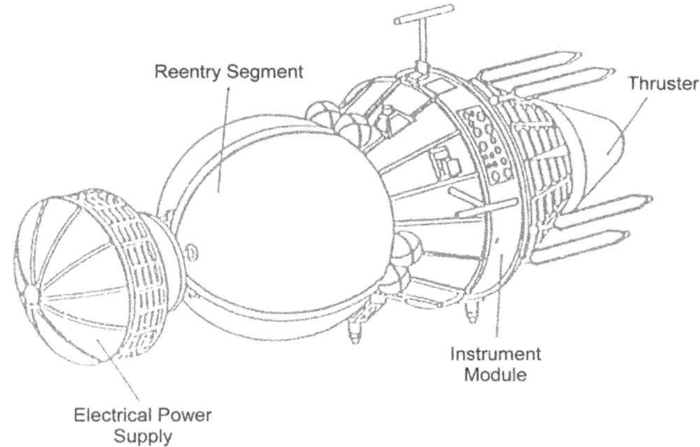

Fig. 8.7. The Foton Capsule System

8.2.4 Space Capsules

Unlike flights involving sounding rockets, space capsules have orbital trajectories. Accordingly, the duration of microgravity is only limited by the power supply or the natural decrease in orbital altitude due to atmospheric drag. The typical mission duration of such capsule experiments hence ranges from a few days all the way up to several weeks.

The Russian Foton capsule (Fig. 8.7) is an example for a regularly used capsule system. After launch into an orbit of about 220–400 km altitude, 14–16 days in microgravity are available for experiments. The capsule has a payload-carrying capacity of up to 700 kg at an average power consumption of 400 W. After the microgravity phase is over, the thrust in the instrument module will be activated, introducing reentry. The spherical reentry segment will finally perform a parachute landing and will be recovered.

The Foton-capsule, among others, is used in the framework of ESA's microgravity program. For example, with BIOBOX, an automated incubator in the interior of the reentry segment was created.

8.2.5 Flight Opportunities

This heading summarizes flight opportunities that, apart from their main missions, are a means of transport for further experiments and payloads. Therefore, via different programs, small, autonomously working experiments are permitted for the upper stages of launch vehicles (Ariane, Delta-II). The experiments are activated after release of the launcher's primary payload. It is also possible to fly experiments in special payload containers, called "Get-Away Special" (GAS) or "Hitchhiker", aboard the US Space Shuttle. For this purpose, experiments must fit into standardized, cylindrical containers. They can either work autonomously (GAS) or maintain com-

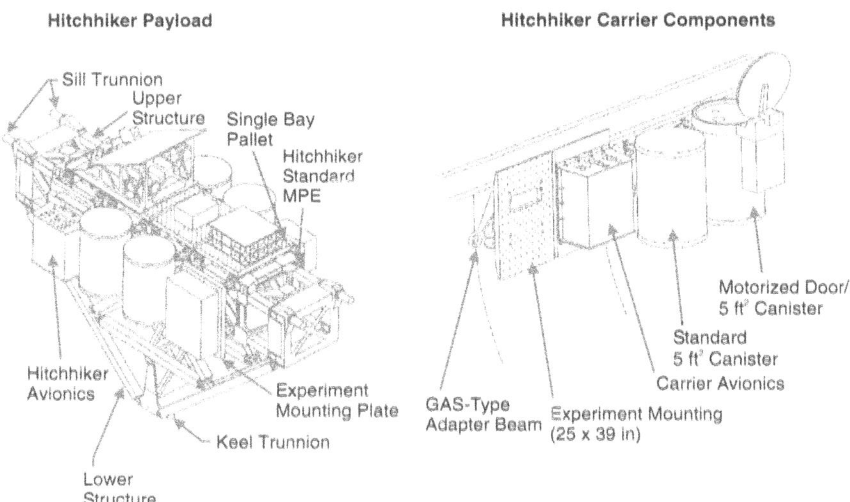

Fig. 8.8. Hitchhiker System inside the Space Shuttle Cargo Bay

mand, power and telemetry interfaces with the Shuttle. During the Shuttle flights, experiment containers are installed either in the cargo bay or on payload pallets (Fig. 8.8), depending on the facilities. As far as small payloads (< 90 kg) are concerned, this is a very cost-effective flight opportunity (about US$ 111–300/kg for GAS-CAN).

Finally, there is also the opportunity to install some of the experiments directly in the Shuttle middeck and to have them performed by the crew. At present, this opportunity is preferred by NASA and US enterprises, one of the reasons being the insecure planning of STS missions. In order to enlarge the storage compartment for Shuttle middeck experiments, the Spacehab was designed. It is a structure that enlarges the pressurized Shuttle cabin in the direction of the cargo bay without significantly restricting the Shuttle's transport capacity. These Spacehab flight opportunities have been commercialized and are offered by the marketing company Spacehab, Inc. (cf. Sect. 2.4.2).

Also the Foton system presented in Sect. 8.2.4 is an interesting option for a flight opportunity. As an example of such an opportunity, it carries the Mirka experimental reentry capsule as a piggy-back payload. On Foton, Mirka is mounted in the place where the power supply segment is normally located (Fig. 8.9). This capsule, weighing 150 kg, is separated from the Foton reentry segment after the deorbit impulse is initiated and can then carry out its reentry mission independently. It serves the purpose of the testing of technology as well as the verification of aerothermodynamic reentry models in the upper Earth atmosphere. As a return-to-Earth vehicle it could carry up to 30 kg of payload.

The microgravity level during the various flight opportunities is, of course, determined by the respective launch system. As for the US Space Shuttle, it is between 10^{-4} and 10^{-5} g.

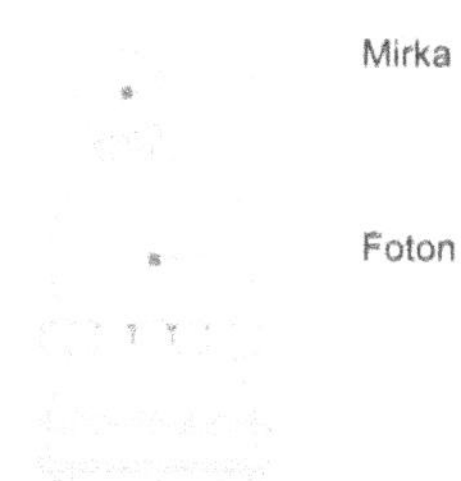

Fig. 8.9. Foton Carrying the Experimental Reentry Capsule Mirka launched in 1997 as a Piggy-back Payload

8.2.6 SPAS

If experiments are to be entirely decoupled from the perturbing influences of the launch vehicle, a free-flying experimental platform should be opted for. The "Shuttle Pallet Satellite" (SPAS) is such a system and was developed by the DaimlerChrysler Aerospace, Germany (formerly MBB), see Fig. 8.10. The Space Shuttle payload pallet has a payload-carrying capacity of up to 900 kg and also carries the necessary supply systems (power, thermal control, communication, etc.). In orbit, the pallet is released by a manipulator arm and, after having carried out the experiments, is recaptured by it. Unlike satellites, which receive their electrical power from huge solar collectors, SPAS is provided with its electrical power by batteries. Thus, the atmospheric braking of the spacecraft is lower, leading to a very good microgravity level of 10^{-6}. The limited battery capacity of 2.6 kWh, however, restricts the autonomous operational duration to about two days.

In 1983, SPAS was successfully demonstrated in orbit and, since then, has repeatedly been sent out, among other reasons, as a launch platform for an infrared spectrometer for atmospheric research (CRISTA-SPAS, Nov. 1994) or as a UV telescope during the Orfeus-SPAS-Mission (Nov. 1996).

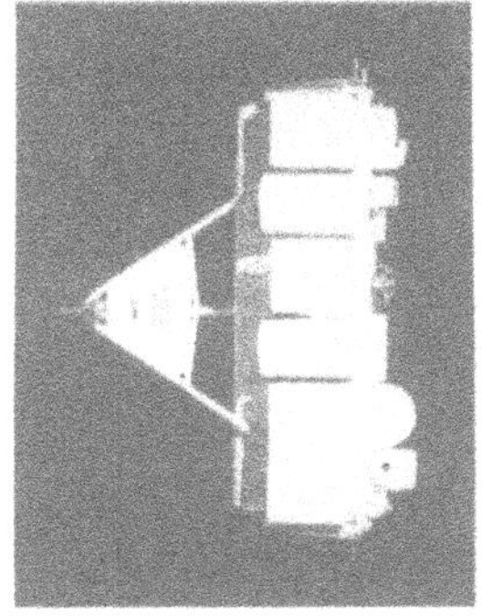

Fig. 8.10. SPAS and EURECA [MBB]

8.2.7 EURECA

As the next logical step towards prolonging μg-duration, the platform EURECA (EUropean REtrievable CArrier, cf. Fig. 8.10) is also carried by the Space Shuttle, but EURECA is independent of launch vehicle regarding the duration of the mission. This is why EURECA has its own supply systems allowing an experimental time in orbit from 6 months to 18 months. Despite the payload-carrying capacity of 1000 kg, EURECA is a compact, uncrewed system making possible a μg-level of $< 10^{-5}$ g.

EURECA was designed as an all-automated microgravity platform for ESA. The initial flight took place from August 1992 to July 1993. On this occasion, however, difficulties in operation of the all-automated systems occurred: due to technical defects, for example, the temperature inside a cooling circuit for a short time exceeded 70°C, which caused the failure of particularly those experiments dealing with protein crystal growth. Improvised human intervention as it would have been possible aboard a crewed system (cf. Spacelab) was not available here.

Although it was conceived for various utilization aspects, subsequent missions, perhaps on a private economic basis, are at present still the subject of negotiations. One reason is mission costs which, despite the fact that it is an uncrewed system, were comparatively high (first mission: about US$ 500 million). On the other hand, in the past, lead times and procedures for access to the EURECA platform were comparable to those of the Spacelab system, which, as an additional advantage, allowed human intervention.

8.2.8 Spacelab

Flights aboard crewed systems are invaluable for tests requiring direct intervention by an astronaut or experimenter in orbit. For this purpose, the Spacelab was developed by ERNO Raumfahrttechnik GmbH, Bremen (now DaimlerChrysler Aerospace AG) for ESA. It is integrated into the cargo bay of the US Space Shuttle and allows microgravity experiments from all disciplines for a duration of up to two weeks maximum. With 17 flights logged to-date, the Spacelab pressurized module has become the most important payload for the US Space Shuttle. Furthermore, it was the only space station-like opportunity for experiments in the West for many years. The Spacelab system is described in detail in Section 2.4. Due to Spacelab's significance as an example of space flight utilization, Chapter 7 ("Utilization") will describe different Spacelab experiments and scientific features in detail.

Nowadays, Spacelab has been replaced more and more by the Spacehab system (cf. Sect. 2.4.2), mainly for reasons of quicker turn-around cycles and its better suitability for resupply and mission flexibility, but less potential for research.

8.2.9 Space Stations

The long-term objective for research in microgravity is doubtlessly a permanently available experimental testbed as can only be optimally offered by a space station.

With the space station Mir, for example, such a space laboratory has been available since 1986, and after the Cold War was over, it was also able to be used by the

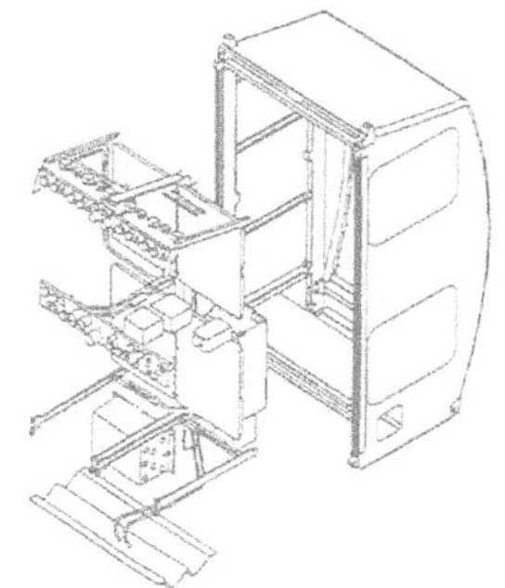

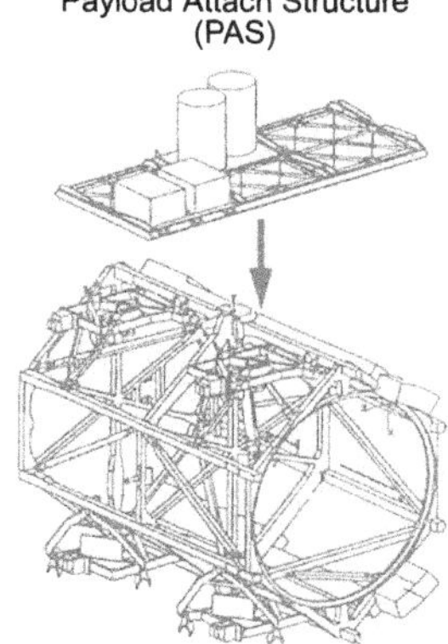

Fig. 8.11. ISPR and PAS [ESA 95]

West. In the year 1999, however, Mir approaches the end of its operational life. At this point, experiment operation aboard the new International Space Station (ISS) is about to begin. Both space stations are presented in detail in Chapter 2 ("History and Current Development").

The International Space Station offers the opportunity for experiments in the pressurized laboratory modules to be conducted, where standardized payload racks (International Standard Payload Racks, ISPR) allow for flexible use of the offered space and resources as well as for reconfiguration in orbit. Additionally, experiments can be installed on unpressurized truss structures. For these so-called "Exposed Facilities", of course, there are higher level requirements regarding autonomy, handling by manipulators or astronauts, and shielding against orbital influences. Apart from an ISPR, Fig. 8.11 shows the way in which a Payload Attach Structure is connected to a truss of the space station. Up to six experiment units, so-called "Express Pallets", can be connected by means of standardized interfaces for power and telecommunication.

However, it must be mentioned that, compared to the Space Shuttle, the ballistic coefficient of space stations is rather unfavorable and, under certain circumstances, may continue to impair the μg-level. Therefore, it is absolutely necessary to pay particular attention to the spatial design of the laboratory modules and their correct positioning within the overall configuration.

8.2.10 Comparison of Flight Opportunities

In order to carry out microgravity experiments, there is a relatively wide variety of flight opportunities ranging from simple drop experiments on Earth to long-term tests aboard space stations (Fig. 8.12).

For space flight experiments, distinctions can be made among capsules, uncrewed platforms and crewed space laboratories in terms of resources and μg quality (Table 8.1).

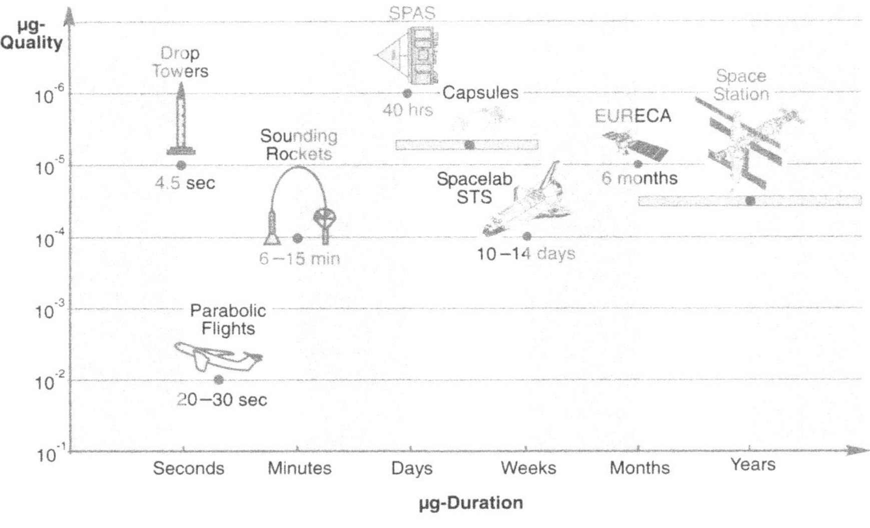

Fig. 8.12. Microgravity Duration and Quality for Different Flight Opportunities

For the selection of an experiment carrier, several factors have to be considered:

- *Duration and Level of Microgravity.* These are quality requirements of the microgravity environment, and they are usually given by the experiment itself.
- *Access to Flight Opportunities.* In the past, the relatively low number of flights, e.g. of Spacelab, in the West, led to the demand for high reliability and the utmost time efficiency during the mission itself. Accordingly, a lead time of several years had to be assumed for the project, all the way from experiment selection, to development of hardware and operation procedures, to experiment qualification and finally to astronaut training. These boundary conditions had a hindering effect particularly on industrial research or technological experiments. With the beginning of operation of the International Space Station (ISS), access conditions will certainly be markedly improved.
- *Crew Intervention.* In the case of unforeseen events, perturbations or defects, on-the-spot human intervention can be essential, and even prevent the loss of an entire experiment under certain circumstances. Particularly, it cannot be generally taken for granted that all-automated systems automatically lead to lower costs and more efficient experiment carriers (cf. Sects. 12.3 and 12.4).
- *Costs.* Last but not least, the relationship between mission costs and experimental boundary conditions is a crucial factor. The International Space Station will give rise to an interesting competition for flight opportunities on the space station segments belonging to the different international partners. This competition can be expected to facilitate the users' access to a microgravity laboratory. Until now, only the Russians have quoted "prices" for the use of their modules.

Table 8.1. Comparison of Different Experiment Carriers for Research in Microgravity

Flight Opportunity	Accommodation Interfaces	Payload Mass [kg]	Power Supply	Microgravity Level [g]	US$/kg Payload
Drop Tower	Standard Module ø 1.5 x 0.8 m	125	0.6 kWh	$\leq 10^{-5}$	
Parabolic Flight	ø 3.0 x 12.5 m	50[a]	< 2 kW	10^{-2}	
Sounding Rockets (TEXUS)	Standard Modules ø 0.404 x 1.0 m	30–80 per Module	28 V DC < 4 kW, < 1 kWh	10^{-4}	12000
Ballistic Capsules (Foton)	< 4.7 m^3	< 700	< 400 W	~10^{-5} [b]	20000[c]
GAS	ø 0.5 x 0.71 m	90	None[d]	$\leq 10^{-4}$	300
SPAS	Standard Mounting Plates	900	28 V DC 2.6 kWh	$\leq 10^{-6}$	
Spacelab	Standard Racks 0.48 and 0.96 m wide	290 per Single Rack 4600 total	28 V DC 2.5 kW Cont. 6.5 kW Peak	$\leq 10^{-4}$	
EURECA	Standard Mounting Plates	1000	28 V DC 1000 W	10^{-5}	
ISS-COF (one ISPR)[a]	10 ISPRs 1.6 m^3 each	400	3–6 kW	10^{-3}–10^{-6}	
ISS-PAS[a]	6 Express Pallet Adapters, 1.3 x 0.9 m	5000 total	3 kW	10^{-3}–10^{-6}	
ISS-ROS				10^{-3}–10^{-6}	30000 [e] 60000/h [f] (m x P) [g] 66000 [h]

[a] For Free Floating Experiments Only
[b] Estimated Value
[c] For External Payloads
[d] To be Provided by the User
[e] Launch
[f] Astronaut Time
[g] Running Cost depend on Mass & Power
[h] Download

8.3 Perturbing Accelerations aboard Space Stations

As already mentioned, the state of weightlessness, particularly insofar as larger space vehicles are concerned, can only be attained to a certain degree. Various effects lead to residual accelerations being inherently present. These effects may impair experiments in microgravity. It is therefore important to both analyze and prevent the causes for the perturbations as far as possible and to know about the specific requirements an experiment has with regard to gravity and/or the acceleration environment.

Perturbation forces having an effect on a space vehicle may be differentiated as follows:

External forces such as atmospheric drag and solar radiation pressure create a *nearly constant* (with respect to time) acceleration of the center of mass and thus of the entire structure. Furthermore, a rotation of the vehicle yields a centrifugal force causing a constant radial acceleration which may impair μg-experiments.

Other forces, however, cause a *transient* field of perturbing accelerations in the space vehicle: During nominal operation of a space station, for example, the attitude and orbit control system, when firing the thrusters or when control torques are applied, gives rise to perturbing accelerations. An operation such as the docking of vehicles also introduces impulses into the space station structure.

Furthermore, the mass distribution of a space vehicle changes due to both the movement of mechanical parts and to crew activities. Consequently, internal forces and/or accelerations are created. These forces, however, do not change the state of motion of the overall system, but they cause an oscillating response of the space station structure itself. The resulting accelerations, called "g-jitter", are characterized by a wide spectrum of frequencies. Although the g-jitter leads to comparatively low perturbation amplitudes, it may considerably impair experiments in microgravity whenever the perturbation frequency lays in the range of the experiments' eigenfrequency.

The spatially extended structure of a space vehicle still evokes further acceleration effects. So-called "tidal forces" may arise, as the balance between gravity and centrifugal force is only perfect in the center of gravity of the space vehicle. Depending on the flight mode of the space vehicle (Earth-oriented or inertial), tidal forces lead to either a constant or a transient field of acceleration. In addition to that, Coriolis accelerations may occur in the rotating space vehicle during linear movements.

Finally, the Earth's oblateness and the irregularities in its mass distribution create a modulation of gravity, also noticeable outside the center of mass.

The most important perturbation effects are characterized in more detail in the following sections. In the order of their influence on µg-quality, they are:

- Atmospheric drag due to the residual atmosphere
- Tidal forces
- G-jitter
- Solar radiation pressure

For reasons of comparison, the acceleration values relate to the Earth acceleration of $g_0 = 9.81 \text{m/s}^2$.

8.3.1 Atmospheric Drag

The frictional drag due to the residual atmosphere can be approximately described by

$$\vec{F}_D = -\frac{1}{2}\rho v^2 C_D A_p \hat{v} \quad \text{and} \quad \vec{a} = \frac{\vec{F}_D}{m} \tag{8.1}$$

Here, ρ stands for the density of the residual atmosphere, v for the orbital velocity, C_D for the drag coefficient, A_p for the incidence area normal to the velocity vector, and $\hat{v}$ for the unit vector of the orbital velocity.

From among this group, the density is the parameter most difficult to describe. Influences such as solar activity, day time, season or geomagnetic activity lead to significant density variations. These factors are explained in detail in Chapter 3 ("Orbital Environment") and Chapter 6 ("Attitude and Orbit Control System"). The

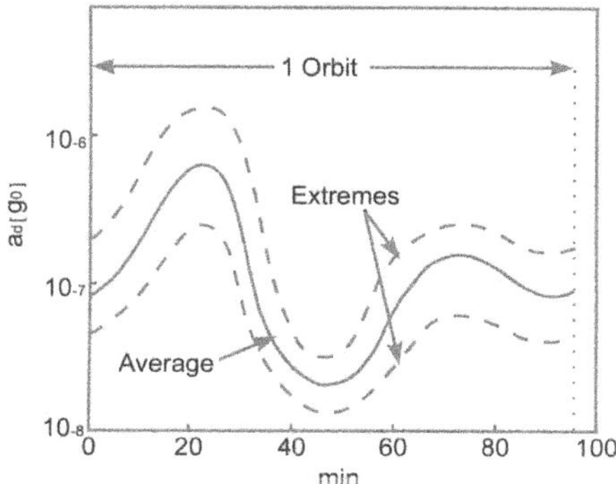

Fig. 8.13. Variations in Atmospheric Residual Deceleration due to Density Fluctuations ($h = 450$ km, $C_D = 2.3$) [Hamacher 87]

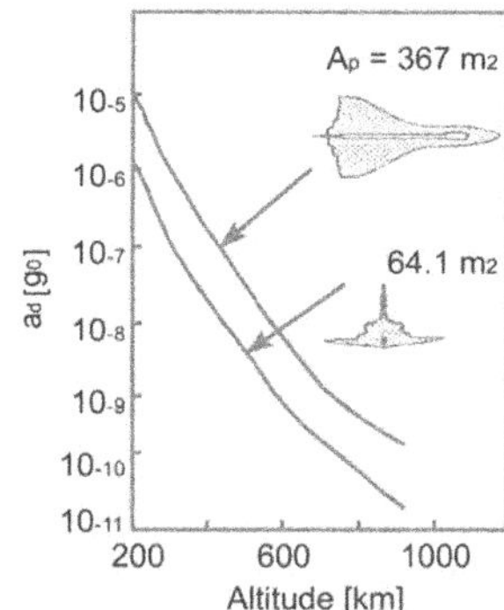

Fig. 8.14. Atmospheric Deceleration of the Space Shuttle Indicated for the Two Extreme Flight Attitudes ($m = 90$ t, $C_D = 2.0$) [Hamacher 88]

dynamics of the density variation for a typical space station orbit, according to which the atmospheric drag varies by about one to two orders of magnitude during one orbital revolution, is shown in Fig. 8.13.

In Fig. 8.14, the atmospheric drag of the US Space Shuttle is shown as a function of the orbital altitude. In the case of the most unfavorable minimum orbital altitude and maximum incidence area, the aerodynamic drag can have a value of up to 10^{-5} g_0.

From the general definition of acceleration (force divided by mass), it becomes obvious that atmospheric braking depends on the relation of surface area to mass of the space vehicle. This relation can vary considerably from one space vehicle to the next. For the US Space Shuttle, it amounts e.g. to between 0.7 and $4.1 \cdot 10^{-3}$ m^2/kg, whereas for the International Space Station, values from 1.7 to $9.2 \cdot 10^{-3}$ m^2/kg are obtained. Temporary changes of the incidence area also lead to a variation of the aerodynamic deceleration. This happens particularly in the case of space stations in Earth-oriented flight mode since their large solar arrays must usually be Sun-tracked (α-tracking). Tracking in this case leads to cyclic modulations of the atmospheric drag (Fig. 8.15).

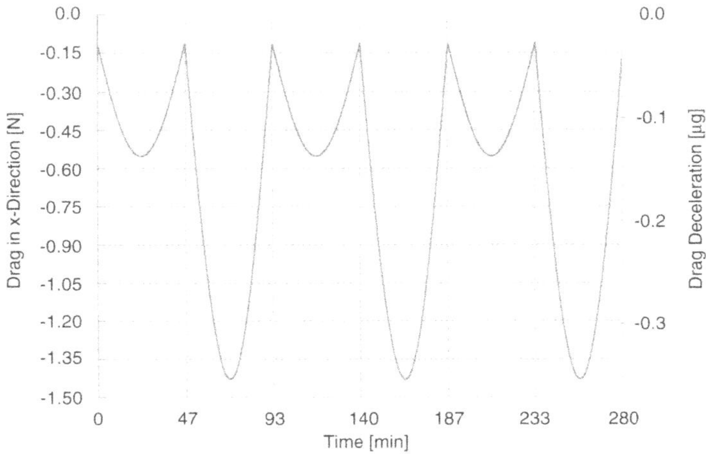

Fig. 8.15. Fluctuations of Atmospheric Drag due to the Tracking of Solar Collectors of ISS ($h = 460$ km, $F_{10.7} = 175$, $m = 415$ t) [Laible 95]

8.3.2 Tidal Forces

Aboard a space station not every experiment can be placed at the center of mass. As the equilibrium of forces between gravity and centrifugal force can only be fulfilled at that particular point, perturbing accelerations caused by the gravity gradient occur.

In the following, the perturbation field that arises will be examined. As a basis, an orbit-oriented space vehicle coordinate system (as presented in Fig. 8.16) is assumed. The experiment be at coordinate $\vec{R}$ relative to the center of Earth. $\vec{R}$ can also be described by the vector sum from the origin $\vec{R}_0$ of the space vehicle coordinate system and the relative experiment coordinate $\vec{r}$.

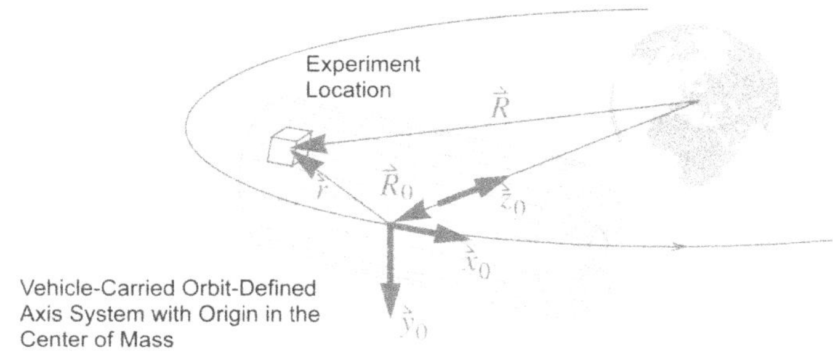

Fig. 8.16. Orbit-Defined Coordinate System of a Space Vehicle

For this estimation, the gravitational force must be considered. Together with Newton's law of gravitation, the momentum equation within a fixed coordinate system results in the following relation:

$$\ddot{\vec{R}}_0 = -\frac{\mu}{R_0^3}\vec{R}_0 \tag{8.2}$$

Assuming that the experiment were able to float freely in space, the following would be valid in an analog way for its acceleration:

$$\ddot{\vec{R}} = -\frac{\mu}{R^3}\vec{R} \tag{8.3}$$

Due to its fixed installation in the laboratory, the experiment is kinematically bound to the movement of the space vehicle. Therefore, the relative acceleration is

$$\ddot{\vec{r}} = \ddot{\vec{R}} - \ddot{\vec{R}}_0 = -\frac{\mu}{R^3}\vec{R} + \frac{\mu}{R_0^3}\vec{R}_0 \tag{8.4}$$

Based on the definition of the square of the vector $\vec{R}$:

$$\begin{aligned} \vec{R}^2 &= (\vec{R}_0 + \vec{r})^T \cdot (\vec{R}_0 + \vec{r}) \\ &= \vec{R}_0^2 \left(1 + 2\frac{\vec{r}^T \vec{R}_0}{R_0^2} + \frac{r^2}{R_0^2}\right) \end{aligned} \tag{8.5}$$

R^{-3} can be described as:

$$\vec{R}^{-3} \approx \vec{R}_0^{-3} \left(1 + 2\frac{\vec{r}^T \vec{R}_0}{R_0^2}\right)^{-3/2} \tag{8.6}$$

The last addend in parentheses in Eq. 8.5 is neglected under the assumption $r^2 \ll R_0^2$. Applying the linearization

$$(1+x)^{-3/2} = 1 - \frac{3}{2}x + \frac{3 \cdot 5}{2 \cdot 4}x^2 - \frac{3 \cdot 5 \cdot 7}{2 \cdot 4 \cdot 6}x^3 + \ldots$$

leads to

$$\vec{R}^{-3} \approx \vec{R}_0^{-3} \left(1 - 3\frac{\vec{r}^T \vec{R}_0}{R_0^2}\right) \tag{8.7}$$

After doing some vector arithmetics we obtain the tidal-acceleration as a function of the experiment coordinate $\vec{r}$.

$$\ddot{\vec{r}} = -\omega^2 \left[\underline{E} - 3\frac{\vec{R}_0 \vec{R}_0^T}{R_0^2}\right]\vec{r} \tag{8.8}$$

Assuming a circular orbit, the term μ/R_0^3 was replaced by the orbital frequency ω^2.

The formulation of the conservation of momentum was made for a fixed coordinate system at the center of mass of the space station. In order to be able to describe the rotation movement in orbit, a rotating reference system is now being introduced. The experiment coordinates in body-fixed coordinates $\vec{r}'$ are calculated with the help of a transformation matrix T:

$$\vec{r}' = T^T\vec{r} \qquad \vec{r} = T\vec{r}' \tag{8.9}$$

After differentiating twice and inserting Eq. 8.8 we finally obtain:

$$\ddot{\vec{r}}' = -\omega^2\left[\underline{E} - 3\frac{(T^T\vec{R}_0)\cdot(\vec{R}_0^T\,T)}{\vec{R}_0^2}\right]\vec{r}' - T^2\ddot{T}\vec{r}' - 2T^T\dot{T}\dot{\vec{r}}' \tag{8.10}$$

where $\underline{E}$ stands for the unit matrix. The terms $T^T\ddot{T}\vec{r}'$ and $2T^T\dot{T}\dot{\vec{r}}'$ describe the Coriolis and Euler accelerations. This expression is increasingly simplified by assuming a circular orbit ($\ddot{T}=0$) and the secure installation of the experiments in the laboratory ($\dot{\vec{r}}'=0$), so that only the first addend is left:

$$\ddot{\vec{r}}' = -\omega^2\left[\underline{E} - 3\frac{(T^T\vec{R}_0)\cdot(\vec{R}_0^T\,T)}{\vec{R}_0^2}\right]\vec{r}' \tag{8.11}$$

By means of this formula, the perturbing acceleration can now be calculated, knowing that the experiment position is relative to the center of gravity ($\vec{r}'$) and to the flight attitude of the space station (matrix T). This will be presented for both the Earth-oriented and the inertial flight modes.

Earth-Oriented Flight Mode: In this case, for the inertial observer, the space station rotates once about the y_0-axis during one orbital revolution. The radius vector $\vec{R}_0$ is always opposite in orientation to the z_0-axis, hence:

$$T = \begin{bmatrix} \cos\omega t & 0 & \sin\omega t \\ 0 & 1 & 0 \\ -\sin\omega t & 0 & \cos\omega t \end{bmatrix}; \qquad R = \begin{bmatrix} 0 \\ 0 \\ -R_0 \end{bmatrix} \tag{8.12}$$

Therefore, the term for the perturbing acceleration Eq. 8.11 is simplified, yielding:

$$\ddot{\vec{r}}'_{\text{Earth-oriented}} = -\omega^2\begin{bmatrix} 0 & 0 & 0 \\ 0 & 1 & 0 \\ 0 & 0 & -3 \end{bmatrix}\vec{r}' \tag{8.13}$$

It is interesting to note that the perturbing field of the tidal forces in this case is time-constant, and that there is no tidal perturbation in x_0-direction (velocity direction). Additionally, the signs show that a y_0-offset from the center of mass (normal direction to the orbit) always leads to a restoring force, whereas an offset in z_0-direction

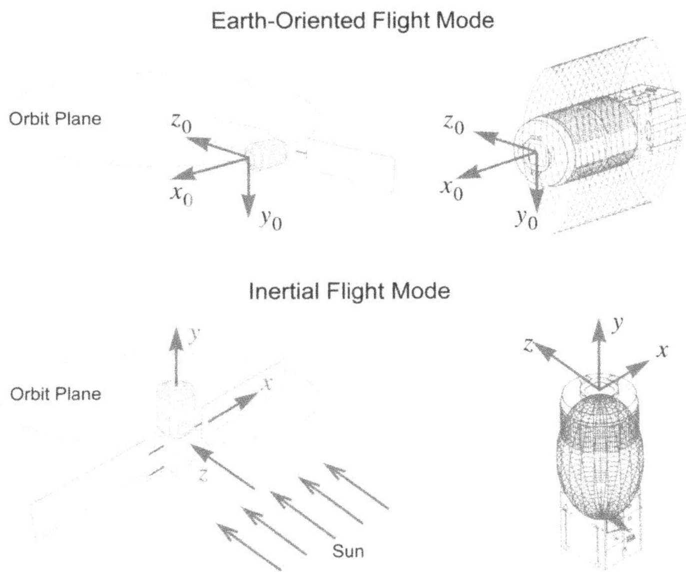

Fig. 8.17. Lines of Constant Tidal Acceleration for Earth-Oriented and Inertial Flight Modes

(direction of the radius vector) tries to enlarge the distance from the center of mass. The isolines of the tidal perturbations thus describe a cylinder with elliptical cross-section whose axis is identical to the x_0-axis (Fig. 8.17).

Direct conclusions for the installation of experiments aboard a space station and for the design of space laboratories can be drawn from the matrix representation and the position of the isolines:

- A shift of an experiment from the center of mass to the x_0-direction (velocity direction) does not lead to any additional tidal acceleration. With respect to the microgravity level, all installation positions lying on a line parallel to the x_0-axis are equivalent.
- Whenever an experiment has the same distance from the station's center of mass in y_0 and in z_0, the z_0-offset will cause, in comparison, triple the perturbing acceleration. Both components become zero when the experiment position lays on the x_0-axis.
- Generally it can be stated that the spatial expansion of a microgravity laboratory in orbit-normal and radial direction must be kept as low as possible. Although, in case of choice, a distance increase in the orbit-normal direction will create less tidal perturbations and should therefore be preferred. In velocity direction, however, a microgravity laboratory can be extended to any degree without increase of the tidal forces.

Inertial Flight Mode: If the orientation of the space station is fixed with respect to the Sun, the transformation matrix T remains constant and can be replaced without loss of generality by the unit matrix. The radius vector $\vec{R}_0$, however, now takes a transient form:

$$\vec{R}_0^T = \begin{bmatrix} \sin\omega t & 0 & \cos\omega t \end{bmatrix} R_0 \tag{8.14}$$

The following is derived from Eq. 8.11 and is valid for the acceleration caused by tidal forces:

$$\ddot{\vec{r}}'_{\text{inertial}} = -\omega^2 \begin{bmatrix} 1-3\sin^2\omega t & 0 & -\frac{3}{2}\sin 2\omega t \\ 0 & 1 & 0 \\ -\frac{3}{2}\sin 2\omega t & 0 & 1-3\cos^2\omega t \end{bmatrix} \vec{r}' \tag{8.15}$$

In this case, the observer in a laboratory module notices a transient perturbation field in x- and z-directions and a constant perturbation field in the y-direction. The isolines of the tidal forces form an ellipsoid (Fig. 8.17). For each component of the experiment's direction vector, a distance as short as possible to the space station's center of mass is of advantage. There is no preferential direction for the spatial extension of a microgravity laboratory, either, so that generally, for the inertial flight mode, a compact and symmetrical topology is favorable.

8.3.3 The g-Jitter

The most complex perturbation is doubtlessly the so-called "g-jitter". This term summarizes all perturbing accelerations that can be traced back to resonances of the space vehicle's structure due to internal or external perturbations. As aforementioned, these can result from control maneuvers, crew activities, unbalanced rotating devices, valves or engines. Due to these various causes it is hardly possible to identify these perturbations one by one. It is rather a wide acceleration spectrum that is caused, mechanically exciting the vehicle's structure. In its upper half, Fig. 8.18 shows the most important causes aboard a space station listed according to their frequency range.

The structure of the vehicle, in turn, reacts to the broadband excitation with a cascade of eigenfrequencies (Fig. 8.18, below). These range from about 0.01 Hz for the primary structure of the space station, to the different structure elements (module cluster/orbiter, pallet/module, rack), and furthermore to the experiment itself with eigenfrequencies of up to 100 Hz. In Fig. 8.19, a perturbation spectrum of the g-jitter during the Spacelab D1 mission is shown, as it was caused by an astronaut who moved abruptly from one wall to another (push off – float – decelerate).

For the description of a disturbing environment caused by the g-jitter, normally an envelope is specified which, with a certain statistic probability, will not be exceeded by the presumably occurring perturbing accelerations. Such a curve is schematically drawn in Fig. 8.19.

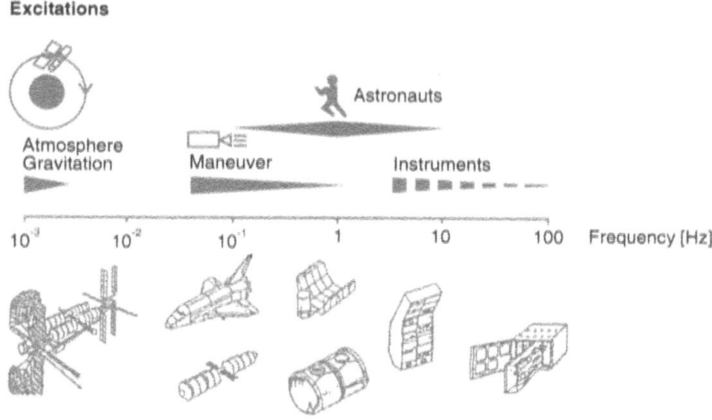

Fig. 8.18. Disturbing Sources and Resonances of a Vehicle Structure shown in the Frequency Range [Feuerbacher 88]

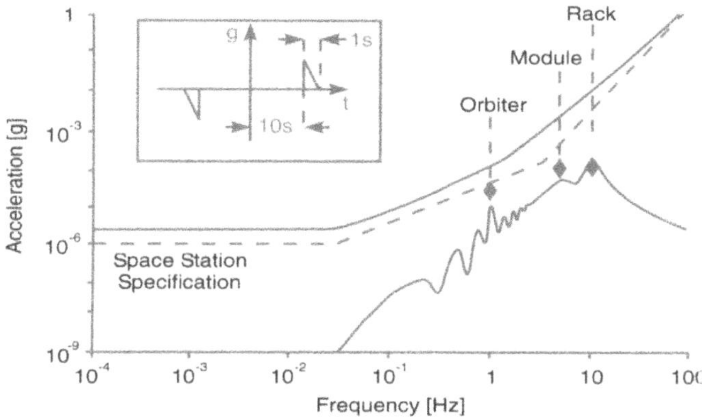

Fig. 8.19. Spectrum of Disturbance of the Spacelab-D1 Mission during Crew Motion and an Example of a "Typical" Tolerance Spectrum of an Experiment [Hamacher 96]

Tolerance Spectrum: Opposed to the specified perturbation spectrum of the space station, there is a so-called "tolerance spectrum" on the experiment side, indicating the maximum acceptable perturbing accelerations as a function of frequency. Reliable experiment operation will be guaranteed if the lines of the perturbation spectrum show a sufficient distance to those of the tolerance spectrum.

The way how to derive such a tolerance spectrum shall be illustrated with the help of a one-dimensional model experiment: Figure 8.20 shows a drop of water in oil as an example from the field of fluid physics.

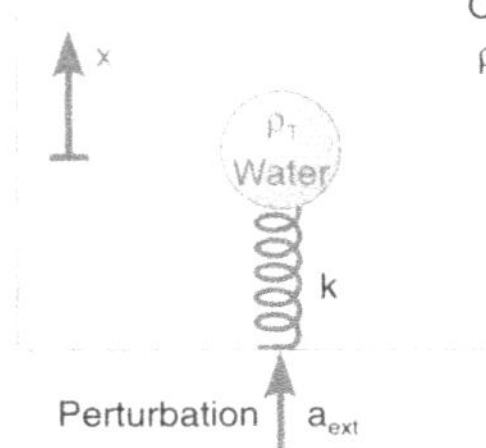

	Water	Oil
Density [kg/m³]	1000	871
Viscosity [10⁻⁶ m²/s]	1	15

Fig. 8.20. Model Experiment: A Water Droplet in Oil

The drop movement is determined by the following forces:

- The buoyancy force resulting from an exterior acceleration influencing the experiment:

$$F_A = (\rho_T - \rho_F) V_T\, a_{ext} \tag{8.16}$$

- The frictional force of the drop with a radius R, here modeled by means of the Stokes' law:

$$F_R = -6\pi\nu\rho_F R\dot{x} \tag{8.17}$$

- A restoring force as yet not clearly defined, shown by Hooke's law:

$$F_F = -kx \tag{8.18}$$

The force equation $\rho_T V_T \ddot{x} = F_A + F_R + F_F$ for the droplet movement leads to the equation of motion:

$$\ddot{x} + \frac{6\pi\nu\rho_F R}{\rho_T V_T}\dot{x} + \frac{k}{\rho_T V_T}x = \left(1 - \frac{\rho_F}{\rho_T}\right)a_{ext} \tag{8.19}$$

Assuming a harmonic course of the exterior (i.e. monochromatic) acceleration, Eq. 8.19 describes the behavior of forced oscillation of the form

$$\ddot{x} + 2\delta\dot{x} + \omega_0^2 x = b\, e^{i\omega t} \tag{8.20}$$

with the damping coefficient

$$\delta = \frac{9\rho_F\nu}{4\rho_T R^2} \tag{8.21}$$

and the eigenfrequency

$$\omega_0^2 = \frac{3}{4}\frac{k}{\rho_T\pi R^3} \tag{8.22}$$

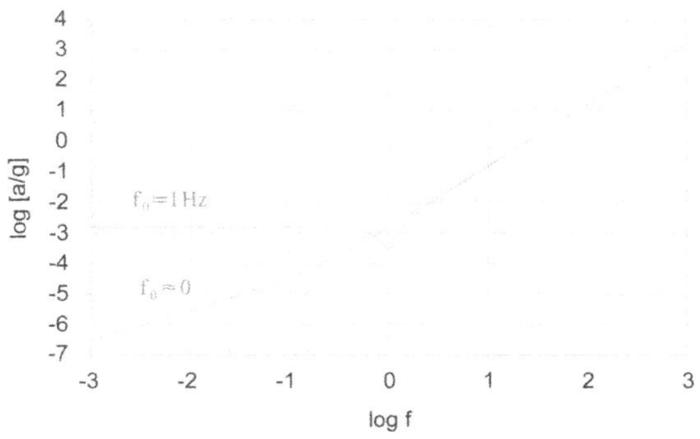

Fig. 8.21. Acceptable Perturbing Accelerations for a Model Experiment

This system has the solution

$$x(t) = a \frac{(\rho_T - \rho_F)}{\rho_T \sqrt{(\omega_0^2 - \omega^2)^2 + 4\delta^2\omega^2}} e^{i\omega t} \tag{8.23}$$

where a stands for the amplitude of the excitation. The largest displacement of the droplet, as a consequence of the perturbing acceleration a, is:

$$\Delta x = a \frac{(\rho_T - \rho_F)}{\rho_T \sqrt{(\omega_0^2 - \omega^2)^2 + 4\delta^2\omega^2}} \tag{8.24}$$

It is now necessary to define a criterion for the maximum acceptable perturbing acceleration $\bar{a}$. Thus, it is an obvious way of expressing it as the maximum displacement of the droplet relative to its dimension:

$$\bar{a} = \frac{\Delta x}{R} \cdot \frac{\rho_T}{\rho_T - \rho_F} \cdot R \cdot \sqrt{(\omega_0^2 - \omega^2)^2 + 4\delta^2\omega^2} \tag{8.25}$$

If a value of 1% is taken as a basis for the allowable relative displacement, the acceptable acceleration can be defined as a function of the excitation frequency. This can be seen in Fig. 8.21 for a droplet of 1 mm diameter.

When initially considering this curve with the eigenfrequency of 1 Hz, this corresponds to an experiment with non-vanishing restoring forces as it comes about, for example, in the case of liquid columns or containerless processing in a levitation field (cf. Sect. 8.4). Such experiments are extremely susceptible to perturbations in the range of their eigenfrequency. Above the eigenfrequency, susceptibility decreases with the square of the frequency, whereas the acceptable perturbing accelerations below the eigenfrequency assume a constant and finite value.

The majority of microgravity experiments are characterized by the fact that no restoring forces are present. In this case, the eigenfrequency disappears ($f_0 \to 0$). At higher excitation frequencies, the susceptibility decreases with the second power as well, whereas at low frequencies it shows a nearly linear course. This simple model

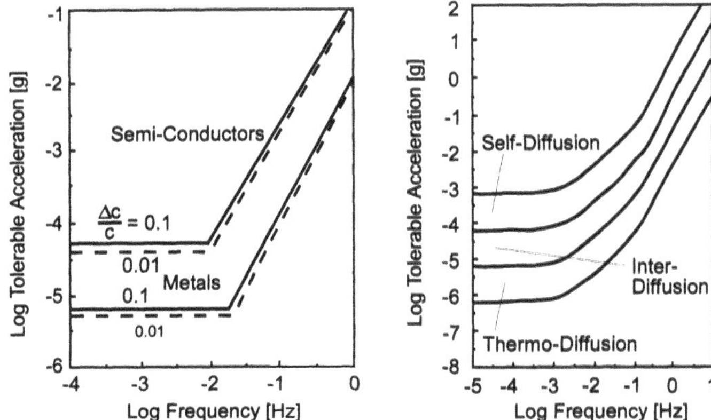

Fig. 8.22. Tolerance Spectra for Different Classes of Experiments [Alexander 90]

leads, however, to a steadily growing perturbation susceptibility for disappearing excitation frequencies. This does not make sense from the physical point of view. In reality, e.g. a frequency-independent, lower limit of sensitivity will be reached if the mean velocity of the free-floating object corresponds to the velocity of the Brownian movement [Feuerbacher 88]. This, however, shall not be discussed in detail at this point.

Figure 8.22 shows the tolerance spectrums for various classes of experiments. The sensitivity peaks superpose the curves by the eigenfrequencies of the experiment (for example, liquid columns), as shown in Fig. 8.21. It remains to be stated that detailed knowledge and careful calculation of perturbation and tolerance spectrums are essential preconditions for the success of experiments in microgravity [Alexander 90, Monti 87].

8.3.4 Solar Radiation Pressure

The perturbing accelerations caused by solar radiation can, to a certain approximation, be described by the relation:

$$a_s = p_s(1+\sigma)\frac{A_p}{m} \tag{8.26}$$

Here, p_s stands for the radiation pressure in Earth orbit given by the solar constant and the speed of light,

$$p = \frac{E}{c} = 4.5 \cdot 10^{-6}\ \mathrm{N/m^2} \tag{8.27}$$

and σ represents the surface reflectivity of the space vehicle ($0 < \sigma < 1$). As in the case of the atmospheric drag, this perturbing acceleration depends on the relationship of the projected incidented area to the vehicle mass. Typical values amount to $2 \cdot 10^{-8}\ g_0$ for EURECA and $2.3 \cdot 10^{-9}\ g_0$ for the US Space Shuttle. When compared to the perturbations mentioned before, these values can be neglected for low Earth orbits. In higher orbits, however, and especially in the geosynchronous orbit, the solar radiation pressure is the major perturbation for the attitude dynamics of a space vehicle (see also Fig. 3.19).

Depending on the altitude and vector of the orbit plane relative to the Sun, the day/night phases related to it, and the flight mode of the vehicle, a perturbation results with both a constant and a periodic contribution (see Fig. 6.6and Fig. 8.13).

8.4 Perturbation Compensation and Levitation

Illustrated in Fig. 8.23 are lines of constant perturbing accelerations as they can be expected to result from the quasi-stationary perturbing influence aboard the International Space Station. It is of particular interest that the isolines, in the side view, are tilted with respect to the longitudinal axis of the service module, the Zarya module (FGB), and the US Laboratory module (cf. Sect. 8.3.2). This is a result of the fact that the perturbation-free flight mode (TEA) of the International Space Station markedly deviates from the LVLH-flight mode ("Local Vertical Local Horizontal", cf. Fig. 6.7).

The station has a set of requirements that define the microgravity environment. The duration requirement details the annual time the microgravity will exist. The disturbance environment is divided into two major classifications: transient and steady-state. Steady-state disturbances are subdivided into quasisteady and oscillatory types. Fifty percent of the ISPR locations inside the US-Lab, JEM, and COF must comply with the duration, quasisteady, oscillatory, and transient disturbance requirements during the microgravity mode. These requirements are defined below:

- *Duration Requirements.* A variety of activities ranging from orbiter docking to EVAs can cause excessive disturbances in particularly sensitive experiments. Therefore, times are set aside throughout the year in which these activities are not conducted, except, of course, during an emergency. ISS is to have at least 180 days a year in "microgravity" mode. The 180 days shall be allotted in increments of no less than 30 days each. During these times, the space station must comply with the quasisteady and oscillatory requirements (see below) simultaneously.
- *Transient Requirements.* Transient disturbances are defined as being neither steady-state sinusoidal nor stationary random within any 100-second measure-

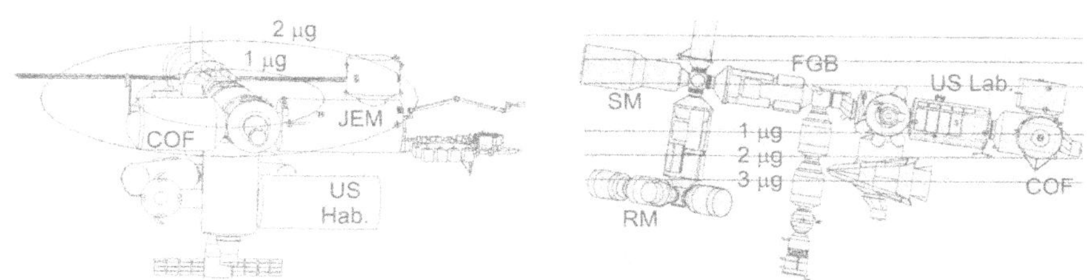

Fig. 8.23. Quasi-Static Microgravity Level of ISS [ESA 95]

Table 8.2. Disturbance Limit for Oscillatory Disturbances

For the Vehicle Subsystems Alone		For the Vehicle Subsystems Plus the Payload Complement	
Frequency Range	Disturbance Limit	Frequency Range	Disturbance Limit
0.01 Hz < F < 0.1 Hz	1.6 μg	0.01 Hz < F < 0.1 Hz	1.8 μg
0.1 Hz < F < 100 Hz	16 μg · F [Hz]	0.1 Hz < F < 100 Hz	18 μg · F [Hz]
100 Hz < F < 300 Hz	1600 μg	100 Hz < F < 300 Hz	1800 μg

ment window. Any single transient disturbance must have a maximum below 1000 μg, and an integrated area of 10 μg-seconds over any 10-second interval.

- *Quasisteady Requirements.* A disturbance is considered quasisteady if at least 95% of its mean-squared value lies below 0.01 Hz as measured over a 5400-second period (the approximate time of one orbit). Examples of quasisteady disturbances are aerodynamic drag and gravity gradient. The entire quasisteady environment must be less than or equal to 1 μg in magnitude and have a component perpendicular to its orbital average vector less than or equal to 0.2 μg.
- *Oscillatory Requirements.* The disturbance limit for oscillatory disturbances is somewhat more complex and covers frequencies between 0.01 Hz and 300 Hz. The root-mean-squared value over any 100-second interval cannot exceed the limits shown in Table 8.2.

The predicted g-jitter perturbations aboard ISS are shown in Fig. 8.24. It can be seen that the perturbing amplitudes are above the specified limit for various excitation frequencies. Depending on the experiment's susceptibility, perturbation compensation may have to be taken into account.

Finally, as a passive means for the compensation of perturbations, the space station architecture itself and the installation of the experiments are to be mentioned. As already shown in Sect. 8.3.2, tidal forces especially can be influenced by the size, attitude and orientation of, e.g, a laboratory module or by the installation of an experiment within such a unit. Furthermore, it is also possible to think of compensating single perturbations, such as tidal forces and the solar radiation pressure, with each another.

Special attention must be paid to the atmospheric drag due to its significant influence in low Earth orbits. A compensation using other perturbations such as tidal forces cannot be applied in this context, so the active way by means of reaction thrusters is the only option. This has already been carried out for satellites performing research of the Earth's gravitational field with residual accelerations of about 10^{-11} g (TRIAD-Project, 1972). By means of adequate control maneuvers, the space vehicle "avoids" a free-floating reference mass inside the space vehicle ("drag-free" concept). For the laboratory operation aboard a space station, this method, however, certainly cannot be applied because of the high propellant requirements. The atmospheric drag is thus determined by the overall configuration of the space station. This once more underlines the necessity to examine keenly during conceptual design phase, the interactions between the space station architecture, the environmental conditions, and the utilization scenarios. To this concern, special attention is given in Chapter 9 ("System Design").

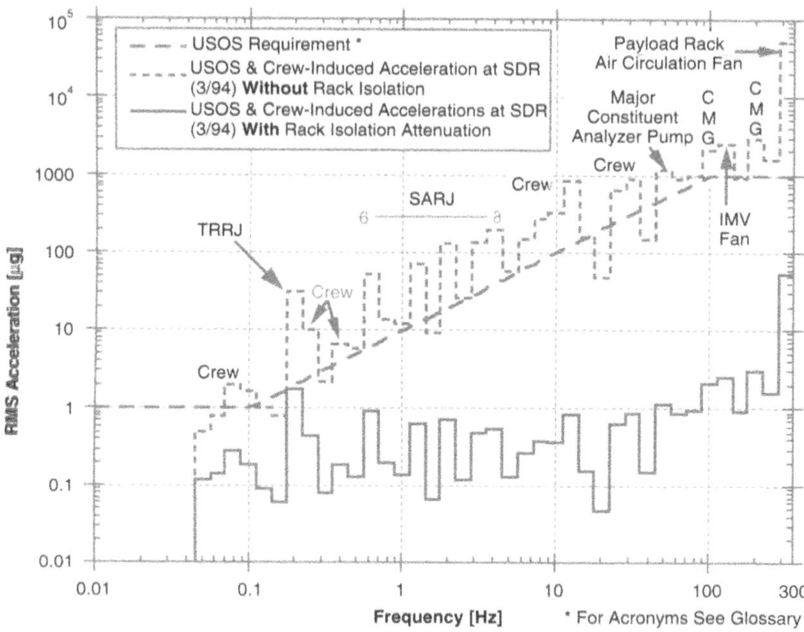

Fig. 8.24. Prediction for the G-Jitter aboard the International Space Station [ISS 94]

From the experiment point of view as well, isolation against perturbations and/or their damping can be considered. It is often sufficient to mechanically decouple the experiment from the space station by means of conventional spring or damping elements in order to decrease the impact of the g-jitter in particular. Aboard the International Space Station, Boeing of USA had the idea of actively decoupling whole experiment racks in order to protect them against vibrational perturbations by installing a considerable number of small actuators. These actuators are driven by a control system which uses accelerometer data to position the rack. The perturbation spectrum obtained from calculations for this so-called "Active Rack Isolation System" (ARIS) is shown in Fig. 8.24, as well. The shown damping effect in the low frequency range will, however, be difficult to achieve because low frequencies involve larger amplitudes. Due to its actuators, the ARIS itself will, in any case, introduce considerable perturbations into the space station structure as well. Their impact on the microgravity level is still unknown. There is reason to fear, however, that the perturbation level will increase considerably within inactively damped experiment racks [Hamacher 96].

An interesting alternative is the "tethering" of an experiment or test item. The prototype of such a buffer element is shown in Fig. 8.25. It consists of a chain of ball joints being tied to one another under pressure by an continous wire. Due to their friction, the ball joints can damp the oscillation movements of the end mass to be fixed, whereby the damping can be controlled by the tension of the tether [Ockels 87].

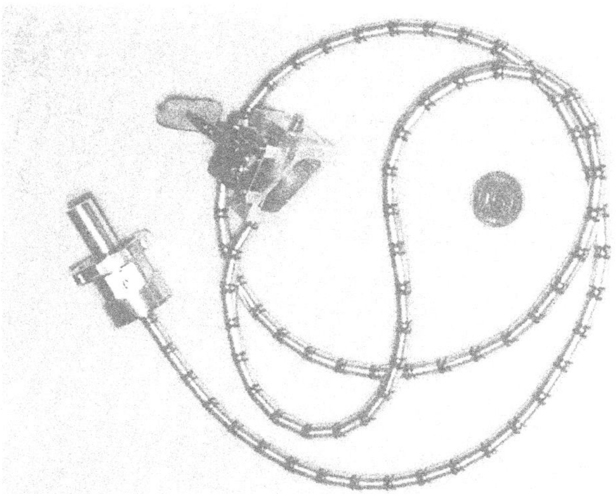

Fig. 8.25. Short Tether for Experiment Attachment in a Microgravity Laboratory

Furthermore, the system damping of existing levitation systems can be used as well. Levitation systems primarily serve the contact-free and low-disturbance positioning of experiments or test items.

Positioning by levitation can be achieved in various ways, through:

- Electrical forces: electrostatic levitation
- Magnetic forces: magnetic and electromagnetic levitation
- Ultrasound: acoustic levitation
- Aerodynamic forces in a controlled flow: aerodynamic levitation

In Fig. 8.26, the principles of electromagnetic and acoustic levitation are explained. A new concept of magnetic levitation is shown in Fig. 8.27. It makes possible the stabilization without any active control of a free-floating droplet in a magnetic fluid. The stabilizing effect is based on the interaction between the magnetic ferrofluid and the non-magnetic drop in a permanent magnetic field. This interaction has initially been calculated by Dr. T. Rösgen and can be used for a considerable number of µg-applications [Huber 96]. Accordingly, such a levitation cell can be used for containerless processing as well as for crystal growth. Vortex-free flow fields may also be caused by magneto-caloric pumping. The first technological experiment (T9, MAGLEV aboard the EuroMir-95 mission) has proved the application potential of this levitation technology.

Common to all levitation and damping systems is the fact that there is an additional perturbation susceptibility from the residual forces due to resonance caused by the restoring forces. This must be considered when designing the hardware. It is absolutely necessary to remember that damping and levitation systems are only suitable for the compensation of transient perturbing accelerations (for example the g-jitter). Constant perturbing accelerations such as the atmospheric drag must be reduced to the bare minimum by designing a space station accordingly.

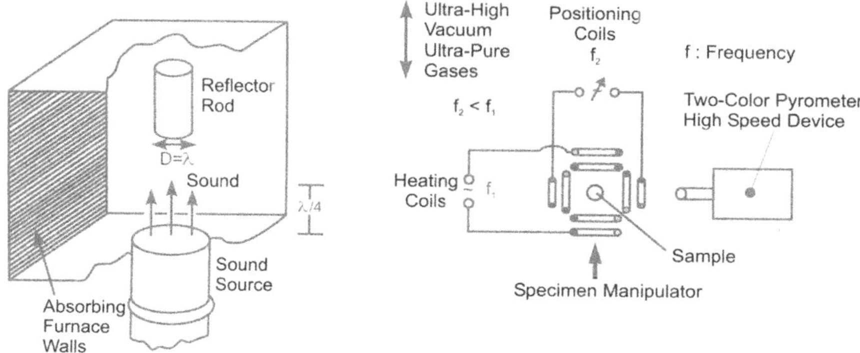

Fig. 8.26. Acoustic Levitation (Left) and Electromagnetic Levitation (Right) [Walter 87]

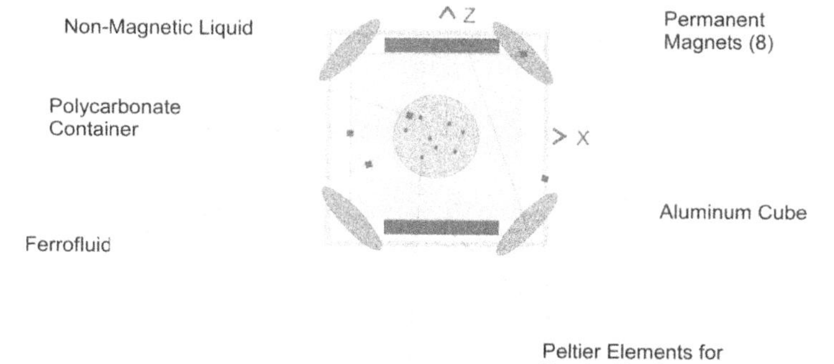

Fig. 8.27. Magnetic Levitation [Huber 96]

9 System Engineering

In the previous chapters, various space stations and space concepts have been presented. In this context, certain elements continue to occur repetitively in one form or the other, such as the following:

- Pressurized modules for experiments and the crew
- Pressurized connection nodes and docking adapters
- Structural elements (e.g. trusses) to connect the modules
- Solar collectors, usually tracking the Sun
- Thermal radiators
- External payloads
- Manipulator systems
- Operational elements such as logistics or rescue vehicles

In principal, any geometric configuration can be formed from these basic components of a space station. For that reason, it is even more astonishing that the results of most of the configuration studies are only a few patterns which are very similar to one another. This observation raises the question of which factors determine the design of a space station in the end – this is a question of configurational system design which will be analyzed in the framework of this chapter. First, the challenges of system design and the environment in which it takes place will be discussed. Second, a possible methodology and possible design tools will be introduced. The third section will cover the conflict between the theoretical-methodical approach and the (sometimes painful) reality of system design and space flight concepts. Additionally, the development of space station concepts from the beginning of the 1980's to the International Space Station (ISS) will be outlined.

9.1 The Life Cycle of a Space Project

All activities necessary for the realization of a technical system, ranging from the initial ideas to operation to final use, are summarized by the term *project*. Its characteristics are the following:

- It is a unique and acyclic sequence of events with a defined beginning and end point.
- There is an objective or goal which has to be reached within specified time, cost and performance constraints.
- Several persons, groups, enterprises or institutions participate in the project.
- There is a certain complexity with respect to technology or organization.

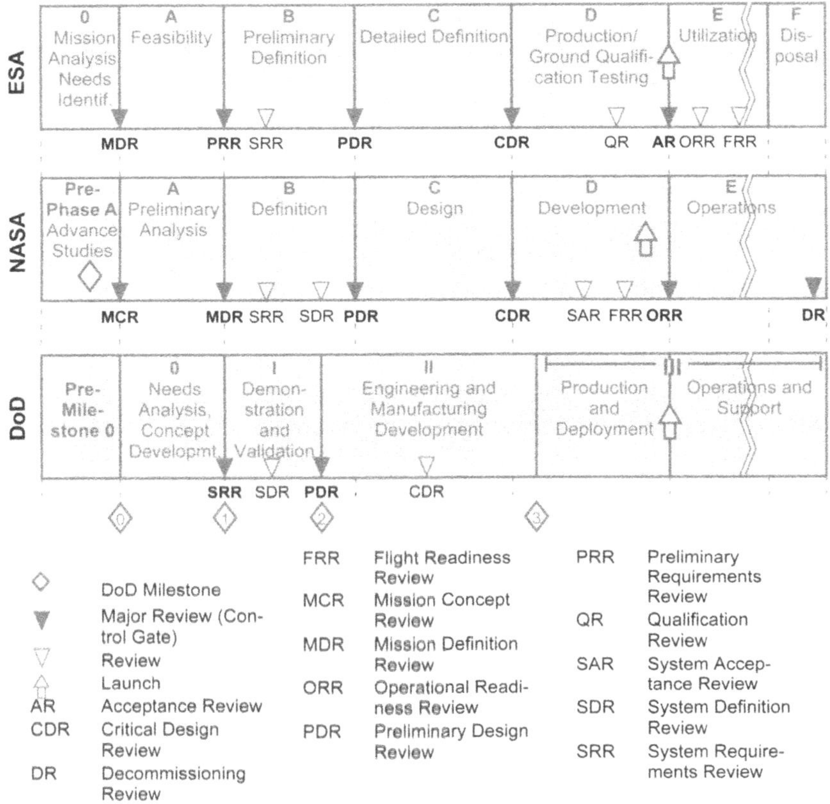

Fig. 9.1. Life Cycle of a Space System

Space projects are commonly structured in several stages or phases, reflecting the different types of activities carried on within a project. The overall process is called the "life cycle" of a space system. Figure 9.1 illustrates the phase models used in different institutions to structure the life cycle. The numbering and naming of the phases is slightly different among these institutions. However, they denote the same types of activities that are needed to conceive, design, build and operate a space system. In Europe the labels 0, A–F denominate the life cycle phases.

Phase 0 – Mission Analysis/Needs Identification ("pre-phase A" or "Advanced Studies", "Pre-Milestone 0"). In this first project phase, the task is to identify the needs or interests that trigger the design activity and to express them in terms of broad objectives. Such broad objectives can be, for instance, not only a telecommunications service or an Earth observation task, but also a scientific or a technological objective. In case of scientific ventures, for example those of ESA, there may be a formal call for ideas inviting the scientific community to submit proposals. In other

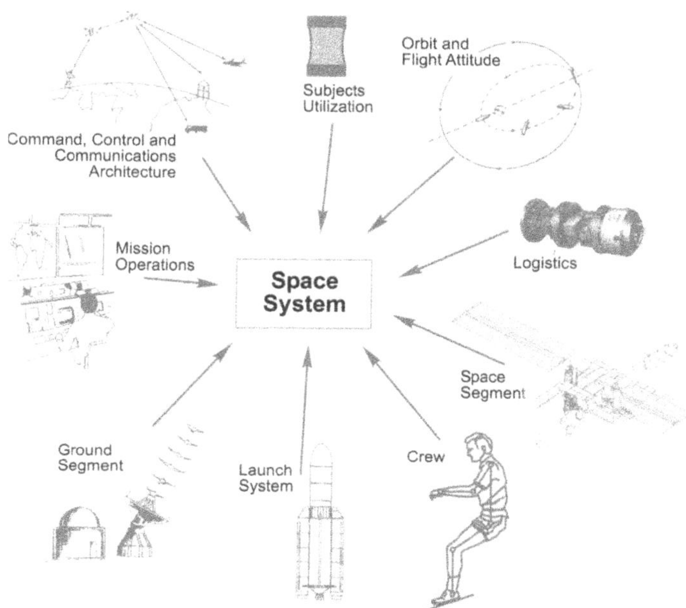

Fig. 9.2. System Elements of a Space System

cases these broad objectives will be named directly by a customer (e.g. a satellite operator) or by political leaders. In the past, crewed missions such as Apollo or the International Space Station have always been driven mainly by political and scientific objectives.

In the course of the conceptual design process, space system solutions have to be found which can both meet the requirements and respect the design constraints. It is important to keep in mind that a space system solution not only concerns the actual space segment, but rather it includes all system elements (Fig. 9.2), such as mission objectives, orbit, launch, flight strategies, space segment, ground segment, logistics concept, etc., in a consistent and balanced combination (system concept).

The most important task in phase A is to compare and evaluate different candidate system concepts in order to identify a small number of concepts (ideally one single concept) that can meet the objectives best. Therefore, meaningful criteria or figures of merit relating the outputs of a system concept with its expenditures must be defined. In the case of an Earth observation mission or an interplanetary probe, an adequate figure of merit could present the amount and quality of scientific data produced by different system variants with respect to their life cycle costs. With complex systems such as a space station, the designer usually encounters some difficulties when searching for adequate evaluation criteria. On one hand, there will be several utilization disciplines aboard a space station; on the other, there is a multitude of system parameters such as equivalent system mass, logistics mass, safety and reliability numbers, efforts for development, etc., which could be used to evaluate the quality of a concept. The situation is further complicated by the fact that political objectives tend to play an important role in large and/or crewed space projects. In the past, such objectives were to demonstrate technological superiority or to manifest leadership. For the International Space Station, cooperation, especially between the USA and Russia, plays an important role. As a matter of fact, predominance of political objectives can add considerable risk to a space system venture. A classic example is the space station Freedom. For nearly a decade it was seen as an outpost for the assurance of a permanently crewed presence of the "free world". The end of the Cold War introduced a new objective of cooperation with the former Eastern Bloc countries for the space station, leading to a complete configurational redesign of Freedom and the resulting new International Space Station. Even so, political considerations will always be a part of large space projects. By nature, these objectives are difficult to quantify, and they can add some extra risk to a project.

To sum up, the activities which are part of phase 0 can be listed as follows:

- Identification of broad objectives
- Translation of objectives into numerical requirements (e.g. an optical resolution, a microgravity level, a pressurized laboratory volume)
- Identification of constraints that are difficult for the designer to change (e.g. launch requirements, political constraints, geographical returns, etc.)
- Establishment of system concepts which are able to meet the objectives
- Assessment of the most important system parameters in system budgets for performance, mass, power, cost, etc.
- Identification of critical points in terms of safety, technology, cost, and other predominant design drivers
- Documentation, evaluation and comparison of system concepts using adequate figures of merit
- Establishment of a market study for commercial ventures
- Marketing of the project to the customer or potential user

Phase A – Feasibility ("Preliminary Analysis", "Concept Exploration"). To conclude phase A, an optimum system concept should be worked out and documented in a system specification. Therefore, the system concepts retained in phase 0 have to be analyzed and evaluated with respect to technical, economic and organizational feasibility. This is why NASA considers phase A (or "preliminary analysis") to be a structured version of the previous phase [NASA SP-6105]. Its activities are:

- To work out and compare alternative system concepts, the goal being to find the one which is "optimal"
- To evaluate technological, economic and organizational feasibility
- To demonstrate solutions for critical points and confirm feasibility with respect to mission, technology, cost and organization
- To establish system specification
- To develop plans for design and development, finances, schedule and organization

Phase B – Preliminary Definition ("Definition", "Demonstration and Validation"). In this phase, system level requirements have to be broken down into subsystem levels. As in phases 0 and A, but on a subsystem level, alternative design solutions have to be worked out, again using critical system parameters such as mass, electrical power, logistics or safety for evaluation and decision making. At the end of this phase, all information needed to initialize the realization phases must be available. This covers mainly a complete set of system specifications, as well as planning documents for development, manufacture, integration, verification and operation, and a detailed description of the organizational structure.

Phase C – Detailed Definition ("Design", "Engineering and Manufacturing Development"). In phase C, the design cycle is brought down to a component level. Furthermore, test and qualification processes, as well as the related facilities, will be established and qualified.

Phase D – Production/Ground Qualification Testing ("Development", "Production and Deployment"). Phase D is the end of the system development. It includes the production, integration, and verification of the flight hardware and its associated test models. In practice, a separation of detailed definition (phase C) from production/qualification (phase D) has proven to be inadequate. For that reason, both phases are generally combined under the label "Phase C/D". In contrast to ESA and DoD, NASA considers launch and in-orbit checkout to be part of the development phase, whereas the operations phase starts once operational readiness has been acknowledged.

Phase E – Utilization ("Operations", "Operations and Support"). For ESA, the launch of the space segment marks the beginning of the system's utilization period. This phase, however, is divided into two subphases: E1, an overall test and commissioning phase, and E2, the utilization phase itself [ECSS-M-30A].

Phase F – Disposal. Once the operational phase of a system has ended, it enters the final utilization phase. One disposal strategy includes a transfer to a graveyard orbit as applied in the case of geosynchronous satellites. Transfer back to Earth or destructive reentry into the Earth's atmosphere are the other options for disposal. The dedicated disposal phase is ESA-specific terminology. There is no distinction between operations and disposal phase on the US side.

Figure 9.1 shows a number of project reviews during the different phases. They are an important tool for the verification and coordination of work progress in the sequence of project phases [ECSS-M-30A, ESA FFP/PS/753]). Consequently, all important project steps in terms of design, development, manufacturing, testing and operation will be formally closed and approved in a review. Therefore external reviewers together with the client (e.g. space agencies) and the industrial contractors will evaluate to what extent requirements and specifications have been met, and whether or not design decisions are justified and complete.

The most important reviews in the project's life cycle are the following [ECSS-M-30A, NASA SP-6105]:

- *Preliminary Requirements Review (PRR).* This review formally terminates the definition of the mission and that of all required system functions. As a result, mission feasibility is confirmed, which means that a functional baseline and a corresponding system concept have been selected as the starting point for the subsequent definition phase. For that reason, the PRR stands at the end of the feasibility phase A. The system is referred to as being in a "functional state".
- *Preliminary Design Review (PDR).* Once all requirements for the system elements and their interfaces have been established, the PDR at the end of phase B is held. This is therefore called the "specified state" of the system. At this time, the system requirements are broken down in detail into all elements of the subsystems. Often the PDR is not a single review, but a number of reviews on the subsystem level, terminated by a system PDR.
- *Critical Design Review (CDR).* The CDR has a prominent position within the design and development flow as it fixes the final design. Therefore, testing and verification must show that all design activities and trade-offs have led to a detailed and complete definition of the system that will meet all the requirements within the planned cost and on schedule. When a system has reached the "defined state", production of hardware and software can start. Therefore, apart from the system itself, production plans, safety aspects, integration, verification and test procedures are also subject to the CDR. Again, the system CDR is preceded by a number of subsystem CDRs.
- *Acceptance Review (AR).* The system has reached an "accepted state" once the customer has confirmed that the completed product conforms to its configuration baseline. This is formally done in the AR. The system has successfully passed through all test and verification procedures and is ready for shipment to the launch site. The AR therefore marks the beginning of the operations or utilization phase.

9.2 The Conceptual Design Problem

When designing a system from scratch, the designer has to deal with a number of special characteristics and requirements:

"Fuzzy" Problem Formulation. At the beginning of the design process, broad objectives and goals are generally hard to come by, and even then they are often only available in "fuzzy" form. Ideal objectives for a space mission can be a telecommunication service, an Earth observation task, a scientific experiment or a technology project. Nevertheless, in many cases the only customer objective stated will be the desire to reuse a certain piece of hardware or launcher, or the political will to cooperate in a technology project, e.g. to enhance international cooperation or to be part of a visible joint venture, etc.

In general, mission objectives should be developed, put into concrete terms, and balanced during conceptual design, together with the corresponding system solution. The availability of precise mission objectives from the very beginning is more the exception than the rule. As a consequence, conceptual design can be characterized by a lack of information as well as the need to process fuzzy information.

Strong Interdependencies among System Elements. System elements are highly interdependent. The complexity of the design task is due to the variety of interrelationships or mutual influences, as the following example shows:

The arrows in Fig. 9.3 illustrate interrelationships originating from the system element "launch system". A launcher can first be characterized by its payload capacity (kg in orbit), its maximum payload dimensions, the realizable launch rate, the launch loads, and the launch system safety and reliability standards. These five characteristics have an influence on other system elements.

- Since the propulsion performance of the launcher is limited, there will be an orbital envelope specifying the borders up to which reasonable payload masses can be delivered. Therefore, orbital altitude and inclination will be limited due to the choice of launch system (arrow ①).
- The launcher performance and payload dimensions have a direct influence on space segment design (arrow ②). For space stations, this generally leads to the constraint of having to implement a modular and orbit-deployable design, with the mass and size of each module being compatible with the launch system.
- The launcher performance together with the realizable launch rate will not only drive the maximum logistics flux (arrow ③), but together with the size of the

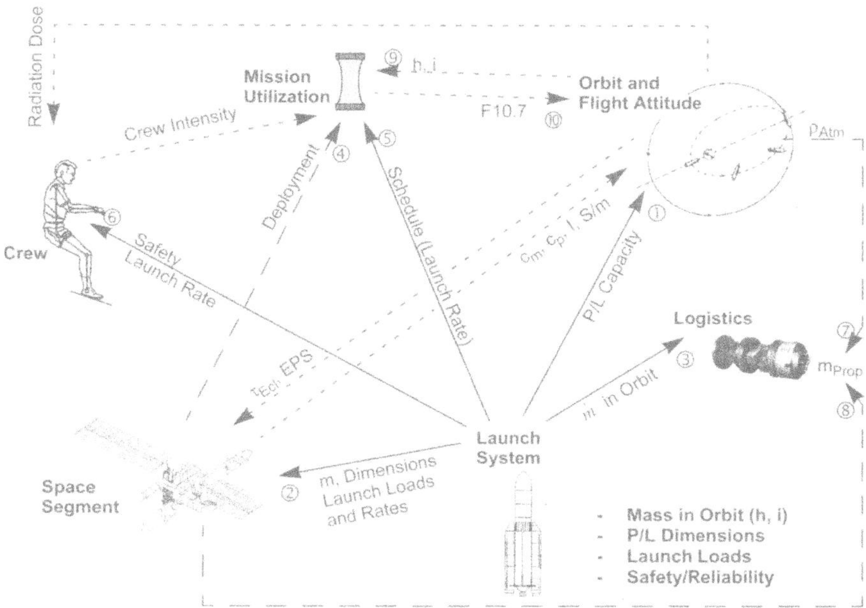

Fig. 9.3. Example of Interrelationships between System Elements

space segment, it will also determine the time required to deploy the system and, therefore, the delay until the beginning of the system's utilization (arrows ④, ⑤).

- Finally, the launch system selection is strongly linked to safety and risk considerations. Besides the general question of whether or not and how humans can be transported aboard a launcher, the launch rate will also limit the crew intensity attainable aboard the space station (arrow ⑥), meaning the number of crew members, permanently crewed or crew-tended, crew rotation, and logistics, etc.

Typically, direct dependencies will induce chains of follow-on dependencies, represented by the dashed arrows in Fig. 9.3. For example, the space segment's reboost propellant, as a major logistical requirement, will be driven by the selected orbit and by the size of the space segment itself (arrows ⑦ and ⑧). However, the orbit will also have a significant influence on the mission if, for example, Earth observation or telecommunication missions are to be carried out (arrow ⑨). Another interdependency occurs among schedule (start of mission), orbit and logistics. Time-dependent solar cycles drive the density of the residual atmosphere in orbit. As a result, both minimum flight altitude (which assures enough margin against accidental reentry of the space segment) and propellant needs for orbit control will vary in time (arrows ⑦ and ⑩).

It is obvious that these chains of mutual dependencies could be extended indefinitely. However, this small example illustrates that the process of designing a space system cannot be described in a linear chain of dependencies with a sequential flow of steps. Space system design is rather a typical *network-type problem* where a change of one element or parameter will always influence a variety of other system parameters. Another characteristic feature which can be emphasized is that a single interrelationship, when taken independently, may appear rather trivial, e.g. "When using launcher A, maximum orbital altitude will be X". The complexity of the design problem, however, arises from the multitude of interrelationships in a network-type structure.

Predetermining Character of Conceptual Design Decisions. Once the most important system elements, such as orbit and flight mode, size and configuration of the space segment, crew intensity, and the operations concept are fixed, the key system parameters such as overall mass, performance, cost or risk will be largely determined. Figure 9.4 illustrates this correlation in a qualitative diagram.

Most of the project expenditures will occur during detailed definition and production phases (phase C/D) when technologies are to be brought to operational readiness and flight models are to be built. The decisions on the type and extent of these efforts, however, are made in the very early design phases, when concepts are developed and selected. As a consequence, the first 5% to 8% of program expenditures determine about 85% to 90% of program development cost, and they also determine the production unit cost, quality, and supportability of the design almost irreversibly [Shaw 92]. In other words, the whole life cycle of a product is present during conceptual design since the life cycle characteristics of a system will inevitably be fixed. Conceptual design therefore plays the key role in the design process, although it is often considered the "play phase", the precursor to the "real" design phase.

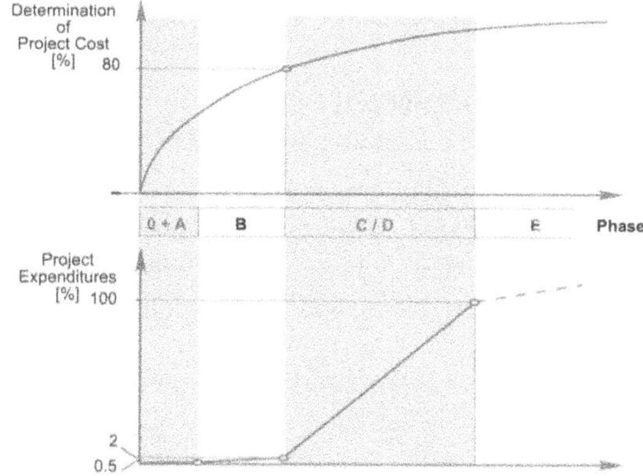

Fig. 9.4. Cost Determination and Project Expenditures through System Design

Extreme Boundary Conditions. As compared to other technical systems with comparable complexity such as nuclear reactor plants or large traffic systems, space systems have to cope with additional boundary conditions:

- Extreme environmental conditions in terms of temperature, pressure and radiation as well as hyper- and microgravity
- Extreme loads during launch and deployment such as acceleration (e.g. during launch), heat loads, vibration or noise
- Extreme requirements for light weight due to restricted launcher performance or high launch cost per kg
- Deployment and operation of a distant system with limited access

In the case of a space system, these conditions collectively cause the mutual dependencies among the system elements to become more obvious when compared with other technical systems. Some of them (e.g. launch, logistics or crew intensity) may even dominate the design, thus becoming *system drivers* which are parameters that largely determine overall cost, performance schedule or risk [Larson 92].

The complexity of the conceptual design problem emerges from the sum of the characteristics mentioned above: "fuzzy" problem formulation, strong interdependencies among system elements, the predetermining character of design decisions, and extreme boundary conditions. Conceptual design is therefore a special problem with more similarities to architectural design than to classical mathematical or natural science problems. In the latter case, the problem can generally be precisely formulated, and the difficulty is related to modeling or experimentation. In conceptual design, problem formulation is already difficult because a precisely formulated design problem will inevitably contain important elements of the solution, e.g. an identification or specification of system elements (space segment, orbit or operations). Therefore, conceptual design largely resembles general planning problems, denoted as "wicked problems" [Rittel 73]: They have no definite problem formulation and their solutions are not true-or-false, but good-or-bad, and they cannot be solved by classical sequential systems engineering approaches, either deductive or stringent in nature. The next section, therefore, presents a methodology for conceptual design of space systems.

9.3 Methods and Tools for Conceptual Design

Considering the special environment and boundary conditions, a methodology adapted for the conceptual system design of space stations has to meet a number of requirements:

- It has to account for the "fuzzy" start-up situation in order to transform it, step by step, into system concepts and numerical requirements, thus bridging the gap between the early conceptual phase and the systematic requirement-driven design definition.
- Boundary conditions and interdependencies between system elements need to be addressed systematically.
- Among the infinite variety of system solutions, realistic and, if possible, optimal system concepts should be worked out within a few weeks.
- It has to deliver the information allowing the "stop" or "go" decision for the further development of the project.
- The methodology should encourage the designer's creativity without being constrained by a rigorous process.

The first requirement leads to the necessity of processing knowledge in different depths of detail. The design process should reflect the non-sequential and non-deterministic structure of the design problem. Besides precise theories and approximate solutions, heuristic methods and iteration techniques will especially be applied in conceptual design. Another tool to support the processing of fuzzy information or to cope with information gaps is the application of arguments by analogy to existing systems. In some cases, even intuition, trial-and-error or educated guess must be applied, if reference data is not available at all.

These requirements characterize the environment of *Systems Engineering* (SE), which is defined as:

> *"...an interdisciplinary collaborative approach to derive, evolve, and verify a life cycle balanced system solution that satisfies customers expectations and meets public acceptability."* [IEEE 1220]

ESA applies a similar definition:

> *"...a mechanism for proceeding from interpretation of the customer's requirements to an optimised product by steadily applying attention to a wide range of product requirements, extending to all details of the user's needs, produceability constraints, and life cycle aspects, essentially through an organised concurrent engineering practise."* [ECSS-E-10A]

Of course, there are many ways to structure the systems design process. However, in recent years, the systems design methodology proposed by Larson [Larson 92] has become the standard for filling the gap between the conceptual design of satellite systems and their detailed specification. This process was also the starting point for the conceptual design methodology for space stations presented hereafter.

9.3.1 Conceptual Design Methodology

Table 9.1 shows a four-step procedure which eventually leads to the formal definition of system requirements and their allocation to system elements (steps H and I). The different steps will be briefly presented. A more comprehensive discussion can be found in [Bertrand 98] and [Larson 92].

Table 9.1. The Space Station Analysis and Design Process

Step	Details	Design Flow
Define Objectives	A Develop Broad Objectives B Develop a Preliminary List of Requirements and Constraints	
Characterize the System	C Develop Alternative System Concepts D Characterize System Elements	
Evaluate the System	E Prepare System Budgets F Evaluate Mission Utility G Select System Baseline	
Define Requirements	H Define System Requirements I Allocate Requirements to System Elements	

The starting point of the process is the definition of the objectives, first in a broad, qualitative manner (step A), and second through an initial "translation" into numerical requirements and constraints (step B). This is the basis upon which the system is characterized: alternative system concepts are developed (step C) representing a range of system solutions suitable for meeting the objectives. For each of the system concepts, a consistent definition of the major system elements has to be found (step D), thus defining in more detail what the system is and what it does. System evaluation should prepare the decision on which concepts are to be retained for further design. Therefore, system budgets for outstanding system parameters such as performance, mass, power, etc. (step E) will be prepared. This also allows the system drivers and the driving requirements of the design to be identified. The cornerstone of the design process, however, is the evaluation of mission utility (step F), since here, all concepts will be compared with respect to the question of how well the requirements and broad objectives are met as a function of either cost or key design choices. The selection of the system baseline (step G) is the formal choice among the different concepts. In the early conceptual design stage, there may be several baselines in order to retain the most promising system solutions for further examination. Definition of baselines is a typical tool for handling complex network-type problems. They provide temporary milestones against which the design progress or success of the project can be measured.

These design steps will be explained in further detail with strong emphasis on characterization since this step contains the core design activity which is the definition of the orbital segment configuration.

Definition of Broad Objectives

The design process starts with an analysis of the mission goals to be met by the space station mission. As mentioned earlier, the specified mission goals may be very "fuzzy". For instance, a given country may have an interest in obtaining access to a small, independent microgravity laboratory, or a satellite operator might wish to access a servicing platform for periodically crewed operations. Therefore, the definition of broad objectives requires a high degree of interaction between the customer and the designer in order to put these objectives into increasingly more concrete terms until the first numerical requirements can be specified.

Primary mission goals for space stations make use of qualities or location factors that are characteristic for the position and situation of a space station (see Fig. 7.5). These issues were presented in Chapter 7.

A space station, in general, rarely serves only one user discipline. However, the type of utilization foreseen for the space station has an impact on overall system design. Therefore, a clear assessment of the utilization profile will be performed beforehand.

Another class of objectives may arise if the space station is part of a higher level mission such as an interplanetary exploration mission. In this case, the overall mission statement has to be reviewed in order to identify subsequent mission objectives for the space station (Table 9.2).

A third class of non-technical, non-utilization objectives is called "secondary objectives" or "the hidden agenda". They may be political, social, economic or cultural in nature, but especially in the case of large programs, they can become very important. For example, technical cooperation with another country may increase political stability, or an ambitious technological project could enhance technological competence, safeguard high quality jobs or demonstrate leadership. Although it is vital to consider these objectives in the design process, too, secondary objectives should not dominate a project. Otherwise, a change in political circumstances can lead to a hold on or cancellation of a program, as it happened, for instance, in the case of the US space station program Freedom.

Table 9.2. Mission-Derived and Utilization-Driven Objectives

Overall Mission Derived	**Utilization Driven**
• *Provide crew accommodations* for habitation, training, laboratory and working areas • *Provide facilities* for berthing, construction, servicing, repair, checkout and launch other spacecraft (transfer vehicles, landers, planetary habitats, equipment etc.) • *Provide logistics* produce, store, transfer *resources* (electrical, thermal, life support, data, propellants, etc.) • *Provide operational support*	• Microgravity Laboratory • Earth Observation • Space Science • Telecommunication • Servicing and Operations Center • Transportation Node

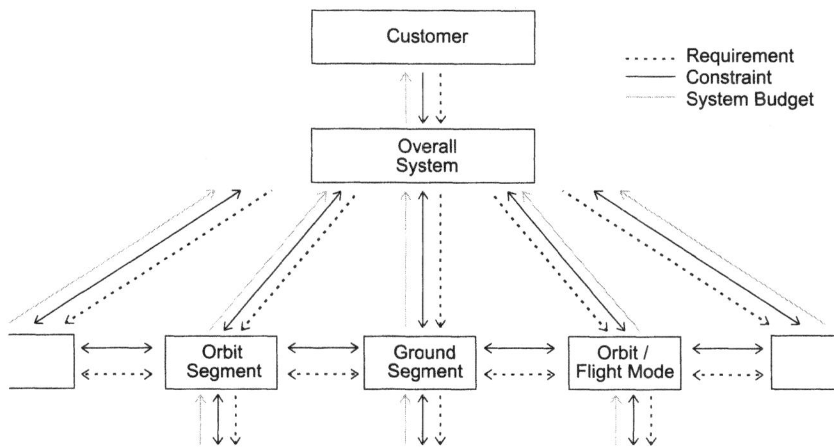

Fig. 9.5. Communication between Mission and System Elements

Development of a Preliminary List of Requirements and Constraints

Having developed a prioritized list of objectives, the next step is to systematically look for constraints on the orbital segment and to translate the broad objectives into preliminary requirements. The "communication" between mission and system elements is shown in Fig. 9.5.

Constraints can be seen as the top-down boundary conditions of the design process. But they may also appear as bottom-up or horizontal constraints from one element to another representing, for instance, technological limits. In all cases, considerable effort is required to change these constraints since this requires major changes in opinion or design for the element which is the origin of the constraint. Constraints can be helpful in the reduction of the initial variety of system solutions, but they should also be subject to critical review in order to assure that there is no unjustified design constraint leading to a suboptimal system.

Requirements are to communicate needs among system elements. They act as design guidelines for the element which is their target. Per definition, they are subject to trade-offs in order to allow for design optimization, especially during conceptual design. Once a complete set of system level requirements for the baseline concept has been established, the requirements for subsequent system levels can be deduced through a top-down process.

It is important to understand that system design does not primarily mean to satisfy given requirements and constraints but to find the best solutions in meeting the client's objectives. Requirements and constraints, therefore, are only communication tools for the support of this process. They are also important for the verification of the design and, later on, the actual system, with respect to the design goals. This can be done by establishing system budgets and by comparing the results of the budgets to the initial requirements and constraints.

As explained previously, the complexity of the design task is due to the fact that these elements are strongly dependent on one another. A good tool to show and analyze this type of network problem is the interference matrix as depicted in Fig. 9.6, where the lines and columns are labeled with the mission and system elements. Interrelationships between the elements can be entered into the boxes of the matrix. Such a matrix can be used to search for subsequent constraints of the previously mentioned launcher selection by entering in the line labeled "Launcher". Each box contains questions which unveil subsequent constraints on the mission element of the corresponding column, for example, launch rate influences on the mission planning (column 1: "Mission/Utilization") or orbit restrictions due to the launching system (column 2: "Orbit/Environment").

A good starting point for this matrix analysis will be the direct constraints imposed by the client or the top-level mission scenario. Interference matrices may also support the translation of broad objectives into preliminary requirements. In this case, when running through a line, the question is whether or not there is a numerical value or a range of numbers that can be linked to the interrelationships in the matrix boxes.

Development of Alternative System Concepts

A system concept in this context is a broad statement of how the mission of the space station could work in practice. In contrast to robotic space systems where missions are mostly concerned with the acquisition, transmission and processing of data, mission outputs of crewed systems like space stations can also be samples, experiment or technology hardware, or a service. The system concept, therefore, outlines the way in which the required mission output can be generated through proposing different options for the system and mission elements. For example, options for the realization of a microgravity test facility could be a small crewed orbiting laboratory, a transfer vehicle with an orbiting power station, an add-on module for the International Space Station, or a robotic platform co-orbiting ISS. It is important to cover a broad expanse of alternative solutions at this point in order not to restrict the design process to a certain concept too early.

The system concept question is linked to the overall class of the mission and associated space segment, which can be described as single element, multi-element with pressurized module backbone, and multi-element with truss backbone as shown in Table 9.3.

The first space stations such as Salyut 1 (1971) or Skylab (1973) were single element configurations designed to accommodate all necessary subsystems needed to allow human presence in one or a few pressurized modules (PMs). This philosophy has basically been applied to the modular Mir space station, too; although the Mir complex currently consists of six major pressurized modules and many subsystem functions are implemented in each of the modules. The strategy of forming the structural backbone of the space segment by the pressurized modules has quite some advantages: From the beginning, there is a fully operational space station. Configuration changes and growth can easily be realized while always keeping good structural stiffness and integrity. However, due to the constraint of mounting pay-

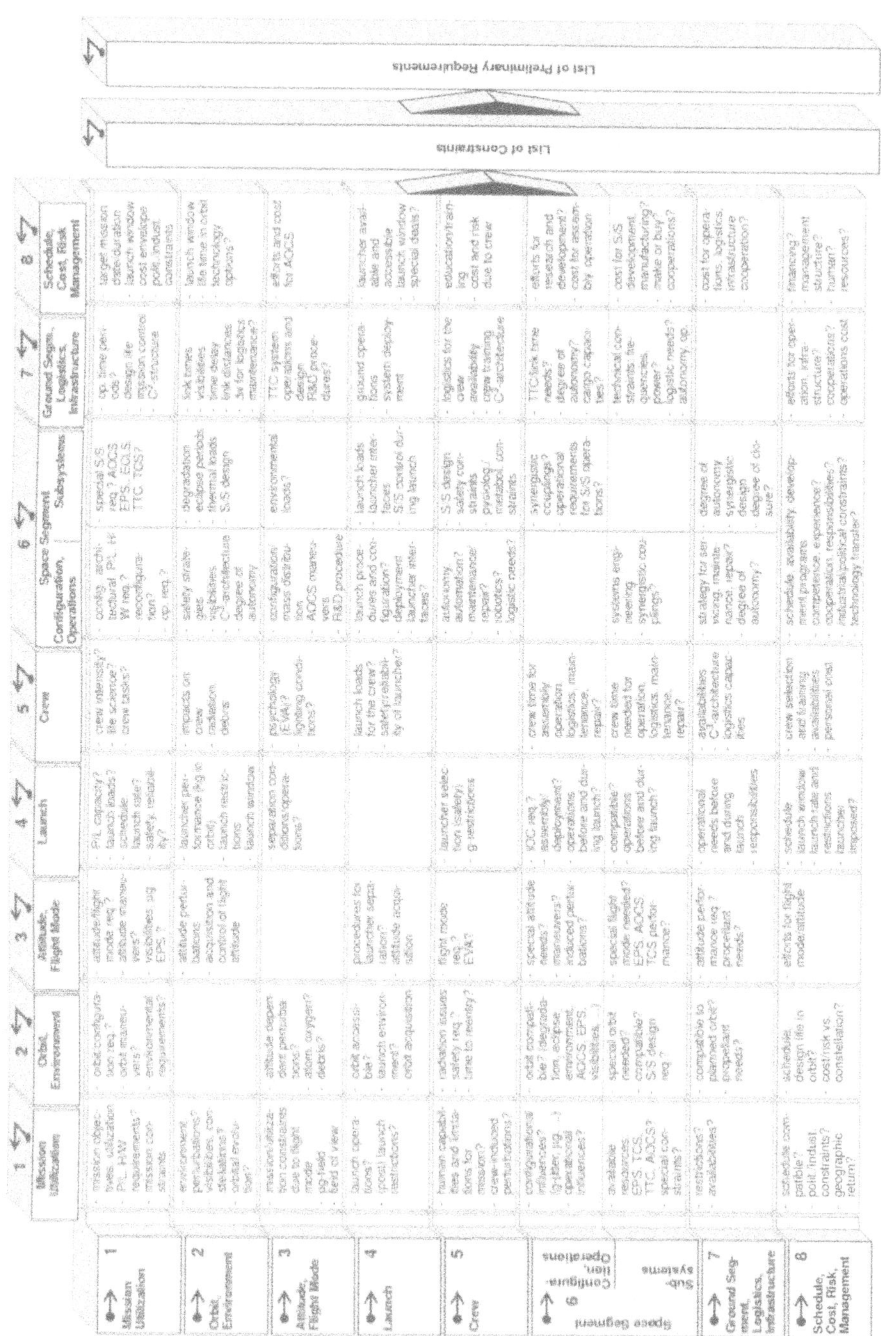

Fig. 9.6. Interference Matrix of System Elements

Table 9.3. Classes of Space Segments

Class of Orbit Segment	Characteristics	Examples
Single Element	One self-contained element + One launch required to deploy and operate + Low cost and risk approach - No growth potential Especially suited for: • One to three crew members • Small power and heat rejection needs • Relatively few years of lifetime • Specialized mission profile • National programs	Salyut 1-6, Skylab
Multi-Element, pressurized module backbone	Subsystems integrated into the PMs with a large degree of autonomy of the individual elements + High structural stiffness + Higher redundancy + Continuous and flexible growth potential + g-jitter at higher frequencies + Good flexibility for reconfiguration on module level - Several launches required to deploy and operate - Limited power density (collector and radiator accommodations) - Limited external P/L accommodations - Reconfiguration inside modules more difficult Especially suited for: • Medium-sized configurations • Three or more crew members • < ten years of lifetime • Sharp mission profile • High configuration flexibility • Moderate power and heat rejection needs • High microgravity requirements • International programs	Salyut 7, Mir
Multi-Element, truss backbone	Centralized subsystems with a large degree of interaction between elements + Good external P/L accommodations for multi-mission scenario + Facilitates the establishment of separate functional areas (power generation, habitation, laboratory, thermal control, etc.) and orbit replaceable units + Higher power densities attainable + Centralized subsystems can save mass + Higher usable volume in a given PM volume - Many launches required to deploy and operate - Limited growth (size and power) - g-jitter at lower frequencies - More efforts to distribute resources (power, thermal, data) Especially suited for: • Diversified mission profile • High power and heat rejection needs • Three or more crew members • > ten years of lifetime • Multinational programs	CDG-Concepts, Freedom, ISS

loads, collectors and radiators directly onto the pressurized modules, payload accommodations, electrical power and heat rejection capacity are limited to a larger extent as compared to another architectural concept: the truss backbone type. Here, a multi-element integrated configuration incorporates a truss backbone. In this case, subsystems such as electrical power, thermal control, or environmental control and life support can be centralized to achieve higher utilization performance inside the modules. Payloads can also find more space with less interference problems on a truss structure. One typical representative of this architectural type is the concept of the US space station Freedom. Therefore, the truss backbone architecture is well suited for diversified mission profiles with different utilization disciplines as well as higher power and heat rejection requirements. On the other hand, pressurized module-backbone concepts favor specialized mission profiles with moderate power needs or high configuration flexibility.

These configurations – single element, and multi-element pressurized module backbone or truss backbone – should be examined for feasibility with respect to mission objectives in the order given. The single-element configuration is inherently superior from a development and deployment standpoint since it can be fully tested and checked-out on ground prior to launch. Furthermore, it can be deployed in a single launch. Operationally, the subsystems can be integrated so that advantage of synergies to the maximum extent desired is taken, thus in that same manner, the station is very efficient in terms of resource consumption. If a single-element configuration is not feasible, most likely due to launch mass or geometrical constraints, the modular configuration is the next option. While resource utilization is degraded by configuration limitations such as shadowing of solar arrays, obstructed thermal dissipation, etc., the development and deployment of this configuration are significantly more straightforward than the final option, the multi-element integrated configuration. This configuration regains much of the operational efficiency of the single-element configuration from a centralization of utilities such as power generation and distribution. However, development and deployment of this configuration are significantly more complex due to the increased functional interdependencies among flight elements and the multiple assembly configurations. Each stage of the assembly must be a stable spacecraft which can survive in the orbital environment pending the next stage of the assembly. This last challenge is particularly demanding in the early stages of assembly.

The selection of a configuration type is appropriate at this point in the design process. Knowing little more than crew size and mission duration, one of the three configurations from Table 9.3 may be selected. Where the criteria are not exact or overlapping, engineering judgment must be exercised. A more precise definition of the configuration, such as quantity of elements and their precise geometrical arrangement, will be developed in the subsequent steps of the design process.

9.3.2 Characterization of System Elements

The characterization of the system concepts is the creative core of the space station design process: all components of the orbital segment have to be arranged in architectural sketches which consider utilization aspects, attitude and orbit dynamics, energy provision and thermal control, as well as operations and safety issues. As a result, the shape, position, and orientation of all components will be fixed, and their key parameters such as size, mass, power and thermal needs will be assessed. Unfortunately, there is no detailed recipe for generating an architecture. It is left in the hands of the designers, to their intuition and creativity, to develop the design, to prioritize and integrate the multiple constraints, requirements and design issues into an architecture. Fig. 9.7 shows a framework of substeps supporting the development of the orbital segment of the architecture.

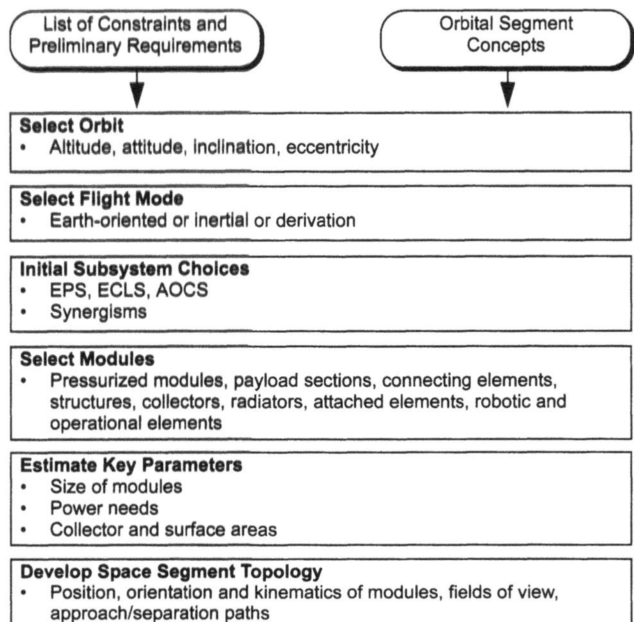

Fig. 9.7. Process for Characterizing the Space Segment Architecture

Selection of Orbit: Space station orbit selection is mainly driven by safety and utilization issues. Operational safety will call for a minimum altitude, dependent on solar activity, in order to prevent accidental reentry due to atmospheric drag. Other safety criteria are radiation and orbital debris. If there are transfer or servicing functions to be provided, access to target vehicles and orbits will be of primary concern. Visibility aspects for Earth observation and telecommunications may lead to requirements in terms of inclination, orbital altitude or eccentricity. Last but not least, the payload performance of the launch systems will be directly dependent on the choice of the orbit. Table 9.4 gives an overview of orbit selection criteria.

Safety concerns are, again, most important for the orbital system. Another constraining criteria still is launch, since it largely determines cost and schedule for system deployment and operation.

Selection of Flight Mode: This issue has already been discussed in detail in Section 6.3 on "Flight Strategies". Table 9.5 summarizes the implications of Earth-oriented and inertial flight modes. Small space stations with a relatively homogeneous mass distribution enable the realization of both flight modes, Earth-oriented and inertial. With growing space station size, attitude stability and controllability issues are becoming more and more important. For large orbital systems like the Mir space station or the International Space Station, it would just be too expensive in terms of propellant to maintain an attitude which needs constant counteracting against the gravity gradient or aerodynamic torques. For that reason, such systems

Table 9.4. Orbit Selection for Space Stations

Criteria	Consequences	Values
Orbital access to target or transfer vehicles	• Propellant needs, mission operations, phasing problems	
Communications performance	• Ground station and satellite coverage, link duration, data rates	
Earth observation performance	• Global coverage, resolution	Higher inclination, orbital altitude
Ground track and swath	• Earth observation qualities (frequency of contact, sight angles etc.)	
Atmospheric drag	• Decline and oscillation in altitude, aerodynamic torque • Microgravity level (quasi-static)	h > 200 km (depending on solar activity)
Launch system P/L performance	• Mass restrictions for system build up and resupply, flight frequency • Latitude of launch site • Orbit restrictions (semi-major axis, inclination) for given launchers	Launcher dependent
Radiation shield due to Van-Allen belts	• Crew and equipment radiation dose	i < 65°, h < 1000 km

Table 9.5. Characteristics of Earth-oriented and inertial flight modes

Earth-oriented	Inertial
+ Favorable for Earth observation and telecommunication + Allows the use of the gravity gradient for attitude stabilization (TEA) + More flexibility for microgravity experiment accommodations (tidal perturbation field) + Earth is a reference for crew orientation (EVA) + Rendezvous and docking operations easier + More mass distribution flexibility for assembly and orbital segment growth - Needs solar array and radiator tracking for best performance - Variable lighting conditions (EVA)	+ Favorable for astronomy + Collectors and radiators simplified (best performance even without tracking) + Constant lighting conditions (EVA) + Constant thermal control conditions - Gravity gradient always a perturbation - Difficult to keep optimal mass distribution during assembly and space segment growth

generally need to fly in an Earth-oriented manner, in a "Torque Equilibrium Attitude", where, over the course of each orbital revolution, no momentum is accumulated in the attitude control system. Typically all criteria for the selection of the flight mode cannot be met at the same time. Therefore, in practice, there will be a compromise depending on the individual mission requirements. The discussion of these factors allows some design requirements on the architectural concept of the space segment to be deduced that will be presented in the paragraph "Development of Orbital Segment Topology".

Initial Subsystem Choices: Some subsystem issues can have a significant influence on the overall configuration of the orbital segment. This is due to the fact that the associated components such as collectors or radiators may represent the largest components in size within the orbital segment architecture, or that the components have special requirements with respect to their design integration. The most important subsystem choices for space stations deal with the following:

- Electrical power supply
- Orbit control
- Thermal control
- Environmental control and life support
- Synergistic couplings between these subsystems

At this design stage, an initial choice of technologies and a draft design of the main parameters (power and mass) have to be performed. These issues are addressed in more detail in Chapters 4, 5, 6 and 10.

Selection of Modules and Estimation of Key Parameters: Before drafting the first architectural sketches, it is necessary to list the components to be integrated into the architecture and to estimate the rough dimensions of all modules. Table 9.6 lists typical modules as recurring elements of space stations. The utilization- or mission-related components such as laboratories, exposed facilities, docking and processing areas, etc. are good starting points for this list. Then, all components which ensure full orbital functionality (AOCS, EPS, TCS, TTC, etc.) can be included.

Next, the size of the modules has to be quantified. In the case of pressurized modules, these numbers may arise directly from the objectives if, for example, a volume for the crew, for experiments or for mission operations has been specified in the list of requirements. However, it is more likely that these numbers will not be immediately available, thus calling for an estimation at that point.

Size, mass and power needs of pressurized modules can easily be assessed by analogy to existing or planned systems. Scaling of this data gives a first, rough order-of-magnitude (ROM) estimate of the configurational items of the space segment. However, assurance that all these components have been part of similar systems must be given. Therefore, compatibility of specific constraints, such as the architectural type, has to be checked before applying these data to a new design. Surely, analogy and parametric methods have their limits in the case of new designs done "from scratch". But since maximum masses and dimensions will still be lim-

Table 9.6. Typical Architectural Components

<table>
<tr><th></th><th>Pressurized</th><th>Unpressurized</th></tr>
<tr><td rowspan="2">Mission P/L</td><td>• Habitats
• Laboratories/working compartments</td><td>• Attached payloads
• Exposed facilities</td></tr>
<tr><td colspan="2">• Transfer vehicles
• Servicing, construction areas</td></tr>
<tr><td>Space Station</td><td>• Habitats
• Laboratories/working compartments
• Interconnecting elements
• Airlocks</td><td>• Truss structures (backbone)
• Collectors (planar or curved)
• Radiators
• External subsystems and payloads
• Manipulator systems</td></tr>
<tr><td rowspan="2">Operational Elements</td><td>• Crew transfer and rescue vehicles</td><td></td></tr>
<tr><td colspan="2">• Logistics vehicles</td></tr>
</table>

ited by the available launch systems in the foreseeable future, the new designs cannot be too far away from what has already been found in the Russian and US programs and studies.

Development of Orbital Segment Topology: Despite the infinite variety of architectural solutions, it is possible to state some general design rules which can be used as guidelines to develop and evaluate a design. Table 9.7 gives a list of guidelines for different design aspects, which have already been partially mentioned in the paragraph on flight mode selection. The column "Relative Importance" is meant to support the prioritization of the different design guidelines since not all design rules can be met at the same time. Category "A" should be fulfilled by all designs. "B"-aspects should be respected in a good design, however, they might be negotiated in some cases. "C"-aspects should be fulfilled if they are not in conflict with "A" or "B" guidelines.

Mass Distribution: First of all, the mass distribution should be such that it inherently favors attitude stability and controllability (see Fig. 9.8). For an Earth-oriented flight mode, this means that the local vertical is the preferred direction to line up masses. This leads to an inertial tensor with its smallest moment of inertia directed toward the Earth, and, consequently, the gravity gradient of the Earth will have a stabilizing effect on the flight attitude. Another important issue is the avoidance of mass distribution asymmetry relative to the orbital plane since, otherwise, the cross-products of inertia I_{yz} and I_{xz} ($I_{yz} = \int_m yz dm$, $I_{xz} = \int_m xz dm$) will cause roll and/or pitch torque. Gravity gradient torque should be generally minimized in the case of the inertial flight mode. This can be achieved with mass distributions leading to an inertia tensor close to a sphere symmetrical shape. In general, secular attitude perturbations in the case of an inertial flight mode can be avoided if one main axis of inertia is oriented perpendicular to the orbital plane (cf. Sect. 6.2.3).

Aerodynamic Drag: Design with respect to aerodynamic drag has two goals: On one hand, the incidence areas normal to the flight velocity vector should be as small as possible in order to minimize propellant requirements for orbital reboost. On the other, space station attitude dynamics will mainly be driven by gravity gradient and aerodynamic torque. As mentioned above, large orbital systems have to respect the torque equilibrium attitude to avoid excessive propellant requirements on the attitude control side. Therefore, aerodynamic incidence areas may intentionally be placed to equilibrate the attitude perturbations. It is also important that large surfaces with changing area of incidence, such as solar arrays, do not cause excessive aerodynamic torque. Therefore, their center of pressure has to be close to the center of mass of the total orbital system. A diagram showing the principles of aerodynamic and gravity gradient torque on an Earth-oriented model space station is displayed in Fig. 9.9.

Mass distribution and aerodynamic drag are permanent concerns for the whole life cycle of the orbital system. This question has to be addressed for all configuration changes, as they might occur due to normal operation (mission or logistics vehicles) or be caused by orbital segment growth.

Table 9.7. Architectural Design Rules for Space Stations

<table>
<tr><th>Aspect</th><th>Earth-Oriented Flight Mode</th><th>Inertial Flight Mode</th><th>Relative Importance</th></tr>
<tr><td>Attitude Stability and Controllability</td><td>• Arrange mass distribution for gravity gradient stability (I_{xx},I_{yy} > I_{zz})
• Avoid mass distribution asymmetries relative to the orbital plane (I_{xz}, pitch instability) and to the plane perpendicular to the velocity vector (I_{yz}, roll instability)</td><td>• Arrange mass distribution to get the inertia tensor close to sphere-symmetric shape (minimize gravity gradient torque)
• Arrange mass distribution such that one main axis of inertia is oriented perpendicular to the orbital plane (no secular torque)</td><td>A</td></tr>
<tr><td rowspan="2">Aerodynamic Drag</td><td colspan="2">• Minimize aerodynamic incidence areas (PM orientations, collectors, radiators)</td><td>B</td></tr>
<tr><td>• Balance the mean aerodynamic incidence areas (aerodynamic center) such that the sum of gravity gradient and aerodynamic torque lead to a TEA
• Keep aerodynamic center of moving surfaces near to the center of mass.</td><td>• Keep aerodynamic center close to the center of mass to minimize secular aerodynamic torque</td><td>A</td></tr>
<tr><td>Growth Potential and Configuration Changes</td><td colspan="2">• Analyze "Attitude stability and controllability" aspect for different orbital stages
• Accommodate frequently changing modules (logistics vehicles) respecting aerodynamic compatibility (center of pressure, see aspect "aerodynamic drag"</td><td>A</td></tr>
<tr><td>Quasi-static µg Level</td><td>• Integrate µg-facilities close to the line parallel to the flight direction vector hitting the center of mass</td><td>• Integrate µg-facilities close to the center of mass</td><td>A*</td></tr>
<tr><td>Transient µg Level</td><td colspan="2">• Avoid long, flexible structures</td><td>A*</td></tr>
<tr><td rowspan="2">Rendezvous and Docking</td><td colspan="2">• Allow for access corridors in radial or orbit tangential direction with sufficient clearance</td><td>A</td></tr>
<tr><td>• Avoid dedicated docking orientations different from nominal orientations</td><td></td><td>C</td></tr>
<tr><td>Pressurized Module Configuration</td><td colspan="2">• Dual egress, redundant access
• Minimize aerodynamic and debris incidence areas</td><td>C</td></tr>
<tr><td>Solar Collectors</td><td>• Need tracking (α and β) for best performance
• Keep the collector's center of pressure close to center of mass of the space station to avoid changes in TEA
• Best location to minimize shading problems: distant in POP direction</td><td>• No tracking needed</td><td>B</td></tr>
<tr><td>Radiator Panels</td><td colspan="2">• Minimize fluid feed lines and feed-throughs in joints
• Check fields of view for payload and sensors
• Check rendezvous and docking, thruster plume interference</td><td>A</td></tr>
<tr><td></td><td>• Need tracking for best performance (no Sun incidence)
• No drag if the panel's normal vectors are perpendicular to the flight direction</td><td>• Panel's normal vector PSL</td><td>C</td></tr>
<tr><td>AOCS Thrusters</td><td colspan="2">• Check contamination, safety and interference
• Self-contained ORU or distributed/refueling by logistics vehicles
• Avoid long propellant feed lines
• Orbit control: line of force parallel to flight direction and close to CM
• Attitude control: maximize CM distance for low propellant consumption</td><td>A</td></tr>
<tr><td>Observation and Communication Payloads</td><td colspan="2">• Check field of view and shading
• May need tracking and steering
• May be sensitive to contamination (thrusters etc.)</td><td>A*</td></tr>
<tr><td></td><td>• Astronomical P/L need α-tracking</td><td>• Earth-oriented P/Ls need tracking</td><td>A*</td></tr>
</table>

* depending on the importance within the utilization profile

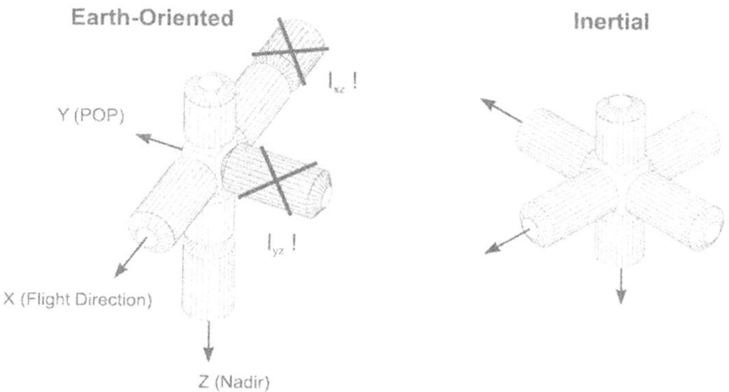

Fig. 9.8. Optimal Mass Distributions with Respect to Gravity Gradient

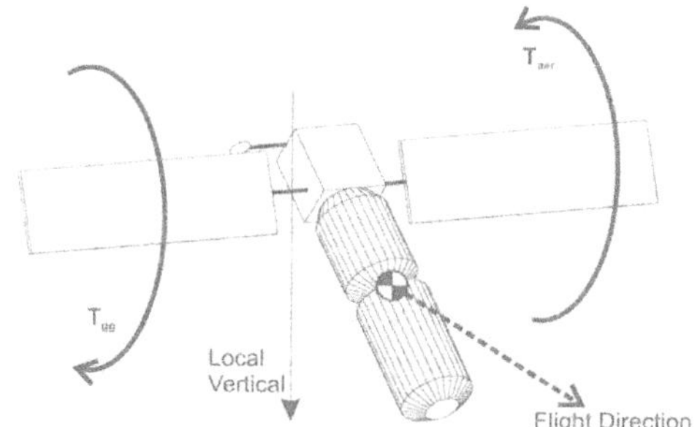

Fig. 9.9. Aerodynamic and Gravity Gradient Torque

Microgravity: Most space station user disciplines make use of the microgravity environment. As mentioned in Sect. 8.3, there are internal and external perturbations on the microgravity environment (drag, tidal forces and g-jitter), which can be partly controlled by the overall architecture. This has already been worked out in Section 8.3.2 for the tidal perturbations: A good architectural design for an Earth-oriented space station will have the modules for microgravity utilization located as closely as possible to the line parallel to the flight direction vector hitting the station's center of mass. Every meter of distance to the space station's center of mass in orbit normal (y_0) or radial (z_0) direction will cause a tidal acceleration of 0.128 µg or 0.384 µg, respectively (orbital altitude: 400 km). Thus, for the accommodation of microgravity-relevant experiments, the x_0 and y_0 directions are preferred. For inertial flight modes, the perturbation field forms an ellipsoid transient in time. Since there is no preferred direction to integrate microgravity sensitive payloads, they should be as close as possible to the center of mass. In-orbit-plane offsets to the center of mass (x, z) will cause transient tidal perturbations < 0.256 µg/m, whereas distances normal to the orbital plane (y) will yield constant accelerations of 0.128 µg/m.

g-Jitter: Undoubtedly, the most complex disturbance to microgravity is the g-jitter. All internal impulses such as control maneuvers, crew movements, pumps or actuators can activate the structural eigenfrequencies of the space station structure itself (see Sect. 8.3.3). The most critical frequencies for microgravity experiments are located in the lower frequency range between 0.1 and a few Hz and are typically caused by oscillations of module clusters, solar array wings, pressurized modules or the shuttle orbiter. An analysis of the g-jitter is quite complex and needs high fidelity structural models which are usually not available for conceptual design. However, it is important to keep in mind that filigree and slender structures will be more sensitive to low frequency structural oscillations and thus should be avoided if microgravity is a key issue in the utilization scenario. Since g-jitter dynamics are strongly dependent on local geometry and transfer functions between geometric entities, both parameters need very accurate modeling and are not commonly available during conceptual design. Therefore, a g-jitter analysis is performed in later design stages.

Rendezvous and Docking: Preferred rendezvous and docking operations require approach and departure corridors in radial or orbit tangential direction with sufficient clearance. Docking and berthing interfaces should be integrated such that there is no need to change the space station attitude for proximity operations.

Collision Risk with Orbital Debris: Whereas micrometeoroids may impact the space station from any direction with about the same probability, the orbital debris flux is dependent on the impact direction: for orbit-mechanical reasons, nearly all debris objects will approach the spacecraft within the local horizontal plane which is called the *debris plane*. Within the debris plane, the object flux, again, is dependent on the direction with two peaks between 45 and 75 degrees and -45 and -75 degrees to the flight direction, respectively (see Fig. 9.10).

This means that for a given cylindrical pressurized module, the impact probability varies with respect to its orientation. The lowest flux will occur if the cylinder axis is perpendicular to the orbital plane (longitudinal arrangement). Table 9.8 shows the relative debris object flux for the orbit-defined coordinate directions calculated for a standard pressurized module of 4.0 m diameter and 12.7 m length.

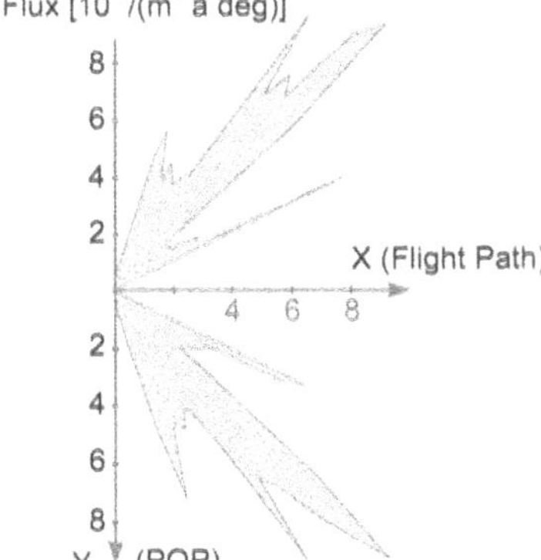

Fig. 9.10. Object Flux per m^2 and year in the Debris Plane as a Function of the Collision Angle, Particle Size > 1cm [Eichler 90]

Table 9.8. Debris Flux for Three Possible Orientations of a Pressurized Cylinder, Particle Size > 1cm (∅ = 4.0 m, l = 12.7 m) [Eichler 90]

	z, x (Flight Direction), y (POP)	z, x (Flight Direction), y (POP)	z, x (Flight Direction), y (POP)
Particle Flux [10^{-4}/a]	1.55	1.88	2.03
[%]	100	121	131

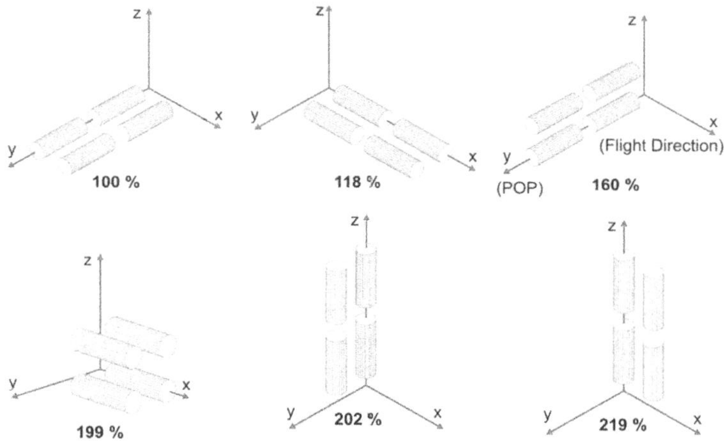

Fig. 9.11. Relative Debris Flux for a Module Cluster with Different Orientations, Particle Size > 1cm (Four Modules, ∅ = 4.0 m, l = 12.7 m) [Eichler 90]

Besides the individual debris flux on a module, a module cluster should take advantage of mutual shading from debris to achieve the lowest possible overall debris flux. An optimum configuration from a debris risk point of view would be to line up the cylinders in flight direction with their cylinder axis oriented POP. The highest debris flux is seen in modules lined up in the radial direction with the cylinder axis pointing to the Earth (see Fig. 9.11).

Module Interconnection Patterns: Figure 9.12 shows some basic interconnection patterns as they were analyzed in the early 1980's by the Boeing company [Woodcock 86]. The way pressurized modules are connected not only influences the overall system aspects of attitude stability and aerodynamic drag mentioned above; in terms of *safety* two escape paths from a module (dual egress) are also just as important as redundant access in case of a failed neighboring module. A planar arrangement of larger size can pose drastic problems for *attitude control* if it is not in gravity gradient orientation; on the other hand, it may be oriented such as to re-

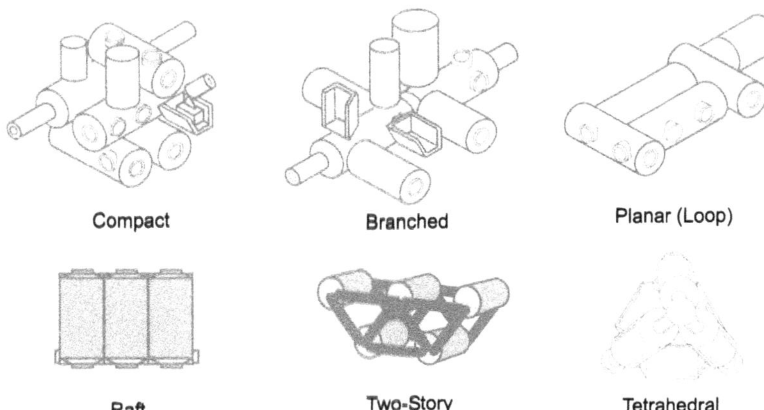

Fig. 9.12. Pressurized Module Interconnect Patterns [Woodcock 86]

ceive the smallest possible *debris flux* (plane of modules POP, cylinder axis perpendicular to the flight direction). *Thermal control* of the modules for view-vector reasons can be managed more easily in the case of planar and branched arrangements than in the case of the compact form. However, this is generally no problem in the case of truss-backbone architectures. For *configuration expansion*, branched and planar arrangements provide a broader range of possibilities but might lead to the already mentioned attitude control problems above a certain size. This is not the case for quasi-spherical compact or tetrahedral concepts which, therefore, are well suited for inertial flight modes. Still, different module arrangements can be conceived of, e.g. a tetrahedral arrangement which features short *transfer paths*, a high *structural stiffness* and balanced *mass distribution*. Another variation of the planar arrangement might be a configuration in two stories, which promises increased access redundancies and the possibility to exchange even the inner modules.

Another interconnection issue is whether or not to use *dedicated connecting elements* (nodes) or to integrate the docking adapters into the modules themselves. Connecting nodes increase the number and complexity of components that have to be built, but they allow for simplification, standardization and better volume efficiency inside the actual pressurized modules. As a drawback, system mass and crew transfer times may be higher e.g. in a raft configuration with the exits and nodes only at the ends of the modules. It is generally believed that a *consistent orientation* of "up" and "down" throughout the whole station supports the visual orientation of the crew. This requirement is difficult to fulfill in the case of branched, compact or tetrahedral concepts.

Solar Collectors: Solar array tracking was already mentioned when the flight mode selection was discussed. Since rotating solar collectors with their large aerodynamic effective surface can cause high cyclic aerodynamic torque, it is very desirable to keep their aerodynamic center close to the center of mass of the overall system. A good strategy to minimize shading problems in the case of an Earth-oriented space station is to arrange the pressurized modules mainly in the orbital plane and to integrate the solar collectors at some distance perpendicular to the orbital plane.

Radiator Panels: Best radiator performance calls for solar tracking in order to avoid solar radiation incidence on the panel. This can be done using one rotating joint only, e.g. using the alpha-joint in concert with solar collectors. If the rotation axis is co-aligned with the flight path vector, the radiator panels will never experience drag.

AOCS Thrusters: In order to achieve maximum performance, orbit control thrusters should be integrated such that their line of force hits the center of mass of the orbital segment. For reboost engines, this line of force has to be aligned with the orbital velocity vector. Attitude control thrusters, by contrast, should have maximum distance from the center of mass in order to generate maximum torque with a given thrust force. These functional requirements have to be balanced with the requirements of propellant storage and feed lines. Furthermore, thruster systems can be a significant source of perturbation of the overall system. Exhaust plumes may contaminate equipment and payloads outside the pressurized volume or may even cause mechanical damage to fragile elements such as solar arrays. Interference and compatibility checks are therefore vital to ensure overall system integrity and performance. Another issue is linked to logistics, maintenance and repair: Since the propulsion system in general will need regular refueling and maintenance, the question is whether or not the propulsion system can be distributed over the orbital segment, or to what degree it should be concentrated in Orbital Replaceable Units (ORUs). Whereas ORUs can reduce system complexity and simplify automated maintenance operations by replacement of self-contained modules, they tend to increase logistics mass. Distributed propulsion system configurations are better suited to minimize logistics and to realize synergistic couplings. However, the integration of propellant tanks and feed lines as well as maintenance and repair operations may be more complicated.

Observation and Communication Payloads may need tracking and steering, too, depending on the flight mode of the overall system. At any rate, the designer has to take care of the free field of view and of potential cross contamination by thrusters or vent lines.

Again, a design cannot fully comply with all these guidelines at the same time. For instance, the minimum drag requirement for a pressurized module is incompatible with the minimum debris flux requirement. That means that the guidelines must always be applied in a prioritized way leading to a compromise solution.

Preparation of System Budgets

Having characterized the orbital segment concept, the next step is to close the design cycle on system level. The goal is to find and assess quantitative criteria, which allow for a comparison of what has been designed against the objectives, constraints and preliminary requirements that were developed along with the customer since the beginning of the design process. Thus, the evaluation of the system, on one hand, is a control unit for the design iterations; on the other, the main task is to prepare the "go" or "no go" decision of the customer. Another important objective for the evaluation is to generate the overall system numbers needed by the subsystem designers.

A system budget is a quantitative assessment of system options using a characteristic system parameter or criterion. Various parameters can serve as budgeting criteria. On a conceptual design level, good criteria have to allow for easy calculation and estimation and have to cover the main aspects with respect to system performance and cost. As mentioned earlier, a lack of information is a typical feature of conceptual design. Therefore, especially at the beginning of the design process, budgets will be established by simplified calculations, analogy or parametric estimations, or even meaningful assumptions and educated guess.

System budgets normally should cover mass, inertias, attitude and orbit dynamics, electrical power, thermal control, logistics, EVA/crew time, and cost.

Evaluation of Mission Utility

Utility evaluation can be done by correlating system budgets with overall system performance data. The goal is to express, using a few parameters, what the system's output is with respect to the efforts spent on different system alternatives. For a space station laboratory, such a measure of effectiveness could be the total crewed experimentation time over life cycle cost, or the pressurized experiment volume over total space segment mass.

In general, it will be difficult to describe the value of a solution as well as the effort to spend it in only one measure of effectiveness. This is why the mission evaluation step, as well as the baseline selection, again requires very close cooperation between the system designers and the customer. Finally, only the customer can state which aspects make up the system's value for him. That person is also the only authority who can specify priorities for different value functions (e.g. development cost, launch cost, total life cycle cost). The same is true for the judgement on system expenditures which may be more or less constraining for the customer. A space agency, for instance, could put emphasis on the technological leadership of its own country and would therefore be willing to pay for national developments but not for external procurement in other countries. In this case, external procurement cost would carry more weight than expenditures in the home country.

Selection of System Baseline

A system baseline is a consistent set of system elements which meets most or all of the mission objectives [Larson 92]. Due to the iterative character of the design process, system baselines in early design phases represent a working hypothesis with the goal being the better understanding and design of the system (heuristic principle). At each design cycle, the number of options is reduced and their description depth will be increased. The more mature a design gets, the fewer changes will occur at the end of the design cycles, and the resulting system baseline will become the planning document for the realization phase.

Definition of System Requirements and Allocation of Requirements to System Elements

The definition of system requirements is more than the formal documentation of the system baseline. At the same time, it is the subject and result of the design process. In the early design phases, requirements are directly drawn from overall objectives and customer needs, representing a first translation into functions and numerical values. Early design requirements are normally linked to the quality and quantity of the system's outputs which the customer desires. Once a design cycle has been completed, the overall system requirements have to be broken down to the next lower system level such that the elements of the subsequent level can meet the higher level objectives with some margin to spare. The level of formality and detail of the requirements framework has to be balanced with project size and the degree of maturity of the design. The designer should avoid letting the requirements become too rigid and overburdening for the project. Instead, they should be treated as an efficient design tool. In order to achieve this, requirements must be subject to negotiation and critical review both between the customer and the system designer, as well as among the different levels of the project hierarchy. Communication and negotiation are vital to assure that there will be no requirements driving the design which are not justified with respect to the mission objectives.

A common method to derive and establish the requirements of a system is the functional analysis, where the overall mission is broken down into a hierarchical structure of functions to be performed [NASA SP-6105, Larson 92, Hunger 95]. Methods to work out and display the functional structure of a system are Functional Flow Block Diagrams (FFDB), Time Line Analysis (TLA) and N^2 Diagrams. A requirement can be seen as a system function allocated to a system element affixed with specifications of how well the function is to be performed and how the functional performance will be verified. All requirements specified at that point have to be verified and adjusted in a subsequent design cycle by bottom-up budgets, simulations or tests.

Control of the Design Process

As mentioned earlier, the systems design process, due to its special problem context, will be iterative by nature. Therefore, the control of the design process is a critical part of the overall design activity. The ideal process will start with broad objectives and derive step by step a system concept in a top-down flow. After each design cycle, design depth can be increased until the whole system is defined down to the component level. However, systems design, in practice, is rarely so straightforward. In general, there will be several candidate concepts suitable to meet the overall objectives; they call for competitive concept development, trade-off studies and baseline selection.

As a consequence, the typical design flow may show many iteration loops (cf. Table 9.1). The need to iterate the design may arise during the evaluation of mission utility (step F) baseline selection or after requirements definition. Anyway, it is reasonable to complete system characterization (steps C and D) as well as system evaluation (steps E and F) before starting an iteration. As an outcome of system evaluation, there may be a need to change the characterization of the system that is, for instance, a change in flight mode or the orbital segment configuration. Moreover, it may become obvious that none of the concepts is suitable to meet the mission objectives sufficiently. In this case, iteration will lead back to the development of new concepts. Finally, it may turn out that the broad objectives cannot be met through a reasonable effort in time and money. This is the most severe iteration case, leading to a challenge of mission requirements, broad objectives, or even the cancellation of the project. In the nominal case however, the process will be followed up by the allocation of requirements (step I) in order to increase the design depth and to restart the design process on the subsequent system level.

9.3.3 Conceptual Design Tools

When reviewing the special environment of conceptual design and the related methodological characteristics, it becomes evident that this environment imposes a number of requirements on adapted design tools. Computer tools for conceptual design should

- be able to take on an initial situation characterized by a lack of information and "fuzzy" requirements and data,
- allow for rapid design cycles in definition, modification, analysis, simulation and evaluation of design variants, given the heuristic and iterative nature of the design process,
- support variable description depths of information starting with "rough order of magnitude" models or assumptions at the beginning. Information generated during conceptual design has to be in such shape that it can be smoothly passed on to detailed design tools,
- be multidisciplinary in methods (design, analysis, simulation, evaluation) and in subject area (e.g. overall system aspects such as attitude and orbit dynamics or microgravity level, as well as subsystem issues such as power systems or thermal control) in order to cover the most important design aspects in a collaborative and concurrent approach,
- be comprehensive in showing a high level of clarity, and
- have consistent and transparent user interfaces.

Large and complex computer tools used during detailed design can generally not be applied for conceptual design problems. For example, the time required to establish an accurate finite element model for structural analysis will easily exceed the time limit typical for a conceptual design study. Furthermore the system parameters (e.g. topology) are not established to a degree where meaningful high-precision models could be defined. Consequently, conceptual design tools need to find a compromise between reasonable simplification and sufficient physical or system-dynamic relevance.

Several institutions and companies have been working on conceptual design tools. NASA-JPL, for instance, established a Project Design Center with the idea to implement concurrent engineering techniques for space mission and system design as well as to create a testbed for early prototyping [Smith 96, Casani 97]. The European Space Agency ESA has developed a computer-aided framework named ESA-

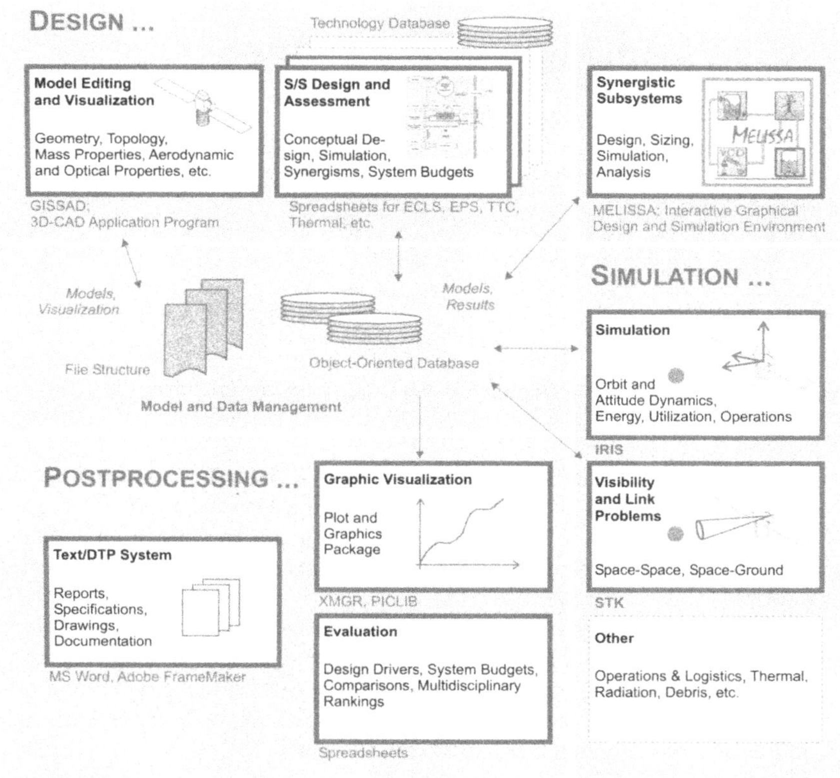

Fig. 9.13. The Space Station Design Workshop Tools

BASE for space systems design and analysis [De Kruyf 91] which is, however, more suited for detailed design. At this point, one example for a conceptual design environment called the "Space Station Design Workshop" developed at the University of Stuttgart [Bertrand 98] is presented (Fig. 9.13). In this approach, all tools needed for the design activity are brought together in a coherent environment while relying as much as possible on already existing or commercially available tools. This enables the designer to focus on software development activities for dedicated functions and interfaces between the applications.

In order to cope with the requirements of lacking and fuzzy information on one side and the need to keep model size and calculation time within reasonable limits on the other, a modeling technique using geometric primitives such as spheres, cuboids or cylinders (Fig. 9.14, center) is applied. The orbit segment is decomposed into geometric primitives which can represent a component's mass properties, its aerodynamic and optical properties, as well as its position and orientation data. Furthermore, kinetic degrees of freedom as well as mission-specific functionalities can be assigned to each primitive. For instance, in the case of a solar panel, its geometry,

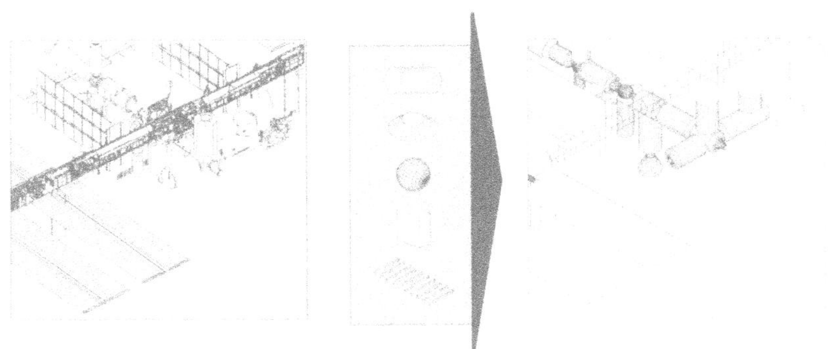

Fig. 9.14. Detailed CAD and Simplified Model Using Geometric Primitives

mass and aerodynamic model can be derived from a flat cuboid; its tracking functionality can be described by a rotational degree of freedom and a certain control law for this movement. And, its electrical functionality can be defined by a figure of merit correlating the panel area and the solar incidence flux with the electrical output power.

This modeling technique with geometric primitives has been implemented in the Space Station Design Workshop with both design and simulation tools. Figure 9.14 illustrates the modeling of the International Space Station using geometric primitives only. Further tools cover on-orbit system simulation, visibility and link analysis, and design and simulation of synergistic subsystems.

Example: Figure 9.15 shows some results of the interaction of these tools within a conceptual design cycle. The two alternative configurations shown have been established by participants of one of the international workshops on space station design carried out regularly at the Space Systems Institute of the University of Stuttgart. The participants (graduate aerospace engineering and architecture students) were tasked to establish a conceptual system design based on only a few broad objectives within five days time while working in a team-centered international environment.

Their mission statement read: "The European Space Agency asks you to conduct a conceptual design study on a joint European-Japanese space station project fulfilling the following requirements:

- The station shall allow for direct and permanent access to microgravity and Low-Earth-Orbit environment in a crewed research laboratory,
- with the main utilization fields of materials science and technology development.
- The proposed space station shall use a minimum configuration with growth possibilities using mainly ISS or other existing hardware.
- The space station shall demonstrate Japanese-European cooperation as a counterbalance to the US/Russian-dominated ISS."

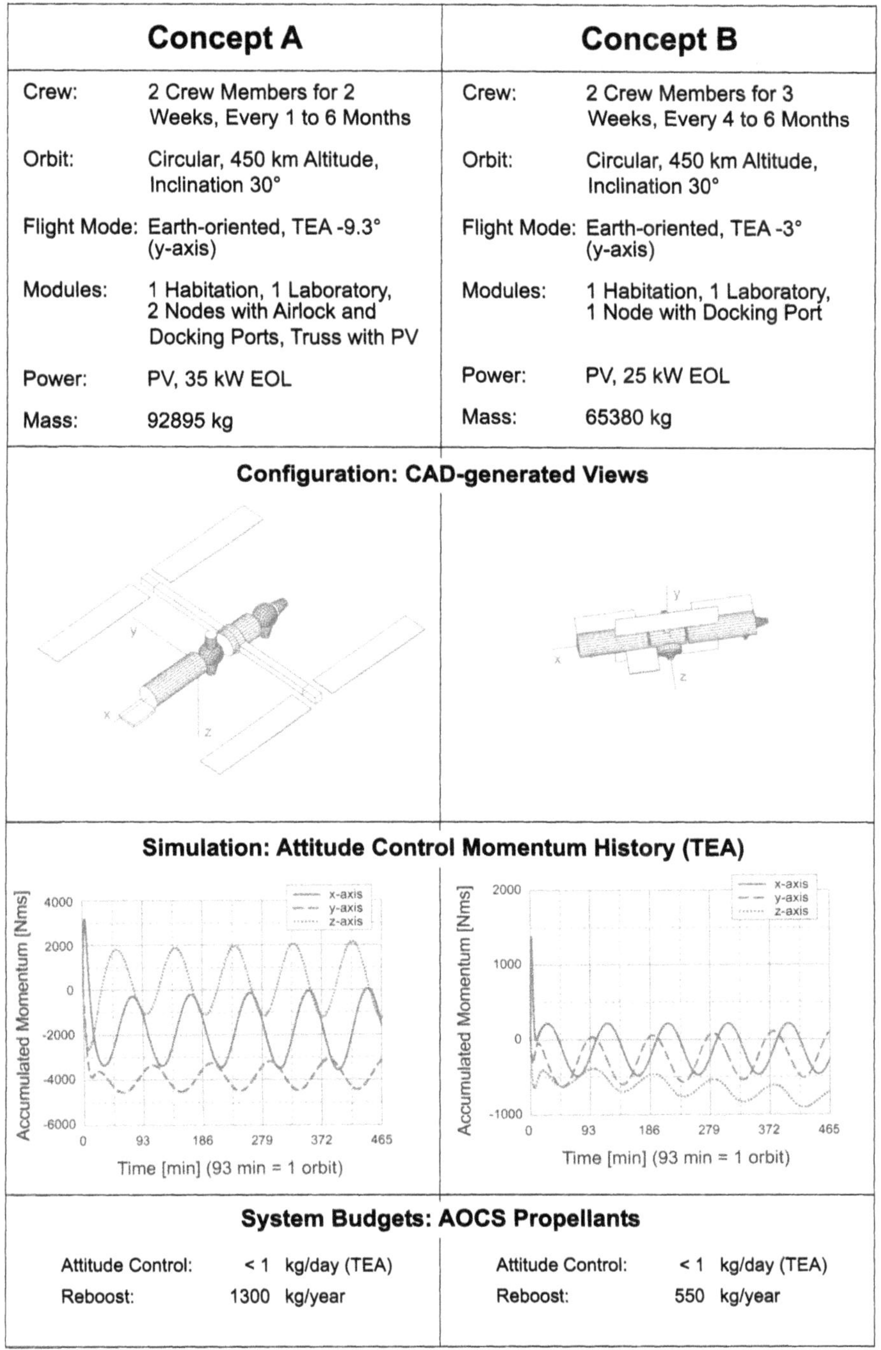

Fig. 9.15. Exemplary Space Station Designs Developed Using SSDW Tools and Methodology

These broad objectives had to be translated by each design team into preliminary requirements in order to establish a system concept suitable to fulfill the requirements. Once the most important system parameters such as orbit, flight attitude, electrical power, crew intensity, number and type of modules etc. had been characterized, the topology of the space segment had to be developed (Fig. 9.15, top and center).

Based on the CAD models of the respective configurations, key system parameters were analyzed using computer simulations. For example, the system parameter shown at the bottom of Fig. 9.15 (attitude control momentum) can be used as a measure of the inherent attitude stability of a configuration. The graphs show the attitude control momentum history over five orbital periods. In both designs, the accumulated momentum oscillates due to the varying angles of attack of the solar panels tracking the Sun, but the mean value increases only slightly. This can be attributed to both stations having a well-balanced torque equilibrium attitude (TEA). Thus, the corresponding requirements for attitude control propellants listed below the diagrams are quite low. The more compact configuration (B) has the smallest oscillation amplitude as well as the lowest reboost fuel requirements due to its smaller drag area.

All system budgets together will enable the refinement or optimization of the overall system, leading in an iterative process to the desired baseline concept.

9.4 Space Station Architectures

In the early 1980's, the so-called "Concept Development Group" (CDG) was commissioned by the US space agency, NASA, to conduct configuration studies for space stations. In the course of these studies, a number of "generic" types of architecture were defined and analyzed [Powell 84, NASA TM 87383]. Due to their fundamental importance also for the International Space Station (ISS), the CDG concepts will be briefly introduced in the next section, with special focus on their configurational characteristics.

9.4.1 CDG and "Freedom" Concepts

The following preconditions formed the starting point of the CDG studies which were aimed at yielding a space station that

- is permanently crewed
- flies at an inclination of 28.5°
- is operated as a scientific and technology platform
- is outfitted with any resource necessary for those facilities whose functions were identified as commercial applications:
  - Testbed for technological developments
  - Transportation node
  - Maintenance facility for satellites and industrial platforms
  - Observatory for Earth and space

Assuming a solar power supply and a zero-g design, the CDG compared different configurations. Since the mission profile implied different user disciplines, truss structures were introduced in order to provide enough space for all the different us-

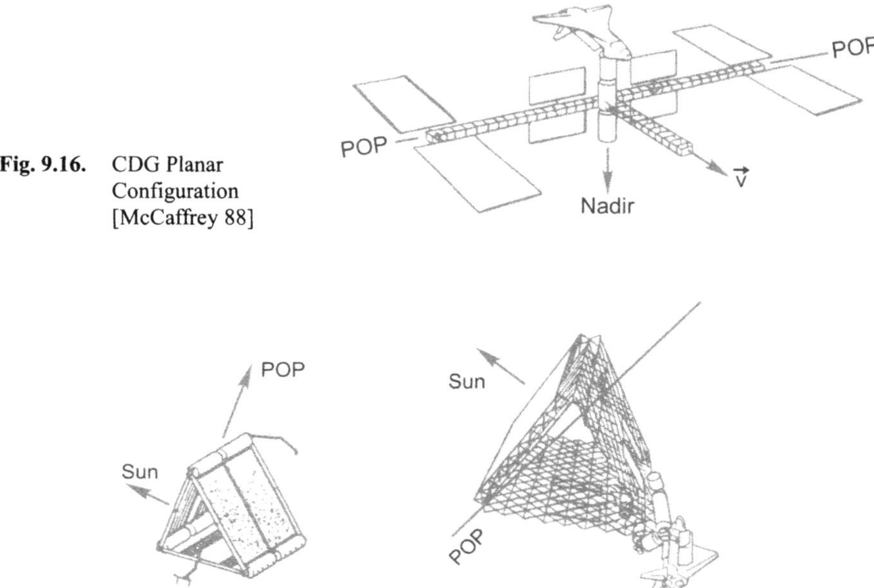

Fig. 9.16. CDG Planar Configuration [McCaffrey 88]

Fig. 9.17. CDG Delta Configuration (Early and Late Version) [McCaffrey 88]

ers and subsystems. The pressurized modules were installed in similar arrangement at suitable locations. This idea was new at the time of the CDG, since up to that point, mission requirements of previous studies had been met by arranging the pressurized modules in the form of a cluster.

The "Planar" configuration (cf. Fig. 9.16) was directly derived from earlier studies and was the predecessor of SSF. With the solar panels being actively tracked (the axis of rotation perpendicular to the orbital plane), the pressurized modules in their orientation could be decoupled from the rest of the station. Thus, both the gravity gradient-stabilized flight mode and the inertial flight mode (for astronomical observation) were possible. Incidentally, an Earth-oriented flight mode, with the modules arranged in a row on the $\vec{v}$ -axis, cf. Chapter 8 on "Microgravity", would have been more favorable for μg laboratory operations than the flight mode presented in Fig. 9.16.

With steadily increasing mission requirements and thus the station's increasing size, it began to be doubtful as to whether such large and flexible structures could be controlled at all (vibration damping).

As a result, the Johnson Space Center suggested the "Delta" configuration (Fig. 9.17), the three-dimensionality of which resulted in high rigidity of the structure. Initially, the pressurized modules were to be arranged at the corners of a triangular cylinder, connected by tunnels. This yielded a symmetrical mass distribution allowing the station to fly in the inertial flight mode, permanently oriented towards the Sun (with one of the three external surfaces serving as solar collector). Moreover, the interior of the prism formed by the modules and the tunnels offered space

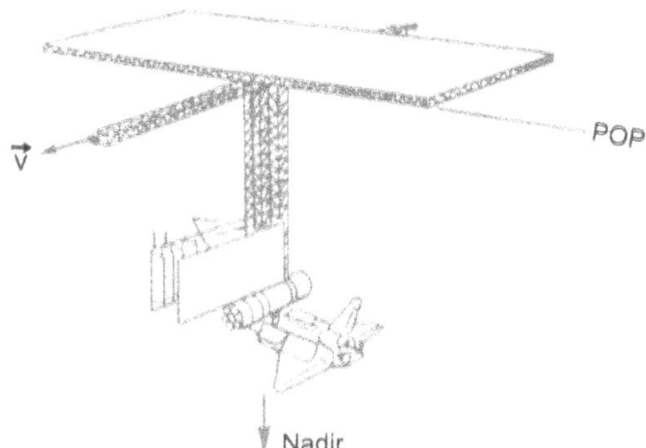

Fig. 9.18. CDG Big T Configuration [McCaffrey 88]

similar to a hangar. Due to increasing mission requirements, the power requirements also increased. As a consequence, the area of the solar array and the triangle became so large that the pressurized modules had to be installed all at the same corner (due to the otherwise excessive length of the tunnels between the pressurized modules). As a consequence, the mass distribution became asymmetrical and that caused problems as to the station's stability. Upon closer examination (e.g. docking of a Shuttle, utilization of the hangar), it was concluded that the resulting perturbations would be too serious, so, finally, the concept had to be abandoned.

The "T" configuration (Fig. 9.18) as an alternative configuration also offered high structural rigidity and was intended to fly in an Earth-oriented mode. In this mode, it was highly gravity gradient-stabilized because of the arrangement of its hangar and the pressurized modules. The advantage of a rigid array (oriented with its edge pointing towards the flight direction and thus yielding a low atmospheric drag) came at the price of oversizing the array by a factor of 2.5 in order to balance the highly fluctuating performance of the solar generator.

Since, in the meantime, doubts concerning large flexible structures had disappeared, a less complex concept was decided on: the "Power Tower" (Fig. 9.19) was a configuration perpendicular to the planar configuration. With articulated solar arrays, this station was to fly in a gravity gradient-stabilized flight attitude. Due to the large distance between the pressurized modules and the solar panels, the aerodynamic torque and the gravity gradient torque could be balanced against one another. Above the generators, there were suitable points for astronomical payloads, where they were far from the polluted surroundings of the pressurized modules. Compared to the planar configuration, expansion was easier (simply by extending the truss in z-direction). Finally, one single uninterrupted track of rails on the truss allowed a mobile manipulator system to reach any point of the entire station.

Contrary to planar configurations, the pressurized modules of the "Power Tower" configuration were not installed near the center of mass. Since the station's µg operations were of considerable importance, this would not have been advantageous

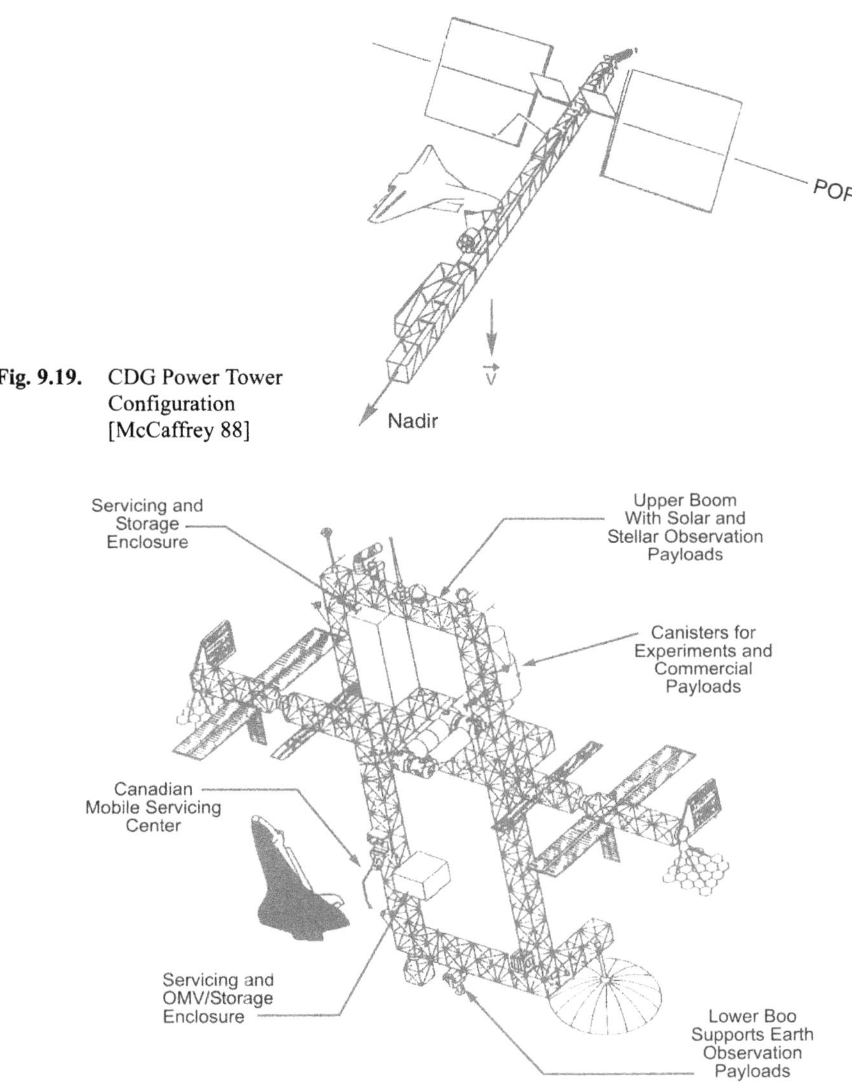

Fig. 9.19. CDG Power Tower Configuration [McCaffrey 88]

Fig. 9.20. Dual Keel Configuration [McCaffrey 88]

for the station. By designing the "Dual Keel" concept (Fig. 9.20), an attempt was made to unite the advantages of both the Power Tower and the Planar Concept, i.e.:

- Low g-levels inside the pressurized modules
- Gravity gradient stabilization by orienting the vertical truss in the z-direction
- Available accommodations for payloads in zenith and nadir directions
- Expansion possibilities (e.g. hangars)

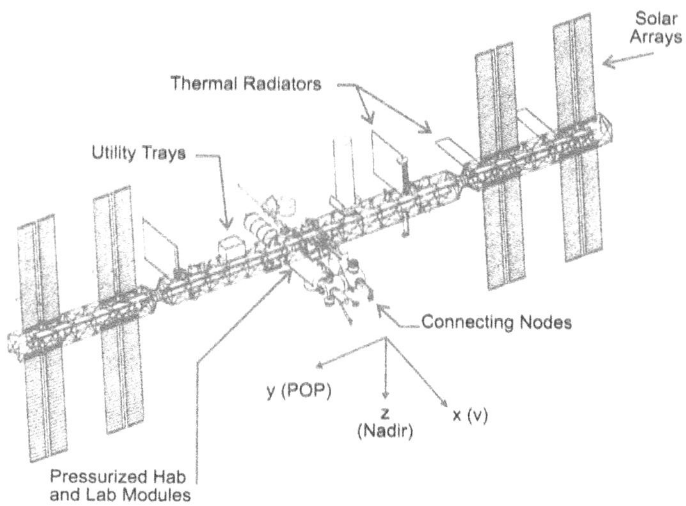

Fig. 9.21. Revised Baseline Configuration [Priest 88]

In the course of taking economic measures, the "Dual Keel" configuration "lost" its vertical dual keel and was renamed "Revised Baseline" (Fig. 9.21). As a consequence, costs for development, assembly and launch were considerably reduced, but on the other hand, some important advantages of the "Dual Keel" concept were lost, for example:

- In the intended flight mode, the gravity gradient no longer had a stabilizing effect. Instead, the balanced attitude would be characterized by a nearly radially-oriented truss similar to the "Power Tower" configuration
- Suitable accommodations along the keels for payloads, especially those for Earth observation, astronomy and servicing have disappeared

As is generally known, neither did this concept reach its realization, but many of its components were directly transferred to the International Space Station unaltered.

9.4.2 The Mir Configuration

Architectures using the pressurized modules as the "backbone" of their structure are a counterpole to those configurations whose backbones consist of a truss (cf. Chapter 2), for example MORL, Skylab, Salyut and Mir. Mir's longitudinal axis consists of the Mir Core Module, the Kvant module as well as the Soyuz and Progress vehicles (cf. Fig. 9.22). At the end of this axis formed by the Core Module, there is the so-called "Multiple Docking Adapter" where up to four more modules can be docked in transverse direction. On one hand, when the longitudinal axis is oriented towards the Earth, this comparatively compact configuration allows a gravity gradient-stabilized flight attitude. On the other, the entire station can change its flight attitude, e.g. for docking maneuvers, so that both active docking adapters - sit-

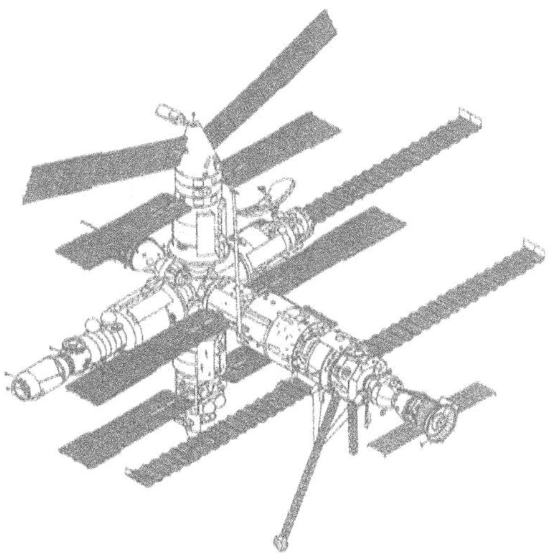

Fig. 9.22. Space Station Mir [NRC 95]

uated at the ends of the longitudinal axis – are oriented in the direction of the orbit tangent. Mir is extremely flexible as to expansion and reconfiguration on the module level since each module carries all necessary subsystems, and indeed, Mir, during its operational life, has existed in various configurations.

A disadvantage of this architectural concept is that the size of the solar collectors directly mounted on the pressurized modules is limited, and, compared to remote solar collectors, they have higher losses due to shadowing. The same applies to the thermal radiators directly installed onto the outer skin of the pressurized modules. Moreover, the module cluster offers less space for installation of external payloads and subsystems, hence limiting extremely diversified mission profiles and the use of the so-called "Orbital Replaceable Units" (ORUs).

9.4.3 The Columbus Free Flying Laboratory

The "Columbus Free Flying Laboratory" (CFFL), a concept temporarily pursued by ESA, consists of only one pressurized module and a service module and is thus a "minimum architecture" for space stations (Fig. 9.23). As the CFFL was designed mainly for microgravity research, it was only temporarily crewed and was intended to dock at the then space station "Freedom" for purposes of maintenance, reconfiguration and repair.

The extremely compact design allowed the Earth-oriented as well as inertial flight mode and, due to its intended application, a very high level of microgravity. The solar collectors were only α-trackable, requiring a prealignment of the station or a partly inertial flight mode for maximum performance. Expansion of the station is theoretically possible; a major prerequisite would be larger investments for an increase in the station's resources (electrical power, thermal capacity, life support system, propulsion systems). Of course, accommodations for external payloads are limited and as a consequence, so is the diversification of mission profiles.

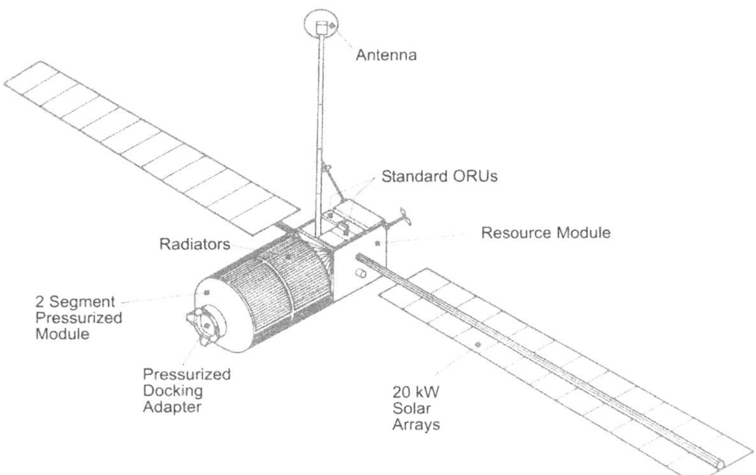

Fig. 9.23. Columbus Free Flying Laboratory (CFFL), ESA Concept [Wöhlke 88]

9.4.4 The International Space Station (ISS)

As to its architecture, ISS is a compromise between Eastern and Western concepts. The Russian Orbital Segment (ROS) includes the pressurized module backbone concept (Zarya, formerly FGB, and Service Module), a method already proven with Mir. In the case of ISS, this design includes additional components for power supply, thermal control and attitude control, all of which contribute to the increase in the performance of the tower-like Science Power Platform (SPP). Especially the integrated design of Zarya and the Service Module allows the fast and flexible beginning of assembly since these components are in a position to provide all resources autonomously. The system will remain autonomous until later on when the United States On-Orbit Segment (USOS) is added, providing efficient and centralized systems for power supply and thermal control. The US, Japanese and European components have all been transferred largely unaltered from the Freedom concept ("Revised Baseline"). For that reason, ISS, in form, looks like as if a Freedom concept was combined with an expanded Mir station. The Russian and US components are connected by the pressurized "Node 1", named "Unity".

Due to this arrangement, some of the advantages of the Revised Baseline (tracking of the solar panels, accommodation of external payloads) are preserved. Moreover, the parallel subsystems in the ROS and the USOS ensure increased safety and redundancy. On the other hand, the mass distribution and the distribution of the incidence areas of the station have some serious shortcomings, such as the following:

- The axis with the smallest moment of inertia is perpendicular to the orbital plane. Thus, as was the case with the Revised Baseline, the gravity gradient has no stabilizing effect on the flight attitude.

- The mass distribution is highly asymmetrical with regard to the orbital plane and the local horizontal plane. Especially the asymmetrical mass distribution with regard to orbital plane causes strong gravity gradient torques around the pitch axis due to a large deviation moment I_{XZ}. As shown in Chapter 6, the amount of propellants required to maintain the LVLH flight mode would be too great. For that reason, the station will fly with about 6°–8° pitch angle in order to achieve a torque equilibrium attitude. This "error" in orientation is clear to see in the illustrations relating to the quality of microgravity (cf. Fig. 8.23), since the isolines of the tidal perturbation are parallel to the orbital plane and also parallel to the direction of velocity.
- Additional disturbances due to the position of the large solar collectors at the truss and the SPP also influence the flight attitude in a negative way. Since their centers of pressure are markedly remote from the station's center of mass, they cause cyclic perturbation torques due to the tracking.

All in all, ISS is characterized by rather conservative technology and system concepts [Foley 96]. Generally, innovative technologies such as gaseous H_2/O_2 propellants, resistojets, solar dynamic power supply, closure of the oxygen loop in the ECLSS, etc. have not been introduced to reduce costs and risk in development and assembly of the station. As a consequence, however, requirements for the station's operation and maintenance will be much higher.

The quasi-static level of microgravity inside the pressurized modules which are arranged in flight direction will be sufficient with 1 to $2 \cdot 10^{-6}$. Those modules of the Russian Orbital Segment which are more remote from the center of mass will have a level of microgravity of 3 to $4 \cdot 10^{-6}$. For optical payloads there will be accommodations (and clear fields-of-view) on the truss, on the "Exposed Facility" of the Japanese segment, and on the Russian pressurized modules. In radial as well as in orbit-tangential directions, sufficient docking ports and approach corridors will be provided for rendezvous and docking.

Despite some very difficult boundary conditions such as:

- The involvement of Russia
- The requirement to use existing Space Station Freedom hardware
- The very short period of time (only six months) in which the station had to be restructured

ISS is nevertheless a very good compromise and will constitute an experimentation and test platform for the different application disciplines until far into the next millennium.

Then, at the latest with ISS's successor station, the task of designing a space station will again be relevant due to changed boundary conditions. Figure 9.24 shows the concept of a future space station in LEO for a crew of 60 [Weiser 96].

This mission scenario includes, apart from research in weightlessness, the first commercial use, e.g. for the production of goods with high added value and for satellite and platform maintenance. The station has a good gravity gradient stability and systematically uses spent upper stages and propellant tanks of launch systems not only to balance the mass distribution, but also to assemble the station's structural

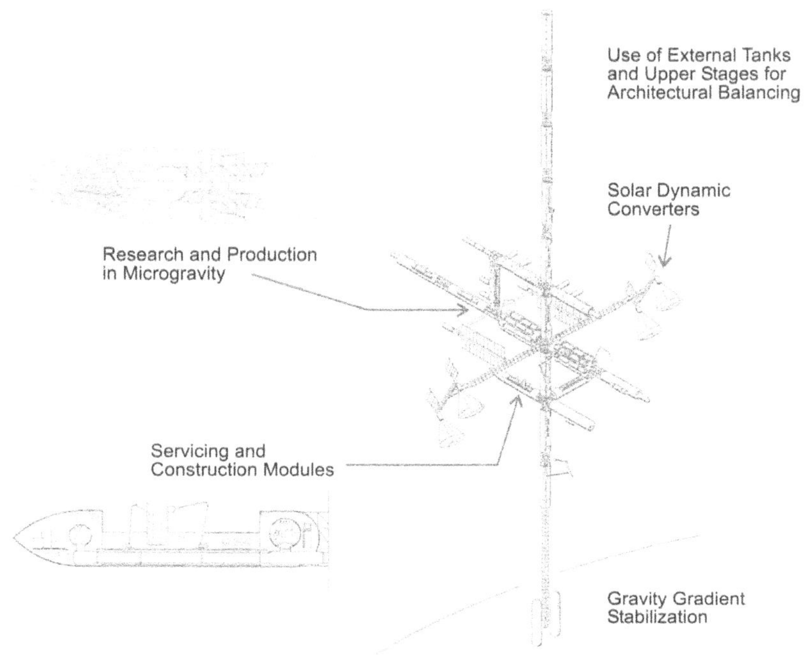

Fig. 9.24. Concept of a Next Generation Space Station [Weiser 96]

backbone. Power will be mainly supplied by solar dynamic converters, the whole concept having been designed for an electrical performance of 500 kW at a total mass of 2500 tons.

Of course, such concepts nowadays are still far from being realized. In the long run, transport into orbit will be a significant problem with every kind of orbital structure since it is and will always be the launch system that determines the mass, shape and size of the single components as well as the related costs of assembly and operation.

10 Synergisms

During the past few years, "synergism" has become a frequently used expression. Terms such as "synergistic effects", "synergetics", and "integrated thought" have become catch phrases, especially in discussions on political, social, or ecological issues. The attempt to use a more or less comprehensive approach is quite common in each of these areas, in order to find an optimum criteria-specific overall solution of a design problem by combining different partial solutions [Webster 83, Haken 83].

Usually, in an Earth environment, economical or ecological requirements lead to synergistic approaches. For space applications though, the omnipresent limitation on transport capacities is the main driver. Logistics during assembly and the operational phase still represent a bottleneck for large space projects like a space station. Thus, the systems engineer will continuously strive to minimize resupply mass, especially with subsystems like the ECLSS or the AOCS. In addition, high safety requirements are typical for crewed systems, demanding high failure tolerance and operational flexibility.

In this chapter, the synergistic approach to the design of technical systems will be presented. First, an introduction to terms, concepts, objectives and methodology of synergistic system design is given. Subsequently, the changes in system layout and key system parameters that result from the gradual introduction of synergistic linkages into a previously unlinked system are demonstrated, using a generic space station as an example.

10.1 Terms and Concepts

According to its definition, "synergism" (Greek *syn-*: together, *ergo*: the work) describes the collaboration of agents or units towards the achievement of a total performance, which is larger than the sum of the individual performances of all agents.

When applied to technical systems, and especially to space stations, "synergism" denotes the linking or partial merging of subsystems, so that, in addition to the required subsystem functions (i.e. supply of power, air, etc.), even higher objectives can be achieved. Synergistic systems design thus represents a counterbalance to classical design approaches, which demand a separation of the main task into subtasks, each of which is then optimized individually.

In Sect. 9.2, the systems design process was also characterized by the fact that the most important decisions about system architecture are made in the very beginning. It is thus obvious that possible synergistic linkages have to be included into the system layout during the preliminary design phase. Once the system components or subsystems with their key properties and interfaces are defined, great effort is required to incorporate configurational changes that include e.g. the merging of subsystems. Consequently, synergisms have to be considered as early as the conceptual design phase.

The following possible objectives may result from implementing synergisms in space system design:

- Increased reliability through the creation of redundancies
- Broader system safety
- Reduction of system and supply masses
- Operational flexibility
- Simplification of logistics (i.e. lower supply material diversity)
- Reduced system complexity
- Reduced construction expenditure

These different objectives generally lead to conflicting requirements (i.e. redundancies vs. mass saving) necessitating assessment and weighing of criteria for the application of synergisms to a specific design. As a rule, however, in synergistic systems design, the main effort is directed at reducing mass – especially resupply mass – and increasing safety, reliability and operational flexibility.

10.2 Coupling of Subsystems

If questions regarding synergisms arise in the design of a space station, the next step is to ascertain which subsystems are fundamentally suitable for being coupled with other subsystems (Fig. 10.1) [RSW 92].

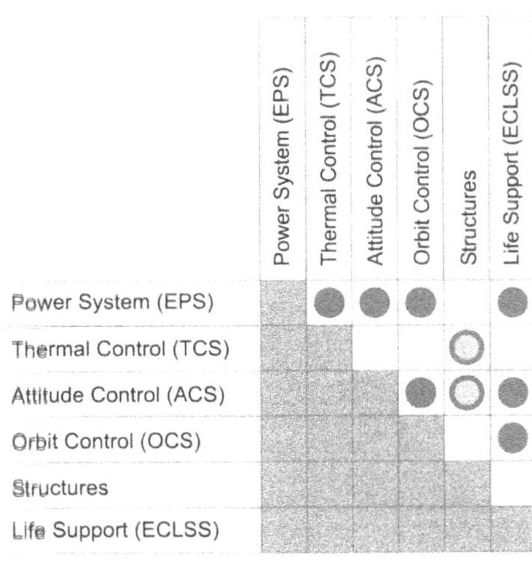

Fig. 10.1. Matrix of Possible Subsystem Linkages

The possible linkages indicated in the matrix above can be illustrated through the following qualitative examples:

Power System	⇔ Thermal Control:	Extraction of thermal energy from waste heat
Power System	⇔ Attitude Control:	Electrolytic H_2/O_2 as propellant from RFC
Power System	⇔ Orbit Control:	H_2/O_2 for fuel cells from RCS propellant tank
Power System	⇔ Life Support:	RFC used as residual contaminant filter for H_2O processing
Thermal Control	⇔ Structures:	Outer structure surfaces serve as a radiator
Attitude Control	⇔ Orbit Control:	Common engines, tanks and propellants
Attitude Control	⇔ Structures:	Design of the station configuration for gravity gradient stabilization
Attitude Control	⇔ Life Support:	Hydrazine for the supply of N_2
Orbit Control	⇔ Life Support:	Wastes as propellants in resisto- or arcjets

Similarly, synergisms can also be identified and used between different functions within a subsystem. For example, significant power and mass reductions may be achieved through the recovery of heat in the thermal control system when using solar dynamic energy production. The following examples, however, focus on synergisms between distinct subsystems.

In the matrix shown in Fig. 10.1, a few markers are filled gray instead of solid black. These cases denote linkages whose character is more static in nature. As a consequence, the respective common function can only be performed by one of either of the subsystems. The coupling rather contributes statically to the achievement of synergistic objectives.

For example, the use of structural parts as a thermal radiator (linkage between structure and TCS) results in direct mass reductions. In contrast to this, the coupling of the ECLSS with the EPS via an electrolysis cell for the filtering of contaminants can be given as an example of a dynamic synergism.

In order to identify from the wide range of possible system couplings those which are useful from the synergistic point of view, one normally has to preselect suitable items, systems, technologies and system components. This selection process requires a broad knowledge of system parameters and properties, and those of all subsystems involved. In the end, the synergistic benefit can only be determined after design and simulation of either the complete subsystem or even the total system have been finished.

10.3 System Balances

A system balance allows a quantitative comparison of various design alternatives in order to find out about their individual advantages and disadvantages with respect to key system parameters. Obviously, differences in required hardware, different operational/logistical concepts and synergistic effects have to be taken into account. The key system parameters normally used for such an analysis are the following:

- Safety and reliability factors
- Equivalent system mass
- System volume
- Crew time requirements
- Cost and Risk
- Scheduling

Within the scope of preliminary design, the equivalent system mass is considered to be of special importance. Mass is one of the drivers during the design process anyway, since parameters such as electrical power, volume or cost are estimated using mass-specific technology factors like e.g. kW/kg, \$/kg, etc. (see also Chapter 9 on "System Engineering"). It is therefore advisable to express all advantages and disadvantages of a certain design concept in terms of mass, namely its total system mass. It is calculated according to

$$m_{total,\,sys} = m_{fix} + \dot{m} \cdot t_{life} + m_{syn} \tag{10.1}$$

from three individual contributors:

- Fixed system mass (m_{fix}). This parameter records differences in hardware. Naturally, the need for spare or exchange hardware over the total design life of the system will have to be taken into account.
- Mass flows ($\dot{m}$). These include consumables like propellants, gases or food.
- Synergistic mass (m_{syn}). This parameter describes all positive or negative effects of a design alternative with respect to the overall system. For example, the additional electrical power that can be provided to the overall system when using an EDC unit for CO_2 filtering leads to a reduction in power system size. In order to estimate the corresponding mass savings (synergistic mass), the mass of a solar array producing an equivalent power can be calculated and then subtracted from the baseline EPS mass. On the other hand, the waste heat produced in the EDC unit will lead to an increase in thermal control system size. This increase can also be stated in terms of radiator and auxiliary system mass, and consequently added to the overall balance.

This methodology of establishing system mass balances results in the equivalent system mass usually being different from the actual system mass, but it directly shows the difference in mass turnover between two design alternatives. As the mass balance usually includes mass flows, by equaling two mass balances, a point in time can be identified at which both corresponding design alternatives have the same system mass:

$$m_{fix,\,1} + \dot{m}_1 \cdot t + m_{syn,\,1} = m_{fix,\,2} + \dot{m}_2 \cdot t + m_{syn,\,2} \tag{10.2}$$

After solving this for t, the break-even point can be calculated:

$$t_{break-even} = \frac{m_{fix,\,2} - m_{fix,\,1} + m_{syn,\,2} - m_{syn,\,1}}{\dot{m}_1 - \dot{m}_2} \tag{10.3}$$

The break-even point shows which design alternative is most suitable for a given duration of operations (design life). In Sect. 4.2, this is demonstrated using an example from the area of life support, comparing two different CO_2 filtration methods (cf. Fig. 4.11 and Fig. 4.29).

The maturity of a design prescribes which methods to apply in determining the masses. Initially, analogies with existing systems can be used. For example, the mass of an experiment module included as a part of a future space station can be estimated from a comparable module used in the Russian space station Mir. In addition to that, masses can be determined with the help of parametric technology data. The mass of a solar array can be estimated, for example, if its collector technology, its surface area, and a corresponding technology factor (kg/m^2) are known. Finally, as the design process is progressing, more and more masses can be determined by summing up known individual component or sub-subsystem masses.

For these estimations, simplifications of system dynamics as well as technological assumptions are unavoidable. This must be taken into account when interpreting the results of such an analysis. In this respect, it is especially important to document all assumptions and simplifications, and to check their appropriateness with respect to conclusions and design decisions continuously.

10.4 Examples for Synergistic Linkages

In Chapter 6 ("Attitude and Orbit Control System"), Sect. 6.3.2, a synergistic propulsion system for orbit control of the ISS has already been presented as an example of such a linkage [Heckert 87, Bertrand 96]. This section will demonstrate more possibilities for synergistic linkages between the subsystems of a space station. In this case, the life support, power supply and orbit control subsystems have been chosen. Initially, a non-integrated system (Fig. 10.2) is presented as the basis for comparison. Subsequently, three increasingly synergistically linked systems are introduced, and their respective effects on the overall system are estimated. For the sake of simplicity, only the equivalent system mass is used for comparison. Safety and reliability criteria as additional parameters for comparison will be briefly mentioned towards the end.

In the remainder of this chapter, the investigation and assessment of synergistic effects will be made with the help of a reference space station. This reference space station shall be designed to permanently accommodate four astronauts and shall have an orbit of 400 km altitude. The design lifetime shall be at least 20 years. All relevant assumptions and space station parameters are compiled in Table 10.1.

An orbital period of 91.9 minutes and the maximum eclipse factor of 39% for an orbit with 28.5° inclination result from the orbital parameters.

An electrical output of 40 kW shall be provided for the operation of the station and the experiments to be conducted therein. The power supply is shared between a photovoltaic system (PV, planar silicon cell) and a solar dynamic system (SD, Brayton process with thermal energy storage).

The physico-chemical life support system is designed for four persons and is of a regenerative nature. Consequently, the losses due to leakage (N_2, O_2, H_2O) and

Table 10.1. Space Station Data

Reference Space Station:	
Crew Size	Four Astronauts
Design Life	$t_{mission} = 20$ a
Required Power	$P_{base} = 40$ kW electrical
Orbit	
Altitude	h = 400 km
Inclination	i = 28.5°
Period	P = 91.6 min
Eclipse Factor	$\tau_{ecl} \leq 39\%$
Subsystems Data	
EPS	
Electrical Power	$P_{base} = 40$ kW
Si PV System	$m_{spec,PV} = 22.5$ kg/kW
	$A_{spec,PV} = 7.5$ m²/kW
Brayton SD System	$m_{spec,SD} = 111.6$ kg/kW
	$A_{spec,SD} = 7.4$ m²/kW
ECLSS	
Physico-Chemical, Regenerative	
AOCS	
Thruster System	GH_2/GO_2
Specific Impulse	$I_{spec} = 380$ s
Reboost Strategy	Permanent Thrust
TCS	
Specific Radiator Mass	$m_{spec,R} = 7$ kg/kW
Specific Radiator Surface	$A_{spec,R} = 1.4$ m²/kW

the non-regenerative substances (additional O_2, food, H_2O sludge) are to be compensated for by resupply.

A propulsion system using GH_2/GO_2 combustion thrusters (I_{sp} = 380 s) is available for attitude and orbit control maneuvers. Orbit control is performed by permanent thrust maneuvers, with the required thrust being scaled by the size of the aerodynamic surfaces of the space station. For attitude control, Control Momentum Gyros are used. For simplicity, the attitude control system will not be included in the following synergistic analysis examples. It is assumed though, that the attitude is kept constant at all times.

10.4.1 Non-Integrated System

In the case of the non-integrated subsystems, the basic system layout will turn out as shown in Fig. 10.2. All consumable goods, especially those for the ECLSS and the AOCS, are to be provided by means of resupply. Subsequently, an analysis of the various subsystems involved will be performed.

Photovoltaic Power Supply. The solar generator, during the Sun phase, has to supply both the basic power for operation of the space station and its experiments as

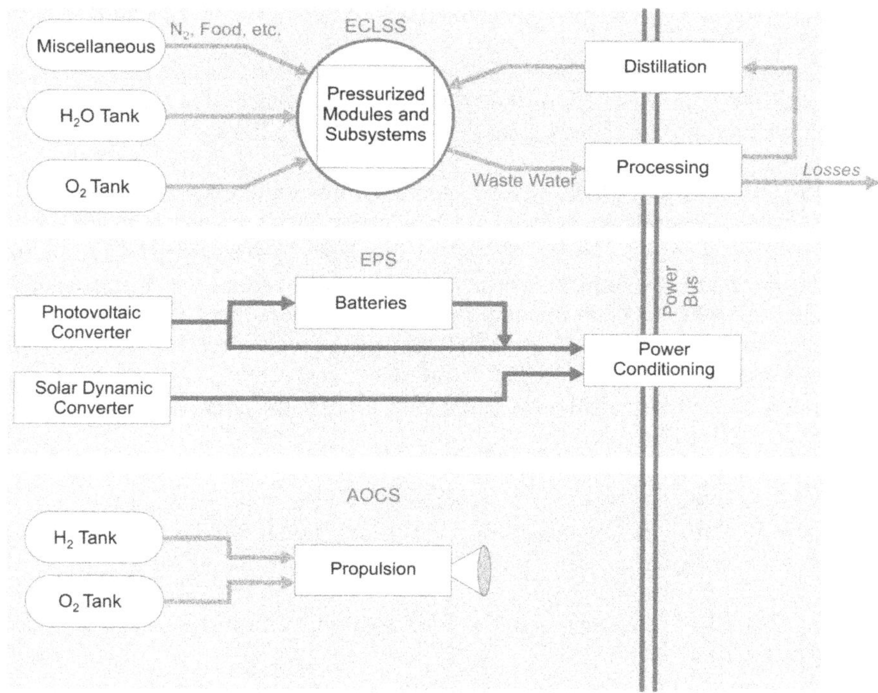

Fig. 10.2. System Layout of the Non-Integrated System

well as power for the charging of the batteries. Consequently, the net performance of the solar generator results from:

$$\begin{aligned} P_{PV,\,net} &= \frac{P_{base}}{2} + P_{store} \\ &= \frac{P_{base}}{2} + \frac{E_{ecl}}{t_{Sun} \cdot \eta_{store}} \end{aligned} \tag{10.4}$$

Assuming a degree of storage efficiency of $\eta_{store} = 0.61$ for the batteries and a solar phase length of $t_{Sun} = 55.9$ min, the PV generators must provide 41 kW net of electrical power (cf. Section 5.2.3).

Ultimately, the solar generator design must take into account the environmentally caused degradation of the solar cells over the operational life of the generator. When using silicon cells, it is safe to assume that after a year, the performance will diminish to $l_{deg} = 98.75\%$ of the original performance level. With a design life of $t_{mission} = 20$ years, the degradation factor Begin of Life/End of Life (BOL/EOL) calculates to:

$$\frac{BOL}{EOL} = \left(\frac{1}{l_{deg}}\right)^{t_{mission}} = 1.286 \tag{10.5}$$

Keeping this in mind, the solar generator must be designed for a performance of 52.7 kW at mission start.

With characteristic dimensions of 7.5 m^2/kW for a specific surface area and mass of 22.5 kg/kW (see Table 10.1), the total generator surface area turns out to be 395 m^2, with an associated total solar generator mass of 1185 kg.

Batteries. NiH_2 batteries were chosen in our example for storage of the photovoltaically collected energy. At a pre-set life expectancy, they are superior to NiCd batteries in terms of energy density, efficiency and depth of discharge (DOD). In the design of the battery system, the primary task is to establish the depth of discharge to be used. A higher DOD means that for the same battery mass, more energy can be taken from the batteries. However, through this, the life expectancy (Design Life, DL) of the batteries is simultaneously reduced, and respectively, with the fixed operating time of the space station set at 20 years, more battery sets are required:

$$n_{sets} = 20a/DL_{batt} \tag{10.6}$$

Table 10.2 illustrates this connection. Columns 2 to 4 show the parameters for a battery with a 55 Wh/kg energy density (ρ_E) at chosen DODs of 75%, 50% and 35%, while column 5 displays a battery with a performance of 80 Wh/kg at a DOD of 35%.

Using Table 10.2, the comparison of equivalent masses for the four options according to the equation

$$m_{batt} = \frac{E_{ecl}}{\rho_E DOD} n_{sets} \tag{10.7}$$

shows an optimum mass for a ρ_E of 55 Wh/kg and a DOD of 35%. This corresponds to a total battery mass of 2474 kg for the provision of 20 kW of electrical power during the eclipse periods, equaling an output of 11.9 kWh of power per eclipse.

Table 10.2. Design Parameters of NiH_2 Batteries

DOD [%]	75	50	35	35
ρ_E [Wh/kg]	55	55	55	80
DL_{batt} [Years]	0.85	2.0	5.4	2.4
n_{sets} (per 20 years)	24	10	4	9
Mass per Battery Set [kg]	289	433	619	425
Battery Total Mass [kg]	6936	4330	2474	3828

Solar Dynamic Power Supply. With the design of the solar dynamic (SD) generator (again for an electrical power output of 20 kW), it shall be assumed that the energy required for power generation during the shadow phase is stored in primary form, i.e. in the form of heat. A detailed discussion of the components (collector, receiver-storage unit, radiator, Brayton thermal power engine, and generator) will not be mentioned here. The SD generator is rather treated as a "black box." Accordingly, characteristic technology parameters as established in Table 5.4 on page 154 are used. Note that the surface areas given there refer to the sun-facing projected surface area of the receiver only, so that allowances for radiator and structural surface area must be added. From the resulting specific surface area of 7.4 m^2/kW and a specific mass of 111.6 kg/kW, a total SD mass of 2232 kg and a total SD surface area of 148 m^2 result.

Life Support System. The life support system is already closed to such an extent that only commodities such as food, clothing, nitrogen, metabolically-bound oxygen and regular water loss need to be compensated for by resupply. These consume a large fraction of the logistics budget; however, they do not play a role in the synergisms to be investigated and will therefore not be considered any further.

A distiller is planned in the water cycle for the recovery of waste water. Since this unit will later on be replaced through the implementation of a synergism, a proper accounting of its contribution is necessary.

A waste water volume of 96.5 kg per day is assumed for the crew of four. Additional 303 kg in system mass must be accounted for the distillation device itself. The electrical performance requirements of 0.7 kW shall be included in the 40 kW base supply of the space station.

Thermal Control System. The thermal control is restricted here to the removal of heat loss from the energy supply. The SD characteristic data already take into account its provision with radiators for disposal of its own waste heat. However, the waste heat of the batteries needs to be considered separately. The waste heat flow which needs to be emitted by radiators calculates to

$$Q_{batt} = \left(\frac{1}{\eta_{batt}} - 1\right) \cdot P_{batt} = 12.8 \text{ kW} \tag{10.8}$$

with $P_{batt} = P_{base}/2$ and $\eta_{batt} = \eta_{store}$. A radiator surface area of 18 m^2 with a mass of 90 kg is thereby necessary. The radiator surfaces are normally oriented in such a way that they cause minimal aerodynamic drag. Therefore, with respect to the propellant balance for orbit control, these surfaces do not need to be taken into account.

Propulsion System. As mentioned at the beginning, the aerodynamic drag shall be compensated for by continuous thrust maneuvers. The required thrust for this calculates to

$$F = \frac{\rho}{2} v^2 C_D A_{Stat} \tag{10.9}$$

using the following values:

- Density of residual atmosphere at 400 km altitude: $\rho = 5 \cdot 10^{-12}$ kg/m^3
- Orbital velocity at a 400 km altitude: $v = 7671.5$ m/s
- Average drag coefficient: $C_D = 2.3$

A_{Stat} includes all surfaces of the space station which are exposed to residual atmosphere. In this example, these are the surfaces of the solar collectors (PV and SD), the radiator surfaces, and the constant surfaces of the space station. The last two contributors remain uninfluenced by the changes caused by the synergisms. Strictly speaking, for the PV and SD surface areas, the surface area should be averaged over the course of an orbit, as they are tracking the Sun and thus present different cross sections with respect to the station velocity vector. For simplicity, though, this is not taken into account here. With respect to the open system, the component surfaces result in:

A_{PV}=395 m^2
A_{SD}=148 m^2
A_{misc}=70 m^2 (various surface areas for pressure modules, structures, etc.).

With this total drag area of 613 m^2, the thrust needed for drag compensation amounts to 0.208 N. The required propellant mass flow then calculates to 35031 kg in 20 years with:

$$\dot{m}_{prop} = \frac{F}{I_{sp} \cdot g_0} \tag{10.10}$$

Balance for a 20-Year Operating Period. On the basis of the assumptions put forth, Table 10.3 lists the equivalent system masses for the investigated subsystems. These values constitute the basis for comparison of the synergistic system couplings computed below.

Table 10.3. System Balance of the Non-Integrated System

EPS	Batteries	2474 kg
	Photovoltaic	1185 kg
	Solar Dynamic	2232 kg
ECLSS	Distiller	303 kg
TCS	Radiator	90 kg
AOCS	Orbit Control	35031 kg
Total		**41315 kg**

10.4.2 Regenerative Fuel Cells for Energy Storage

In this section, the synergistic effects of a Regenerative Fuel Cell (RFC) will be investigated. On one hand, this device is coupled with the life support system by hydrogen and oxygen flows; on the other, it will assume the task of energy storage in place of the batteries, thereby also making it a part of the power system (Fig. 10.3).

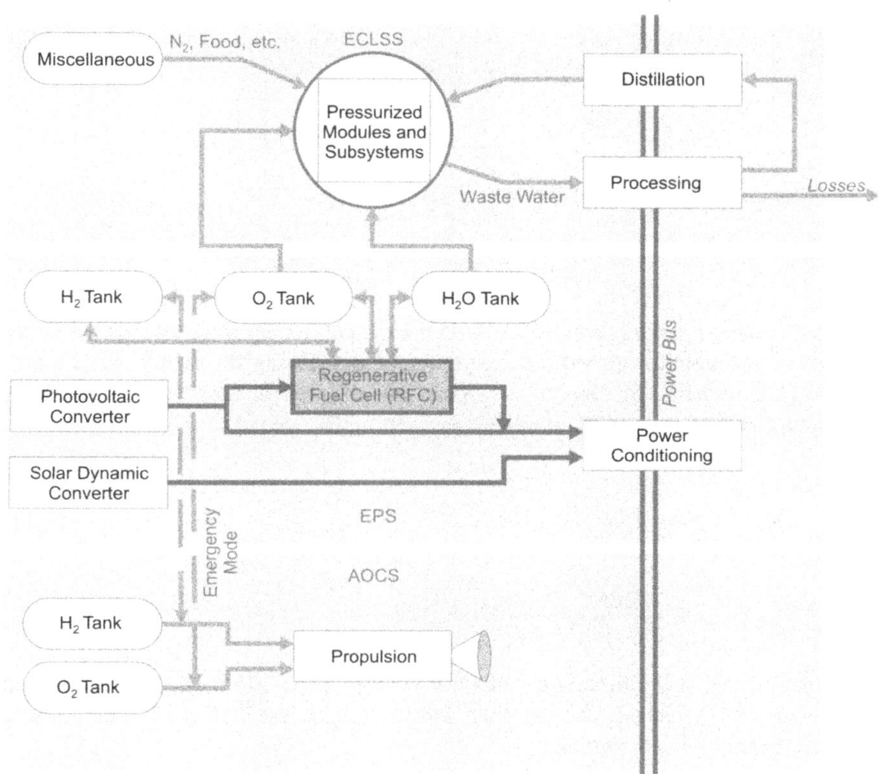

Fig. 10.3. System Layout: RFC for Energy Storage

Table 10.4. RFC Data

Fuel Cell	
Specific Mass	$m_{spec,FC} = 15$ kg/kW$_{el}$
Design Life	$DL_{FC} = 3.4$ a
Efficiency	$\eta_{FC} = 0.65$
Specific Energy	$e_{FC} = 2.85$ kWh$_{el}$/ kg$_{H_2O}$
Electrolyzer	
Specific Mass	$m_{spec,EL} = 3$ kg/kW$_{el}$
Design Life	$DL_{EL} = 1.7$ a
Efficiency	$\eta_{EL} = 0.8$
Specific Energy	$e_{EL} = 5.5$ kWh$_{el}$/ kg$_{H_2O}$

Power System. As in the case of the batteries, the fuel cell shall provide 20 kW of electrical power during eclipse periods. With the technological assumptions from Table 10.4, the following system mass is directly obtained:

$$m_{FC} = \frac{P_{base}}{2} \cdot m_{spec,\,FC} \cdot n_{sets} = 900\ kg \tag{10.11}$$

In the process, the number of sets was determined by:

$$n_{sets,\,FC} = \frac{t_{mission} \cdot \tau_{ecl}}{DL_{FC}} = 2.29 \Rightarrow 3 \tag{10.12}$$

The electrolyzer must be designed in such a way that during solar periods the entire amount of propellant required for the power supply during the eclipse can be produced. Thus, during the electrolysis operation, the device receives the following electrical power:

$$P_{EL} = \frac{P_{base}}{2 \cdot \eta_{EL} \cdot \eta_{FC}} \cdot \frac{t_{ecl}}{t_{Sun}} \tag{10.13}$$

$$= \frac{20kW}{0.8 \cdot 0.65} \cdot \frac{\tau_{ecl}}{1 - \tau_{ecl}} = 24.6\ kW$$

If a conventional low temperature technology with a specific mass of 3 kg/kW and a life span of 1.7 years is chosen, then for an eclipse factor of 0.39, the resulting amount of required sets will be

$$n_{sets,\,EL} = \frac{t_{mission} \cdot \tau_{Sun}}{DL_{EL}} = 7.18 \Rightarrow 8 \tag{10.14}$$

and the life-cycle mass of the electrolysis unit will be:

$$m_{EL} = P_{EL} \cdot m_{spec} \cdot n_{sets} = 590\ kg \tag{10.15}$$

During operation of the RFC, the following amounts of water are put through during each charge and discharge cycle:

$$m_{H_2O} = \frac{P_{EL} \cdot t_{Sun}}{e_{EL}} \tag{10.16}$$

$$= \frac{24.6kW \cdot 0.9343h}{5.5kWh/kg_{H_2O}} = 4.2\ kg$$

In a similar manner as with the non-integrated systems, the required power of the PV-system calculates to:

$$P_{PV} = \left(\frac{P_{base}}{2} + P_{EL}\right) \cdot \frac{BOL}{EOL} = 57.4\ kW \tag{10.17}$$

Accordingly, the system mass of the PV system rises to 1290 kg with a collector surface area of 430 m^2.

Thermal Control System. The waste heat flows to be removed result from

$$\dot{Q}_{FC} = \left(\frac{1}{\eta_{FC}} - 1\right) \cdot P_{FC} = 10.8\ kW \tag{10.18}$$

for the fuel cell, and

$$\dot{Q}_{EL} = \left(\frac{1}{\eta_{EL}} - 1\right) \cdot P_{EL} = 6.2\ kW \tag{10.19}$$

for the electrolyzer. Since the fuel cell and the electrolyzer are not operated at the same time, only the greater heat load has to be considered, in our case 10.8 kW for the fuel cell. This leads to a radiator surface area of 15.1 m^2 and mass of 75.4 kg.

Orbit Control. Lastly, the enlarged collector surface has to be included into the propellant balance for orbit control. Due to the overall aerodynamic surface of 648 m^2, a propellant requirement of 37031 kg results over the operational lifetime.

Balance for a 20-Year Operating Period. In the remaining subsystems, there will be no change in the system mass with respect to the non-integrated configuration. The mass balance for a 20-year period is compiled in Table 10.5.

Table 10.5. System Budget: RFC for Energy Storage

EPS	Photovoltaic	1290 kg
	Solar Dynamic	2232 kg
	Fuel Cell	900 kg
	Electrolysis Unit	590 kg
ECLSS	Distiller	303 kg
TCS	Radiator	75 kg
AOCS	Orbit Control	37031 kg
Total		**42422 kg**

With this variant, lower system masses actually result, roughly 15% less due to the replacement of batteries by RFCs; however, this savings is overcompensated through the propellant requirements for orbit reboost, so that over the entire life span of the station, an additional 2000 kg of supplies needs to be provided. However, the following advantages can be found:

- Additional safety in case of loss of the energy supply (possible use of ECLSS gases for emergency power supply)

- Additional safety in the case of logistics problems: If, for example, the resupply vehicle were grounded, propellants for the essential orbit reboost could be obtained by water electrolysis.
- When compared with battery systems, an important advantage of the RFC consists of the fact that the stored energy does not increase linearly proportional to the system mass. A capacity increase is possible through overfilling or by bringing in additional H_2/O_2 tanks. Also, the control of the solar generators' initial output is simplified, since output peaks no longer have to be disposed of in dissipation resistors, but rather can be stored through increased propellant production.
- Lastly, more flexibility in nominal operations is achieved by the RFC. With appropriate refueling of the gas buffers, an additional 20 kW of power is available in the sunlit phase so that, for example, special payloads can be operated. By oversizing the fuel cells, more peak output could be produced without having to enlarge the solar generator.

This simple example illustrates that even if the fix system masses are not necessarily reduced by implementing synergisms, substantial advantages can still be attained for the flexible and safe operation of the station.

10.4.3 Regenerative Fuel Cell for Pollutant Filtering

In addition to the use of the RFC unit for energy storage, the splitting and synthesis of water in the unit can be used for the processing of waste water. In this case, the waste water is fed into the electrolysis unit, while purified water is taken from the fuel cell (Fig. 10.4). The task of energy storage for station supply during shadow phases shall be performed, as in the aforementioned case, by the RFC unit. This section describes an additional synergistic linkage of the life support and power subsystems.

As a basis for this example, we choose an integrated water cycle (see Chapter 4 on the "Environmental Control and Life Support System"). The entire waste water amount produced daily (96.5 kg) shall be processed in the RFC unit. The distillation unit for water processing, mentioned in Sect. 10.4.1, can thereby be omitted; however, in this case, the electrolysis and fuel cells must be adjusted for the waste water mass flow. In the last section, the sizing of the RFC for energy storage resulted in a water throughput of 4.2 kg per cycle or 66 kg per day. The water throughput must thus be increased by an additional m_w = 30.5 kg per day for the processing of all waste water.

Power System. Due to the losses in the RFC, the energy required for the conversion of the additional amount of water of m_w = 30.5 kg per day into hydrogen and oxygen calculates to:

$$\begin{aligned} E_w &= m_w(e_{EL} - e_{FC}) \\ &= 30.7\ kg \cdot (5.5 - 2.85)\ kWh/kg = 81.4\ kWh \end{aligned} \tag{10.20}$$

This energy must be additionally delivered during the solar period by the PV system. However, as the distillation unit was removed, its power requirements of

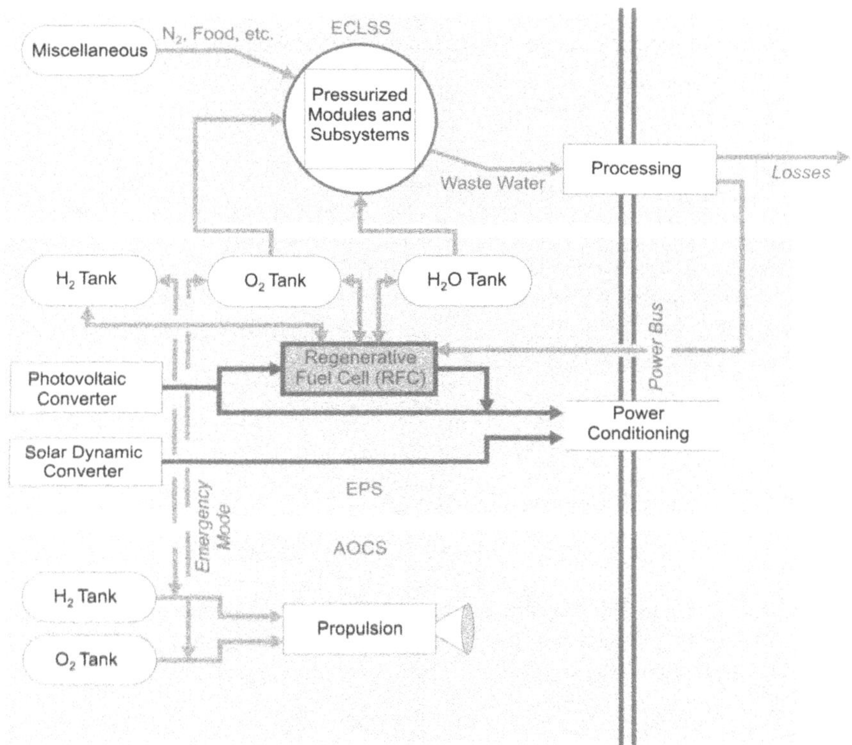

Fig. 10.4. System Layout: RFC for Energy Storage and Pollutant Filtering

P_{dist} = 0.7 kW can be subtracted from the PV design power. From the power balance for one day we obtain the new design performance of the solar generator:

$$P_{PV} = \left[\left(\frac{P_{base}}{2} - P_{dist}\right)\left(1 + \frac{1}{\eta_{FC} \cdot \eta_{EL}} \cdot \frac{\tau_{ecl}}{1-\tau_{ecl}}\right) + \frac{E_w}{t_{Sun}}\right] \cdot \frac{BOL}{EOL} \tag{10.21}$$

$$= \left[(20\ kW - 0.7\ kW)\left(1 + \frac{1}{0.65 \cdot 0.8} \cdot \frac{0.39}{0.61}\right) + \frac{81.4\ kWh}{14.64\ h}\right] \cdot 1.286$$

$$= 62.5\ kW$$

This corresponds to a solar generator mass of 1405 kg with an active collector surface of 468 m^2.

The electrolysis cell has to be designed in such a way that the total amount of 96.5 kg of water per day can be processed during the Sun phases. We obtain:

$$P_{EL} = \frac{m_{H_2O} \cdot e_{EL}}{t_{Sun}} = \frac{96.5\ kg \cdot 5.5\ kWh/kg}{14.64\ h} = 36.3\ kW \tag{10.22}$$

$$m_{EL} = P_{EL} \cdot m_{spec} \cdot n_{sets} = 871\ kg$$

The fuel cell capacity does not necessarily have to be extended. For the management of the additional 30.5 kg per day it is enough to operate the fuel cell for a longer period of time during the Sun phase of the orbit. This, however, decreases the effective life span and increases the number of sets which are required during the mission from three to four in our example:

$$n_{sets} = \frac{t_{mission} \cdot \tau_{ecl}}{DL_{FC}} \cdot \frac{96.5\ kg_{H_2O}}{65.8\ kg_{H_2O}} = 3.3645 \Rightarrow 4 \tag{10.23}$$

This results in an equivalent system mass of:

$$m_{FC} = P_{base} \cdot m_{spec,\,FC} \cdot n_{sets} = 1200\ kg \tag{10.24}$$

The alternative would be an increase of the FC output by a factor of $(96.5/66=)$ 1.462, thereby causing it to be 29.24 kW, in order to handle all of the water accumulated during the eclipse. This, however, leads to an equivalent mass of 1316 kg.

Thermal Control System. Also in this case, the synergistic effects on the thermal control and power systems should be estimated. If the electrolyzer and the fuel cell work simultaneously, the thermal control system will have to be able to manage both heat loads at the same time by

$$\begin{aligned} \dot{Q}_{RFC} &= \left(\frac{1}{\eta_{EL}} - 1\right) \cdot P_{EL} + \left(\frac{1}{\eta_{FC}} - 1\right) \cdot P_{FC} \\ &= \left(\frac{1}{0.8} - 1\right) \cdot 36.3\ kW + \left(\frac{1}{0.65} - 1\right) \cdot 20\ kW = 19.8\ kW \end{aligned} \tag{10.25}$$

which corresponds to a radiator mass of 139 kg and a radiator surface of 27.8 m^2.

In the case of an increased fuel cell capacity, the electrolyzer and the fuel cell again do not have to be operated at the same time, which would result in their respective waste heat fluxes occurring at different times. Consequently, the thermal control system may be designed for the higher heat load, in this case that of the fuel cell:

$$\begin{aligned} m_R &= \left(\frac{1}{\eta_{FC}} - 1\right) \cdot P_{FC} \cdot m_{spec,\,R} \\ &= \left(\frac{1}{0.65} - 1\right) \cdot 29.3\ kW \cdot 7\ kg/kW = 110\ kg \end{aligned} \tag{10.26}$$

But since mass savings with respect to the radiator cannot compensate for the increased fuel cell mass when implementing the second strategy (see previous subsection), the first option is chosen.

Propulsion System. Based on the space station's total aerodynamic surface of 686 m^2, the required propellant for orbit control turns out to be 39212 kg over 20 years of operation.

Table 10.6. System Balance: RFC for Energy Storage and Pollutant Filtering

EPS	Photovoltaic	1405 kg
	Solar Dynamic	2232 kg
	Fuel Cell	1200 kg
	Electrolysis Unit	870 kg
TCS	Radiator	139 kg
AOCS	Orbit Control	39212 kg
Total		**45058 kg**

Balance for a 20-Year Operation Period. Compared to the exclusive use of the RFC unit for energy storage, the balance (displayed in Table 10.6.) has shifted again to the disadvantage of the system mass.

But this again contrasts to the following more relevant synergistic advantages:

- The intensive purification effect of the electrolysis process was used for the life support system.
- At the same time, all synergistic advantages concerning the flexibility, safety and reliability of the station's operation discussed in the previous example are still valid. Due to the slightly higher amount of water stored aboard the station, the energetic flexibility of the station is increased.

However, the technological problems linked to the use of an RFC as a water processor were left out up to this point, especially those regarding the electrolysis of waste water (contamination of membranes and electrodes). Probably, at the current level of technology, a partial integration of the water cycle would be appropriate, e.g. by processing only the condensed water from the life support system.

10.4.4 Electrolytically Produced Propellants

Another obvious possibility for the synergistic couplings of subsystems is to obtain the required amount of propellants for orbit control by electrolysis in the already existing regenerative fuel cells (Fig. 10.5).

When scaling the subsystems we are confronted with the problem that the design parameters depend on each other in a cyclical way: the propellant requirement for

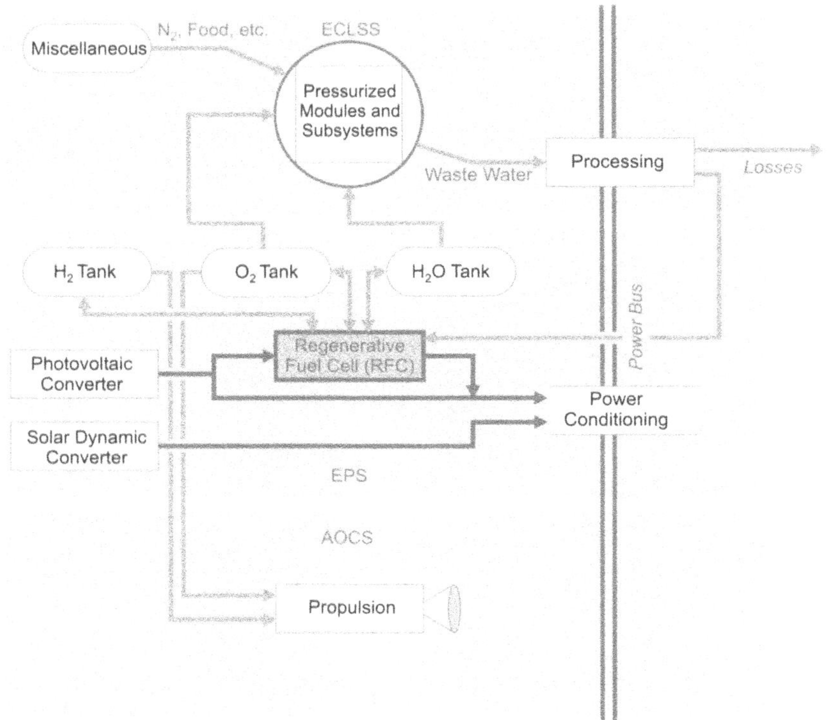

Fig. 10.5. System Layout: Electrolytically Produced Propellants

orbit control is dependent on the size of the aerodynamic surfaces and thereby on the collector surface of the photovoltaic generator. This generator needs to be proportioned in such a way that the electrolysis cells can split a sufficient amount of water into propulsion gases. Again, a reference configuration will be varied in order to find a solution. In this case, this reference configuration will be the one described in the previous example (RFC for energy storage and contaminant filtering).

Power Supply and Thermal Control Systems. Based on the results from the reference calculations, a design propellant requirement of 39500 kg over 20 years will be assumed. This corresponds to a daily H_2O requirement of 5.4 kg, which must be electrolytically processed for propulsion purposes.

According to the following equation, the additional power requirement for the electrolysis process calculates to:

$$P_{EL,prop} = \frac{m_{H_2O} \cdot e_{EL}}{t_{Sun}} = \frac{5.4\ kg/d \cdot 5.5\ kWh/kg}{14.64\ h/d} = 2\ kW \tag{10.27}$$

A total of (24.6 kW + 2 kW =) 26.6 kW would have to be provided for electrolysis for energy storage and propellant preparation. However, as it is irrelevant whether the propellant is obtained from potable water or from waste water, the design capacity of the electrolysis cell from the previous example (36.6 kW) can remain unchanged. This assumption is valid as long as the water requirement for energy storage and propellant production does not exceed the amount of polluted water to be filtered. In the latter case, caused by the cyclic dependencies among propellant needs, photovoltaic surface area and electrolysis units, an iterative approach to find a viable design would have to be used.

Orbit Control. Up to this point the propulsion system has been operated with incomplete combustion, i.e. with a mass ratio of $m_{H_2} : m_{O_2} = 1:4$, in order to attain maximum specific impulse.

A quick estimate, however, shows that this does not lead to a desirable result with regard to electrolytic propellant production:

The estimated 39500 kg of propellant would contain 7900 kg of hydrogen, which would require 71100 kg of water for its electrolytic production. Correspondingly, 63200 kg of oxygen would result from the electrolysis, from which, however, only 31600 kg would be required for propellant purposes. Even after deducting the metabolically bound oxygen of 0.11 kg per crew member per day, an oxygen surplus of almost 28400 kg would remain over the course of the design life of 20 years. As long as the oxygen cannot be used elsewhere (e.g. refueling of transfer vehicles), this option is unfavorable with regard to the supply levels.

For that reason, the following balance will be based on a stoichiometric combustion, with the slight decrease of the specific impulse being neglected within the framework of this estimate.

Balance for a 20-Year Operating Period. Compared to the previous example, no change in the mass balance occurs under these assumptions. As already mentioned, this is effective as long as the amount of waste water is sufficient for energy storage and propellant production. Again, synergistic advantages can be stated as follows:

- The logistics are significantly simplified, since now only water and not hydrogen and oxygen in gaseous or cryogenic form need to be supplied for propulsion purposes. The corresponding mass savings for tanks, valves and feed lines in the resupply vehicles have not been considered in this mass budget.
- Through the additional integration of the power, life support, and propulsion systems, the total system is further simplified, with a simultaneous gain in redundancies.

10.4.5 Safety and Reliability

The safety and reliability of a system cannot be directly assessed by merely considering the system parameters, as it is possible with the system mass criterion. In a detailed analysis, the failure probabilities of the individual components would have to be estimated, and these would then have to be projected onto the corresponding values for the entire system via the established interactions.

Under the simplified assumption that obtaining redundancies increases safety and a higher number of components reduces it, a comparative qualitative statement about system variants can be made. The number of redundancies (here defined as links between subsystems) and the number of components can be taken straight from the system circuit diagrams. If we define a safety factor according to the equation

$$f_s = 1 + \frac{\Delta R - \Delta K}{\Sigma K} \tag{10.28}$$

with ΔR = Growth of the number of redundancies with respect to a design with separated subsystems,
ΔK = Growth of the number of components with respect to a design with separated subsystems,
ΣK = Total number of components,

a rough comparison of three examples is possible. The safety factor for the non-integrated system is fixed at 1. The number of its components corresponds to the number of component boxes in Fig. 10.2, and hence amounts to 13. Redundancies which are based on synergistic couplings are not visible.

In the third example (electrolytic water preparation), for instance, the number of components also amounts to 13 while the number of redundancies has grown to 3, which is due to common oxygen and water tanks for EPS and ECLSS and to the employment of RFC for energy storage and water preparation. With K = 13, R = 3 and K = 0 the safety factor calculates to 1.23. The results for all examples are compiled in Table 10.7.

Table 10.7. Safety Factors for all Four Examples

Example	ΣK	ΔK	ΔR	f_s
1	13	0	0	1
2	14	1	2	1.07
3	13	0	3	1.23
4	11	-2	10	2.09

10.5 Summary

Figure 10.6 and Fig. 10.7 allow a comparison of the results of the mass balances and safety factors. In each case, the stated results are with reference to the non-integrated system.

We see that – even though total system mass increases to some extent – significant improvements in system safety and also in operational flexibility result from the integration of subsystems aboard a space station. Especially peak loads in both the power and life support subsystems can thus be covered.

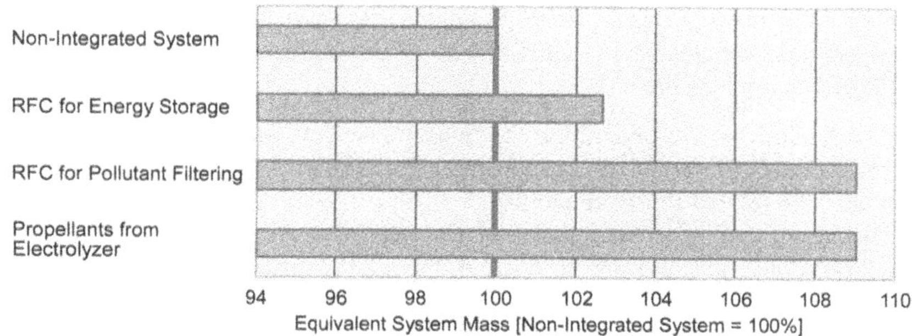

Fig. 10.6. Synopsis of Mass Budgets

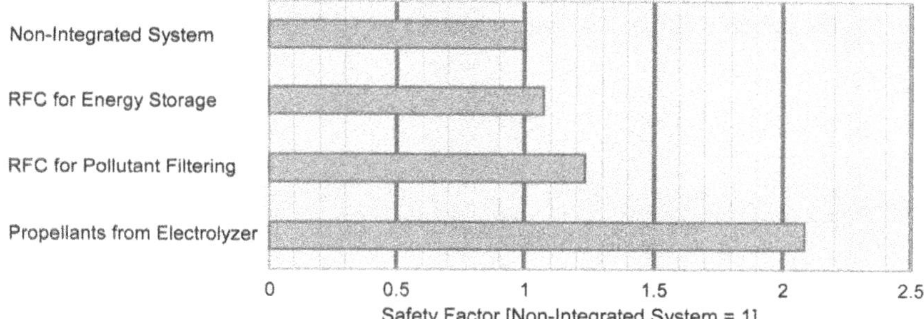

Fig. 10.7. Synopsis of Safety Factors

A standardization of components will also be of interest for the system operator since the total number of spare parts that have to be held in storage could be reduced. For that reason, and since the number of species to be handled decreases, the logistics are simplified. A further consequence is an increased maintainability of the station, which reduces the workload of the crew during operation and reduces training expenditures during preparation.

The significance of synergy for the system design process lies in a less complex hardware layout and in the possibility of creating redundancies without mass increase through use of common species. This, combined with a lower susceptibility to failure due to the reduced number of components and the resulting lower system complexity, increases safety and reliability. On the other hand, interfaces between subsystems may become more complex and the boundary between subsystems can change depending on the operational mode. This will require more flexibility and demand a global perspective on part of the designer.

Nevertheless, the decision as to which synergisms are to be used for a specific systems design must be made in each individual case, taking into account system mass and operational requirements. Only the development and comparison of different concepts will make such a decision possible.

To arrive at new options it has been realized that advanced technologies must be developed, thus enhancing the capabilities of a space station and reducing costs [ERTD 96]. Examples are:

- Solar dynamic power generation for "high power/low drag" energy generation
- Hydrogen-oxygen propulsion to eliminate the need to transport hydrazine propellant and the related safety concerns
- Resistojets for supplemental propulsion
- Electric propulsion for drag make-up (arcjet, MPD thrusters)
- Reclamation of water and oxygen in a closed loop environmental control system for use by the crew or as a propellant for hydrogen-oxygen propulsion

Many of these technologies, which often were studied in detail before being discarded or delayed indefinitely, could still be added to the ISS. Some could result in benefits to future space systems, and some could make a significant contribution toward reducing the operating costs and enhancing the capabilities of the ISS.

11 Human Factors

At the beginning of Chapter 4 which deals with life support systems, human beings have been defined as a "black box". In order to maintain their vital functions, different mass and energy flows as well as environmental conditions such as atmospheric pressure and air composition, identified in Chapter 4, must be guaranteed by the life support system.

Even if this system-level approach had been appropriate for deriving the requirements of a life support system, it would mean an inadmissible short-cut with regard to the crew's tasks. Aboard a space station, humans are at the center of the mission and have diverse functions: observer, worker, experimenter or test subject. Additionally, human beings are exposed to various factors influencing them (see Fig. 11.1).

When conceiving or planning a space station, attention must be paid to those special needs, skills and restrictions of humans that exceed physiological aspects. This must be done in order to ensure the success of the mission. These factors are generally summarized under the concept of "Human Factors".

11.1 Terms and Historical Development

Sanders & McCormick [Sanders 87] define "Human Factors" as follows:

> *"Human factors discovers and applies information about human behavior, abilities, limitations, and other characteristics to the design of tools, machines, systems, tasks, jobs, and environments for productive, safe, comfortable and effective human use."*

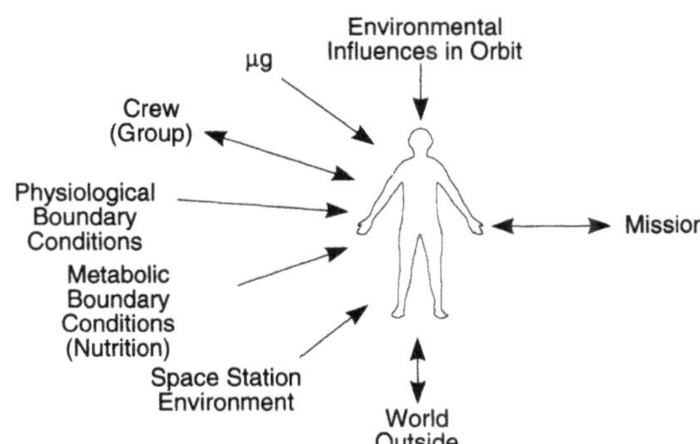

Fig. 11.1. Factors Influencing Humans Aboard a Space Station

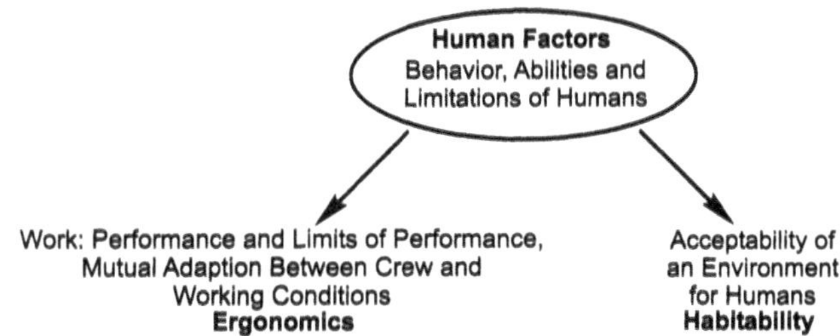

Fig. 11.2. Possibility of Structuring the Notion of "Human Factors"

The notion of ergonomics, e.g. with regard to the automobile sector or tools, is far more common than the term Human Factors. According to the Encyclopaedia Britannica [Britannica 90], Human Factors

> *"...refers to the design of machines, machine systems, work methods, and environments to take into account the safety, comfort, and productiveness of human users and operators."*

While the concept of "ergonomics" focuses on a working human being, the definition of Human Factors is wider and addresses not only the working world but also important parts of the environment in a space station (systems, environment, tasks). Likewise, it aims at a particular objective, namely the mission itself.

Figure 11.2 shows one way of structuring "Human Factors". One possible aspect is to aim at ergonomics; another is to address wider aspects of how acceptable the environment surrounding the space station is for humans. The latter is summarized under the concept of "habitability". The notions "Ergonomics" and "Habitability" also refer to different periods in time: Ergonomics mainly applies to work tasks of limited duration, whereas habitability takes into account long-term consequences for the crew aboard or in the vicinity of a space station. The latter subject will be addressed separately in Sect. 11.5.

Consequently, by taking human factors into account, the efficiency of a certain procedure can be increased, or the use of a device or tool can be made more user-friendly. Misuse of a machine must be excluded to increase the level of safety of the entire system and to prevent early tiredness of the crew. A comfortable work environment, however, also increases the operator's acceptance thereof and hence has a positive effect on the general psychological state of mind of the crew and, subsequently, on the success of the mission.

At present, Human Factors are mainly being taken into account when it comes to the relationship human working hours/costs (production in large-scale industries) or when human error, i.e. wrong decisions due to environmental stress, must be prevented, e.g. in the case of military systems, cockpits, cars. Both factors come into play in the case of space stations, where additionally the operators, i.e. the crews, are permanently exposed to the work environment for long periods of time.

Although the concept of Human Factors, though not literally mentioned, can be traced back to prehistoric times, there is a three-step process of its historic development, whose different periods run parallel to the steps of industrial development.

Phase 1 – Machine Age (about 1750–1870). Individual and, at that time, rather controversial ideas emerged on the relationship man/machine (e.g. "L'Homme Machine" by La Mettrie, 1748; "The Art of Directing the Great Sources of Power in Nature to the Use and Convenience of Man", Thomas Tredgold, Institute of Civil Engineers, 1828).

Phase 2 – Development of the Power Engine (about 1870–1945). After several individual contributions (such as investigations by F. Taylor on optimized shovel sizes), two centers formed during the 20th century dedicated to research in the field of human factors: the University of Cambridge during the 1930's, and the US Army Air Corps Medicine during the 1940's. During the Second World War, the knowledge acquired was mainly applied to the selection and education of Air Force pilots, and it was not until after the end of the war that it grew into wider application.

With the idea that work processes should not result in machine-like repetitions but adapt to human nature (N. Wiener, 1950), the knowledge about "Engineering Psychology" now could grow in almost any industrial field. The US Department of Defense then started to actively support research in the field of Human Factors. This came about due to the understanding that the increasing complexity of weapons systems used exceeded their operators' limits. It was no longer that the human being had to be adapted to the device through selection and skills, but that the device had to be adapted to human needs by designing it correspondingly (forces, illumination, acoustic, etc.).

Phase 3 – Information Age (since 1945). Up to this point, Human Factors had only covered the investigation of manual and receptive processes. Due to steadily increasing development in computer technology and the appliance of artificial intelligence as well as knowledge-based systems, the range has been broadened to the planning of a cognitive process. The design of the whole complex process of transmitting information, control and communication, all taking place at one single place of work, is being investigated.

Future development in the field of Human Factors will mainly address differences in the personality of individuals and the improvement on teamwork by taking into account aspects of group dynamics. Certainly, the smooth integration of computer technology into personal life, already announcing itself, will be one of the topics involved.

11.2 Humans in Space

At the starting point of crewed space missions, weightlessness was thought to have catastrophic consequences for the human organism. In the meantime, stays in space for durations of up to a year have become absolutely common. What is true, however, is whenever leaving the familiar field of Earth's gravity, human beings are af-

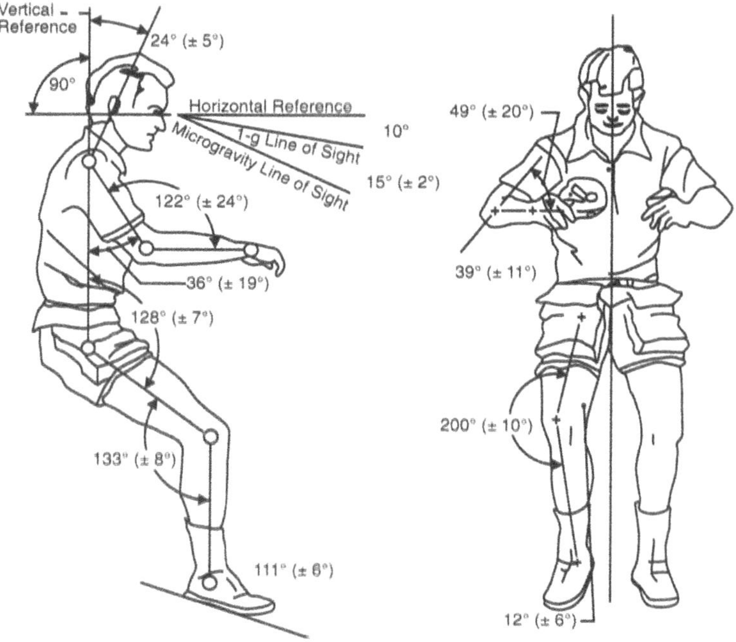

Fig. 11.3. Neutral Posture in Microgravity [ESA PSS-03-07]

fected by considerable changes, as are presented in the following sections (see also Sect. 7.3).

Neutral Posture. In microgravity, a bent posture will appear due to the natural tonus in joints, muscles and tendons deviating considerably from the neutral posture when sitting or standing in gravity (cf. Fig. 11.3).

The most important changes are the following:

- The actual body height is reduced combined with an up-shift of the human center of gravity.
- The legs are in a position somewhere between sitting and standing, the sole of the foot is angled by 21° with respect to its position on a horizontal floor.
- Compared to the 1g posture, the line of sight is angled downwards by about 15°.
- In a neutral posture, the shoulders are elevated, and the arms are slightly bent.

For these obvious reasons, it is absolutely necessary that the design of work stations, tools or furniture must correspond to this special posture.

As far as posture is concerned, it has to be considered that (unlike other activities in a 1g-environment) certain activities or movements are more difficult to perform. For instance, bending forwards is more strenuous, as gravity no longer pulls the body down and hence, the abdominal muscles are used more. The same applies to sitting straight and standing in a 1g-fashion when performed in microgravity.

Impairment of the Sense of Balance. During the first days in space, the sense of balance may be influenced. Minor symptoms are moving or posture illusions as well as dizziness. If, additionally, symptoms like bouts of perspiration, nausea and vomiting occur, this disorder is called "Space Adaptation Syndrome" (SAS). Similar to kinetosis on Earth, this disorder is supposed to result from a sensorial conflict among visual stimulus, senses of touch and distorted signals of the vestibular apparatus.

Approximately 50% of the astronauts are more or less affected by SAS. However, after a maximum of 3–5 days in space, the process of adaptation is completed, making the crew virtually immune to this disorder. During the process of adaptation, however, the performance of the crew and hence the entire mission may be considerably impaired. Medications such as Scop-dex, a combination of scopolamine and dex-amphetamine, can suppress most of the SAS symptoms, though [Kohl 86].

Fluid Shift. In gravity, blood circulation and the natural tension in muscles must prevent accumulation of fluid in the legs. In missing gravity, the 1g-compensation mechanism shifts body fluid from the legs to the upper part of the body and the head. The face appears to be slightly swollen (the so-called "puffy face"), wrinkles are smoothed and thus astronauts appear to look younger (Fig. 7.28). At the same time, the volume especially of the calves is reduced and therefore, the expression "chicken legs" has been established to characterize this effect of the shift in body fluids.

Muscle Atrophy. As a consequence of the disuse in the course of the stay in space, muscle tissue is reduced. This reduction, however, can be limited by regular workouts on ergometers and treadmills.

Decalcification. Bone atrophy due to missing gravity is to be considered the most serious physiological consequence. For the time being, apart from the risks due to radiation, this is another limiting factor for long-duration missions in space. Up until today, the exact mechanism could not be completely understood. It is supposed, however, that on one hand, by the absence of mechanical strain on the body and by psychological stress, and thus a change in the hormonal level on the other, the production of bone mass is being slowed down. Recent research also discovered reduced rates of calcification with isolated bone cells in microgravity, so that, additionally, there has to be a direct influence of the microgravity environment on every single cell.

Usually, weight-carrying bones such as leg or back bones are mainly affected by loss of calcium. As far as these bones are concerned, a reduction of bone mass by up to 20% has been noticed with some astronauts. Astronauts on long-duration missions can lose about 100 mg to 800 mg of calcium per day which means that after a 8-month stay, 5% to 10% of calcium contained in the body is excreted (Fig. 7.30).

Although bone atrophy can equally be slowed down by physical training, it is nevertheless a serious problem whenever a mission exceeds six months, since even after post-mission convalescence on Earth, bone weakness may subsist.

Most of the bio-medical changes disappear after the return to Earth within a few hours or, maximum, a few weeks. When considering Human Factors, however, there are also changes under microgravity influencing the performance of a human being in a positive way:

Improved Mobility. After a short period of acclimatization, astronauts are in a position to easily bridge distances by using the newly obtained third-dimension degree of freedom. Even working positions unusual in a 1g environment (e.g. inverted standing or lying) are no longer a restriction.

Handling of Weights. The crew is able to handle and transport fairly massive objects. The precondition, however, is grips or foot loops in order to fix the astronauts in a stable position.

Improved Vision. Predominantly Russian cosmonauts reported on improved eyesight during long-duration missions, particularly in the case of Earth observation.

The main task of Human Factors Engineering (HFE) is to creatively include these changes and boundary conditions. The following section will illustrate some basic aspects.

11.3 Human Factors Engineering (HFE)

In a project of crewed spaceflight, the main purpose of HFE (which could also be referred to as applied ergonomics) is to integrate knowledge acquired in the field of Human Factors into that project. This happens on system, sub-system and module/ unit level.

11.3.1 Organization and Integration

Due to its nature, HFE can be found in all areas and at all levels of a space mission. For that particular reason, it cannot only be classified as a subordinate task within a project (cf. Fig. 11.4), but also comes up in every discipline. In this very flexible way, HFE can deal with questions on system level such as space station operations and system safety and also influence subsystem development, for instance, designing an operator's console.

During preparation and conduct of a space mission, the HF group (if existing) must keep close contact with the other organizational units. The HF group's task is to verify the correct functioning of the overall system of a space vehicle and its crew and to restore it, if necessary, by changing the Man/Machine-Interface. Usually, the HF engineer acts within a field of controversy: from the engineering department's point of view, for example, the function of a unit or the technical difficulties of realization are at the center of interest, whereas aspects dealing with Man/Machine-Communication or Habitability are only considered second priority.

But especially during the stages of development, the participation of HFE in the design of a space vehicle or device is absolutely necessary, as the most important decisions are made at that point, for example, vehicle and device configuration, interior design or division of labor between man and machine. After having deter-

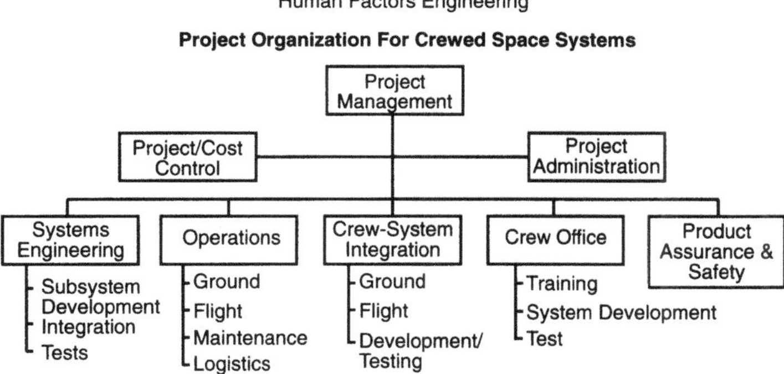

Fig. 11.4. Example of an Organizational Structure for a Crewed Space Flight Project

mined the design, the emphasis of HFE inevitably shifts to the verification of the systems as well as to mission preparation and the support of crew training.

During the mission, all systems should be permanently checked as to ergonomics and efficiency to make sure they meet the expectations. If necessary, proposals for changes in the course of the mission or in the use of devices must be made immediately in order to guarantee the mission's success. The experiences acquired during a mission must be collected and, together with new technological development, will influence future changes in the design (e.g. spare parts).

11.3.2 Methods of HFE

As already mentioned, HFE is also responsible for the design of the so-called "Man/Machine-Interface" (MMI). This interface can be either active (control devices, displays, terminals, etc.) or passive (grips, foot loops), of low complexity (e.g. a simple switch) or high complexity (computer, software). The concrete aspects that have to be taken into consideration when designing the MMI are the following:

- Anthropometrics
- Physiology and ergonomics
- Psychology
- Sociology
- Environmental conditions
- Occupational medicine
- Training
- Crew selection
- Medical support
- Preventive measures
- Performance
- Tasks and work-routines

Consequently, HFE starts with scientific investigation of human beings and their reaction to environmental influences (objects, surroundings, situations, etc.) from which conclusions can be drawn about human skills and limits, behavior and motivation. In the next step, acquired information can be systematically applied to designing products, tools or procedures as well as to the environment in which they are used by human beings. An important step is the comparative testing of different design concepts in order to ensure that they result in the desired influence. The realization of Human Factors by means of a three-step method including investigation, application and verification, comprises, apart from scientific research, mainly empirical or heuristic procedures. In this context, lessons learned during former missions or simulations play a central role.

Standardization. Part of the implementation of Human Factors Engineering is the consistent recording of experiences and their establishment in standards. As a result of Western research, particularly the NASA Standard "Man Systems Integration Standards" (NASA-STD-3000) and the ESA standard "Human Factors" [ESA-PSS-03-70] can be recommended here.

First of all these documents show in detail anthropometrical boundary conditions such as height, operator forces (Fig. 11.5) or ranges of motion. Furthermore, concrete principles of configuration for space vehicle architecture such as volume requirements per crew member or space requirements at work-desks (Fig. 11.6) are stated. This information was obtained by systematically evaluating Skylab, Space Shuttle and Spacelab missions.

Analytical Methods. Often, an efficient design of MMI is hard to establish by standardization alone. Figure 11.7 shows an example of an analytical method for designing the MMI for an experiment in the field of biomedicine.

Force in Newtons			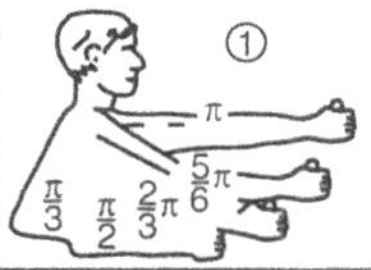 (1) Degree of Elbow Flexion (Radians)				
Motion		**Hand**	π	$\frac{5}{6}\pi$	$\frac{2}{3}\pi$	$\frac{\pi}{2}$	$\frac{\pi}{3}$
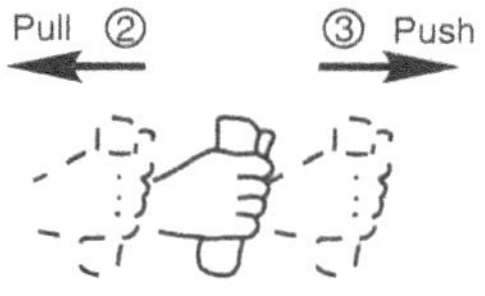 Pull ② ③ Push	(2) Pull	L	222	187	151	142	116
		R	231	249	187	165	107
	(3) Push	L	187	133	116	98	96
		R	222	187	160	160	151
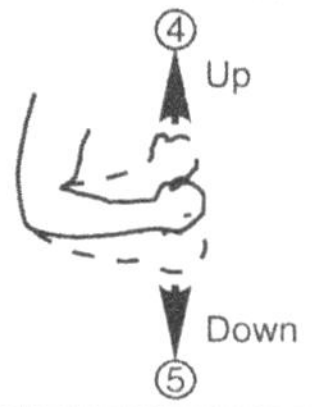 ④ Up ⑤ Down	(4) Up	L	40	67	76	76	67
		R	62	80	107	89	89
	(5) Down	L	58	80	93	93	80
		R	76	89	116	116	89

Fig. 11.5. Arm Strength Valid for 95% of the European Population (Male) [ESA PSS-03-07]

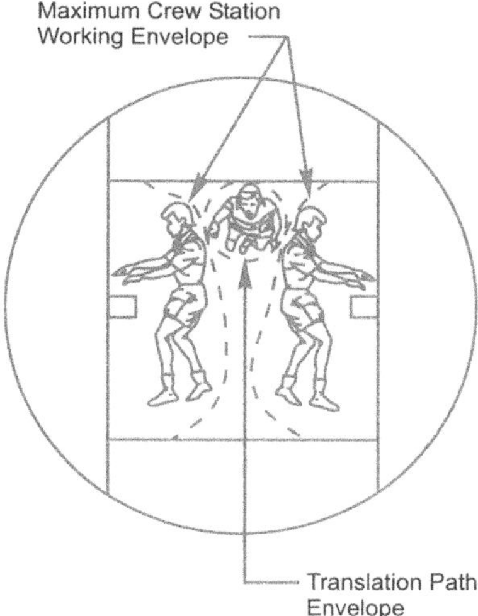

Fig. 11.6. Volume Requirements for Work Areas and Translation Paths

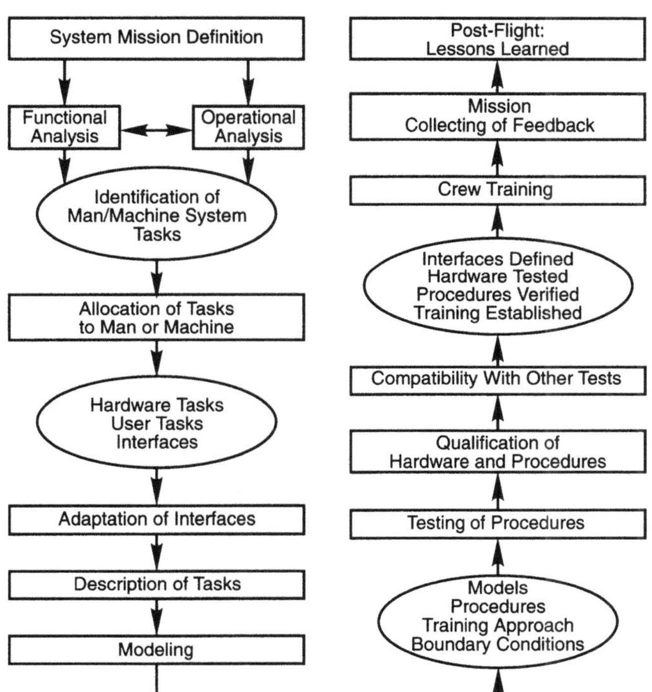

Fig. 11.7. Methodology for Man-Machine-Interface (MMI) Design [Friess 91]

In this context, the starting point is the analysis of the functions within the system man-machine. The crucial point lays in the repartition of the functions between man and machine. To underline that the discussion is hereby often more diverse than it initially appears to be, complete automation shall be taken as an example. First of all, we have to keep in mind that a fully automated system relieves the operator by 100% and hence constitutes the easiest and thus ideal solution. However, this solution stands in contrast to the experience that newly introduced all-automated systems do not routinely produce the expected results. Especially the breakdown of an all-automated system causes a considerable loss of confidence for the operator. This leads to the crew avoiding, whenever possible, the operation of devices they do not trust. And obviously, this certainly does not result in a reduction of their work load.

As a consequence, all-automation often increases the effort in training instead of reducing it: the crew must be able to operate a unit both during its automated operation as well as manually. The nominal all-automated operation leads to a lack of training in manual operation in case of malfunction; automated systems often have more functions than their conventional counterparts and thus increase the complexity of the system.

Once the functions of the hardware and those of the crew are identified, the interfaces can be adapted. Generally, first of all, the machine is adapted as much as possible to the human being. This is achieved by the following steps:

- Design of the working environment for the human being
- A clear separation of tasks between humans and machines
- Dynamic adaptation of the device to different working situations
- The operator's degree of information (intelligent, i.e. context-dependent and selective)
- Task-related controls

Yet, in order to complete the MMI, an additional effort to adapt every single operator to the machine will be necessary. This is accomplished as follows:

- Preliminary crew selection
- Bridging differences in personality
- Human ability to physiological adaptation
- Training
- Safety requirements

The ideal MMI does its best, of course, to keep this second block of adaptation as small as possible.

11.3.3 Means of HFE Support

In order to fulfil their tasks, several tools or support devices are at the HF engineers' disposal:

Computer-Aided Tools. Computer-aided tools are especially applied in the design phase. CAD-programs, for example, make the efficient design of space station architecture possible because design modifications in the framework of those iteration cycles typical of design can easily be managed. Figure 11.8 shows the mid-deck of

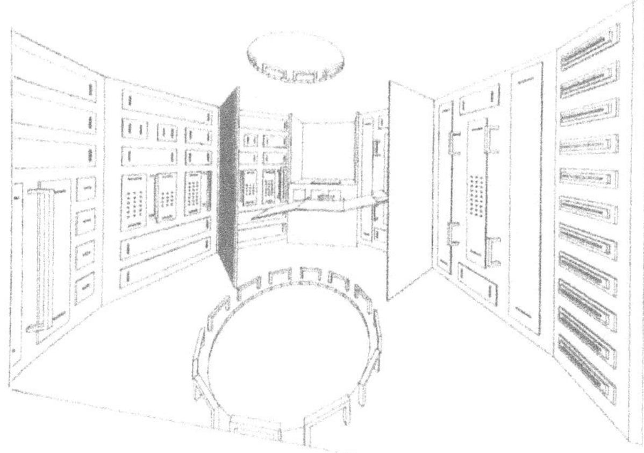

Fig. 11.8. CAD-generated View of the Mid-Deck of a Model Space Station [ESA 93]

a Skylab-like habitation module as it was set up during the ESA study "Habitability Mini-Laboratory" with the help of the CAD-system "STAR Architecture UX".

Apart from its application as a flexible design tool, the CAD program could also be applied to visualize the spatial impression, for example, on a virtual walk through a pressurized module. In addition to that, analyses of the astronaut's field of vision are possible.

Furthermore, there are also programs for graphical-interactive simulation of the kinematics of mechanical systems and their interaction with the human being. For example, the DYNAMAN program [ESA 93] set up a model of human posture and motion in a 1g and 0g environment through which it was possible to analyze procedures or MMI by simulation (cf. Fig. 11.9).

Mock-Ups. Mock-ups are used in reduced scale or on the scale of 1:1. Even if they (compared to computer-aided methods) are less flexible with regard to the design process, they allow three-dimensional representation of architecture or spatial impression in a markedly superior quality. By using adequate cameras (based on fiber-optics, for example), it is also possible to take up any desired angle of view or to take a virtual walk through the room.

In the so-called "Space Station Mockup and Trainer Facility" (SSMTF) of the NASA Johnson Space Center, there is a 1:1 mock-up of pressurized parts of the International Space Station (cf. Fig. 11.10).

This facility, first of all, is to support the different project reviews of the systems and facilitate the development and verification of devices and procedures for the crew quarters aboard the station. In this context, systems improving the standard of living, storage compartments, EVA systems, maintenance facilities, telecommunication systems, display and control systems (D&C), and data management are of particular interest. Emphasis is put on the crew training prior to a mission, as well. As soon as operation of ISS has started, the SSMTF will be used for mission support

Fig. 11.9. Computer Simulation of Activities at a Payload Rack [ESA 91]

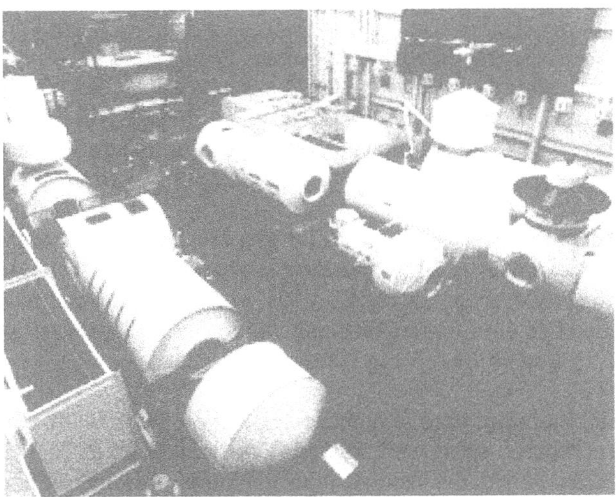

Fig. 11.10. Space Station Mock-up and Trainer Facility (SSMTF) at NASA Johnson Space Center [SSMTF 96]

in real time, e.g. when modifying and verifying operating procedures or reacting to anomalies in operation.

Every single module is equipped with the so-called "high fidelity outfit" with individual systems (depending on the requirements) mirroring either entirely or partly the functions of the original. The replicated systems include food preparation (wardroom/galley), waste disposal, hygiene, showers, sleeping areas, closets for crew equipment, safety systems, medical support systems, workstations and maintenance systems. Also the necessary basic functions of subsystems such as power supply,

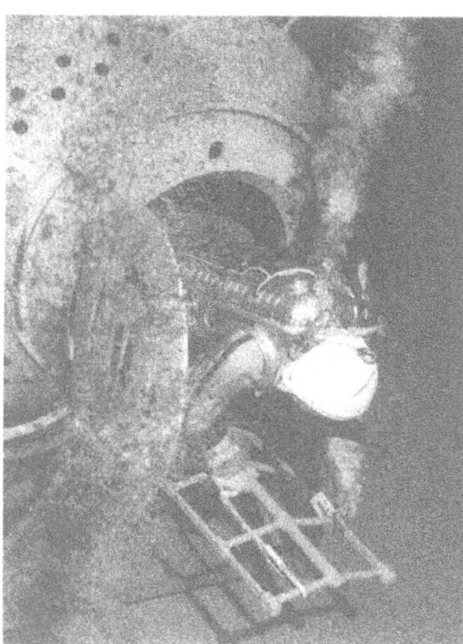

Fig. 11.11. Astronaut Leaving the Air Lock Mock-Up of the Space Station Mir [DLR 95]

life support and telecommunication are provided either identically to the original or by suitable Ground Support Equipment (GSE).

A former training model of the Mir space station can be seen in the "Europapark" theme park in Rust near Freiburg, Germany.

Neutral Buoyancy Facilities. Apart from a faithful copy of the spatial arrangement, also the microgravity environment must be simulated in order to develop, verify or optimize MMIs or work procedures. One possible option are tests in neutral buoyancy facilities. By trimming with ballast, it is possible to compensate for the weight by buoyancy. This method is mainly applied in crew training. For that reason, entire mock-ups, for example, of the Space Shuttle cargo bay or the space station Mir are installed in enormous water-filled pools, where complex installation or repair procedures are developed and practiced (cf. Fig. 11.11).

Parabolic Flights. This way of compensating for gravity for 20–30 seconds with the help of jet aircrafts has already been discussed in Chapter 8 dealing with microgravity (cf. Sect. 8.2.2). Due to comparatively frequent flights, easy access and a moderate price, parabolic flights are of great importance to HFE. Apart from astronaut training, procedures and devices are often tested as to their basic suitability with regard to application in microgravity, as shown in the following examples:

- Handling of electrical or mechanical connectors in microgravity with and without a pressurized suit
- Exchange of components such as experiment racks (Fig. 11.12)

Fig. 11.12. Exchange of a 1:1 Model of an "International Standard Payload Rack" during parabolic flight

- Testing of new installation methods, devices or equipment (crew restraints, space furniture, tools, etc.)
- Verification of operation and maintenance procedures for experiments and subsystems

Tests In Space. Sometimes it is essential that tests be conducted directly in space. Prior to the development of the International Space Station, for example, various assembly procedures for large truss structures were tested several times. Once ISS is put into operation, flight opportunities are at the crew's disposal more often and more directly. In the field of engineering science and technology demonstration, e.g. it might be considered to replace some of the test and qualification procedures on Earth (which are very complex and comprise several phases) by one single test aboard the International Space Station.

11.4 Design of a Workstation

Workstations aboard a space station are needed for various tasks, for example, to control and verify experiments or supply systems, or to operate a manipulation system outside the pressurized modules. Thus, they usually form the interface between the technical systems and the crew. The detailed design of such a work station not only depends on the changing boundary conditions of the corresponding task, the environment and the crew, but also general design guidelines and experiences must be taken into account with respect to a safe and efficient operation. These are outlined with the help of the following examples:

Anthropometrical Range. Firstly, it must be guaranteed that a workstation is height-adjustable. Figure 11.13 gives an impression of the range of heights of crew members. In Europe, for example, the minimum height of 1.46 m is defined by the fact that only 5% of all female Europeans are smaller than 1.46 m, whereas the maximum height is fixed at 1.92 m as it is exceeded by only 5% of the male population.

Apart from the adaptation to the height of the potential user group, the boundary conditions specific to microgravity already mentioned at the beginning must be considered: Direct requirements for the design of operating or working surfaces result

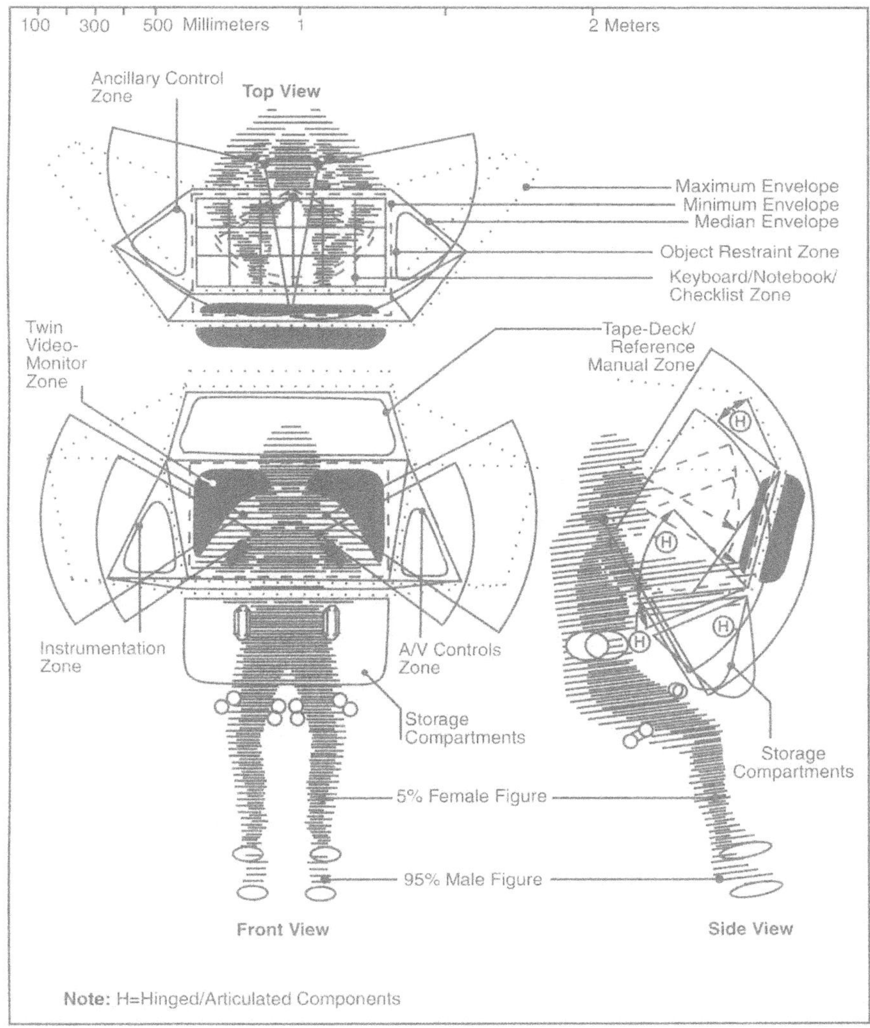

Fig. 11.13. Anthropometric Boundary Conditions for a Workstation [NASA CP 2426]

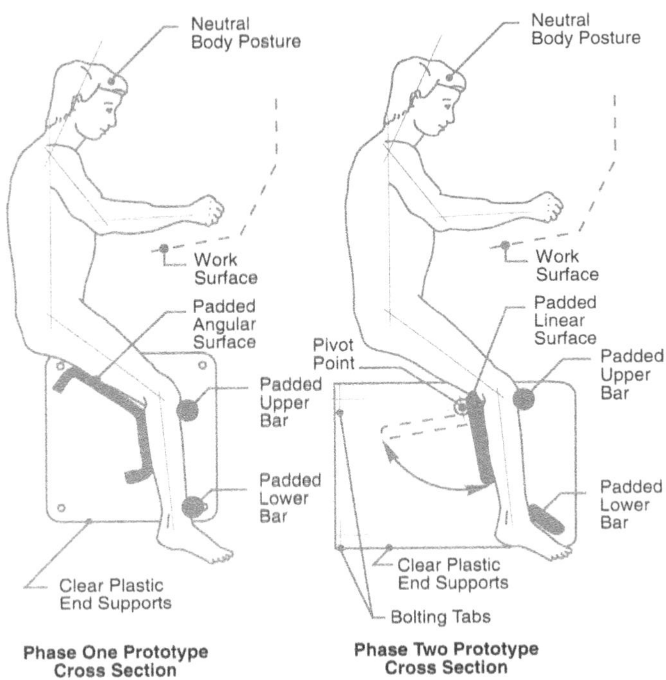

Fig. 11.14. Leg Restraint [Nixon 87]

from the altered neutral posture (cf. Fig. 11.3). The lowered field of vision likewise requires a special adaptation of displays.

Body Restraint. Various facilities such as ropes, grips and foot or hip loops can be installed to fix the operator in front of a workstation. In any case, it must be guaranteed that, especially during mechanical work, occurring forces or moments can be drawn off, and the task is not interfered with by the restraint. The user's comfort and the effort necessary to put the grips on and take them off are of great importance, as the acceptance by the crew largely depends on these factors – and so does their work efficiency. Figure 11.14 shows phases of development of a leg restraint that allows free-handed operation of workstation controls.

Before being able to design single controls, the functions and tasks have to be divided between humans and machines. In this respect, the functional analysis mentioned in Sect. 11.3.2 gives support. When separating the tasks, the special skills and limits of man and machine alike must be taken into consideration, as well. Repetitive, time-consuming functions (such as routine maintenance works, leakage testings, calibrations, etc.) as well as dangerous tasks should be translated into automated machine functions. Human beings, on the other hand, can generally make those decisions that require the overview for the overall system or intuition more efficient and safer. The same applies to the flexible or fine manipulation of experiments and devices as well as to repairing tasks.

When controls are to be designed in detail, we can distinguish between general and element-specific design guidelines. Among the general characteristics are the following:

- *Orientation.* In microgravity, the number of possible control surfaces is usually growing compared to 1g. Different "local verticals" within a workstation might lead to disorientation or slowed operation. For this reason, the orientation within a workstation should be adhered to as consistently as possible. For example, the Human Factors engineer can define horizontal or vertical reference planes or lines by corresponding coloring or shaping of the workstations. Also the corners of a window or of a display or their illumination are suitable for supporting the astronaut's orientation.
- *Consistency and Standardization.* Controls that can be found on more than one workstation should feature identical design. This improves the user friendliness and the safety level, and, at the same time, reduces the training effort of the crew. Same attention should be paid to similarities between different groups of controls, system units and vehicles. Whereas mirror-image arrangement of controls, for example, should be avoided in these cases.
- *Surfaces.* At every workstation, sufficient space for tools, manuals, portable data terminals, etc. must be provided. Here, too, special requirements due to the microgravity environment must be taken into account by providing suitable means of restraint (mainly Velcro®).
- *Decor, Coloring and Illumination.* These are to improve efficiency, safety and comfort during operation. Also in this case, numerous recommendations are already on hand at ESA PSS-03-70 and NASA STD 3000. Table 11.1 shows an extract of color ranges for different work areas.
- *Ventilation.* Chapter 4 on the subject of life support systems already dealt with the importance of cabin ventilation. The designer of such a workstation must know that this function has to be implemented as a part of the MMI.
- *Encoding.* In order to safely locate and handle equipment and experiments especially in a large orbital system, a uniform encoding system is essential. Encoding can be carried out by brightness, dimensions, surface pattern, shaping and coloring. Colored and numbered codes facilitate the location and identification of a required device. Furthermore, defined color-codes are provided to mark "front/back" orientation, mechanically movable components, or those relevant to safety.

Apart from the general requirements for the design of a workstation, there are numerous special considerations which have to be taken into account when designing the corresponding controls. At this point, we will only discuss switches and displays.

Usually, and also in this case, the most important criteria of configuration are mainly aspects of safety. Every safety-related switch must be secured against inadvertent activation/deactivation by, for example, recessing it into the front panel or by installing switch guards or covers. Additionally, the state of the switch must be clearly visible for the operator, and, when switching, there must be a noticeable feedback (for example through the action point, optical or audible signals). Already

Table 11.1. Color Recommendations for Work Areas (Selection) [ESA PSS-03-70]

Space Module Areas	Use the Colors in Small Amounts Only or as Trim																									Use in any Amount													
	Pale Pink	Reddish Pink	Red	Red Orange	Maroon	Orange	Apricot	Yellow Orange	Chamois	Yellow	Yellow Green	Green	Blue Green	Blue	Violet	Royal Blue	Purple	Magenta	Reddish Brown	Buff	Burnt Sienna	Brown	Olive Drab	Grey	Dark Grey	Champagne	Cinnamon	Beige	Salmon	Peach	Straw	Ivory	Cream	Maize	Pale Yellow	Pale Green	Pale Blue	Lavender	White
Work Areas																																							
General Workstations							★	★	★	★	★	★	★	★		★			★	★	★	★		★		★	★	★			★	★	★	★	★	★	★		★
Data Processing						★	★	★	★	★	★	★	★	★		★			★	★	★	★		★	★	★	★	★	★	★	★	★	★	★	★	★	★		★
Communications						★	★	★	★		★	★	★	★		★			★	★	★	★		★	★	★	★	★	★	★	★	★	★	★	★	★	★		★
Maintenance													★	★		★											★	★			★	★	★	★	★		★		★
Mechanical Equipment Power Generation													★	★		★			★	★	★	★		★			★	★			★	★	★	★	★		★		★
Security									★	★	★	★	★	★		★			★	★	★	★		★			★	★	★	★	★	★	★	★	★	★	★		★
Logistics									★	★	★	★	★	★		★			★	★	★	★		★	★		★	★	★	★	★	★	★	★	★	★	★		★
Administration					★	★	★	★	★	★	★	★	★	★		★		★	★	★	★	★			★	★	★	★	★	★	★	★	★	★	★	★	★		★
Service Areas																																							
Laundry						★	★	★	★	★	★	★	★	★		★				★					★		★	★	★	★	★	★	★	★	★	★	★		★
Health Maintenance										★	★	★	★	★														★	★	★	★	★	★	★	★		★		★
Assembly Areas																																							
Conference/Briefing			★		★	★	★	★	★	★	★	★	★	★		★			★	★	★	★		★	★	★	★	★	★	★	★	★	★	★	★	★	★		★
Training						★	★	★	★	★	★	★	★	★		★			★	★	★	★		★		★	★	★	★	★	★	★	★	★	★	★	★		★
Storage Areas																																							
Food Storage			★				★	★	★	★																		★				★	★	★					★
General Storage/Supplies			★									★	★															★				★	★	★					★

Notes:

The Use of Saturated (High Chroma) or Dark (Low Value) Colors Shall be Restricted to Small Amounts

Black is not Recommended for Use Anywhere

when choosing a control, utmost attention must be paid to the required operating force, the necessary precision in operating, or the field of application (coarse or fine adjustment). This is emphasized by Table 11.2 using the example of computer input devices.

Finally, a workstation must not be regarded as being isolated from a space station's environment. Aboard a space station, there is a wide range of activities which, naturally, are mainly performed simultaneously in a very confining work environment. For that reason, not only functional points such as access; transfer times; logical sequence of activities; contamination (for example food or body hygiene); and restrictions caused by traffic, noise, incidence of light, or vibration are of impor-

Table 11.2. Advantages/Disadvantages of Different Computer Input Devices [ESA PSS-03-70]

Advantages	Disadvantages
a) Joystick	
• Can be used comfortably with minimum arm fatigue • Does not cover parts of screen in use • Expansion or contraction of cursor movements is possible	• Slower than a light pen for simple input • Must be attached, but not to the display • Unless there is a large joystick, an inadequate control and display ratio will result for positional control • The displacement of the stick controls both the direction and the speed of cursor movement • Difficult to use for free-hand graphic input • Not good for option selection
b) Track ball	
• Ball excellent for three-dimensional rotation of objects • Efficient use of space • Allows user to concentrate attention on VDU screen • Unaffected by microgravity if properly designed	• May need two devices to accommodate handedness
c) Mouse	
• Relatively fast • Has low error rates for large targets • Allows user to concentrate attention on VDU screen	• Must be adapted for microgravity use • Requires flat work surface • Difficult to use for free-hand graphic input • High error rates with small targets • Lost time when mouse held backwards or sideways • Some training needed • Wheels slipping sometimes a problem
d) Light pen	
• Fast for simple input • Good for tracking moving objects • Minimal perceptual motor skill needed • Good for gross drawing • Efficient for successful multiple selection • User does not have to scan to find a cursor somewhere on the screen • May be adaptable to bar coding	• May not feel natural to user, like a real pen or pencil • May lack precision because of the aperture, distance from the CRT screen surface, and parallax • Contact with the computer may be lost unintentionally • Frequently required simultaneous button depression may cause slippage and inaccuracy • Must be attached to terminal, which may be inconvenient • Glare problem if pen tilted to reduce arm fatigue • Fatiguing if pen is held perpendicular to work surface • If pointing to dark area, may require user to flash the screen to find pen • One-to-one input only (zero order control) • May be cumbersome to use with alternating, incompatible entry methods, like the keyboard • Tends to be used for purposes other than originally intended, e.g. for key depression • Tends to be fragile • Hand may obstruct a portion of screen when in use • Care must be taken to provide adequate "activate" area around choice point • Cannot be used on gas panel

tance for the installation of a work station aboard a space station, but "human" aspects as well. Working at a workstation may disturb, on one hand, the privacy of a living or dining area, and on the other hand, the work itself may in turn require a certain separation for reasons of safety or confidentiality. In Fig. 11.15, different activities aboard a space station are listed in a coordinate system, divided into private/public and individual/group activities, respectively. Correspondingly, the larger the distance between two activities, the larger is their potential incompatibility.

When regarding the variety of influences and interaction which arise from the Human Factors point of view, it becomes obvious that a good Human Factors design cannot be considered isolated from hardware design. The attitude often prevailing

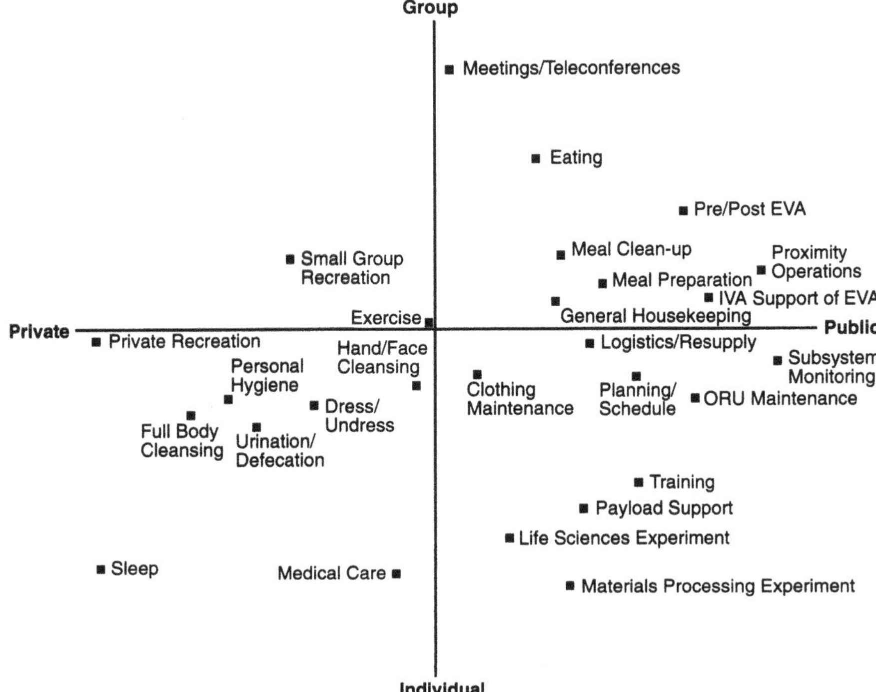

Fig. 11.15. Social Classification of Different Activities

in engineering departments that Human Factors engineering has to confine itself to the colorful design of switches, in practice often leads to complicated system modifications and additional improvements or to less than optimal user interfaces. These consequences may later on considerably influence the course of the mission. Usually, the definition of the MMI deeply influences the design of the subsystem due to the necessary distribution of the functions and the choice of controls, and, therefore, the definition of the MMI must be seen as the central requirement to the design, particularly in the case of space stations.

11.5 Habitability and Crew Performance

An application that exceeds the heretofore discussed area focussing on the direct work environment is described by Human Factors Engineering with the notion of "habitability". This general term describes a certain degree of acceptability of an environment to a human being. Habitability includes all qualities that medically and psychologically support a crew's stay aboard a space station, both at work and during leisure.

At the lowest level, the degree of habitability is determined by direct environmental factors such as temperature, humidity, illumination, noise level and food

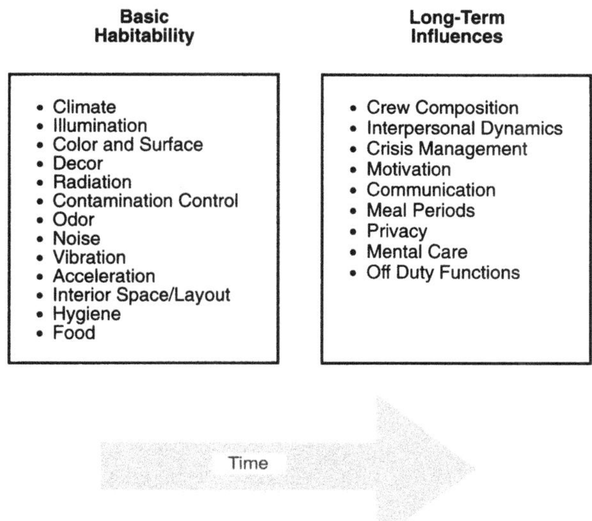

Fig. 11.16. Factors of Habitability

quantity/quality (cf. Fig. 11.16). All these factors influence the well-being of the crew particularly by supporting physiological functions. With increasing duration of a mission, the influence a stay in an isolated and limited room has on the crew, grows drastically. Habitability is now additionally determined by the maintenance of the crew's health and psychological balance.

Optimal utilization of all available space aboard a space station helps to improve habitability. The interior design with visual width, a separation of the modules into public and private areas, and the possibility for the crew to change the interior layout support the reduction of stress situations in narrow rooms considerably.

The degree of habitability directly influences a factor which is important for the mission: crew performance. This term includes the crew's ability to fulfill its tasks correctly and in a reliable manner. Table 11.3 presents a list of factors influencing the crew performance, divided into three areas: physiological conditions, working capacity and psycho-social aspects.

As crew performance during the mission cannot directly be measured quantitatively, it has to be discussed by physicians, HF engineers and program coordination groups who constantly evaluate the course of the mission. In the same way in which a defective subsystem can be repaired in orbit, it has to be decided for the subsystem "crew", which countermeasures can be taken if the crew does not meet the required performance objectives.

Russia, in particular, acquired considerable experience in the field of long-duration missions and was the first nation to take the above-average growing psychological strain of its crews into account. These experiences constituted a valuable basis for following missions. However, projects such as the International Space Station or a Mars mission will introduce additional complications, namely because of larger

Table 11.3. Factors Influencing Crew Performance

Physiological State	Work Capacity	Psycho-Social Aspects
• Circulation • Muscles and Skeletal System • Vestibular System • Sight • Immune System • Dose of Radiation	• Ergonomics under Microgravity • Wake/Sleep Cycles and Work/Rest Cycles • Workload Distribution • Isolated Work Environment • Desynchronization (Loss of External Day/Night Cycle as Time Indicator) • Disturbed Sleep	• Group Composition - Astronaut Image and Function - Sex, Age and Cultural Background - Personal Attraction - Emotional Stability - Readiness to Cooperate - Social Adaptability - Similarities with other Members of the Group - Degree of Complementing other Members of the Group - Size and Agreeable Nature of the Group • Interpersonal Dynamics - Leadership - Team Spirit - Conformity - Group Performance - Changes of Group Behavior over Time • Behavior during Crisis - External Influences (Training for Dangerous Situations, Response to Threat, Post-Situational Reactions and Review & Evaluation) - Internal Crises (Hallucinations, Drug Abuse, Grief, Death aboard the Station) • Motivation • Communication • Privacy • Leisure and Entertainment

crew size (four to eight persons). A new factor in this field on which there has only been little research performed, will increasingly gain influence; that factor being group dynamics.

11.6 Astronaut Selection

In spite of past accomplishments easing the physiological and psychological load on the crew, and regardless of potential human factors engineering achievements in the foreseeable future, astronauts must still be carefully selected from the large group of applicants to ensure they have "the right stuff" required for making their mission a success. Traditionally, applicants outnumber actual astronaut positions by about 100:1, and even with the increased number of astronauts required for the International Space Station program, that ratio is not going to be reduced significantly.

This final section of Chapter 11 will give a brief overview of astronaut tasks and duties, present exemplary astronaut selection (and rejection) criteria, and describe the actual selection process used to match the right person with the job. Much of the following is based on European Space Agency guidelines [ESA 90], which in turn closely follow NASA and RSA equivalents.

11.6.1 Astronaut Tasks and Duties

Candidate astronauts must be able to cope, in an efficient and reliable way, with a high level of occupational demands which they may not have previously faced in their careers. Their activity during training and space flight will have to be conducted under a high degree of stress and in close cooperation with other crew members. Astronauts are expected to be generalists, with a quick understanding of specialized topics and a large background of scientific and technological experience.

Specifically, astronauts must be able to handle the following tasks which may occur during a flight:

- Scientific payload operations: activating, using and deactivating laboratory equipment, communicating with ground-based project scientists, conducting experiments
- Flight safety and contingency operations
- Spacecraft and space station maintenance
- Extravehicular activities

In addition, during both mission preparation and post-mission activities, astronauts must be ready to withstand rigorous physical training and demanding scientific training, participate in systems and mission planning and development, and interact with the public, especially through the media.

11.6.2 Selection Criteria

Applicants for an ESA astronaut position must be nationals of an ESA member state or of an ESA associated state involved in an ESA Manned Space Programme. NASA astronauts must be US citizens. Although no mandatory age limits will be applied, the preferred range of age is 27 to 37. Applicants should be able to speak and read English, which is the international *lingua franca* of the aerospace world.

Applicants must be prepared to provide a full medical history, covering all current or past physical and mental problems, and permit the collection of further information as deemed necessary by the examining medical body. They must also be prepared to participate in extensive medical screening, including internal and x-ray examinations. In addition, certain tests will be performed to evaluate the applicant's (muscular, cardiovascular and vestibular) body system using facilities such as rotating chairs, centrifuges, pressure chambers, and special airplanes.

Comprehensive questionnaires force the applicants to give detailed insights into their life, e.g. by asking them to describe their individual working, sleeping and eating habits. They are also asked to indicate all cases of mental or physical illness in their family tree.

Medical Requirements

Applicants must comply with medical criteria (see also Sect. 11.6.3). They should have a satisfactory medical history and be in a sound state of health, have normal weight, and be of normal psychiatric disposition.

Psychological Requirements

General characteristics expected of the applicant include:

- Outstanding logical reasoning and mental calculation
- Good memory and power of concentration
- Rapid perception and selective attention
- Good spatial orientation
- Manual dexterity, fast reaction
- Sense of responsibility, alertness
- Ability to work in a team
- Communication skills, including languages
- Decision-making ability, methodical approach to tasks
- Sense of humor, enjoyment of physical activities

The applicant's personality should be further characterized by high motivation, mental flexibility, gregariousness, empathy with fellow workers, low level of aggressiveness, and sound emotional stability.

Previous practical experience of the applicant in certain areas is also deemed desirable. Examples include:

- High-risk or emergency situations, including situations dealing with dangerous substances, or in dangerous locations
- Execution of procedural tasks
- Management of complex systems
- Teamwork
- Communications and robotics operations
- General practical scientific and technological work, including computer-based work
- Survival training

The existence of certain mental/psychological traits or conditions may exclude an applicant from further consideration (see Sect. 11.6.3).

Scientific and Technical Requirements

Applicants must possess a university degree (or equivalent) in Natural Sciences, Engineering, or Medicine, and preferably at least three years of post-graduate related professional experience. In general, applicants must be well versed in the scientific disciplines, and have demonstrated superior capability in their respective fields, preferably including operational skills.

11.6.3 Rejection Criteria

Maybe even more interesting than the listing of "positive" requirements in the previous section are some examples of what constitute rejection criteria for aspiring astronauts (Table 11.4). In contrast to the rather "fuzzy" requirements, these are generally stated very precisely in the space agencies' astronaut selection guidelines, where the respective lists cover dozens of pages. Not surprisingly, the table includes

Table 11.4. Mandatory Rejection Criteria

a) Anthropometric Parameters	
• Overall body height less than 1.53 m or exceeding 1.90 m • Eye-seat distance less than 0.625 m or exceeding 0.763 m • Knee joint angular mobility of less than 0° forward or 140° backwards	• Body weight above or below specified limit (e.g. for a 1.85 m tall 30-year old male: lower limit 61 kg, upper limit 96 kg) • Ankle flexibility of less than 20° proximal and 45° distal, measured from neutral position
b) Metabolic Parameters	
• Leucocyte count of less than $4.3*10^9$/l or of more than $10*10^9$/l • Coulter platelet count of less than $140*10^{12}$/l or of more than $440*10^{12}$/l • Detectable blood, nitrites, proteins, glucose, or bacteria in urine	• Total blood cholesterol level above 6.2 mmol/l • HDL cholesterol of less than 0.9 mmol/l or of more than 2.2 mmol/l • Blood testosterone level above 9 ng/ml or below 4 ng/ml (male); above 0.9 ng/ml or below 0.2 ng/ml (female)
c) Audiovisual Parameters	
• Bilateral hearing impairment of more than 25 to 40 dB (depending on frequency) • Reduced tympanic mobility • Inability to perform Valsalva maneuver (depressurization of mid-ear)	• Astigmatism of more than 2 dioptries • Myopia of more than -2 dioptries • Color blindness, insufficient night vision • Vision of less than 1.0 (20/20) • Unilateral reduction of field-of-view of more than 15°
d) Physical Performance Parameters	
• Abnormal heart function during treadmill exercise • Abnormal heart function or too high abdominal volume change during negative-pressure test • Abnormal respiratory volumes during rest or exercise phases	• Abnormal heart function or blood pressure, substandard tissue oxygenisation, reduced field of vision during centrifuge test • Abnormal heart function or vomiting during rotating-chair test
e) Psychological Criteria	
• Abnormal interpersonal behavior • Lack of maturity • Anti-social stress reactions • Current or past alcohol or other drug dependency	• History of psychotic states • Significant anxiety, sleeping, or sexual problems • Detectable affective or personality disorders
f) Other Issues	
• History of severe allergic reactions • Recurrent or chronic infections • Nasal septum deviation • Paralyzed vocal chord • Asthma • Anemia • Chronic gastritis	• Severe history of motion- or air-sickness • Hypertension of more than 140 mmHg (systolic) or 90 mmHg (diastolic), respectively • Missing or additional finger • Length of legs differing by more than 2.5 cm • Dental problems affecting the facial contour • Disorders of the nervous system (Migraine, Narcolepsy)

many items that have no effect on an applicant's suitability for demanding occupations on Earth, or their everyday health, or their life expectancy under normal conditions. Nevertheless, someone suffering e.g. from one of the listed diseases, or someone not meeting the specified parameter ranges, is – in the eyes of the selecting agencies – more likely to pose a risk to a space mission than someone conforming to all criteria. This additional risk is thus eliminated by disqualifying all applicants who meet even one of those rejection criteria.

11.6.4 Selection Process

The European astronaut selection process, which is outlined here as an example, starts with an "Announcement of Opportunity" distributed to national space agencies, which outlines the requirements and the selection process that will be used.

Each national agency then places job ads in major newspapers. Applicants must submit their documents to national preselection committees who conduct national-level screening using questionnaires, interviews, medical examinations, and psychological tests. Three to five applicants per country are then preselected and recommended to ESA for further examination.

More psychological suitability tests are performed there, and an additional questionnaire dealing with the scientific background of the applicants is answered. In a first round of selections, the ESA "interview committee" prepares a shortlist of promising applicants who then undergo more medical tests and a face-to-face interview with the committee. The outcome determines who is picked by the ESA "astronaut selection board" for a second round of in-depth interviews, which finally leads to the selection board recommending the applicants to be selected by the ESA director general.

After their selection, the candidate astronauts are given four-year contracts, with six-month probationary periods. European astronauts are based at the European Astronaut Center in Cologne, Germany, but their training and mission preparation requires them to travel frequently. Astronaut candidates have no right to fly on a space mission, neither can they insist on being used only for a specific type of activity. Selection of mission participants from the European corps of veteran and candidate astronauts is done on a mission-by-mission basis.

In the March 23/24, 1998, ESA Council Meeting a decision was taken to revise the European Astronaut Policy leading to a single European Astronaut Corps. This means that the national astronaut offices existing in some European countries (France, Germany) are dissolved in a couple of years and some of the astronauts or astronaut candidates will be taken over by the European Astronaut Corps.

12 Logistics, Communications and Operation

The operation of space stations and payloads requires enormous effort concerning the transportation of astronauts, equipment/goods and data from Earth to the station and vice versa. This chapter will address logistics, data processing and transmission, automation and also effective maintenance systems. The logistics and operational systems are important subsystems and largely relevant for determining the station's orbit (altitude, inclination, attitude), the design of the station and its peripheral systems, as well as integration and ground support. They are hence relevant for system design, too. The importance of these systems becomes clear when considering the following: despite the size of the International Space Station, its operation will be ensured by only six to seven onboard astronauts. Compared to the operation of Spacelab, the time available for handling a payload element or an experiment is reduced to about 10%–20%. It is thus apparent that operations and maintenance systems play a crucial role. Fortunately at present for terrestrial and space flight applications, data and communication systems (e.g. mobile radio communication and multimedia broadband networks), automation and robotic systems, as well as competing space transportation systems develop extremely fast. As a consequence, it is sometimes very difficult to quickly integrate the spin-on advantages resulting from this development into the plans for a space station and to make use of them. For that reason, the operation at the end of the operational life of ISS will be markedly different from its initial operation; i.e. when designing the subsystems and defining the interfaces, it is important to allow for the exchange of components and the integration of improved systems.

12.1 Logistics

The operation of a space station and, to a limited extent, also that of a maintainable platform, require the resupply of goods, the return of valuable equipment, and also the disposal of defective equipment and waste products. Once the station takes on the function of a "space port" or a service center, this also comprises the transportation requirements of orbit transfer vehicles and subsatellites. Above all, crews must be transported to the station and subsequently returned to Earth after their mission is completed.

The consumables needed aboard a station are, apart from propellants for attitude and orbit control, mainly goods for the crew's daily use such as breathing air, water, food and clothing, as well as safety and survival equipment. In addition to that, further raw materials for later processing, spare parts/instruments, and also extension modules must be transported to the station. Waste products and defective instruments, or instruments which are no longer needed, must be disposed of for reasons of hygiene or of space efficiency; products of space manufacturing must be returned to Earth. When there is no sufficient transportation capacity (such as the Space Shuttle) for return flights, parts which are to be disposed are put into a capsule or ATV and brought on a reentry trajectory where they will burn upon re-entering Earth's atmosphere.

The following considerations will not take the transportation requirements for the assembly of a space station into account; they fully concentrate on the stationary operation of the completely assembled station.

12.1.1 Transportation Requirements

The requirements for transportation to and return from the station depend on the following:

- Number of astronauts and the duration of their stay aboard the station
- Propellant requirements for attitude and orbit control
- Utilization scenario
- To what extent space products shall be manufactured

Additionally, these requirements also strongly depend on the efficiency of the life support system (ECLSS) as to treatment and reusage of e.g. breathing air and water. The overall system mass of a space station during its entire operational life, which is composed of the mass of the completely assembled station and the total amount of astronauts and goods to be transported during this time, determines the life cycle costs. In 1994, the Russian Space Agency RKA indicated the following rough distribution of the average costs over the life cycle of the space station Mir: 30% for manufacture and transportation of the station's modules, 45% purely for operational costs for the station including launch of Soyuz and Progress vehicles for resupply, 25% for experiments and their operation. This means that 75% of the costs are allotted for assembly and operation (infrastructure) of the station and "only" 25% of the costs go toward its actual utilization. This ratio infrastructure/utilization of 3:1 is expected to be typical for the International Space Station as well, and might even be higher.

Which is the maximum resupply mass per astronaut and year? Based on the following assumptions

- the basic supply for a human being as presented in Table 4.6,
- a maintenance and degradation-resistant station,
- no cyclic orbital reboost maneuvers and
- an open ECLSS, etc.,

about 5.34 tons/year resupply mass per astronaut will be necessary. Due to the clothing requirements and LiOH filters, this is 23% more than the minimum resupply mass needed for the support of a human being in space (4.318 tons, see Table 4.2). When taking a closer look, it can be seen that the amounts of washing water and LiOH resupply for the life support system, in the case of a primitive station, amount to about half of the resupply mass, literally "weigh heavily". If the water consumption is lower or, as was the case with the Salyut stations, a large part of the water and the breathing air is recycled, markedly lower masses will result.

A first reference value can be derived from the Russian Salyut 6 and Salyut 7 stations which were the first two space stations operated over a duration of several years including crew exchange and well-established logistics (Salyut 6: 1977–82, 16 crews, a total of 4.07 man years; Salyut 7: 1982–87, 10 crews, 5.06 man years). Both stations housed long-duration crews with two, sometimes three astronauts, and needed 20 tons (Salyut 6) and 25 tons (Salyut 7) of resupply mass including propellants for orbital reboost (cf. Table 2.1), i.e. nearly 5 tons per year per astronaut. This relatively small resupply mass can be explained by having economized on water and the partial recovery of water having resulted from transpiration (about 1 kg per day per astronaut).

A second reference value results from the reported Mir data (up until 1994) of 10 to 12 tons per year [NRC 95]. During the years before the docking with the Space Shuttle, the Mir logistics were mainly handled by about 5 Progress-M flights per year (cf. Table 2.6). In 1993 it became known that one astronaut consumed about 2.3 kg of water per day, of which 66%–80% was recycled water, and that during the first seven years of Mir's operation, 7.5 tons of water were able to be recycled. In addition to that, during the same period of time, about 2.5 tons of oxygen were recovered from urine; that corresponds to the Mir requirement of 750 days for an average crew of slightly less than three astronauts and a recovery rate of 29% [Space News 2-8.8.93]. As a consequence, about 5.5±0.5 tons/year resupply mass per astronaut including propellant consumption during orbital reboost can be assumed for the operation of Mir.

The following values can be assumed for the International Space Station: For potable water, rehydration of food, teeth brushing, 2.81 kg/day is needed, and for washing water 6.80 kg/day. This results in 9.61 kg per day per astronaut, i.e. 57.66 kg per day in the case of six astronauts. When adding 2.18 kg/day for payload operation and 3.33 kg/day for life science experiments, the result is an additional 5.51 kg/day, leading to a total of 63.17 kg/day. When assuming a recycling rate of 80%, as it is planned for the ROS, an end result of 12.6 kg/day is obtained. When considering this value for water resupply and the remaining values in Table 4.6, about 18 tons per year are obtained for six astronauts. An increase in the water recycling rate from 80% to 91%, as planned for the USOS, will reduce this mass by about 2 tons. Thus, an average of 17 tons per year (USOS + ROS) can be assumed. If we now take into account that CO_2 removal aboard ISS will be partially regenerative, another 0.5-1.5 tons/year can be saved. In order to obtain a better comparison to the Mir station, the propulsion requirements for attitude and orbit control ought to be added, resulting in an additional average mass of 10.5 tons/year (cf. Table 12.1). In the case of the assumptions made for ISS, this would result in a total of 26–27 tons/year, i.e. about 4.4 tons per year per astronaut. Compared to the Salyut station, this is a decrease in mass of 14%, which amounts to even 25% in comparison with that of Mir.

By now, it is clear that increasingly improving comfort and thus increased exchange of subsystem parts subject to wear can be compensated for by gradually increasing closure of the process cycles by recycling the consumables. By further developing these subsystem synergisms, and averaging over a long period of time

Table 12.1. Planned Annual Transportation Masses and Volumes from Earth to ISS and Vice Versa

		Resupply		Return	
		Mass [kg]	**Volume [m³]**	**Mass [kg]**	**Volume [m³]**
Internal Cargo	Ambient Stowage [a]	19936	85.09	14114	64.27
	Thermally Conditioned [b] +4/-20 Degrees	2374	7.36	13.59	0.057
External Cargo	Ambient Stowage	6117	19.11	6117	19.11
	Survival Electrical Power [c]	680	2.13	680	2.13
Oxygen [d]		2633	N/A	0	N/A
Nitrogen		794	N/A	0	N/A
Propulsion		10421 ± 2266	N/A	0	N/A
Total Cargo		42955 ± 2266	113.69	20925	85.57

[a] Internal ambient stowage cargo includes crew, system maintenance, and user payload resupply/return logistical items and packaging volume.
[b] Accommodations for +4°C and -20° C thermally conditioned cargo are provided by system to support crew resupply logistical items and packaging volume, and can support return transportation of user experiment products.
[c] Some external cargo items require survival electrical power for thermal heaters during transportation or temporary on-orbit storage.
[d] This may be offset by on-orbit oxygen generation.

Table 12.2. Summary of Annual Resupply Masses per Astronaut

Space Station	**Resupply Mass in tons/astronaut year**	**Assumptions**	**Remarks**
Minimum Resupply Mass (on Earth)	4.3	No Air Revitalization, No Clothing	Input Mass Flow Based on Table 4.2
Maximum Resupply Mass (in Space)	5.4	With Open ECLSS (LiOH Filters), With Clothing, *Without Orbit Reboost*	
Salyut 6 and 7 1977 - 1987	≤ 5.0	Partial Water Recovery, With Orbit Reboost	Including Payload H/W
Mir 1986 - 1994	5.5 ± 0.5	Water Recovery (66%-80%), Oxygen Recovery (≤ 29%), With Orbit Reboost	Including Payload H/W and Replacement Items
ISS 2004 - 2013	4.4 ± 0.6	Water Recovery (80%-91%), Oxygen Recovery (27%-81%), With Orbit Reboost	*Not* Including Payload H/W and Replacement Items

(solar cycle), a value of 4.4±0.6 tons per year per person is computed (reference value). According to these calculations, about 20–27 tons/year for payload operation (payload elements, experiments, consumables, further expansion, etc.) remain in the case of the International Space Station with a crew of six astronauts and a planned logistics requirement of a total of 50 tons/year. Recent figures for the ISS transportation masses at low and high solar activity (i.e. in the years 2005 and 2010) confirm these values (cf. Fig. 12.1).

Table 12.2 gives the summary of the estimated annual resupply masses per astronaut for past (Salyut 6, Salyut 7 and Mir) and future (ISS) space stations operations.

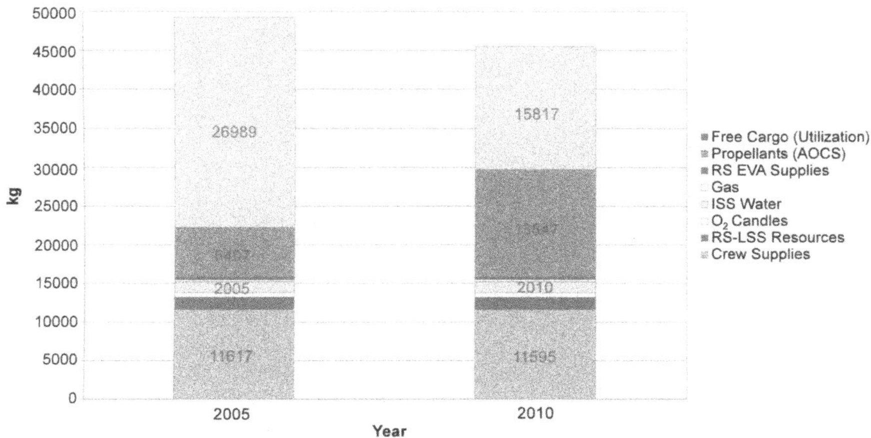

Fig. 12.1. Transportation Masses to ISS at Low (2005) and High (2010) Solar Activity [Boeing 96]

There is little knowledge concerning replacement or replenishment requirements for subsystems, components, experiments or complicated instruments. This applies even more so to the resupply requirements for raw materials for future manufacture of products in space.

The masses which must be returned to Earth can, of course, in a manner similar to the method applied with Salyut and Mir, be markedly reduced by stowing all instruments, materials and substances which are no longer needed into non-reusable logistics return capsules. These capsules are then decoupled from the station and, by means of a suitable braking maneuver, are brought onto a reentry trajectory. Thus, independent of technological progress, the mere logistics return mass can be kept very small. Of course, this does not apply to experimental instruments, samples and valuable equipment items.

According to Table 12.1, a yearly transportation requirement from Earth to the station is indicated at an average mass of 43 tons and a volume (not including oxygen, nitrogen and propellants) of 114 m^3. The transportation requirement from the station back to Earth amounts to 21 tons and a volume of 86 m^3. The more recent values of transportation masses from ground to the station indicated in Fig. 12.1 are somewhat higher.

12.1.2 Launch Systems and Transportation Capabilities

The transportation requirement determines, for the given payload capacity of the launch systems available, the frequency of resupply flights. This section will compare the space transportation systems available for the next 10–15 years and their payload capacities.

In Fig. 12.2, the payload masses to typical LEOs are presented as a function of the orbit's inclination. This is given for launch vehicles which might be put to use with the International Space Station, e.g. the US Space Shuttle, the European

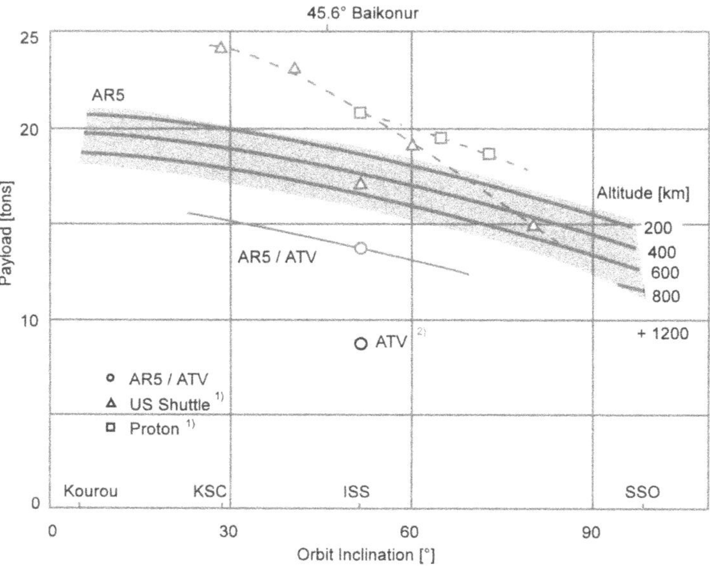

Fig. 12.2. Payload Masses to Low Earth Orbits

Ariane 5, the Russian Proton D-1 and the Japanese H-II. In the case of missions to higher orbits and in target orbits with increasing inclination relative to the equatorial plane, the transportation capacity is reduced (since it is mainly dependent on the geographical latitude of the launch site). The only orbits that can be reached in a direct way are those with an inclination corresponding to the degree of latitude of the launch site. Target orbits with lower inclinations can generally only be reached with the help of additional stages and a drastic loss of payload capacity. For that reason, the payload curves in Fig. 12.2 have their lower limit at 28.5° (for the Space Shuttle, corresponding to the northern latitude of the Kennedy Space Center (KSC)) and at 45.6° (for the Russian Proton rocket, corresponding to the latitude of the launch site Baikonur). Compared to these two, the European launch vehicle Ariane 5 has an advantage due to its near-equatorial launch site at 5.0° in Kourou, French Guyana. The Ariane 5 can reach practically all orbit inclinations, not counting the local limitations of the launch direction. Some useful data on typical launch sites of the launch systems mentioned here, as well as some remarks concerning the maximum dimensions of the payload bays (and thus the permissible size of payloads) are given in Table 12.3.

The only vehicles available for the transportation of crews are the US Space Shuttle (seven astronauts, if necessary ten) and the Russian Soyuz-TM system (three astronauts). However, only the Space Shuttle can transport large payload masses (maximum 25 tons) from a station back to Earth. Compared to that, the Soyuz capsule return mass of less than 50 kg is negligible.

Table 12.3. Transportation Capabilities of Launch Systems Relevant to ISS [Isakowitz]

Type	Ariane 5	US Space Shuttle (STS)	Proton D-1	H II
Launch Mass	710 tons	2040 tons	689 tons	260 tons
Orbit				
Type	70 x 300 km	407 km [a]	200 km [a]	185 km [a]
Inclination	51.6 °	51.6 °	51.6°	30°
Payload				
Mass	18 tons	17.1 tons	20.9 tons	10.5 tons
Dimensions[b]	4.57 m x 10.3 m	4.6 m x 18.3 m	4.1 m x 14.7 m	3.7 m x 9.1 m
Launch Costs	$120 M	$130–$245 M	$50–$70 M	$150–$190 M
Launch Site	Kourou 5.2°N; 52.8°W	KSC 28.5°N; 81.0°W	Baikonur 45.6°N; 63.4°O	Tanegashima 30.2°N; 130.6°O

[a] Circular Orbit [b] Diameter x Length

12.1.3 The Automated Transfer Vehicle (ATV)

The payload masses indicated in Fig. 12.2 relate to transportation to Low Earth Orbits with no requirements concerning orbit maneuvers. The Space Shuttle, as it was shown several times with Mir, can fly to a station and dock to it. All other launch systems, when performing rendezvous or docking maneuvers, depend on transfer vehicles which are separated from their launch vehicle in an Earth orbit and have to carry out the transfer to the station. In Russia, this task is assumed by the Progress system (for uncrewed resupply flights) and the Soyuz-TM capsules (for crew transportation to Mir).

In Europe, the Automated Transfer Vehicle (ATV) is being developed for the supply of cargo to the International Space Station. According to Fig. 12.3, the ATV can, in different configurations, carry different kinds of freight to the station in both evacuated and pressurized containers. The dimensions of the whole ATV, however, are limited to the maximum dimensions of the fairing of the space transportation system Ariane 5.

Generally, these resupply vehicles are limited with respect to their launch mass by the payload capacity of the launch system used. The mass which can be carried to the station is reduced by the masses of structure and equipment as well as by the necessary thruster and propellant masses for attitude and orbit control. For example, the ATV orbital transfer stage carried into a LEO of 30 x 300 km altitude by the Ariane 5-E system can, at an initial mass of 20.5 tons, carry only up to 7.5 tons of payload for resupply of the station (cf. Table 12.4). The maximum mission operation is about 6 months.

The ATV will provide the following services to the International Space Station:

- Delivery of dry and liquid cargo to the Station, such as experiments, food, compressed air and water
- Refuelling of the Station, i.e. the transfer of propellant to the Zarya (FGB) module, built by the Russian company Khrunichev under Boeing contract

- Reboost and attitude control during reboost of the whole Station, i.e. orbit corrections using the ATV propulsion system to compensate for the continuous loss of altitude by the Station
- Removal of waste from the Station followed by controlled destructive reentry of the ATV

As a result, the ESA share of the ISS operating costs will be reduced by such non-financial (in-kind) contributions. Instead of contributing financial means to the joint ISS operation fund, the partners of ISS have the opportunity to "pay" their share in the form of operational and logistical, i.e. in-kind contributions, for example, in the form of space transportation [Feustel 96]. An internal ESA calculation is based on an annual maximum transportation mass of 54 tons/year. Since the ESA part of the total ISS operating costs amounts to 8.3%, the ESA logistics part will be 4.5 tons/ year. As a consequence, with the currently planned frequency of one ATV launch every 15 months, this target will be exceeded for the benefit of ESA's own cargo

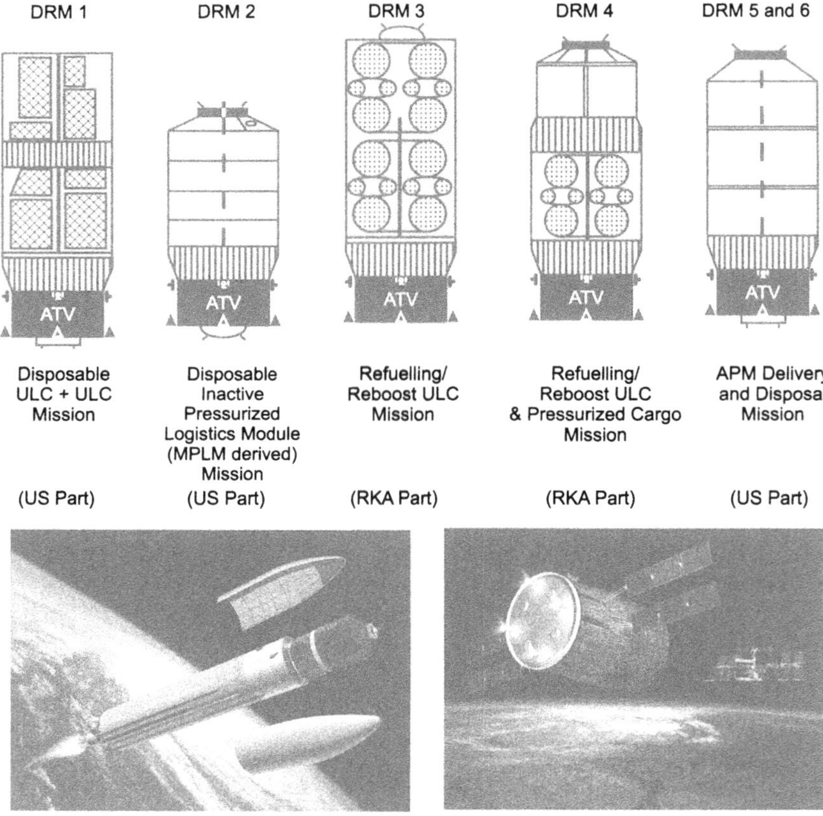

Fig. 12.3. Configurations of the ATV for Different Reference Missions. The payload containers MPLM and ULC described in Sect. 13.4 are accommodated in the Integrated Cargo Carrier (lower part).

Table 12.4. ATV Mission and Transportation Capabilities [Amadieu 98]

ATV	Maximum Mass at Launch	20.5 t
	Maximum Diameter and Length	4.5 m x 10.1 m
	Span with Solar Arrays Deployed (3.8 kW after 6 Months)	18.3 m
	Initial Orbit (Ariane 5 Evolution)	30 x 300 km
	Spacecraft Dry Mass	4.6 t
	Cargo Carrier Dry Mass	3.9 t
	Net Cargo Mass for ISS Orbit	≤ 7.5 t
Mission Duration		
	Orbit Adjustment + Transfer (incl. Spare Time)	≈ 50 Hours
	Final Approach and Docking Maneuver to ISS (automatic or manual by ISS crew)	5 Hours
	Docking Duration to ISS	6 Months
	Decoupling and Deorbiting Maneuver	9 Hours
	Destructive Reentry	1 Hour
Cargo Upload/Download Capability (400 km, 51.6°)		7.5 t/6.5 t
Upload Cargo can Consist of:		
	Dry Cargo Carried in Pressurized Container	≤ 5.50 t
	Water	≤ 0.84 t
	Air, Oxygen or Nitrogen	≤ 0.10 t
	Propellant*) for ATV (306 kg MMH and 554 kg Oxidizer)	≤ 0.86 t
	Propellant*) for ISS Reboost and Attitude Control	≤ 4.00 t
*)ATV Propulsion: 4 Main Thrusters (490 N each), 10 AOCS Thrusters (220 N each)		

(i.e. 1.5 t/year). After the ATV mission is completed, the vehicle, under consideration of safety zones, will be brought onto a reentry trajectory which will cause the vehicle, along with its on-board disposable items, to burn up in the Earth's atmosphere.

The ATV has been designed as an expendable transportation vehicle and hence, similar to the Progress capsules, it cannot return payloads from ISS back to Earth. This task will be accomplished by both the Space Shuttle and by return capsules yet to be developed.

In order to increase the future lifting capacity of the Ariane 5 in favor of large, geostationary communication satellites (i.e. an increase of the GTO payloads from 5790 kg to 7400 kg) and also in favor of the ATV, and in order to reduce costs at the same time, the ESA Ministerial Council in October 1995 in Toulouse, France agreed on a series of measures leading to the development of a so-called "Ariane 5-E" (E = evolution). As a consequence, the ATV payload mass can be up to 7.5 t for a 400 km circular orbit. The ATV contract was signed with European industries (prime contractor is Aerospatiale's Les Mureaux establishment, prime contractor for production is DaimlerChrysler Aerospace, Bremen, Germany) on 25 November

1998, the production phase will probably begin in 2000. The maiden launch is planned for late March 2003. Until 2013 or longer, depending on the station's lifetime extension, there will be at least 8 servicing flights.

Since the beginning of SSF/ISS cooperation with NASA, ESA has tried to integrate its launch vehicle Ariane into the transportation scenario. Although until the present time, some European logistics systems have been considered during the planning, the real breakthrough did not take place until the so-called "Mixed Fleet Scenario" and the new concept of in-kind contributions to the operating costs came about by Russia joining the international partnership. The Mixed Fleet Scenario in the new operational concept includes the idea of the partners not paying their part of the operational costs to NASA monetarily, but instead directly contributing actual transportation activities through the use of their own vehicles for the joint effort for ISS's operation. For that reason, the ATV is the ultimate precondition in order to be able to integrate Ariane 5 into the logistics concept at all. That would enable Europe to "pay" the largest part of the European share of the total operational efforts through non-financial contributions instead of paying monetary sums to NASA.

The six reference missions illustrated in Fig. 12.3 were closely examined in 1995/96 as to their suitability for the design of the ATV, depending on the payloads to be transported and the launch structures belonging to them.

The traffic model for the supply of ISS upon which NASA and the Russian Space Agency RKA had initially agreed included the following annual flights:

- Five US Space Shuttle flights, four of which are carried out with the pressurized payload container MPLM and the remaining one with two of the so-called "Unpressurized Logistics Carriers" (ULCs)
- Four Russian (crewed) Soyuz launch vehicle/Soyuz-TM flights
- Five Russian (uncrewed) Zenit launch vehicles with Progress-M capsules

In the framework of the Mixed Fleet Scenario, the reference missions DRM 1 and DRM 3/4 presented in Fig. 12.3 offer the best prospects for taking part in the logistics flights in a way that is reasonable and feasible for all partners involved. A typical traffic model could, for example, include two Ariane 5/ATV flights per year (cf. Fig. 12.4) as follows:

- One Ariane 5/ATV flight including two ULCs for the transportation of unpressurized payloads (reference mission DRM 1)
- One Ariane 5/ATV flight including propellant tanks and, according to whether or not it is necessary, one pressurized transportation container. After having delivered the resupply to the station, the ATV would be used to carry out a reboost maneuver for the entire station (reference mission DRM 3/4).

The DRM 1-type mission, which has already been accepted by NASA can replace (compared to the initial NASA/RKA traffic model) one Shuttle flight per year, and the DRM 3/4 mission two Zenit/Progress flights per year. As a consequence, NASA itself does not need to develop a ULC transportation vehicle, and Europe has an autonomous and exclusive role in an important transportation task for the entire space station. The mission DRM 3/4, i.e. transportation and reboost instead of two Progress flights, is currently favored by NASA. Since the propellant and propulsion capacity of the ATV is considerably higher than that of the Progress-M capsule, re-

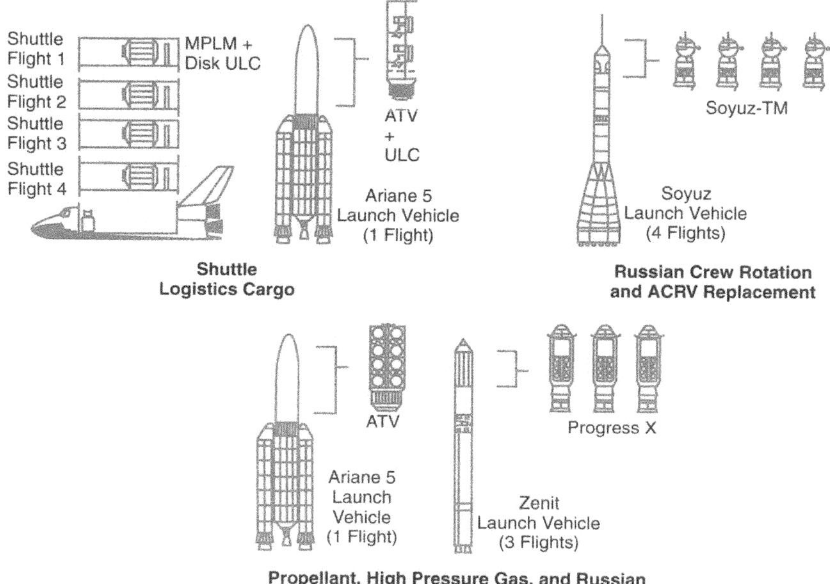

Fig. 12.4. Annual Traffic Model for the Logistical Supply of ISS with Participation of European Vehicles

boost maneuvers will be necessary less often when carried out by the ATV. The reboost function will perhaps not be counted as a contribution to transportation, but as an increase of the European infrastructural contribution and that means that additional utilization options for ISS could be bartered for it. However, an agreement between the space agencies ESA and RKA concerning the reboost task and its accounting in the ISS infrastructure or operation fund has not been reached yet.

NASA officials are concerned that Russia will not be able to build and launch enough Progress vehicles to fulfill its commitments. Therefore, NASA is spending money to build an Interim Control Module (ICM), which will serve as a temporary back-up if Russia's long-delayed Service Module is not ready for launch in late 1999. The ICM, combined with the ATV and a Japanese transfer vehicle (HTV), would provide complete back-up to Russian vehicles for station attitude control and reboost after 2001 [AW&ST, Aug. 24, 98].

12.1.4 Return Vehicles

Return vehicles can also perform the task of a "life boat" for the rescue of astronauts and remain docked to the station until they are needed. For this purpose, European planners intend to participate in the development of a Crew Return Vehicle (CRV). The CRV could be carried by the Space Shuttle to ISS or launched into an Earth orbit by an Ariane 5 and transported to the space station by means of the ATV propulsion module (see Fig. 12.5).

Fig. 12.5. Ariane 5 with CRV

On the condition that the European Council of Ministers approves (a decision which is expected not earlier than 1999), the CRV could, together with the laboratory module COF and the ATV, become one of the pillars of the European crewed space flight program. Europe has its own capabilities and know-how in the field of technology and operation of crewed space laboratories (Spacelab) and uncrewed space transportation (Ariane program). However, where crewed space transportation is concerned, Europe is far behind the two big "spacefaring" nations, USA and Russia, and currently totally depends on the use of their transportation vehicles.

Contrary to COF and ATV, the CRV to date is not an integral part of the ISS logistics concept. NASA, however, is greatly interested in obtaining a means of rescue for the space station crew in emergency situations. The CRV is conceived as a more capable successor to the Russian Soyuz-TM spacecraft, which will be the first lifeboat docked to ISS in the early stages of assembly. Presently, NASA and ESA are studying a Crew Return Vehicle based on the experimental X-38 reentry vehicle concept. A test flight of the X-38 vehicle is scheduled for early 2001. The X-38 design uses a lifting body concept (with short delta stub wings) providing aerodynamic maneuvering and control capability during atmospheric flight to guide the vehicle to a designated landing site.

The CRV would weigh up to 15 tons and will be conceived to carry seven astronauts (with an overload capacity of eight persons). Two CRVs are planned to be attached to ISS at all times, with two others back on Earth being refurbished. It should be able to dock to the station for a duration of six months to three years, depending on the deployment phase. The spacecraft will require the Resource Module developed for the ATV with some additional, safety-increasing redundancies and a reinforced structure. During its aerodynamically controlled return flight it is planned to achieve a smooth landing with a landing accuracy of less than 5 km radius.

Both the Space Shuttle missions to ISS and those of the Ariane 5/CRV (which will perhaps be available in the future) are complex and costly and will, for that reason, take place at intervals of several months. In order to frequently return and recover material samples or small payloads, small, semi-ballistic return capsules are a possible option. They can be aerodynamically maneuvered within the Earth's atmosphere and can thus reach a pre-determined landing site with a high degree of accuracy. As a consequence, the logistical efforts and the operating costs of returning urgent payloads can be minimized.

In order to initiate the return flight, the deorbit maneuver can be performed in the conventional way by chemical rocket thrusters. During the operation of Mir, such a capsule (called "Raduga", with a mass of 350 kg) was used several times for the return of up to 150 kg of samples. Another option frequently discussed for deorbiting spacecrafts but not yet tested is the use of tether systems. Due to orbit and tether dynamics, the capsule can be decelerated with the help of a long tether (e.g. about 35 km in length). As illustrated in Fig. 12.6, two procedures of dynamic and static release could be employed, the former requiring shorter tether lengths. However, deployment control appears to be more challenging. Tether control might be possible by using the interaction of a conductive tether with the magnetic field of the Earth. Since parts of the tether system remain aboard the space station and can be reused, tether concepts can be expected to yield lower total masses [Ockels 95]. Such tether systems can also be used for the dropping of discarded space station modules, the return of which to Earth is not worth the effort. As a consequence of the impulse transfer from the dropped object to the space station, the station will be

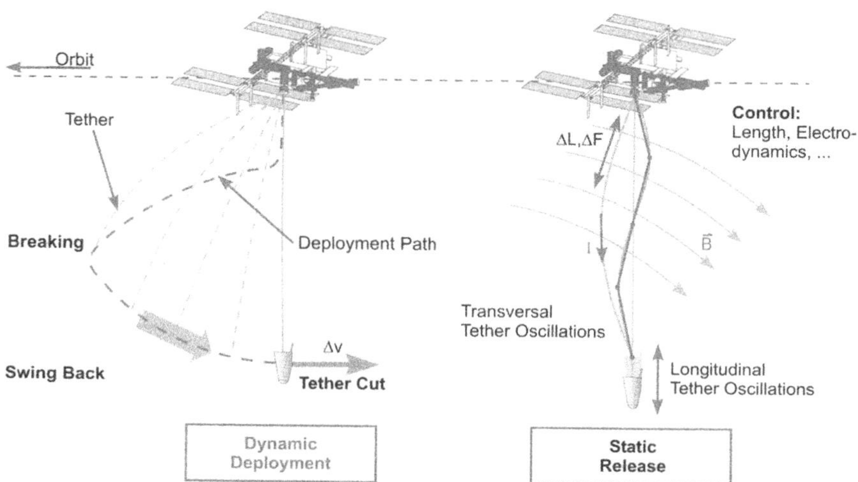

Fig. 12.6. Deorbit Maneuvers by means of Tethers: Dynamic Deployment and Static Release of the Return Spacecraft

lifted to a higher, elliptic orbit during the separation, and thus, propellants for orbit control may be saved, especially when heavy modules are being dropped [Zimmermann 96].

12.1.5 Extravehicular Activities

In the past decade and a half, some very ambitious and spectacular Extravehicular Activities (EVAs) have been performed in LEO from the Salyut/Mir orbital complexes and from the Space Shuttle, like satellite retrievals, maintenance and repair, welding, assembly of structures, and repair and refurbishment of the Hubble Space Telescope. Soviet/Russian EVA experience from Salyut to Mir (1977–February 1995) represents 112 cosmonaut-sorties of 452 hours duration. US Space Shuttle EVAs (1983–February 1995) total 59 astronaut sorties of 348 hours duration [IAA EVA 96].

The assembly and maintenance of the International Space Station at the end of this century will require up to 40 EVAs (50 to 80 astronaut sorties), or 300 to 500 hours per year. Over the projected lifetime of ISS, this represents an increase in extravehicular activity of more than one order of magnitude when compared with the total Soviet/Russian effort over the course of 18 years, or the US effort over the course of 12 years.

This significant increase in the number of EVAs requires space suit systems that provide a maximum of interoperability for safe, efficient, and flexible EVA operations. Ideally, if different suit systems are used, they should be operable from all of the station's airlocks, by crewmembers of different countries, and with space station transfer vehicles being fully supportive of all systems in case of an emergency. At a minimum, different space suit systems must be operationally compatible with exterior translation, worksite restraint, and free-float safety tethering and/or rescue provisions; with communications systems; and with equipment interfaces and tools necessary for station assembly, repair, and maintenance.

Unfortunately, no interoperability exists between current US and Russian EVA suits (STS EMU and Orlan-DMA/M, respectively, see Fig. 12.7) due to incompatibility of communications systems and the Airlock Interface Control Panel/Umbilicals. With respect to airlock hatch dimensions, the largest STS EMU size is too large for use in the Mir emergency airlock.

Fig. 12.7. EVA Space Suit Systems: Russian Orlan DMA (left) and US STS EMU (right) [IAA EVA 96]

Prebreathing protocols present another problem. Oxygen prebreathing is required before decompressing the airlock to vacuum due to reduced suit pressure to avoid decompression sickness symptoms. For STS EMU and Orlan DMA/M, different prebreathing protocols are used due to different conditions such as suit and cabin pressure requirements. This is an inconvenience for nominal operations, but for emergency situations a harmonization of prebreathing protocols is mandatory to enable immediate response.

An EVA space suit system provides not only routine operations capability outside a spacecraft or space station, but it also serves as emergency back-up for the life-support system of a spacecraft or of other space suits. Thus, in space projects such as the International Space Station which require international cooperation, an EVA space suit system of any national origin must be able to support any other space suit system or spacecraft, regardless of its origin. Space suit interoperability is mandatory to warrant a maximum of safety.

12.2 Data and Communication Systems

The operation of space stations and payloads requires carefully designed data and communication systems. This is due to the large amounts of data to be processed, to the variety of data sources and data sinks, and also to the interaction of on-board and ground systems under constantly changing transmission conditions. For that reason, mobile radio communication for both very short and long distances is required, under integration of satellite systems and ground-based communication systems. All this should happen with utmost accuracy and, if possible, permanent availability.

When comparing a space station (where astronauts are present during operation) to uncrewed space systems, it seems, at first glance, possible to do this without some of the functions of classical satellite transmission technology, since a crew can directly perform important monitoring tasks and intervene, if necessary. The following factors, however, in fact increase the overall requirements and thus call for a high level of automation [Hartl 88]:

- Safety requirements
- Complexity and extent of certain control functions for the space station and payload systems
- The necessary relief for the astronauts of their actual, mission-relevant tasks which cannot be automated

The coming sections will address the following subjects: Sect. 12.2.1 describes the typical requirements for the data management system of a space station, Sect. 12.2.2 and Sect. 12.2.3 will deal with the transmission paths between space stations, ground control facilities and distributed data systems. The design of radio communication systems will be discussed in Sect. 12.2.4. Section 12.2.5 and Sect. 12.2.6 will describe antennas and modulation and coding techniques. Finally, in the Sect. 12.2.7 and Sect. 12.2.8, data relay satellite systems and those data and communication subsystems important for the International Space Station will be addressed.

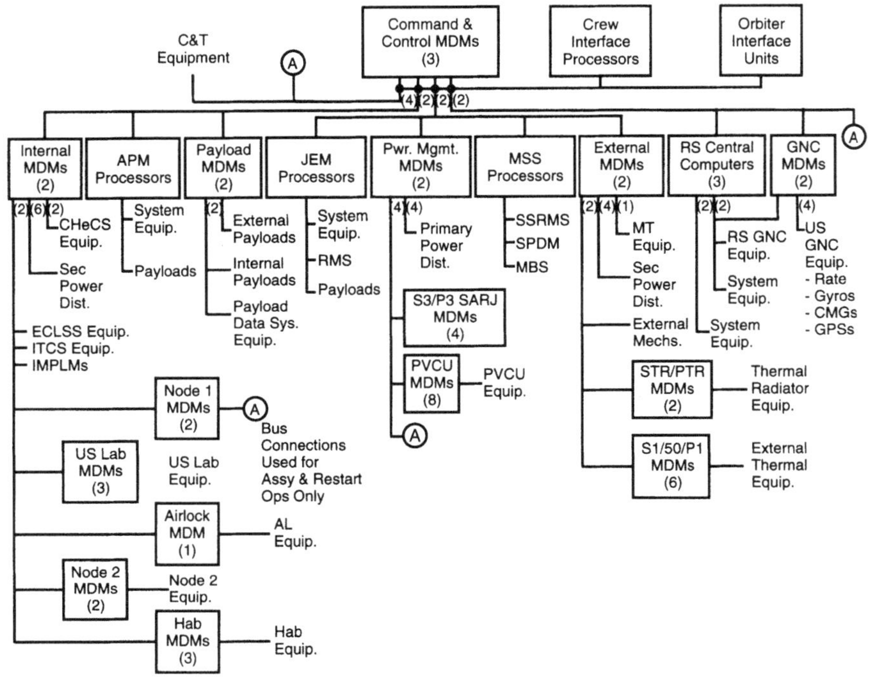

Fig. 12.8. Illustration of the Data Management System of ISS (Figures in Parentheses = Number of Implemented Systems)

12.2.1 The Data Management System

In order to get an initial impression of the complexity of the subsystem "Data Management System" (DMS), Fig. 12.8 shows the DMS for the International Space Station. The important DMS subsection of Guidance, Navigation and Control (GN&C) has already been discussed in Chapter 6, "Attitude and Orbit Control System", (cf. Fig. 6.18).

The DMS supports the monitoring of the course of operation both aboard the station and in the ground control center by the following actions:

- Recording, formatting, multiplexing and distributing (data acquisition) the measurement data of sensors, actuators and other systems (including voice and video data)
- Establishing and maintaining the radio link between the space station and the ground control center and – if nearby – the Space Shuttle (signal acquisition)
- Determining the relevant orbit data (Doppler shift, distance, angle)
- Determining the angular position and the direction to the Sun in order to measure position and velocity
- Receiving the system data which supply information on the space station and its instrumentation (telemetry)

- Receiving payload data and processing it for further signal conditioning, displaying this data (monitoring) and transmitting it; moreover, transmitting commands in the form of command data to the instruments of the space station (telecommand).

The following telecommunications aspects are characteristic for space stations:

- Relatively long and complex transmission paths (strong signal degradation, long propagation time, command delay)
- High speed (Doppler effect, antenna tracking)
- The largest possible data rate for very different systems and users and, at the same time, limited line capacity
- High specific costs for electrical power and data transmission
- Large variety of interfaces to be integrated (OSIs = Operator/System Interfaces) including displays and terminals
- Technological requirements of space (weightlessness, vacuum, radiation)

12.2.2 Transmission Paths to Space Stations

Possible transmission paths to space stations can be shown with the examples of Mir and the International Space Station (see Fig. 12.9). First, there are the direct transmission paths from numerous ground stations. These stations have been placed either in a strategically favorable way all over the world or via favorably positioned radio relay ships, when space flight began to develop in the USA and in the then Soviet Union. The duration of visual contact (and thus favorable transmission conditions at high frequencies) between these ground stations and low-flying space stations or platforms is only a few (≤ 10) minutes. For that reason, the USA established the so-called "Tracking and Data Relay Satellite System" (TDRSS), a system of geostationary data relay satellites with the help of which markedly longer contact times can be achieved. The Russian space station operators, in the case of Mir and in the initial phase of the International Space Station, still relied on their ground network. After the dissolution of the Soviet Union, this network was limited to the area of what is today Russia, and at present it comprises seven ground stations. The frequencies for direct transmission are in the S-Band while those for transmissions via relay satellite are in the Ku-Band. The reception areas for the International Space Station including two TDRS satellites and the RKA ground network are shown in Fig. 12.10. Perhaps later on, Russian geostationary data relay satellites of the "Luch" type will be added. In Table 12.5, the usual frequency bands for satellite communications are displayed; this table does not, however, contain frequencies for short-range transmission for EVAs, rendezvous and docking (RvD). For the latter transmissions, civilian frequencies in the UHF range are used (300-1000 MHz).

Another option for space-ground communications is the use of current or planned satellite-based mobile radio communication systems such as Iridium, Globalstar, ICO (former INMARSAT), Teledesic, etc. that use frequencies in the L/S-band. Moreover, Ka-band satellite systems planned for world-wide linkage of multimedia systems might also be used. The International Space Station in its 2010 state will be only one of several million data sources. Just as on Earth, its operators and users will book and pay for (and will thus also wish to determine) bandwidths, transmission

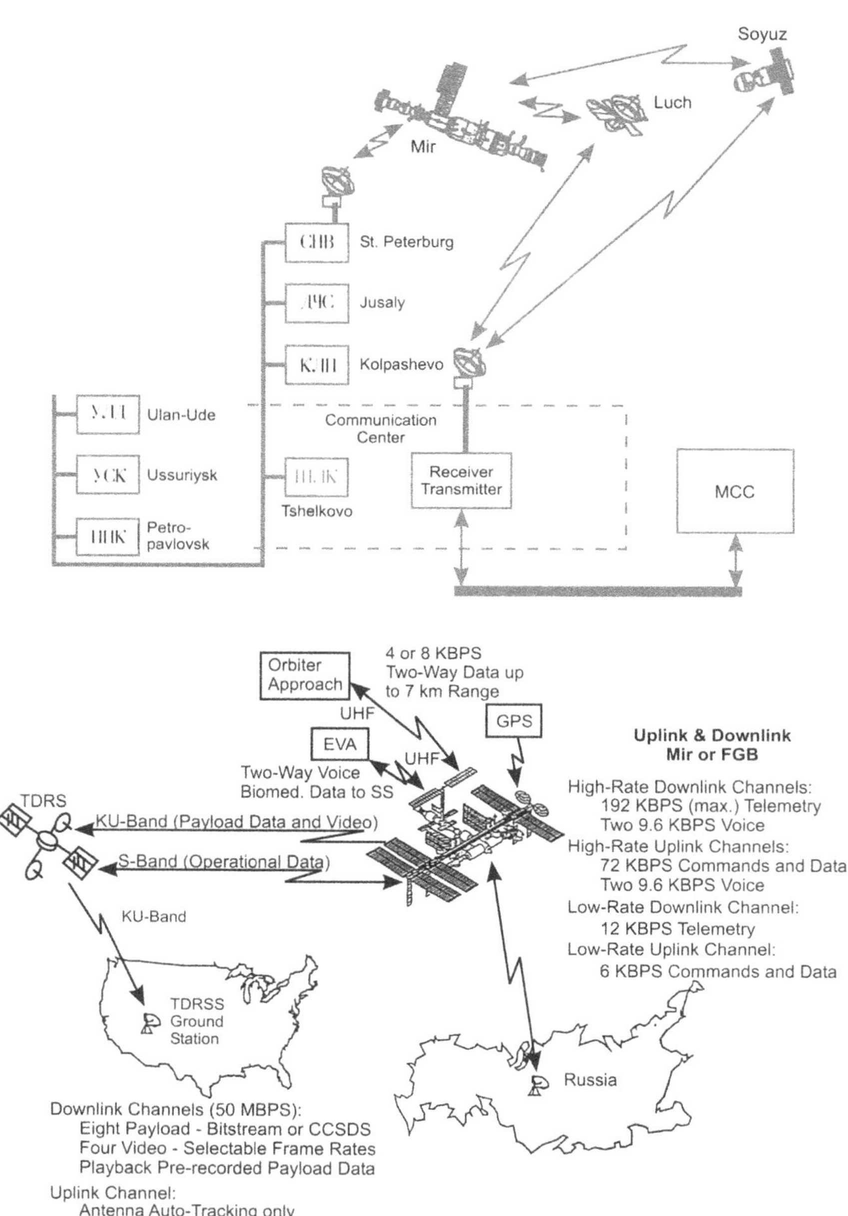

Fig. 12.9. Transmission Paths to the Space Stations Mir and ISS

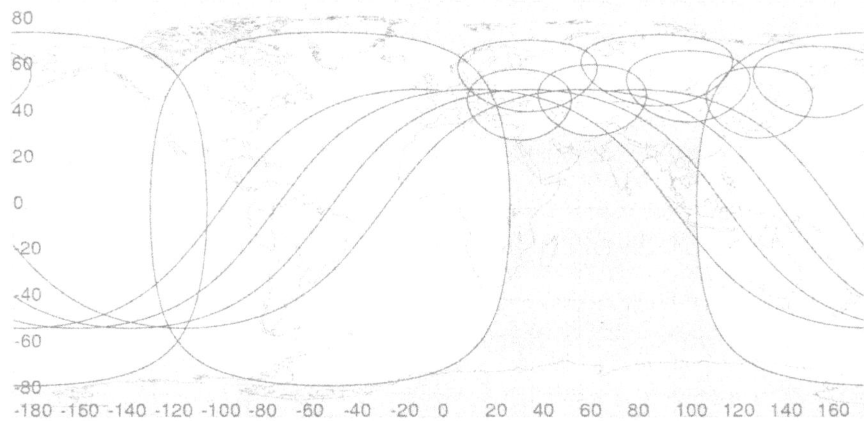

Fig. 12.10. The Reception Areas Planned for ISS including Two TDRS Satellites and Seven RKA Ground Stations

Table 12.5. Frequency Bands for Satellite Links

Frequency Band	Frequency Range [GHz]		Service	Limits of Downlink Power Flux Density
	Uplink	Downlink		[dBW/m²]
UHF	0.2–0.45	0.2–0.45	Military	–
L	1.635–1.66	1.535–1.56	Maritime/Navigation	-144 / 4 kHz
S	2.65–2.69	2.5–2.54	Commercial TV	-137 / 4 kHz
C	5.9–6.4	3.7–4.2	National Communications Satellite	-142 / 4 kHz
X	7.9–8.4	7.25–7.75	Military Communications Satellite	-142 / 4 kHz [a]
K_u	14.0–14.5	12.5–12.75	National Communications Satellite	-138 / 4 kHz
K_a	27.5–31.0	17.7–19.7	National Communications Satellite	-105 / 1 MHz
SHF/EHF	43.5–45.5	19.7–20.7	Military Communications Satellite	–
V	~ 60		Inter-Satellite Links	–

[a] No Limits in the Military Band from 7.70 to 7.75 GHz

time, operational resources and also their links. This will happen in a decentralized way and will be defined on short notice. The astronaut will, if no other opportunity seems favorable to him, call his or her family at home on a cellular telephone at $ 1/ minute, and in the case that a "business connection" to an experimenter should not be available at short notice, the astronaut will connect the corresponding experiment to the experimenter's laptop with the help of an on-board multimedia-capable laptop, via a publicly available multimedia satellite systems.

12.2.3 Distributed Data Systems

Until recently, space systems consisted mainly of centralized systems, where data processing and transmission were carried out by one single computer, i.e., centrally. The first step in the direction of distributed systems consisted of the use of spatially separated receivers and multiplexers. Both the Space Shuttle and the Spacelab have initial approaches of distributed systems. In their case, more than one computer carries out the same task, mainly for reasons of redundancy. In the case of the International Space Station, the DMS tasks listed in Sect. 12.2.1 will mainly be fulfilled in a distributed mode. In the context of DMS architectures, often the terms network topologies, physical transmission paths, and the components necessary for it as well as software and programming languages are used.

Network Topologies: Distributed Systems are connected with one another by means of data lines. Apart from 1:1 connections such as between computer and printer, the smallest system is called a bus. Three or more components are connected in parallel to the same line. A signaling protocol determines the way in which the devices must communicate with one another, how they signal their readiness and how the transmission and data storage is organized among them. Some typical examples are the data and address bus of a microprocessor or the SCSI bus for hard drives and other peripheral devices.

The unit next largest in size is the Local Area Network (LAN). A LAN can show different forms of connection which are often adapted to the situation present in each special case. Typical constellations are again the bus (Ethernet) or the star and ring topologies. In the case of a LAN, all stations are principally "known" among one another, so the routing of data does not take place here. The transmission path can be considered loss-free; disturbances are normally only caused by component failure and are immediately detected. The signaling protocol is mainly limited to the allocation of access to the network for the single stations.

Wide Area Networks (WAN) have the longest paths. Here, the different systems are connected to one another by sometimes redundant lines. The individual ways cannot be considered loss-free anymore, i.e. data may be lost or disturbances may occur which will change the data. This must be prevented by suitable error checking in order to guarantee the data's consistency. The most flexible networks that are connected in several ways to one another sometimes use seemingly arbitrary connections to send the data on the fastest possible way. If this way is blocked or one subsystem is not fully functioning, the next best way is calculated (routing). In this context, each station along the way must be able to guarantee storage and forwarding of the data. Modern telephone systems, to which the future satellite-based mobile radio communication systems such as Iridium and Globalstar (Satellite Personal Communication Networks, SPCN) also belong, work according to this principle.

Such systems are characterized by the fact that they can be flexibly extended and operated with device-independent signaling protocols. The "International Standards Organization (ISO)" suggested a corresponding standard for networks called "Open Systems Interconnect" (OSI) standard. This standard will certainly gain ground aboard space stations and platforms since it can display details in a transparent way and the user must only know the higher levels of the signaling protocol which is structured in a hierarchical manner. It is thus possible that experimenters on ground can reach and control their experiments via commercial computer networks. Much work remains to do be done for the space station operators and safety experts in order to define and monitor the operational and safety standards resulting from this development.

Physical Data Links consist of twisted pairs of wires, coaxial cables or fiber-optic cables. Twisted pairs are used for star topologies. They are inexpensive, allowing data rates of up to 100 Mbit/s and cable lengths of up to several hundred meters. Coaxial cables are used in bus topologies. They are better shielded and their advantage is that connecting further components is relatively easy. Fiber-optic cables transmit modulated light signals. They are immune to electromagnetic interferences and their maximum data rate is limited mainly by connected components such as light-emitting diodes and lasers (on the transmitter side), as well as photodiodes or other components (on the receiver side). The maximum data rate might well reach several Gbit/s. Aboard space stations, fiber-optic cables are increasingly replacing the common coaxial cables for short distances and low data rates.

The only possible means of data transmission for mobile users, either directly from ground to the space station or from ground via a satellite to the station, is radio transmission. The transmission capacity largely depends on the transmission frequency and this, in turn, on the transmission window. The latter is defined by the attenuating effect of the ionosphere at low frequencies and by that of the atmosphere at high frequencies, and it is in the range of several hundred MHz up to several GHz (see Fig. 12.11). One characteristic property of radio links is that the signal-to-noise ratio decreases directly with the square of the distance, thus requiring increasingly large transmitter and receiver antennas. The duration of a link between the ground antenna and the space station antenna is given by the orbital parameters, especially by the station's orbital altitude and the smallest elevation angle ε of the ground station in which a reliable radio link can still be established. The Fig. 12.12 and Fig. 12.13 contain some important data which will later on be needed for designing the transmission system. The radio link with a ground station at space station altitudes of 200–500 km, i.e. orbital periods of about 90 minutes, will last for a maximum of 10 minutes and the Doppler frequency shift, e.g. at 1.6 GHz (L-Band), will be at a maximum of 40 kHz.

Software and Programming Languages. Over the past few decades, costs for computers and components for data transmission, storage and presentation, etc. have drastically decreased. For that reason, the software and programming mainly determine the costs of the DMS. Until the present time, assembler programming has been wide-spread. It is essential everywhere where complete access to hardware or high speed processing is required. In the future, a combination of assembler programming and higher programming languages such as C or Ada that allows both direct hardware access and user-friendly, higher-level programming will be more common. Higher programming languages have considerably improved the productivity of small programs and allowed large complexity. However, as the complexity increased, productivity again decreased since design, testing and verification as well as debugging, i.e. the search for errors, reduced the productivity and, as a conse-

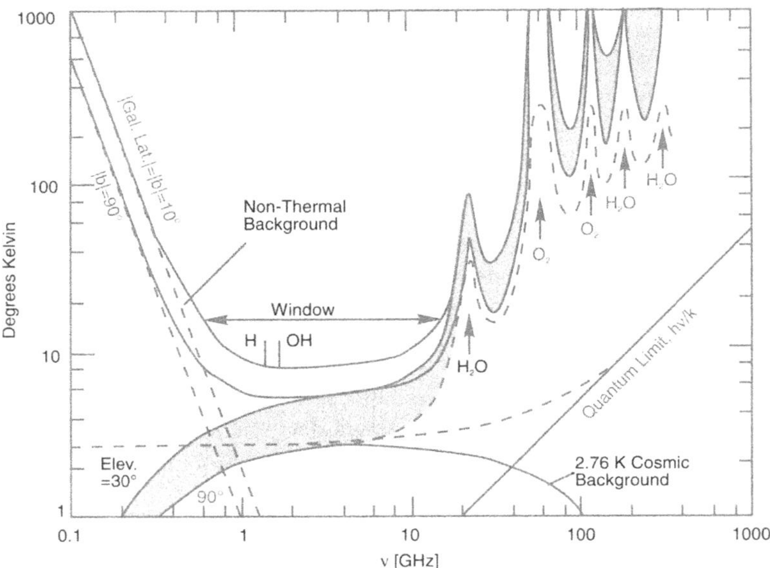

Fig. 12.11. The Window for Transmission Frequencies from the Earth to the Space Station is Limited by the Different Noise Sources (see also Fig. 12.14)

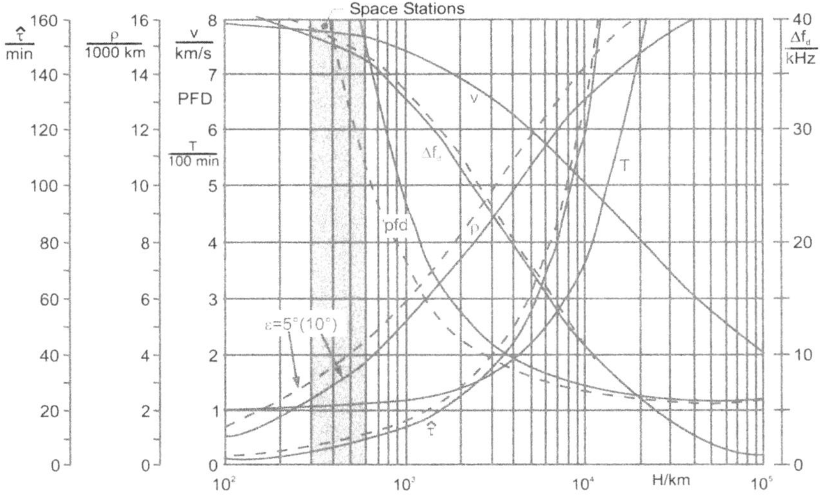

Fig. 12.12. The velocity v, orbital period T, coverage area ρ per satellite during equator passage and the maximum period of visibility τ for satellites on circular orbits. *PFD* is a Power Flux Density normalized to $H \to \infty$ for radio signals, and Δf_d is the Doppler frequency shift at 1.6 GHz.

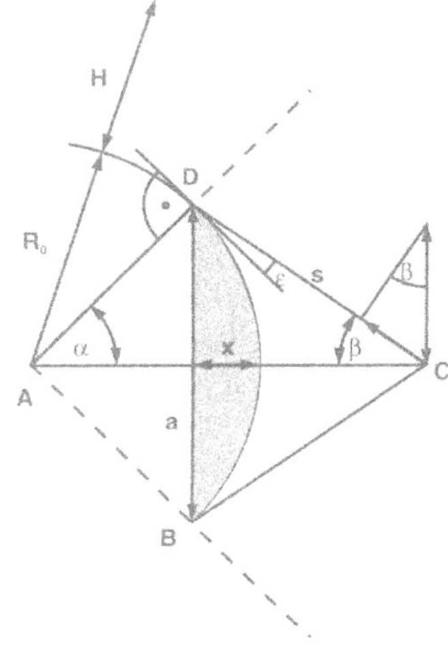

$$\beta = \sin^{-1}\left[\frac{\cos\varepsilon}{1 + H/R_0}\right]$$

$$s = \frac{R_0 \cos(\varepsilon + \beta)}{\sin\beta}$$

Revolution Time $T = 2\pi[(R_0 + H)^3/(R_0^2 g_0)]^{1/2}$

Velocity $v = 2\pi(R_0 + H)/T$

Maximum Time of Sight (Zenith Passage)

$\ddot{\tau} = [\pi/2 - \varepsilon - \beta] \cdot T/\pi$

Arc Section $\rho = \widehat{BD} = 2R_0[\pi/2 - \varepsilon - \beta]$

Normalized Power Flux Density

$PFD = [s(H)\beta_0(H)/s(H \to \infty)\beta_0(H \to \infty)]^{-2}$

with $\beta_0 = \beta(\varepsilon = 0)$

Doppler Frequency Shift (at Pole)

at $f_c = 1.6\ \mathrm{GHz}$: $\Delta f_d = f_c \cdot v_r/c$

with $v_r = (R_0 g_0)^{1/2}(1 + H/R_0)^{-3/2}\cos\varepsilon$

Elevation Angle: ε

Satellite Altitude: H

Earth Radius: $R_0 = 6.37104 \cdot 10^6$ m

Earth Acceleration: $g_0 = 9.80665\ \mathrm{m/s^2}$

Earth Angular Speed: 0.00437527 radians/min

Fig. 12.13. Radio Link Geometry

quence, also the advantages. This problem will be largely avoided by structured programming where brief and modular routines are used. Structured programming avoids the use of "goto" commands and makes the program's flow more clear. A typical developmental environment for these programming languages comprises analysis tools, editors, compilers, debuggers, libraries of standard routines, and integrated test routines, and can also be used for automated code development. An example of such a structured programming language is Ada, a language initiated by the US Department of Defense (DoD). In the meantime, the use of Ada was stopped again: this language employs a very strict type checking, and thus the processing and casting of data packets easily becomes tedious, because, especially in the case of limited bandwidths, it is common practice to reuse data fields. The object-oriented programming is carried one step further: the code and the data form one unit and thus feature better protection against inconsistent data. Object-oriented programs hide the internal method of operation of their modules and thus allow the exchange of routines without the necessity to change the rest of the program.

ISS Data Handling and Communication. ISS provides an onboard command and data distribution network associated with a forward and return antenna communication link to transfer research date and video to the payload developer. A 72 kbps S-band forward link is used to send commands for payloads. Data onboard is distributed from the payloads to a prime and redundant (44 input, 36 output) Automated Payload Switch (APS). The payload MIL-STD-1553B bus, the Ethernet, and high data links are connected to the APS. The APS selects 8 outputs for the High Rate Frame Multiplexer (HRFM). The HRFM distributes the data to the high-rate modem for distribution to the 50 megabit per second (Mbps) Ku-band system. The estimated coverage for the Ku-band system is approximately 50%. Later in the assembly sequence, the data rate of the Ku-band system will be increased to 75 Mbps. When assembly is complete, the data rate will be 150 Mbps. The orbital coverage of the Ku-band system is predicted to be just over 60%. However, several initiatives are in place to increase this orbital coverage to continuous availability. Included in these data rates at ISS assembly complete configuration are four compressed channels of video downlink, and three channels of video uplink. The Payload Multiplexer/Demultiplexer provides 300 megabytes of nonvolatile mass storage for payloads. A 216 Gigabit communications outage recorder is provided to record research data during loss of signal with the communication system. Video onboard is distributed from the payload to one of several switches. Each switch routes the signal to a recorder or monitor, or distributes it to the downlink. Three video compression units will be onboard the station (one each in the US Lab, JEM, and COF) to allow video to be downlinked simultaneously with the research data.

12.2.4 Radio Communication System Design

This section will address only those fundamentals of communications systems via satellite radio links that are important for the understanding and the preliminary design of space stations. For the design of radio systems, the physical transmission path between transmitter and receiver will be investigated as it is presented in Fig. 12.9 between ground antenna and space station antenna (Mir or ISS), both down- and uplink. Based on the transmitting power, the following will have to be calculated: the power emitted by the transmitter antenna, its distribution in the solid angle according to the radiation pattern, the part of the transmitting power collected by the receiver antenna, and the noise power which additionally occurs in the receiver, thus determining the signal-to-noise ratio.

The result of such a calculation (called link calculation), shall be presented at this point, whereas the derivation leading to this result will be presented later on. This calculation forms the relation between data rate, antenna size, transmission path, and transmitting power. It is indicated for digital data as follows:

$$\frac{E_b}{N_o} = \frac{P L_1 G_t L_s L_a G_r}{k T_s R} \equiv \text{Link Equation} \tag{12.1}$$

Here, E_b/N_0 is the relation between the energy stored in one bit and the noise power density where P is the transmitting power, L_l is the loss between transmitter and antenna, G_t is the antenna gain of the transmitter antenna, L_s is the so-called "free-space losses" that depends on the radio link, L_a is additional transmission (attenuation) losses, G_r is the gain of the receiver antenna, k is Boltzmann's constant, T_s is the system noise temperature and R is the data rate. In order to obtain acceptable bit error rates, the ratio E_b/N_0 must be between 5 and 10. Once the data rate R and the orbital altitude are determined (and thus the maximum distance between the transmitter and the receiver antenna), most of the variables that will determine design and costs, such as P, G_t, G_r and R, can be calculated. The main transmission losses are caused by the geometry of the antenna, because the area of the receiver antenna can only "see" and receive a small fraction of the radiation spread into space. Beside these main losses, there are also absorption losses due to ionospheric and atmospheric effects. In clear weather, they are relatively small within a range of 200 MHz–20 GHz (cf. Fig. 12.11 and Fig. 12.14).

For the derivation of Eq. 12.1, it can be seen as a product of two parts of different nature:

$$\frac{E_b}{N_0} = \frac{PL_1}{kT_sR} \cdot G_tL_sL_aG_r \tag{12.2}$$

The first term on the right-hand side represents the transmitter power PL_l in relation to the inevitable noise power (related to one bit) of the transmitter. The second term describes the spread mechanism from antenna to antenna, by which the transmitting power is reduced during the transmission. This derivation is based on a transmitter in the center of a sphere with a radius s. PL_l is the transmitting power which is isotropic and thus homogeneously illuminates the sphere's surface. Thus, the power flux density W is $PL_l\,/\,(4\pi s^2)$. When the beam of the transmitter antenna is narrow, the power flux density must be multiplied by the gain of the transmitter antenna. Transmission losses between the transmitter and the sphere's surface, for example, by absorption of the atmosphere or by rain, are considered with a factor L_a. The product PL_lG_t is called the "Equivalent Isotropic Radiated Power" (EIRP); i.e., $(EIRP)L_a\,/\,(4\pi s^2)$ [W/m^2] are transmitted onto the sphere's surface. Now, this value must only be multiplied by the effective area of the receiving antenna A_r in order to obtain the received power C.

The effective area of the receiving antenna is $\eta\ (\pi D^2/4)$. Here, the efficiency η is between 0 and 1, and it is also a function of the surface roughness and other deviations of ideal antenna properties. For parabolic antennae, η is in the range of 0.55 and 0.7. Thus, the received power is:

$$C = \frac{PL_1G_tL_aD_r^2\eta}{16s^2} \tag{12.3}$$

A close relationship exists between the antenna gain, the geometric size of the antenna and the wave length due to the electrodynamics. The antenna gain can also be defined as the relation between the effective receiving area A_r and the effective area $\lambda^2/4\pi$ of a (hypothetical) isotropic antenna; in this context, λ is the wavelength of the transmitted signal. Hence, the following is valid for the receiving antenna:

$$G_r = \left(\frac{\pi D_r^2\eta}{4}\right)\left(\frac{4\pi}{\lambda^2}\right) = \frac{\pi^2D_r^2\eta}{\lambda^2} \tag{12.4}$$

When we now put Eq. 12.4 in Eq. 12.3, the following is obtained:

$$\begin{aligned} C &= PL_1 \cdot V = PL_1 \cdot G_t L_a G_r \left(\frac{\lambda}{4\pi s}\right)^2 \\ &= PL_1 G_t L_s L_a G_r = (EIRP) L_s L_a G_r \end{aligned} \tag{12.5}$$

Here, C is the received power, V the reduction factor due to the spread mechanism and $L_s = (\lambda/4\pi s)^2$ is the so-called "free-space loss". In the case of digital data, the energy received per bit E_b equals the received power C multiplied by the duration of a bit $1/R$, thus $E_b = C/R$ with a data rate R in bps and E_b in Ws or J.

The noise power in the receiver normally has a homogeneous spectral noise power density $N_0 = kT_s$, with T_s as system noise temperature. The total noise power in the reception area with a bandwidth B (B being dependent on the data rate, the modulation and the coding) is

$$N = kT_s B = N_0 B \tag{12.6}$$

with N_0 in W/Hz, N in W, k as the Boltzmann's constant = $1.381 \cdot 10^{-23}$ J/K, T_s in K and B in Hz. Now with Eq. 12.5 and Eq. 12.6, the link equation Eq. 12.1 indicated in the beginning is obtained.

Normally, link budget calculations are carried out in decibels (dB) since this allows the simple addition or subtraction of the parameters. The gain or loss of one element of the link equation is indicated by the ratio P_0/P_i; expressed in a logarithmic scale decibel (dB), this becomes $10 \cdot \log_{10}(P_0/P_i)$, where P_i is the input power to an element such as to the antenna or to the transmission path and P_0 is the output power. Even dimensional quantities as, e.g. the power, are expressed in dB, like dBW.

Knowing this, the link equation can be expressed in decibels as follows

$$\begin{aligned} \frac{E_b}{N_0} &= P + L_1 + G_t + L_s + L_a + G_r + 228.6 - 10\lg T_s - 10\lg R \\ &= EIRP + L_s + L_a + \frac{G_r}{T_s} + 228.6 - 10\lg R \end{aligned} \tag{12.7}$$

where E_b/N_0, L_l, G_t, L_S, L_a and G_r are given in dB, P in dBW, $10 \lg k = -228.60$ dBW/(Hz K), T_s in K and R in bps. The second equation is used when $EIRP$ and G_r/T_s are indicated in dBW or dB.

The ratio of carrier-to-spectral noise density ratio is as follows:

$$\frac{C}{N_0} = \frac{E_b}{N_0} + 10\lg R = EIRP + L_s + L_a + \frac{G_r}{T_s} + 228.6 \tag{12.8}$$

The system noise temperature T_s depends on the sum of all parts of noise sources which are collected by the receiving antenna. In Fig. 12.14, noise temperatures and the equivalent signal attenuation resulting from them are shown on a large frequency range covering radio communications for the space station. As a consequence, it is advisable not to orient the receiving antenna with a narrow lobe to the Sun. The frequency range between 1.0 and 10 GHz is above the range of considerable iono-

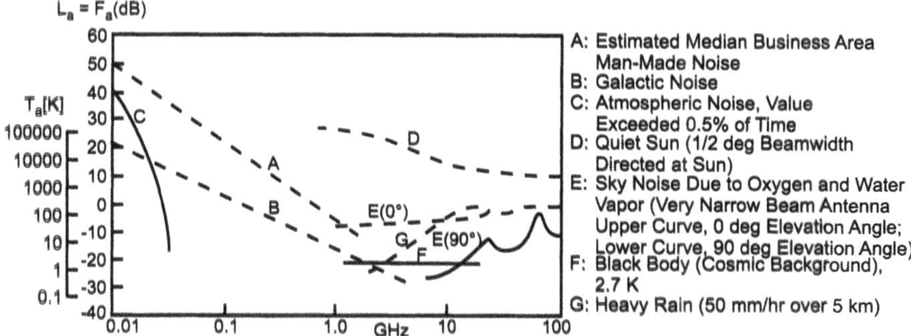

Fig. 12.14. Signal Attenuation by Natural Noise Sources and Others [Wertz 91]

spheric or man-made noise and less influenced by atmospheric and rain attenuation. The total attenuation for elevation angles over 10° is only a few dB and thus is not problematic. For that reason, this range is called "radio communications window".

All noise sources between antenna input (with T_{ant}, containing all noise sources received) and receiver output contribute to the receiver noise temperature T_r. These can be caused by lines, filters, amplifiers and other components. If the noise levels of both the antenna and receiver are considered, the system noise temperature can be calculated as follows:

$$T_s = T_{ant} + \frac{T_0(1-L_r)}{L_r} + \frac{T_0(F-1)}{L_r} \tag{12.9}$$

Here, L_r is the transmission loss between antenna and receiver. The second term on the right-hand side of the equation corresponds to the noise of this transmission path, and the third term is due to the noise of the receiver with a reference temperature T_0 and the noise figure $F=1+T_r/T_0$. For a low-noise receiver, $T_0 = 290$ K, $L_r = 0.89$ or 0.5 dB and $F = 1.1$ or 0.4 dB. Hence, the following results for the noise levels in Eq. 12.9: $T_s=T_{ant}$+36K+33K.

12.2.5 Antennas

In Fig. 12.15, the most common antenna types for space flight application are illustrated. Taking the horn antenna as an example, some of the characteristic properties of antennas will be shown. The transmitted power radiated from a wave guide horn in consideration of the radiation wave properties are presented in Fig. 12.16. The main lobe and the side lobes are due to the fact that in direction (A) there is no extinction due to the identical phases of all radiation, whereas in direction (B) there is complete extinction including intermediate levels in between. The direction (B) is characterized by the fact that for each point on the wave front there is a corresponding point in the distance $D/2$ with a phase shift of $\lambda/2$. Therefore, both sources make one another extinct. Half the cone angle of the main lobe is given by $\alpha/2=\lambda/D$ [rad],

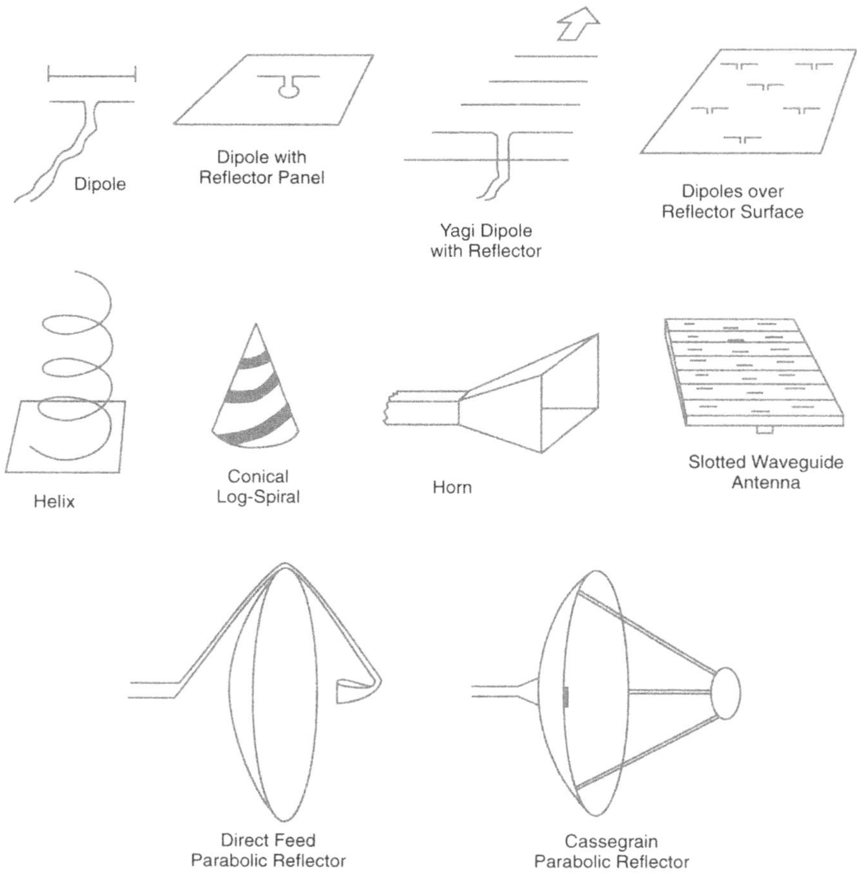

Antenna Type	Gain G	Effective Area
Isotropic Antenna	1	$\lambda^2/4\pi$
Elementary Dipole	1.5	$1.5\lambda^2/4\pi$
$\lambda/2$ Dipole	1.64	$1.64\lambda^2/4\pi$
$\lambda/4$ Dipole over Conducting Surface	3.28	$3.28\lambda^2/4\pi$
Horn	$10(A/\lambda)^2$	0.81A
Parabolic Reflector	$6.2\text{-}7.5(A/\lambda)^2$	0.5A to 0.6A
Ideal Phased-Array Planar Antenna	$L_1\pi(A/\lambda)^2$	A

Fig. 12.15. Typical Space Station Antennas with their Gain and Effective Antenna Area

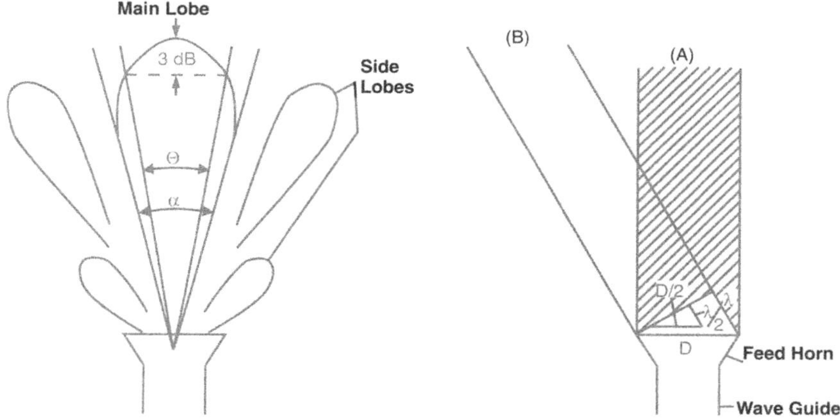

Fig. 12.16. Transmission Characteristics of a Horn Antenna in Consideration of the Wave Properties of the Radiation

whereas the technically interesting angle Θ up to which at least half of the maximum transmitting power is available makes up about 60% of the full angle:

$$\Theta_{3dB} \approx 0.6\alpha = 1.2\frac{\lambda}{D}[\text{rad}] \approx 70\frac{\lambda}{D}[°] \tag{12.10}$$

A horn antenna with an aperture of 10 cm and a wavelength of 3 cm (corresponding to a frequency of $f = c/\lambda = 10$ GHz), thus has a beam width of 21°. If the transmitted signal were reflected by a parabolic reflector (as presented in Fig. 12.15, lower left) with a diameter of 1 m, the diameter D of the initially planar wave front would be increased to 1 m and the beam width would be reduced to 2.1°.

The gain of a beam antenna puts the radiated power of the main lobe in relation to the radiated power of a (hypothetical) isotropic antenna of the same power. According to Eq. 12.4 and Eq. 12.10, the following is generally valid:

$$G = \frac{\pi^2 D^2 \eta}{\lambda^2} \quad \text{and} \quad \Theta_{3dB} \approx 70\frac{\lambda}{D}[°] \tag{12.11}$$

The following can be directly derived from it: $G\Theta_{3dB}^2 \sim \eta = \text{const.}$ for every antenna of a certain type. Some typical values for the gain and the effective area of antennas are given in Fig. 12.15 as they are used in the field of space stations.

The link equation Eq. 12.5 deserves special attention with regard to the antenna properties and the selection of the frequency. As can be seen from Fig. 12.15, a distinction must be made between two antennas: beam antennas (e.g. horn and parabolic antennas with $G \sim 1/\lambda^2$) and antennas that radiate at a constant angle (dipoles with G independent of the wavelength λ). Thus, for a transmission path with a transmitter antenna (gain G_t), for the transmission path s, (free-space losses $L_s = (\lambda / 4\pi s)^2$) and for the receiver antenna (gain G_r), the following options for the product $V = G_t L_s G_r$ result:

- Option A: Two beam antennas, i.e. $V \sim 1/\lambda^2$ and a frequency $f = c/\lambda$ as high as possible
- Option B: One beam antenna and one antenna with a constant angle, i.e. $V = constant$ and thus, the selection of the frequency is not influenced
- Option C: Two antennas with a constant angle, i.e. $V \sim \lambda^2$ and thus, the frequency should be as small as possible.

As a consequence, in the case of transmissions from the Earth to the space station, options A and B within the frequency window (cf. Fig. 12.11 and Fig. 12.14) are to be chosen; the disadvantage in the case of option B, i.e. a smaller receiving power can often be balanced by the advantage of only one antenna needing to be tracked. When dealing with large transmission paths outside the atmosphere, for example, between a space station and a geostationary relay satellite, option A is preferred; with LEO or HEO relay satellites, option B might be the more favorable from a technical point of view. For short-range transmissions, however, i.e. rendezvous maneuvers and EVAs, option C is technically the simplest, but an absolutely sufficient solution.

The polarization of the antennas used in pairs should always be identical, i.e. in the case of linear polarization, always vertical or horizontal, and in the case of circular polarization, always oriented towards the left or the right. If this rule is not observed, there will be significant polarization losses. A linear/circular mismatch, for example, would mean a loss of 3 dB. Further losses might be due to disturbances indicated in Fig. 12.14 and also due to intermodulation, interference, multi-path and shading effects, antenna mispointing and surface roughness and other effects [Hartl 88, Maral 86 and other textbooks on radio engineering and telecommunications]. So-called "link margins" from 5–15 dB, according to the application, must be taken into account when considering the effects mentioned above.

12.2.6 Modulation and Coding

Around 1980, digital technology totally replaced the analog signal and transmission technology in the field of space systems. In a digital system, usually at first the analog signal, e.g. the voltage for a servo motor, must be sampled with at least double the maximum signal frequency f_m. As early as 1928, Nyquist had already found out that a signal could theoretically be reconstructed from the digital signal if the following is valid for the sampling rate:

$$f_s \geq 2 \cdot f_m \qquad \text{Nyquist Theorem} \qquad (12.12)$$

For example, the human voice has a bandwidth of about 3.5 kHz, i.e. the signal must be sampled at least 7000 times per second. In reality, due to technical limitations, a factor of 2.2 or more has to be included into Eq. 12.12 for a good sound reproduction, i.e. a sampling rate of $f_s = 7.7$ kHz. In the case of modern telephone systems, language is sampled with indeterminate losses of sound quality with a sampling rate of between 4 and 8 kHz. For music CDs, the sampling rates are 44.1 kHz, for TV-

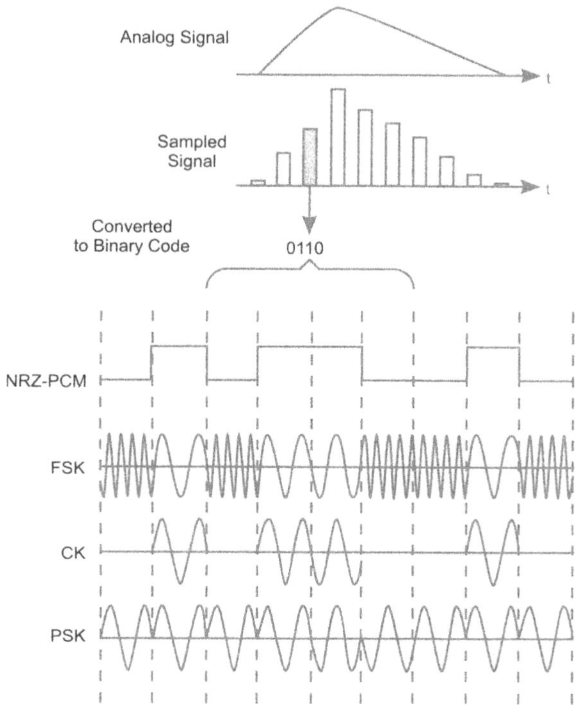

Fig. 12.17. Sampling and Conversion of an Analog Signal into Binary Code with Subsequent Modulation of an RF-Carrier Frequency

video transmission in black/white or color they are up to several MHz. Thanks to modern techniques for data compression and coding, factors of about 1 in Eq. 12.12 are obtained by means of consciously accepted limitations in the signal quality. Future satellite systems for multimedia applications are currently being planned for a channel bandwidth of about 2 MHz.

In Fig. 12.17, the way in which an analog signal is digitized by sampling and how it is converted to a binary code (generally 2^n steps per analog value) is shown. The data stream between microprocessor and modulator/demodulator consists of a series of binary codes. For radio transmission, the coded signal must be modulated onto a high-frequency carrier. The carrier frequency is always strictly assigned and fulfills the requirements discussed above.

There are principally three ways to modulate a data signal onto the HF-carrier: either by amplitude, by frequency or by phase of the HF-carrier. In the case of analog signals, these three modulation methods are called amplitude modulation (AM), frequency modulation (FM) and phase modulation (PM). In the case of digital signals where the carrier can only take discrete states, the terms Carrier Keying (CK), Frequency Shift Keying (FSK) and Phase Shift Keying (PSK) are used. In Fig. 12.17, all three modulation methods are illustrated. Carrier Keying (CK), due

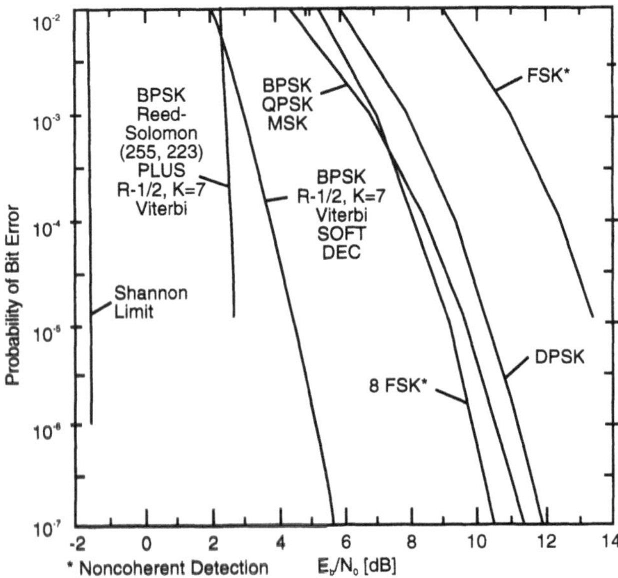

Fig. 12.18. Probability of Bit Error as a Function of E_b/N_0 [Wertz 91]

to its low transmission safety, no longer plays a role. Today, mainly the PSK modulation in different variants is used (binary: BPSK, quadriphased: QPSK, differential: DPSK, etc.) mostly in connection with error-correcting codes (e.g. the Viterbi-Code with $R = 1/2$, i.e. every second bit is a correction bit) [Renner 88]. The probability of bit error as a function of E_b/N_0 is shown in Fig. 12.18. The typical values of the probability of bit error that were actually confirmed are, for example, from 10^{-5} to 10^{-7} for $E_b/N_0 = 3$ dB to 6 dB.

The channel capacity of a radio link as an important resource must be used in an efficient way in order to transmit maximum data from many data sources. This can be carried out simultaneously for different transmitters with different carrier frequencies (Frequency Division Multiple Access – FDMA) or by means of a carrier frequency and different time-slots (Time-slot Division Multiple Access – TDMA) or, thirdly, by means of different coding by an HF Signal (CDMA with Spread-Spectrum (SS) modulation). The line capacity of a radio link can, as shown in Fig. 12.19, be represented in the form of a cuboid, where, with the help of cuts made perpendicular to the axes (frequency, time, signal-to-noise ratio), the channel capacity is apportioned.

12.2.7 The Tracking and Data Relay Satellite System (TDRSS)

The concept of the US TDRS system is presented in Fig. 12.20. Frequencies and other factors influencing the transmission path can be seen from Table 12.6. Data of the satellites which fly at altitudes of up to several thousand kilometers and that of space stations at lower altitudes are transmitted via two TDRSS satellites which are

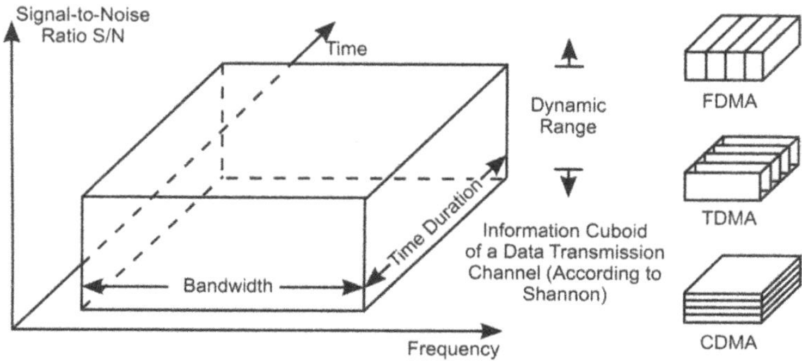

Fig. 12.19. Channel Capacity of a Radio Link in the Form of a Cuboid with the Axes Frequency, Time and Signal-to-Noise Ratio

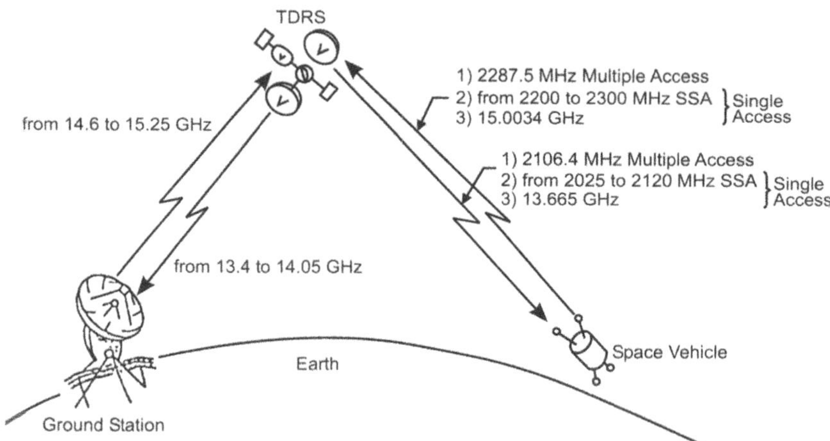

Fig. 12.20. The TDRS System

positioned on geostationary orbits at a longitude of 41° West and 172° West. These TDRSS satellites, viewed from Earth, always have the same angle with respect to Earth. They receive telemetry, telecommand and other data signals from satellites and space stations, amplify those signals and send them, after a frequency conversion, to a central ground station in the USA. Vice versa, this station sends telemetry (TM), telecommand (TC) and other data signals to the TDRSS satellites, which then transmit the data to satellites or space stations. Under the given conditions, there is usually radio transmission via TDRSS to the central ground station as can be seen in Fig. 12.10. For that reason, in many cases, data storage aboard the space station is no longer required.

The European Space Agency ESA, as well, had planned that a data relay satellite DRS would be used from the 1990's onwards. It had been intended to be compatible with the TDRSS and to have additional broadband channels for high frequencies.

Table 12.6. TDRSS Units for Radio Transmissions to Near-Earth Satellites

Function	S-Band Multiple Access		K_u-Band Single Access		S-Band Single Access	
Frequency [MHz]	TM [a] 2287.5	TC [a] 2106.4	TM 14896–15121	TC 13750–13800	TM 2200–2300	TC 2025–2120
Antennas	28 dB	23 dB	52.6 dB	52 dB	36 dB	35.4 dB
EIRP [dBW]		34		43–49		43.4–46
Bandwidth [MHz]	5	5	88/225	50	10	20
Modulation		SS/PSK		SS/PSK		free
p_E		≥ -154		≥ -152		
Capacity	20 Satellites Simultaneously		2 Satellites Simultaneously		2 Satellites Simultaneously	

a For Acronyms See Glossary
EIRP: Equivalent Isotropically Radiated Power = Antenna Gain · Transmission Power
SS/PSK: Spread Spectrum with PSK Modulation
p_E: Necessary Power Density with SS/PSK at TDRS in $dBW/(m^2 \cdot 4\ kHz)$

Such relay satellites are necessary for handling the large data transmission requirements of space stations and Earth observation satellites. For budgetary reasons, the development of this satellite was abandoned by ESA. Instead, the technology satellite Artemis, which was built by ESA and which is to be launched by Japan on an H-II rocket in 1999, will presumably be used in order to test relay services, including those for ISS. As time goes on, it would seem quite certain that commercially operated data relay satellite systems will eventually be used to guarantee the immense data flow between the space station and the Earth.

Table 12.7 shows a simple link calculation for two examples:

a) Signal from a tracked parabolic antenna (1.5 m diameter) on Earth is to be transmitted in the L-band to a space station and shall there be received by a hemispherical antenna ($G_r = 3$ dB).
b) Signal from a parabolic antenna (1 m diameter) aboard a space station is to be transmitted to a geostationary TDRS data relay satellite in the Ku-Band.

12.2.8 Data and Communication Systems for the ISS

Some of the most important data and communications systems for the International Space Station are presented here. Fig. 12.21 shows the command and data handling system together with the payload data system, and Fig. 12.22 illustrates the US part of the telecommunications and antenna operating system.

12.3 Automation and Maintenance

Often, crewed space flight is at the center of criticism due to the relatively high costs involved. Alternative suggestions usually entail automated experimental facilities [DPG 90, LuR 91, LuR 93] with the help of which experiment costs are supposed to

Table 12.7. Link Calculation for
a) Narrow-Band TC Signal from a Ground Station to a Space Station and
b) Broad-Band Multimedia Signal from the Space Station to the TDRSS Satellite

a) Uplink: Ground to ISS			**b) Downlink: ISS to TDRSS to Ground**
1.6	Frequency	[GHz]	14.0
10.0	Transmitting Power P_t	[dB W]	4.77
25.4	Antenna Gain G_t	[dB]	40.72
35.4	EIRP = P_tG_t	[dB W]	**45.49**
-158.34	Free Space Losses	[dB]	-207.53
3.0	G_r Receiving Antenna	[dB]	52.0
-25.5	T_r Operation Noise Temperature	[dB1/K]	-25.55
228.6	Boltzmann's constant	[dBHzK/W]	228.6
-10.0	Additional Attenuation + Margin	[dB]	-10.0
73.16	(C/kT)r	[dB Hz]	**83.0**
-10.0	E_b/N_0	[dB]	-10.0
63.16 $\hat{=}$ 2 MHz	Data Rate	[dB Hz]	73.0 $\hat{=}$ 20 MHz
	Data Used:		
1.5 m	Antenna Diameter		1 m
10 W	Transmitting Power		3 W
10°	Minimum Elevation		10°
290 K	System Noise Temperature		290 K

be considerably reduced. This chapter aims at discussing possibilities for and limitations of automation technology. Experimental facilities, as they are used in the fields of materials science and biology, will serve as examples.

12.3.1 Payload Operation aboard a Space Station

Compared to experiment platforms and laboratories, the operation of payloads and experiments aboard a space station is characterized by some important differences which will be outlined in the following.

In the case of relatively briefly operated systems such as Spacelab, Spacehab or EURECA, the total experiment facilities can and have to be transported into orbit at the beginning of a new mission. A space station, on the other hand, offers the opportunity to make better use of the transportation capacities available: the emphasis is on resupply of experiment and test hardware, whereas the resupply of payloads can be limited to an absolute minimum. This leads to a markedly longer operational life for scientific instruments, typically in the range from 2 to 5 years. However, it is unrealistic to expect that operation over such a long period of time will be free of failures and thus of maintenance work. For that reason, in the future, maintenance and repair tasks will be a firm part of the crew's tasks aboard a space station.

On the other hand, astronauts can dedicate markedly less time to scientific operations due to the complexity of a space station with its variety and high numbers of

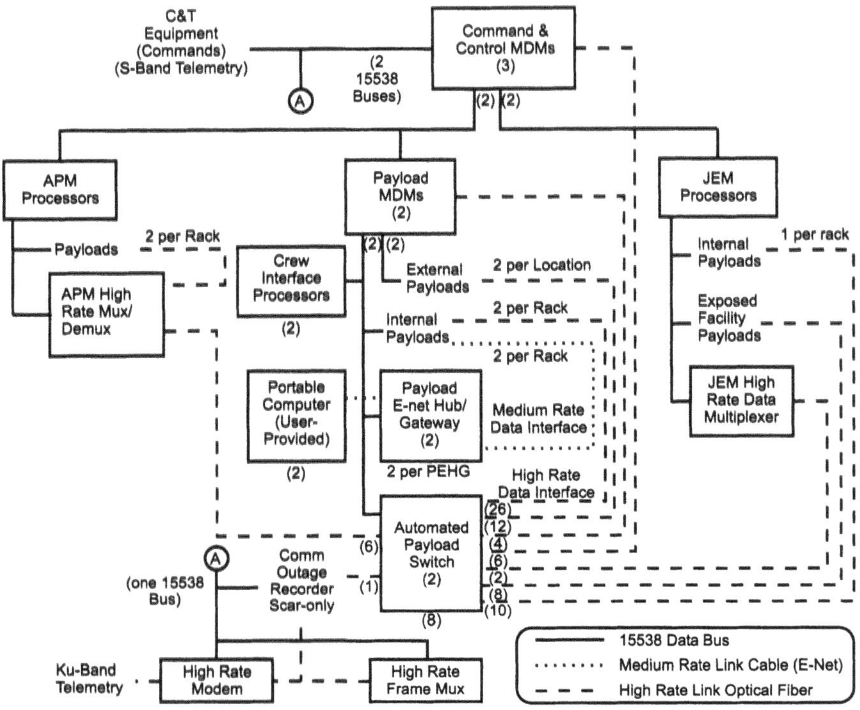

Fig. 12.21. Command and Data Handling System, Payload Data Subsystem Summary. Figures in Parentheses = Number of Implemented Systems

experiments and subsystems. It can be assumed that, compared to the Spacelab, only about 20% of crew time will be available per experiment facility. In order to ensure the efficient operation of the space station, it will be absolutely necessary to operate experiments and subsystems in an automated mode whenever possible – and to limit intervention by the crew to those cases where a human being can still work more efficiently than an automated system.

In any case, the overall operation scenario must be taken into account before deciding on the type and extent of automation. For that reason, this chapter is not intended to offer general, cure-all solutions, but it will show connections with the help of solution concepts for experiment facilities and will illustrate these by means of concrete design examples.

Some indispensable elements of experiment facilities' operation are the performance of experiments as well as the exchange of experiment samples. During long-duration operations, this additionally includes maintenance and repair tasks. Since the performance of experiments is, independently of crew availability, for the most part automated anyway, it shall not be addressed in detail in the following sections.

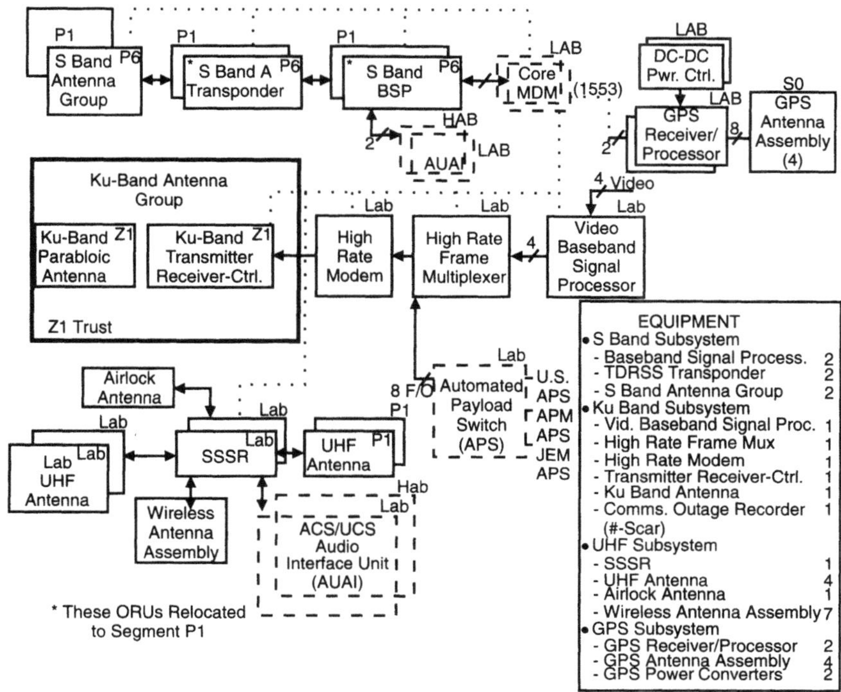

Fig. 12.22. The USOS Communication and Antenna Operating System

Sample exchange, maintenance and repair tasks are basically not very different from one another. Some typical examples of maintenance tasks are the following:

- The functional testing of safety units or the calibration of sensors or instrumentation amplifiers; in these cases, the connection of external instruments is necessary.
- The replenishment of consumables such as process gases, cooling water lost by leakage, or lubricants. Usually, in these cases, storage containers are either replenished or exchanged.
- The exchange of components which are limited as to their operational life, such as high-temperature heater elements

These tasks have in common their need to be carried out relatively frequently, i.e. several times during their operational life. Usually, maintenance activities are limited to a small group of components. For that reason it is an obvious option to support individual solutions for each experiment in order to simplify these tasks, as it has been done for numerous Spacelab experiment facilities. From the point of view of automation technology, however, the exchange of experiment samples is a special case for facility maintenance.

Contrary to maintenance tasks, repair work can become necessary for all components susceptible to wear and tear. They are necessary, for example, in the following cases:

- Statistical failures, e.g. electronic circuits
- Material defects, e.g. elastomer seals
- Overstressing; e.g. in the case of prototype facilities. This is mostly due to mistakes in design, manufacturing tolerances or inexact assembly.

As a consequence, the instruments' structure must allow the on-orbit exchange of all components and sub-segments that might be affected by damage, failure, etc. First reasonable thing to do is certainly to determine whether units are exchangeable or not. If this decision process were mainly influenced by the optimal use of transportation capacities, it would be decided that the smallest possible units be exchanged. Exchange units that are too large (e.g. complete experiment racks) will lead to excessive transportation requirements, exchange units that are too small will lead to other limiting factors such as available crew time, means of assistance (e.g. tools, measuring instruments), and also the required necessary expertise and training needed to carry out the repair. Detailed examinations [ESTEC 92] have recommended that exchanges be carried out on the component level, e.g. valves, pumps, Printed Circuit Boards (PCBs), etc., since this would achieve a balanced use of all resources available.

The exchange of relatively small parts aboard the space station and also the opportunity to connect external test instruments for functional testing causes a number of design requirements to be satisfied by the payload design. These requirements will be discussed in the following sections.

12.3.2 Design of Payloads that are Subject to Maintenance and Repair

A payload suitable for long-term operations mainly has to support the "easy" exchange of all components sensitive to malfunctions. From this requirement, some concrete design instructions can be derived concerning the accessibility of components and the simplification of interfaces between them. First, these constructional characteristics are independent of whether the unit is operated by astronauts, robots or a combination of both.

Accessibility of Components. Payloads are nearly always installed into racks that are only accessible in a limited way either from the back or from the side (see Figures 13.3 and 13.4). For that reason, the following is recommended for payloads:

- All instruments are principally installed from the front side.
- The further disassembly of the instruments is designed in the form of drawers that can be drawn out for repair work.
- Infrastructural lines such as electric cables, cooling water or vacuum lines are positioned in the back of the rack.
- The functional connection of the drawers among one another is preferably carried out by self-coupling connections at the back side.

- Supply units for fluids (gas, water) are to have a two-dimensional structure, for example, mounted onto a base plate that can be drawn out to the front. As a consequence, easy access is guaranteed from both the top and the bottom.
- The interior of furnace-process chambers is also easy to access if the main assembly plane for the guiding rails is the back chamber wall and the components are mounted from inside the chamber. After removal of the front cover and perhaps also in-between structures as well as separation of the interface, the entire interior structure, e.g. the heater together with the cooling jacket, can be drawn out towards the front and be removed from the chamber.
- Electronic boxes, after removal of the cover, should allow the exchange of whole cards. This will require the laying of all the connectors of the cards to that side which is located opposite the cover.

Generally and for safety reasons, no uncontrolled parts may be released during disassembly. This can be achieved by using secured screws or recessed grooves for sealing rings. Moreover, screw connections should be designed as self-locking. Adhesive tapes or securing by wires, however, are not suitable with regard to maintenance and repair tasks.

Simplification of Interfaces. In order to position the components, mechanical adjustments should be avoided in any case. The auto-centering and/or precision adjustment or stop functions can be transferred to the subsystem structure.

In the case of electrical connections, the use of manually-coupled plugs on the component level is advantageous. Correspondingly, PCBs should have plug boards only on one side which then establish electrical contact during installation.

In the case of gas systems, special attention should be paid to establishing all sealing connections during the installation. This can be achieved, for example, by installing them on modular blocks and by arranging the outlets and inlets on one side. Pipe segments intended for disassembly should be designed in a U-shaped form.

12.3.3 Automation of Payload Operation

A precondition for an automated payload operation is that not only experiment samples but also defective parts can be exchanged without crew intervention. In Fig. 12.23, a schematic of the task-division of such an exchange process is shown. Individual tasks such as "Withdrawal from Storage" and "Transportation" form a sequence that is tightly scheduled. Correspondingly, partial solutions for automation are only useful under certain conditions since the timing and spatial coordination of automated operations with "manual" operations usually leads to unnecessary risks, stress and strain for the crew and thus would not achieve the intended resource savings. Other tasks such as reconditioning and verification of the facility are also suitable for a partial automation strategy since, for the execution of these two tasks, there is no immediate coupling with regard to time. This part of automation – analog to the performance of experiments – is preferably carried out by the facility itself (cf. Sect. 12.3.4).

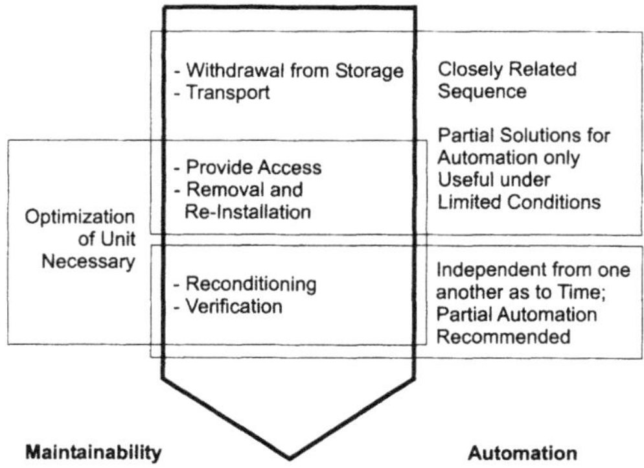

Fig. 12.23. The Individual Steps of an Exchange Activity with Regard to Automation and Maintainability

Internal and External Automation/Operation. Compared to its environment, a typical experiment facility is an at least partially closed system. Every external intervention into this facility, e.g. in a vacuum system or in a thermostatic volume, leads to a disturbance of the experiment conditions and, after the intervention, requires more or less complex procedures for the reconditioning and verification of the facility and the experiment environment. For the exchange of experiment samples, or, in special cases, also for the exchange of parts subject to wear such as light bulbs, *internal automation* is considered suitable. With internal automation, the exchange is carried out by a relatively simple internal mechanism during the performance of which the system can remain closed. In Fig. 12.24, the cartridge for storing

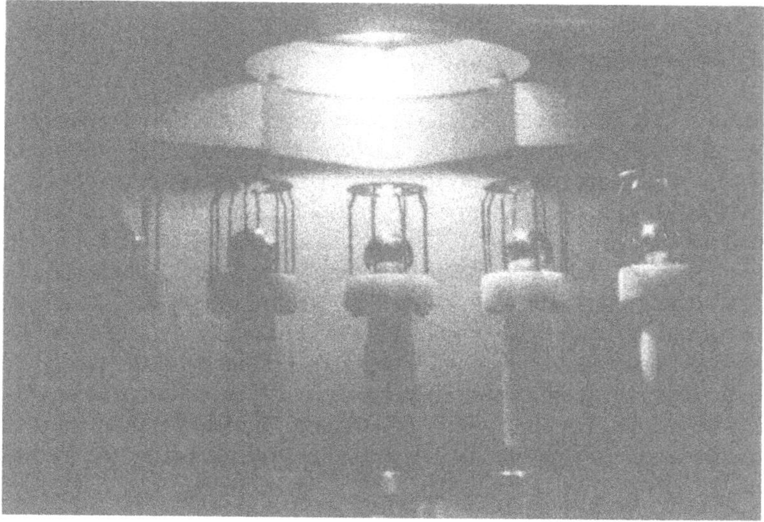

Fig. 12.24. The TEMPUS Cartridge

Table 12.8. Internal vs. External Automation

Internal Automation (Specialized Mechanism)	**External Automation** (Robots)
+ High Level of Precision + Advantageous where closed systems are required (handling/processing of toxic materials; susceptibility to contamination (sterility, ultra-high vacuum); storage at stable temperature levels, e.g. biological samples) - Minimal Level of Flexibility - No possibility of accessing external units; therefore, only partial solutions (exchange/ replenishment of magazines remains necessary)	+ Suitable for Transportation and Exchange + High Level of Flexibility - Limited Degree of Accuracy - High-Level Hardware Requirements

the experiment samples of the Spacelab high-temperature furnace TEMPUS is shown as an example. Here, the material samples can be exchanged by simply turning the cartridge without having to open the experiment facility. External interventions are only carried out for exchanging the cartridge or, in the case of disturbances, for installing spare parts.

External automation, on the other hand, requires the opening of the system during each exchange and, consequently, also the subsequent reconditioning and verification of it. The exchange of parts must be carried out either by an astronaut or by a relatively complex robot. The latter, however, can also be used for performing maintenance tasks.

In Table 12.8, a summary of all characteristics and evaluation criteria of both the internal and the external automation strategies are presented.

In summary, it can be said that internal automation is only suitable for routine exchange activities of, e.g. experiment samples or parts subject to wear. Exchange tasks in the field of logistics for maintenance and repair work should be carried out, according to the tasks' complexity and the flexibility required, either by an external robot or by an astronaut.

Automation Including Robots. When hearing the word "robot", the first idea that comes to mind is the all-automated assembly lines as they are used, for example, in the automobile industry for body assembly in mass production. These are highly specialized systems that, as far as repetitive tasks are concerned, outdo human beings in terms of reliability and speed. In our case, however, the problem is a totally different one: time only plays a subordinate role; what is more important is high flexibility in order to carry out all necessary exchange activities. But also in this context, when taking a closer look at it, modern robots are still inferior to the human being in the following essential ways:

- The end-effectors (grippers) currently available are by far less flexible than the human hand. For example, the end-effectors of the ROTEX experiment robot [Hirzinger 91] had two degrees of freedom, whereas the human hand features more than 20! The absence of flexibility on the robot side must thus be compensated by the facility that is to be operated by the robot. For example, specially designed depression points must be included into *all* individual parts which are intended to be manipulated by the robot. On the other hand, these gripper interfaces can be designed for relatively easy and universal use. For example, the DLR recently presented the most complex four-finger robotic hand that has ever been constructed. For the first time, the integration of all 12 degrees of freedom into a hand and the outfitting of the fingers with three active degrees of freedom was achieved. As a consequence, this hand features a degree of movability comparable to that of a human hand. NASA's Johnson Space Center is also developing comparable concepts with regard to future "robonauts".

- A sensor-based control which a human being can perform by doing corresponding movements because of the direct feedback of human sensation (tactile sensation, sensation of temperature, sight), has not been the industrial standard up to the present day, but it was successfully demonstrated with the robotic arm ROTEX. Without this kind of sensor-based control, however, at any point in time the robot must have a precise internal model of reality including positions, orientation and degrees of freedom of all its individual parts to be in a position to perform maintenance and repair tasks in a practically "blind" manner. Inevitably, the problem of positioning accuracy occurs due to the manufacturing tolerances and yieldingness of the robot. Moreover, such an automation concept is very inflexible with regard to unforeseen events such as the release of a screw.

For that reason, modern telerobotic concepts (as they were nationally developed, e.g. named MARCO) are based on the assumption that the robot in a virtual environment on ground is shown its tasks, taking into consideration all sensor patterns that can be expected to occur in the real world. As a consequence, once aboard a space station, the robot will be able to carry out its tasks in an autonomous and flexible way by adapting to, for example, slightly changed conditions. Prerequisites are, as in the case of ROTEX, multi-sensory gripper systems including e.g. force-torque sensors in the wrist, laser distance measuring devices, and miniaturized stereo cameras.

Typically, four hierarchical levels are taken as the basis of such a task-oriented remote programming concept. The two lower levels (elementary operations and sensor-based reflexes, e.g. the turning of a bayonet lock) are used by robot experts on ground in order to define complete operations (changing the position of the bayonet locks) or in order to combine complex tasks in a whole sequence of operations. What is decisive is the fact that during the operational phase of the station, it will be the payload expert (i.e. generally no robotic expert) who is present at the telerobotic ground station. And it will be the payload expert who, in this "virtual world", is "clicking" the objects to be handled (drawers, doors, bayonet locks, etc.) with the help of a 3D cursor or a data glove and is "dragging" them to pre-defined places.

All leading space flight nations and organizations are intensively preparing the use of robotics in space. The International Space Station will include a 3-robotic system manufactured by Canada. The system is called "Mobile Servicing System" (MSS) and will be used for changing out and transporting exchange modules and is movable outside the station on a track structure. At the end of a completely symmetrical arm, 17 m in length and with seven degrees of freedom, there is the so-called "Special Dexterous Manipulator System" (SPDM). The SPDM is a two-arm robotic system with a total of 14 degrees of freedom and a turnable base platform for the

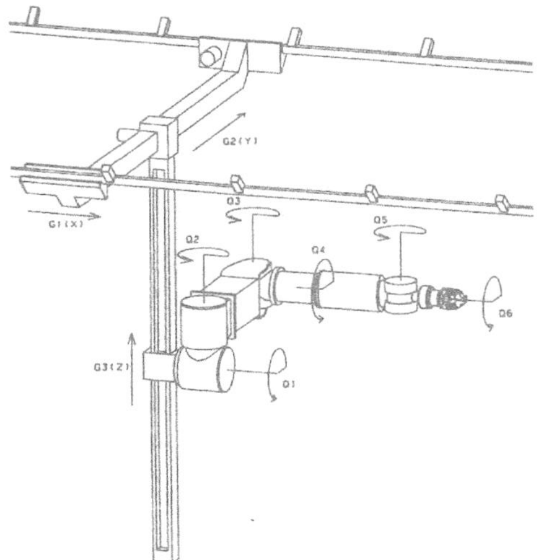

Fig. 12.25. EMATS – Concept of a Robotic System for Space Stations

temporary positioning of exchange modules. Force-torque sensors and TV cameras at different places on the arms support this system. In principle, the system can be controlled remotely from the ground, but it will certainly not reach the level of accuracy demonstrated by ROTEX.

In 1999, NASA is preparing the testing of a multi-arm robotic system called RANGER in the Shuttle cargo bay. Compared to the RANGER system, the ERA manipulator (a development of FOKKER and DASA) and the Japanese JEM-RMS are rather simple. They are used to transport loads within the Russian SPP and Japanese JEM-EF parts of ISS. Also in 1999, ESA plans to fly an experimental robotic system named JERICO that will be coupled with the external structure of the Russian space station Mir. In the long run, ESA plans to install the so-called "Technology Exposure Facility" (TEF) at the external structure of ISS. The TEF includes a remotely controlled robotic system for the handling of experiments in the free space environment. At present, final decisions concerning the use of robotics inside the laboratories of ISS have not yet been taken.

The necessary movability of this manipulator, e.g. inside a laboratory, could be ensured by a mobile base in the form of a sled with three degrees of freedom [ESTEC 93] (cf. Fig. 12.25).

The performance of maintenance and repair works by such a system are quite possible when all devices and instruments are designed especially for automation ("Design to Automation"). Over and above those points explained in Sect. 12.3.2, this would have the following consequences:

- The existence of small depression points for the robot's end-effectors which can compensate positioning and manufacturing tolerances, for example, by means of autocentering guidance.

- The mechanical fastening of the components must be designed in such a way that an end-effector can disassemble and at the same time hold on to them.
- During assembly/disassembly, all components must at any point in time be in a state that is clearly known to the robot. This means for example, that there must be no lose parts whatsoever.
- In the case of fastening elements, it must be taken into account that comparatively small linear forces but relatively large torques can be applied.

These boundary conditions usually lead to devices with high constructional efforts and complex operational procedures even for relatively simple maintenance work. In Fig. 12.26, some steps necessary for exchanging a heater in an experiment rack are shown.

Here, the task is additionally aggravated by the necessity of stabilizing flexible parts, e.g. cables, and of using special installation tools for mechanically demanding activities such as the installation or removal of a heater element. The example clearly shows that, with the state-of-the-art robot technology, the automated performance of maintenance and repair tasks can only be obtained with relatively large efforts. Also in this case, the maintenance and repair concept can only be decided upon through considering the specific requirements of experiments and instruments.

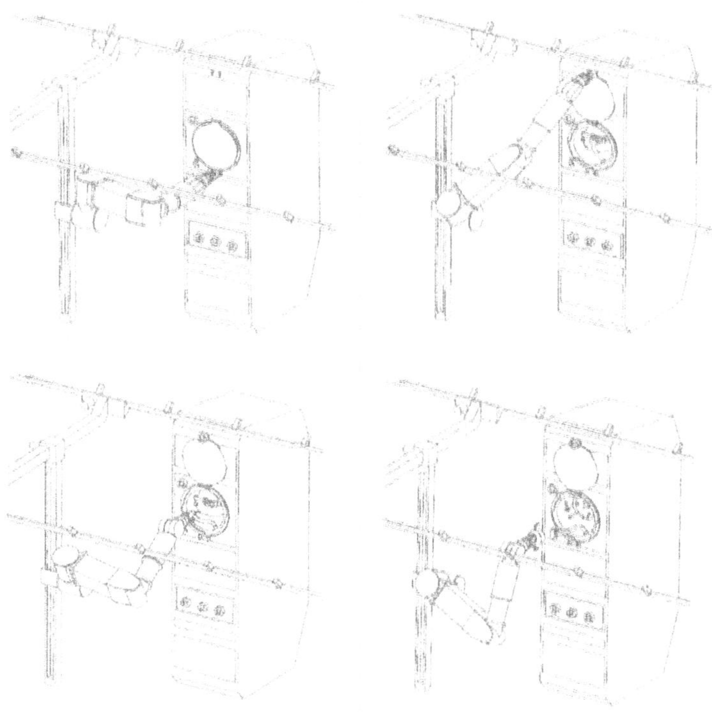

Fig. 12.26. Heater Exchange By Means of a Robot

For these reasons, the following suggestion of the so-called "Payload Tutor" (PT) which the Italian space agency ASI presented to NASA is quite promising: in order to perform the experiments, a small robot moving on tracks does not move through the entire space laboratory (for example, the COF), but only alongside one single rack. At any time, the robot can be removed from this rack and installed in front of another.

At the end of 1997, for the first time, the Japanese space agency NASDA sent a free-flying telerobotic system called ETS VII into space. As to autonomy and flexible remote control, this system does not overtake the ROTEX level, but it demonstrates free movability. Japan attaches great economic importance to the servicing, maintenance and repair of the space flight system by free-flying robots. Similar to ROTEX, two robots of different size will carry out prototype tasks at task boards. The docking to a piggy-backed "training satellite" as a precondition for practicing satellite repair is also planned. During this project, there will be cooperation between the NASDA and European institutes and agencies, including those of Germany.

12.3.4 Testing and Verification

After completion of each exchange activity, the correct installation of the corresponding component must be proved. The tests that have to be carried out to that end will not only verify the function of the installed components, but also all interfaces which were separated during the exchange process. The time necessary for this verification can by far exceed the time that was necessary to carry out the exchange. A high degree of automation is imperative in this context. The correct functioning of an instrument can be verified comparatively easily: the electronics system uses a corresponding self-test function. Additional sensors will generally not be necessary if verification and test runs have already been taken into account during the development of the facility. The evaluation of these data can be carried out either on ground or by the instrument itself.

The necessary leakage test, however, is a special challenge when designing experiment facilities. Since in the closed environment of a space station, a leakage rate determination test by means of helium or other trace gases is not possible, pressure drop or pressure rise methods must be used. The measuring time necessary for these methods increases proportionally with the volume to be tested and its interior pressure and can last up to a few days. For that reason it is recommended that small test volumes by means of double seals at connections that will frequently be opened and closed be generated. Most of the time, pressure measurement requires additional sensors with relatively high degrees of accuracy, but then can take place automatically.

12.3.5 Summary

Compared to the laboratory systems operated before (e.g. Spacelab, Spacehab), the operation duration of a space station has become increasingly long, and thus, automation and maintainability have become crucial aspects for efficient payload and subsystem operation. The special requirements for the long-duration operation of a system in orbit necessitate that aspects of maintainability and repair must be taken into account already during the design phase of a payload. This can come about by, for example, guaranteeing access, simplifying interfaces, and possibly performing the necessary facility test in orbit.

Although payloads that were especially designed to facilitate maintenance and repair will have a volume that is 1.5 times higher (compared to the highest density design) and a mass that is higher by 20%, these payloads will have several advantages during operation. For example, an expensive experiment facility can be repaired in a routine manner without having to return the whole experiment rack back to Earth.

The complexity and size of a space station with its high number of subsystems and limited number of astronauts will undoubtedly require a high level of automation. The question as to whether a certain task should be carried out by automation or by crew intervention must be investigated in each individual case. In this context it will be decisive to discern whether this task can be automated with an acceptable degree of effort, or whether human beings with all their flexibility and their cognitive abilities will work more efficiently.

12.4 Telescience

Experiment operation aboard a space station differs significantly from that aboard a crewed, short-duration mission (e.g. Spacelab) or an uncrewed, long-duration mission. The nearly unlimited experiment time together with only a limited availability of the astronauts aboard calls for a new concept of scientific experimentation. Telescience offers such a concept that is based on the experiences acquired during scientific research in terrestrial "hot" laboratories. The term "telescience" was coined in 1985 by Peter Banks on the occasion of the Space Station User Panel in Stockholm, Sweden, and it means "the integration of telepresence and teleoperations in the operating environment of scientific research activities" [Schmidt 87]. Today, we may define "Telescience" as the acquisition of information through remote experimentation and observation. The concept is hence based on interactive communication between the experiment aboard the space station and the scientist on Earth. For that reason, telescience makes heavy demands on the experiment, including remote control (teleoperation) and the transparency of all activities (telepresence). In order to achieve its realization, the on-board Data Management System, the communication between station and ground, and also the data distribution on Earth, must fulfill special requirements.

During the earlier TEXUS, Space Shuttle and Spacelab missions, the concept of telescience had already been tested in a rudimentary way, with experiments in the field of crystal growth and fluid dynamics. The future telescience operation will combine the experiences of uncrewed space flight missions (mainly acquired in the classical disciplines astrophysics, planetary research, Earth observation, etc.) with those obtained during Spacelab missions, and with those acquired during advanced experiments on Earth as well as those obtained during operation of non-accessible experiments with particle accelerators, reactor plants, and plasma fusion reactors.

12.4.1 Crew Time – A Critical Resource

In comparison to Spacelab missions of one to two weeks duration (with the Extended Duration Orbiter Columbia or Endeavour) but also compared to earlier space stations, the International Space Station as a year-round operated research facility offers a very high utilization potential. In Fig. 12.27, the result of an estimation carried out in 1987 by NASA can be seen. The subject of this estimation was a comparison between the relative utilization potential of space stations operated over a long period of time (Skylab, Mir 1990 and the then planned Space Station Freedom) and that of Spacelab missions of a one-week duration (e.g. the Spacelab mission D1). In fact, the ratio of ISS/SSF in terms of total volume, crew, and power, with the help of Table 2.10, yields 50% more ISS utilization capability as compared to SSF. Even if the considerations only included the best imaginable case of the use of a space station (= electrical power, number of racks, crew time and payload operation available), Table 12.9 clearly shows that, above all, the availability of astronauts will be very limited for the experiment operation of the International Space Station [Messerschmid 92].

Also in the case of a more realistic estimation of crew availability and in comparison to Spacelab, crew time aboard ISS will be one of the most critical resources. This becomes clear when taking a look at Table 12.9. It is interesting to note that the number of rack units aboard Spacelab is nearly the same as the number planned for

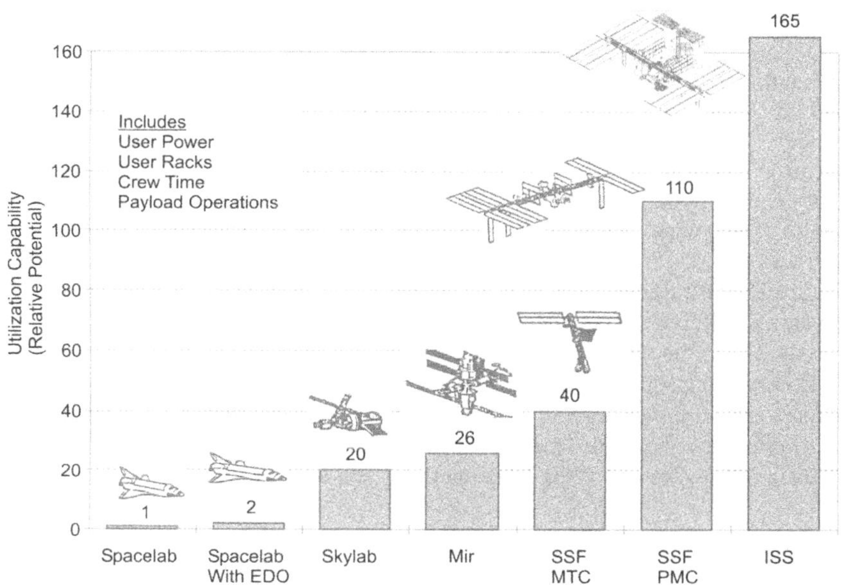

Fig. 12.27. Comparison of Utilization Potentials: Space Stations vs. Spacelab Missions; NASA Dunbar Commission, ISS estimation given by authors

Table 12.9. Crew Availability

	Spacelab (SL) 1 Shift/2 Shifts	International Space Station (min./max.)
Astronaut per Module	5/3	1/2
Working Hours per Day	12/2 x 13	8/10
Working Days per Week	7	6
Racks per Module	20	20/24
Crew Availability in Hours/(Rack x Days)	3.0/3.9	0.29/0.86
Crew Availability Compared to SL	1	0.1–0.22

any of the ISS laboratory modules (US Lab: 24, JEM: 23, COF: 20; these racks have half the size of the ISPRs of which there are 13 in the US Lab, 10 in JEM and 10 in COF, i.e. 33 ISPRs total; cf. Sect. 13.2). The main difference, however, is that aboard Spacelab, five astronauts worked in one shift or three astronauts in two shifts every available day, whereas aboard ISS, on average there will only be one astronaut per module. Additionally, due to the long mission duration, the work hours will approach terrestrial conditions. In the most unfavorable case, the "intensity of attention" available for one ISS rack will be on a 10% to 20% level, when compared to a Spacelab rack.

According to NASA planning, the *crew time for research* is only about half of the total available crew time: 120 crew-hours per week will be available for research purposes at assembly complete. This number assumes the nominal crew size of six. These hours must be apportioned among all science, technology, and commercial investigations onboard the station. An additional 40 hours per week for research would be available if a seventh person joined the crew. Hence, with a crew of six to seven, the crew time for research to be apportioned only among the 67 USOS racks (US Lab, JEM, COF) results in 0.26 to 0.34 crew-hours per rack (½ of an ISPR) and per day. This corresponds well with our lower estimate given in Table 12.9, i.e. an attention level per rack on a 10% level when compared with Spacelab operations.

It is hence apparent that this difference can only be compensated for by automation of experiments and telescience. It is expected that the utilization of the individual racks aboard ISS will not reach the Spacelab level by any estimation (this would require a large number of experiment samples, exchange and spare parts, etc.). Thus, the actual difference between Spacelab's and ISS's operation may still be a factor of only 2–4 to the disadvantage of the latter.

The first consequence is that an astronaut, more than ever before, should only carry out a task if it really cannot be fulfilled by automation. Second, experimenters on Earth must be enabled to monitor and perform their experiments as far as possible on their own by means of the latest methods and with the help of telescience.

12.4.2 Teleoperation and Telepresence

Telescience operation requires a number of prerequisites [Feuerbacher 90, Schmidt 87] listed as follows:

- Transparency of the performance of an experiment: all relevant data of the rack and the experiment should be available to the user on Earth, if possible, as near to real time as possible. In many cases, this also includes the transmission of video data. Problems to be dealt with are, for example, the reliability of the facilities, interrupted or flawed data transmission and signal delay.
- The users must have the possibility to evaluate their experiments promptly in their immediate environment (Payload Operations Control Center POCC or a User Home Base UHB). This provided, they will be able to judge the quality of the results obtained, compare them with reference data of the on-ground experiment, and perhaps discuss them with their colleagues. Distributed intelligence on ground as well as electronic intelligence both on ground and in space are advantageous.
- The possibility of direct intervention in the course of the experiment must be provided. This can be done with the help of a feed back channel to the experiment and a set of well-defined, trained commands. These enable the experimenters to influence their experiments within a limited range of physical parameters and to profit from the status and safety check on ground (POCC and UHB).
- During the defined experiment time, the users must be able to use the resources agreed upon in advance in a flexible way, i.e. changes in real time within the cycles, with the option of new planning for new cycles.

Telescience operation should be strived for in the case of automated operation as well as in the case of astronaut availability. In any case, the astronauts will be released from routine tasks and are thus able to directly and immediately react to unforeseen events. During the telescience operation, only changes of those parameters will occur which are usually monitored by astronauts during on-board operation and which can be changed by commands or adjustments via computer, switch or rotary knobs. The long-term objective of telescience is to also master more complex tasks involving robots.

During the Spacelab mission D2 in 1993, the operation of an experiment robot from the ground was achieved for the first time. This ROTEX experiment is shown in Fig. 12.28 and the following various modes of operation were successfully demonstrated with it:

- Automated sequence with pre-programming on ground and re-programming from the ground
- Teleoperation aboard via stereo-TV monitors
- Teleoperation from the ground via predictive computer graphics

Moreover it was shown that despite the considerable delay in time between the command given on ground and the actual performance in the space laboratory, ROTEX was able to capture an object floating freely inside the workroom of the ROTEX gripper arm (cf. Fig. 7.38) [D2 95].

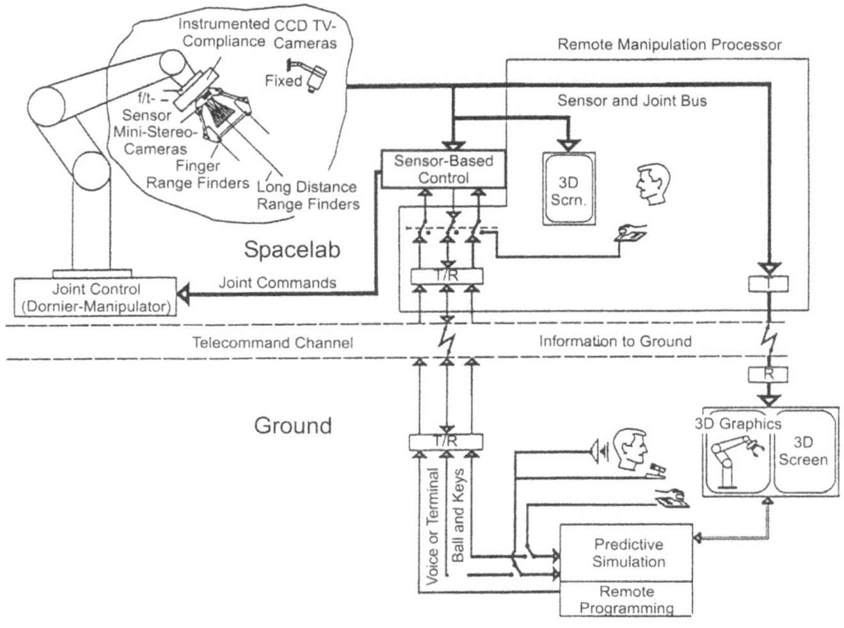

Fig. 12.28. ROTEX Teleoperation: Onboard via TV Monitors and from the Ground via Predictive Computer Graphics

During the operation of the International Space Station there will be various options from which to choose, that, in the order of their appearance, incorporate an increasingly technical complexity with simultaneously increasing flexibility for the experiment performance:

$$\text{Astronaut} \rightarrow \text{Automation} \rightarrow \text{Teleoperation} \rightarrow \text{Telerobotics}$$

The advantages of telescience can be summarized as follows:

- Interactive operation
- Full transparency in real time
- Interfaces integration into experiment
- No knowledge transfer necessary
- Immediate reaction to unforeseen events
- Optimal use of expertise

In the case of ISS, for the first time in space research, scientists will be the primary operators of their payloads. Researchers will have the choice of where their ground operations will be based. Whether located in a Telescience Support Center at one of NASA's Field Centers, in the mission control center of an International Partner, or in a remote site, i.e. a research laboratory, researchers will be able to manage and operate their experiments. A Telescience Support Center is a NASA facility equipped to conduct telescience operations onboard the International Space Station. Remote sites are non-NASA facilities such as commercial space centers, universities, private industry, and other government agencies.

Telescience will maximize the ability of researchers to communicate with the crew, to direct the "commanding" of their experiment hardware and software, and to manage the receipt, processing, and analysis of their research data from laboratory desktop computers. To support researchers in remote locations, NASA has developed the Telescience Resource Kit (TReK). TReK is a PC-based telemetry and command system which, when networked to either a Telescience Support Center or to the Payload Operations Integration Center (POIC), will give researchers access to their payload onboard ISS in addition to ground-based planning and information management systems.

Researchers will define their science planning requirements and provide their science expertise on payload and experiment operations. Investigators will develop the detailed operations plans for their experiments, will participate in training the crew, and will help develop the procedures the crew will use while in space. Researchers' planning inputs will also be used to develop an operational schedule for onboard activities.

Since ISS is designed for extensive communications, including voice, video, telemetry, and downloads of real-time and stored data, researchers will easily be able to assess the success or shortcomings of their on-orbit research protocols. Investigators will be able to discuss problem areas with mission planners and will be better able to devise new procedures. For example, a procedure that resulted in a limited sample yield could be modified to improve future test results. ISS operations are designed to give scientists the flexibility to modify or enhance their experiment operations as long as they remain within their resource constraints. Detailed payload operations plans will be developed one week in advance, thus allowing researchers time to review previous data and revise their research approach, thereby improving the quality of data collection on later experiment runs.

13 The International Space Station

As a complement to the previous chapters, this chapter on the International Space Station (ISS) will describe the special characteristics relevant to the mission, planned laboratory facilities, and user support as well as important technical and programmatic aspects of the station's operation.

Previous chapters have already dealt with the assembly of ISS, its most important components and parameters (Sections 2.5–2.6) and the station's orbital environment and subsystems (Chapters 3–6). Important aspects of utilization have been discussed in Chapters 7 ("Utilization"), 8 ("Microgravity") and 12 ("Logistics, Communications and Operation"). As this chapter is the conclusion of this book, supplemental information and an overview of the current space station plans will be given, thus it may also serve as a guide for users that are planning experiments for research and technology demonstration [ESA-Guide 96, ESA-Guide 98, NASA-Guide 98].

13.1 Station and Mission Elements

This first section deals with those characteristics of ISS, and the entire mission, which are of interest especially to users. These characteristics will help us realize the vast possibilities of ISS in terms of multilateral cooperation based on concerted action and bartering. ISS draws upon the resources and scientific and technological expertise of at least 16 nations and will provide unprecedented access to a crewed, long-duration space research facility. The governments of the United States, Canada, Europe, Japan and Russia are named "International Partners", in short "Partners". Each Partner can establish its own bi-lateral agreements with other countries to supply portions of the partner's contribution, often in exchange for research access to the station once it is operational. The United States, for example, has additional agreements with Brazil and Italy to this end.

Each partner nation is allocated a share of station resources in accordance with its contribution to the program. The partners use their own internal mechanisms to apportion space and research capability to their scientific, technological, and commercial interests. To coordinate their research, the partners have established bi- and multilateral working groups. The United States will have the use of the US laboratory module in addition to 46.7% of the European and Japanese pressurized laboratory space. Not including the Russian laboratory space, this constitutes over 75% of available station research accommodations. Thus, the Partner's share, at present, is NASA (75.3%), Italy (0.85%), Brazil (0.45%), Japan (12.8%), ESA (8.3%) and Canada (2.3%) [NASA-Guide 98]. The utilization and distribution of these accommodations and resource capabilities is achieved at the International Partner level rather than at a specific laboratory or attached site level. This approach provides each International Partner with broader opportunities for planning to accomplish its specific payload mission objectives.

13.1.1 Characteristics of ISS

The International Space Station will be assembled from several basic components: modules, nodes, truss structures, solar arrays, and thermal radiators. Modules are pressurized cylinders and represent habitable space onboard the space station. Modules may contain research facilities, living quarters, and any vehicle operation stations and equipment to which the astronaut may need access. Nodes connect modules to each other and offer external space station access for purposes such as docking, EVA access, and pressurized payload access. Truss structures are the erector-set-like girders that link the modules, the main solar power arrays, and the thermal radiators. Together, the truss elements form the Integrated Truss Structure. Solar arrays collect solar energy for conversion into electricity for the operation of ISS and its payloads. Thermal radiators bleed off excess heat into the vacuum and cold of space.

Once its assembly has reached completion, ISS will feature eight solar arrays and two large radiators mounted on the 108 m long truss structure. The US modules will be directly connected to the center segment of the truss; behind them there will be the Russian modules and the vertical truss including the solar arrays. The European COF and the Japanese JEM module will be situated at the "front" of the station (with respect to its flight direction) in front of the US Laboratory module (cf. Fig. 13.1).

The Canadian manipulator system SSRMS can move along the truss between the first (inner) pair of solar arrays. The station's dimensions are 108 m x 74 m and it will have a total mass of 415 tons. Moreover, 110 kW of electrical power will be generated aboard ISS, of which 47 kW are allocated for research purposes. After the first phase of assembly, three astronauts will live and work aboard the station. The number will be increased to six or seven after completion of assembly. Each crew member will stay in orbit for at least three months – this is the minimum period possible, corresponding to the length of the intervals between two Space Shuttle flights to ISS.

The attitude and orbit control strategies defining the station's flight direction, altitude and attitude were covered in Sect. 6.2. Perturbations aboard the station in the form of quasi-static microgravity levels and g-fluctuations (g-jitter) were explained in Sect. 8.4, whereas the self-induced environment around the station (mainly influencing external sensors and observation instruments) was covered in Sect. 3.9.

The International Space Station will include the following orbital elements (the acronyms in parentheses designate the corresponding parts shown in Fig. 13.1, see also Table 2.8 and Fig. 2.26):

- The functional and payload module "FGB" (acronym derived from Russian designation): The FGB, also named Zarya (Russian for "dawn") is manufactured in Russia and financed by the USA. It is an autonomous space vehicle including

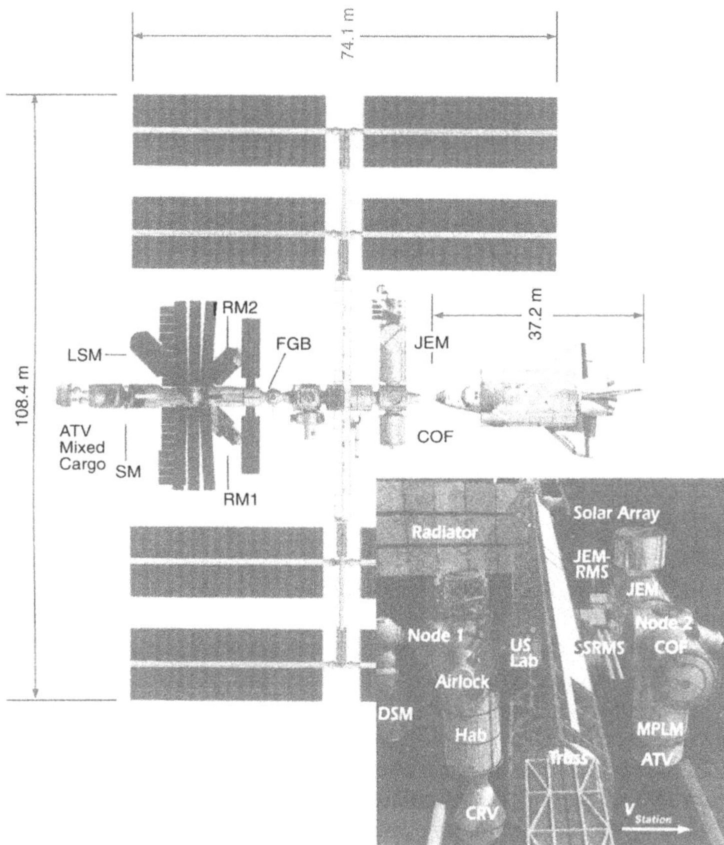

Fig. 13.1. The view from above shows, with respect to flight direction, the rear part of the station (on the left) with the fundamental elements of the Russian Segment and a European Automated Transfer Vehicle (ATV) docked at the Russian Service Module (SM) and carrying Mixed Cargo. The forward part of ISS (in the lower right corner) comprises the European laboratory module COF with an ATV docked to it, with the ATV carrying an MPLM (Mini Pressurized Logistic Module) [ESA-Fakten 96, DARA-ISS 96].

power supply and thermal control as well as navigation, propulsion, and communication equipment.

- The Service Module (SM), provided by the Russian Space Agency (RKA): It includes habitation, work and sleep compartments for up to three crew members and complements the propulsion and attitude control functions of the FGB. At the rear part of the SM there is a docking port for uncrewed Russian Progress vehicles. The Mixed Cargo-version of the European Automated Transfer Vehicle (ATV) also can dock at this port.
- Six pressurized laboratory modules for scientific research:
  - A laboratory module (US Lab) from NASA

  - A US-contributed Centrifuge Accommodation Module (CAM)
  - Two Research Modules (RM1, RM2) from the Russian Space Agency RKA
  - A Japanese Experiment Module (JEM) from the Japanese Science and Technology Administration (STA)
  - A European laboratory module, the "Columbus Orbital Facility" (COF), from the European Space Agency (ESA)
- Various unpressurized accommodation sites on the US, Russian and Japanese elements for installation of scientific and technological external payloads
- A habitation module (Hab) from NASA, offering space for four crew members
- A Life Support Module (LSM) from RKA, complementing the life support functions of the Service Module (SM)
- The Russian Universal Docking Module (UDM), where the Russian research modules RM1 and RM2, the LSM, and the docking segment will be mounted (the UDM also being used for docking with Russian transportation vehicles)
- The Russian Docking and Stowage Module (DSM), an additional docking possibility for Russian space transportation vehicles
- The Russian Docking Compartment (DC), also serving as an air lock for the crew during EVAs
- The US air lock
- The Node 1 and the Node 2 as well as adapters between the various pressurized modules
- A docking port for the US Space Shuttle at Node 2 and at the nadir end of the US Habitation Module (US Hab)
- The Crew Return Vehicle (CRV) which is permanently docked at the station, a crewed reentry vehicle serving as "life boat" for the station, thus ensuring the safe return of the crew to Earth in case of emergency (as yet undecided as to which of the International Partners will furnish the CRV.
- The following three manipulator systems will be available outside the station:
  - The Canadian remotely operated space station manipulator system (SSRMS)
  - The European Robotic Arm (ERA), moving along the Russian "Science Power Platform" (SPP)
  - The Japanese remotely controlled manipulator system (JEM-RMS), mounted on the JEM
- Two large support structures: One truss from NASA and the SPP from Russia, together comprising the structure that links the different station elements.
- Photovoltaic solar arrays, radiators for temperature control, attitude control systems, communication equipment, and corresponding distribution systems for power supply and communications.

13.1.2 The Gradual Assembly of ISS

The first element of ISS is the FGB module, launched on November 20, 1998 by a Russian Proton rocket. During the following six months, three further flights by the US Space Shuttle and Russian launch vehicles are to take place, transporting Node 1 (manufactured in the USA, launched December 4, 1998), two adapters and the Russian Service Module (SM) into space. All these components together make a "core station", which can be permanently crewed from July 1999 onwards with a crew of three, who is to live and work in the Russian Service Module (cf. Table 2.8).

The subsequent installation of the first segments of the US truss and the Russian SPP, as well as that of the solar arrays, radiators and communications equipment which are to be mounted onto them, considerably increases the resources this "core station" is able to offer. The European Robotic Arm (ERA) plays an important role in the assembly process of the Russian SPP and the different Russian modules.

In late 1999 or early 2000, the US Lab, the Airlock and the Russian Docking Compartment are to be added. Over the years 2000–2003, the scientific capacities will be gradually expanded by launching and installing the two Russian research modules RM1 and RM2, the Japanese Experiment Module (JEM), and the European Columbus Orbital Facility (COF). The outfitting of the research modules with racks (ISPRs) and other equipment is done during intermediate Space Shuttle flights.

Assembly of the station will be complete in 2004. From this point onwards, the total number of astronauts aboard the station can be increased to six or seven, since the US habitation module (Hab) to be launched in January 2004 can accommodate up to four persons.

13.1.3 Mission Characteristics

After the Assembly Phase, the Utilization Phase will begin in 2004. This phase is projected to last 10 years, and it is divided into sections of three months each. At the beginning or end (depending on the point of view) of such a section, a Space Shuttle flight to ISS will take place; the crew, parts of payload elements, and equipment will be exchanged. Moreover, the Space Shuttle will deliver supplies to the station and return experimental test items to Earth. After each Shuttle "visit", the station will be lifted by a reboost in order to compensate for the loss of altitude during the preceding three months caused by the decelerating effect of the residual atmosphere. Each reboost maneuver will increase the station's altitude by about 10-40 km, the exact altitude increment being dependent on the influence the solar cycle has on the atmosphere.

ISS will be operated in several different modes. The primary modes of operation for research are "Microgravity" mode and "Standard" mode. Other modes of operation include Shuttle docking, Extra-Vehicular Activity (EVA), and orbital reboost (see Sect. 6.3.2 "Orbit Control Strategies" and Sect. 8.4 "Perturbation Compensation and Levitation"). The crew time available for research is 120 h/week for six astronauts, a seventh astronaut would add another 40 h/week. When fully operational, ISS will provide 26 kW minimum continuous and 30 kW average power and thermal conditioning to payloads during standard and microgravity modes of operation. During various other ISS operating modes, a minimum of 6.5 kW of continuous power is provided to the integrated set of payloads.

The nominal atmospheric pressure onboard ISS is maintained between 97.9–102.7 kPa, with a minimum of 95.8 kPa. Carbon dioxide levels are maintained below 0.7% to ensure the medical safety of the crew. The relative humidity is maintained between 25% and 70%. To support specific biological research, agreements have been reached to reduce the atmospheric carbon-dioxide content to 0.37% (with the goal of reaching 0.3%) for two 90-day periods each year. This is accomplished by running several carbon dioxide removal assemblies simultaneously. The oxygen partial pressure is maintained in the range of 19.5–21.1 kPa. Potable water for payloads and experiments is provided at a rate of 5.5 kg per day.

Continuous operation and supply cycles require great effort on ground. For example, all activities related to the launch of the first payload have to be repeated for all subsequent payloads, sometimes even overlapping one another. At the end of the first phase of experiment selection (and coordination on the ISS Partner level), and in the case of the subsequent selection phases alike, a "Partner Utilization Plan" (PUP) is agreed upon for each year of the steady-state station utilization. The composition of the payload and the sequence of the scientific schedule for each increment have been coordinated and fixed far in advance by the "Space Station Strategic and Tactical Planning Process". If this planning includes an exchange of payload elements, the "Payload Integration and Operations Preparation Process" will have to be started in time. For the European COF operation, for example, this process has to be continued quasi-permanently in order to continually "move" the payload through all the different phases of preparation up until launch.

Payload flight operations, firstly, comprise the new payload's onboard installation, testing and verification. The next steps are the scientific tasks that the payload is to perform, its maintenance and repair (if necessary) and finally, its removal and return to Earth. In the case of payloads situated in one of the pressurized parts of the station, astronauts will perform installation on rack or drawer level. Usually, this requires a connection with standard elements such as screws, latches and joints as well as electrical, fluid-mechanical and vacuum tube connectors. External payloads are installed with the help of the SSRMS robotic system which is controlled by the astronauts. Small payloads like the Express Pallet Adapters are installed with the help of the small Special Purpose Dextrous Manipulator (SPDM), which, for this task, is mounted at the end of the larger SSRMS robotic arm. After successfully checking the complete installation, all internal or external payloads are cleared for routine operation. Either astronauts, autonomous control and robotic units, or ground control (or a combination of all three) will operate the payloads and intervene, if necessary. Inside each research module, astronauts can control the payloads either by means of laptops, (which are connected to the Local Area Network (LAN)) or directly by means of switches, displays and experiment computers. Procedures which have to be processed automatically, are usually monitored by the experiment computer; alternatively, this can be performed by the Payload Control Unit (PLCU).

An experiment can also be controlled by the original Payload Investigator (PI). In this context, resources must not be exceeded (electrical power, cooling, data systems and astronaut time are available only for a limited period of time); this is laid out in a short-term plan on the basis of commands which were validated before the flight. As a consequence, the scientist can monitor all relevant experiment and housekeeping data and intervene, if necessary. The scientist, in turn, is supervised by the Control Centers (CC) and Payload Operations Interface Control (POIC) (cf. Fig. 13.2), but they will only intervene if the User Home Base (UHB) operation turns out to have any negative influence on the crew, on other payloads, or on re-

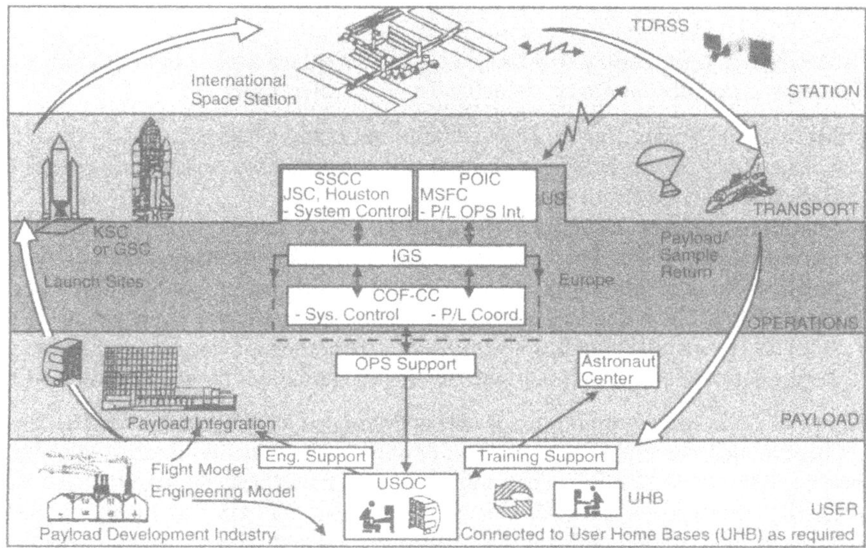

Fig. 13.2. Schematic Representation of the Onboard/Ground Operation with the Different Control Centers

sources not assigned to the PI, or if the previously defined operational window is exceeded or has to be limited.

All the different ways of transmitting commands, voice or video between space station and ground via TDRSS or other relay satellites are described in Sect. 12.2. At present, the uplink data rate is rather small (72 kbit/s); the downlink data rate of 43 Mbit/s is likely to be increased to 150 Mbit/s by adding further satellite transponders (TDRSS, Artemis, other DRS capacities). For payload data, the MIL-STD-1553B is applied where the LAN protocol corresponds to the IEEE 802.3 convention.

13.2 Pressurized Modules and Payload Structures

Seven pressurized station elements are planned as laboratories or can be used for experiment operations: the laboratory modules US Lab, JEM, COF, RM1 and RM2 mentioned before, and the modules US Hab and US Centrifuge. These modules and their utilization will be described in the following sections. Facility descriptions may vary in level of detail dependent upon the level of definition and stage of development of a particular facility at the time of the writing of this book.

13.2.1 The US Laboratory Modules

United States Laboratory (US Lab). Figure 13.3 shows the interior of the US Lab. Elements inside this laboratory which carry experiments are:

- Multi-User Facility Racks: They contain, for example, furnaces of the Material Science Laboratory (MSL) type, which are one of the Microgravity Facilities for Columbus (MFCs) from ESA.
- Laboratory Support Equipment (LSE) racks
- Express Racks (ER): They can easily and quickly be exchanged and can carry payloads in Standard Interface Rack Drawers (SIR-Drawers) or Mid-Deck Lockers (MDLs). Their feed lines are established automatically by insertion (back side) and by manual connection (front side). This concept was successfully tested during Spacelab missions and with the Spektr module of the space station Mir.

Figure 13.3 also shows an Express Rack containing eight MDLs and two SIR-Drawers in extended position. These types of double racks are called International Standard Payload Racks (ISPRs). Their dimensions and interfaces are standardized with slight deviations for NASA, ESA or NASDA racks. The US Lab offers 13 ISPR-Locations, each of them 203 cm high, 105 cm wide and with a maximum depth of 97 cm. Each ISPR has a volume of 1.6 m^3, a mass of 100 kg and can carry 700 kg of payloads. The rack has internal mounting provisions to allow attachment of secondary structures. The ISPRs will be outfitted with a thin center post to accommodate sub-rack-sized payloads, such as the 19 inch Spacelab Interface Rack (SIR) Drawer or the Space Shuttle middeck locker. Utility pass-through ports are located on each side to allow cables to be run between racks. Module attachment points are provided at the top of the rack and via pivot points at the bottom of the rack. The pivot points support installation and maintenance. Seat tracks on the exterior front posts allow mounting of payload equipment and laptop computers.

The 33 ISPR slots onboard ISS (13 in US Lab, 10 in JEM, 10 in COF) provide a common set of interfaces regardless of location. Nonstandard services are also provided at selected locations to support specific payload requirements. Their electrical power supply provides between 3 and 6 kW with 114.5–126 VDC and one data bus can transmit 50 kbps. Additionally, there is a high rate link with a transmission rate of 32–100 Mbps (see Sect. 12.2.3 "ISS Data Handling and Communication"). Lines for cooling water (for water temperatures between 16–20°C) and for vacuum connections (for pressures below 0.13 Pa) will remain closed if an Express Rack is installed in these places. More detailed information on ISPR for power, command and data handling, video, thermal control, vacuum exhaust system (for waste gas), vacuum resource, supply of gases (carbon dioxide, argon, helium) can be found in [NASA-Guide 98].

The US Lab will be the first laboratory module to be launched and will carry mostly US and a few European Early Utilization Payloads.

US Centrifuge Accommodation Module (CAM). The main task of the CAM is to offer a space laboratory for animal experimentation and general biological experiments by using the Centrifuge Facility (CF). The CAM offers additional stowage space for payloads. For that reason it contains, apart from the CF, four racks. As

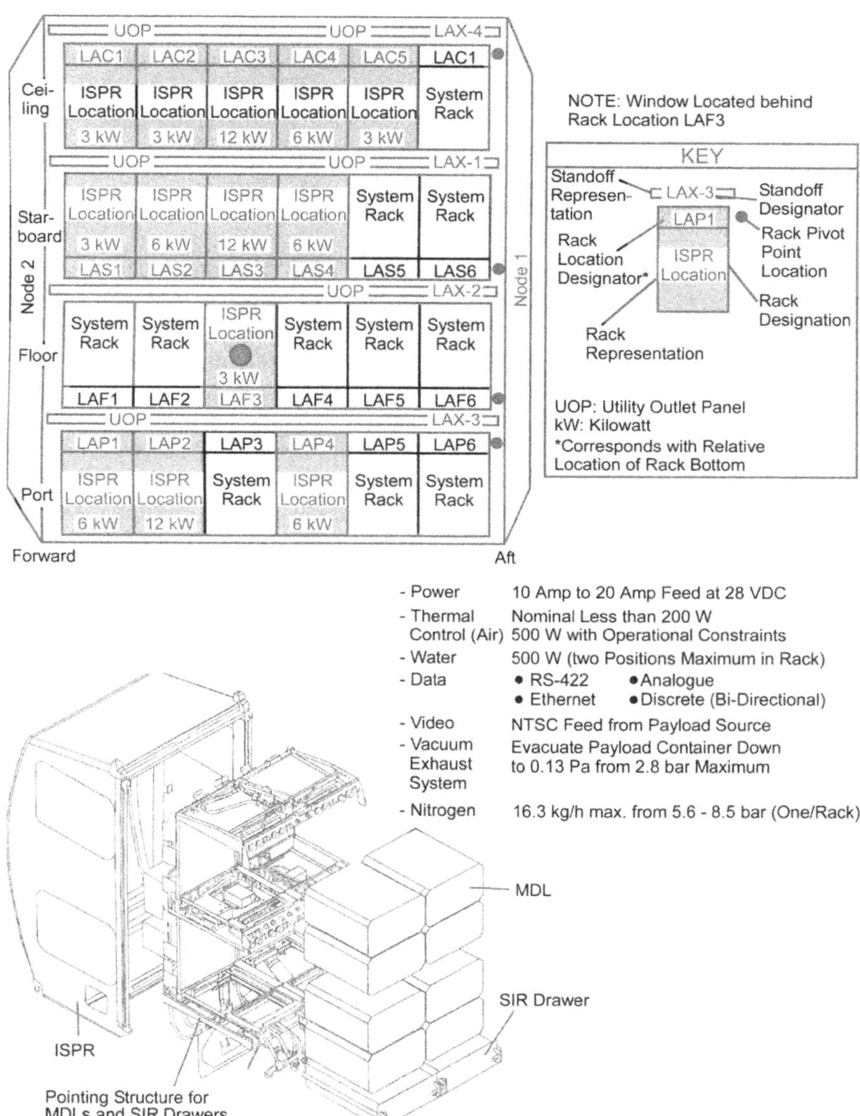

Fig. 13.3. Rack Distribution in the US Lab and an Express Rack with Eight Mid-Deck Lockers (MDLs) and two Standard Interface Rack (SIR) Drawers in Extended Position [ESA-Guide 96]

shown in Table 2.8, the CAM will not be launched until the end of the assembly phase.

US Habitation Module (US Hab). This module offers a sleeping area, space for leisure activities, cooking, eating, and it also contains training facilities with monitors for maintenance of physical fitness. Toilet, shower and handwashing apparatus, etc., are also included. The US Hab does not offer any space for ISPRs.

13.2.2 The Columbus Orbital Facility (COF)

The Columbus Laboratory is designed as a general-purpose laboratory to accommodate payloads in the pressurized module with experiments in life sciences, physical sciences and technology development. External attachment sites are available on the end cone for payloads requiring celestial pointing, Earth pointing or other space environment.

The cylindrical COF module is 6.7 m long and has a diameter of 4.5 m. Its estimated launch mass will be about 10 tons, plus the initial payload mass (corresponding to five ISPR plus payload) installed before launch, in total 12.7 tons. The COF on-orbit mass should not exceed 18 tons. The COF has its own Environmental and Life Support System (ECLSS). Figure 13.4 shows the arrangement of the racks as well as a cross-section of the modules. The quadrants on the ceiling and side walls accommodate payload and storage racks. A total of ten ISPR locations is reserved, of which five are allocated to ESA and the other five to NASA, who, in exchange, will provide transportation and resources in their facilities.

The subfloor as well as the end cones accommodate subsystems and also one storage rack. The following elements are payload-carrying elements:

- Multi-User Facility Racks containing the COF-MFCs
- European Drawer Rack for drawers of different size, which are compatible with the MDLs and SIDs of the Express Racks, including power supply (voltage between 28 V and 120 V), data and video links, and connectors for nitrogen, waste gases, and thermal control
- Stowage Racks: Usually, these are passive, but it is possible to provide limited resources for the drawers.

The outer structure and interfaces of all racks correspond to the ISPR and must not exceed a payload mass of 400 kg (for launch with the COF) or 700 kg (for a separate launch with the MPLM). The ISPR basic accommodation units support the large pressurized facility-type payloads in complete racks. A range of European Standard Experiment Drawers (SEDs) and Mid-Deck Lockers are available for smaller pressurized payloads to allow for easier integration and short flight access. The Columbus resources available to ISPR payloads are provided through a Utility Interface Panel. This is located under the ISPR and, together with the use of flexible utility lines, allows the ISPR to be tilted without disturbing the interface and thus supports "all-round" rack servicing and maintenance requirements. The resources provided to the Columbus ISPR by the Utility Interface Panel are compiled in Table 13.1. Containing an integrated European payload, the COF will be launched by the Space

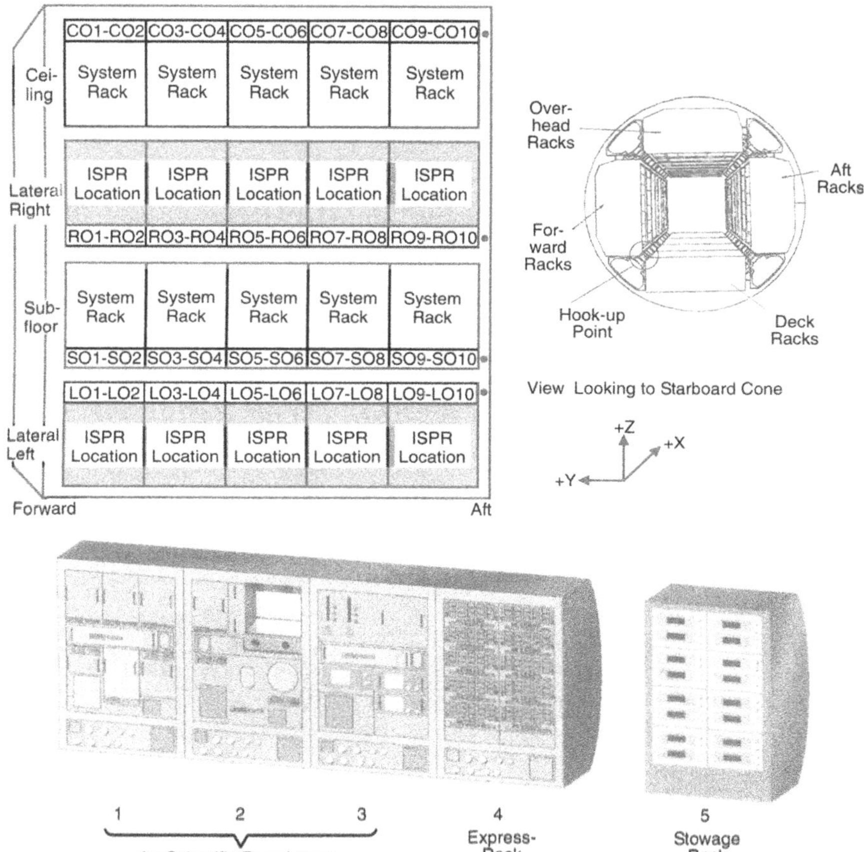

Fig. 13.4. Rack Distribution in the COF and Side View of the Racks as well as Presentation of the Different Rack Types [ESA-Guide 96, ESA-Fakten 96]

Shuttle in February 2003. The COF will be attached to the Node 2 element which connects it to the station, and by this connection, the COF will rely on the station's resources, such as electrical power (≤20 kW of 120 VDC, for payloads ≤13.5 kW), excess heat rejection (≤14 kW through medium-temperature water cooling loop and cabin air cooling), Trace Contaminant Control and Monitoring (TCC and TCM), and also data transmission to the ground via the TDRSS satellites.

Through its contributions to the ISS program, Europe has acquired the following permanent access to user accommodation and utilization resources:

- Five ISPR racks in the Columbus laboratory
- Two Express Pallet Adapters on the Columbus External Payload Facility
- 2.5 kW of electrical power (average, 8.3% of 30 kW)
- 13.25 hours of crew time per week (8.3% of 160 h/week)

Table 13.1. Resources Provided to Columbus ISPR by the Utility Interface Panel

Resource	Nominally Available	Comments
Main Power Auxiliary Power	6 kW or 3 kW @ 120 VDC 1,2 kW @ 120 VDC	6 kW=medium power, 3kW=low power. For backup in case of main power loss.
Vacuum	0.13 Pa under a gas load of 0.1 Pa·liter/s	
Venting	1000 hPa to 0.13 Pa within 2 hours	Only 1 ISPR vented at any one time to prevent cross contamination
Cooling Water	Inlet between 16°C and 20°C Outlet < 49°C	Flow rate fixed/calibrated during payload integration between 30 kg/h and 190 kg/h
Nitrogen Gas	Max. flow rate 5.43 kg/h Operating pressure 5170 hPa to 8270 hPa Max. pressure 1378 kPa Temperature 15.6°C to 45°C	
Video	NTSC Standard	Capability for video from ISPR. Can be configured for high rate data upto 32 Mbit/s
High Rate Data	Upto 32 Mbit/s	High rate data to/from US Lab. APS
Fire Detection System		Payload signals required for credible fire risks (incl. smoke sensor & fans where necessary) and safing commands from Columbus Laboratory vital systems
EWACS		Emergency & warning signal interfaces and commands with direct link to Columbus Laboratory vital systems
1553B Bus	51.2 kbit/s data rate	Primarily intended for payload low rate data and payload commanding
802.3 LAN	< 1.25 Mbit/s data rate for all payloads	Primarily intended for payload medium rate data (via HRM)

In addition, a yearly average of 1.13 tons of Shuttle pressurized payload mass upload, 0.6 tons pressurized download, 0.6 tons Shuttle upload/download for external payloads, and 8.3% of TDRSS capability can be purchased by ESA from NASA.

13.2.3 The Japanese Experiment Module (JEM)

Apart from payloads for Intra-Vehicular Activities (IVA) which correspond to those of the COF, the JEM module includes the possibility of deploying and mounting external payloads on the JEM-EF (Exposure Facility) via an air lock. During such an operation, the module's autonomous robotic arm JEM-RMS (Remote Manipulator System) as well as JEM's own logistics module (ELM = Experiment Logistics Module) can be put to use (cf. Fig. 2.16). Similar to the COF, JEM contains Multi-User Facilities, Life Support Equipment (LSE) and Express Racks. Due to its larger size, JEM can accommodate 23 racks, of which 10 will be ISPRs (five to be used by NASA, cf. Fig. 13.5). JEM will be launched by means of the Space Shuttle earlier than the COF.

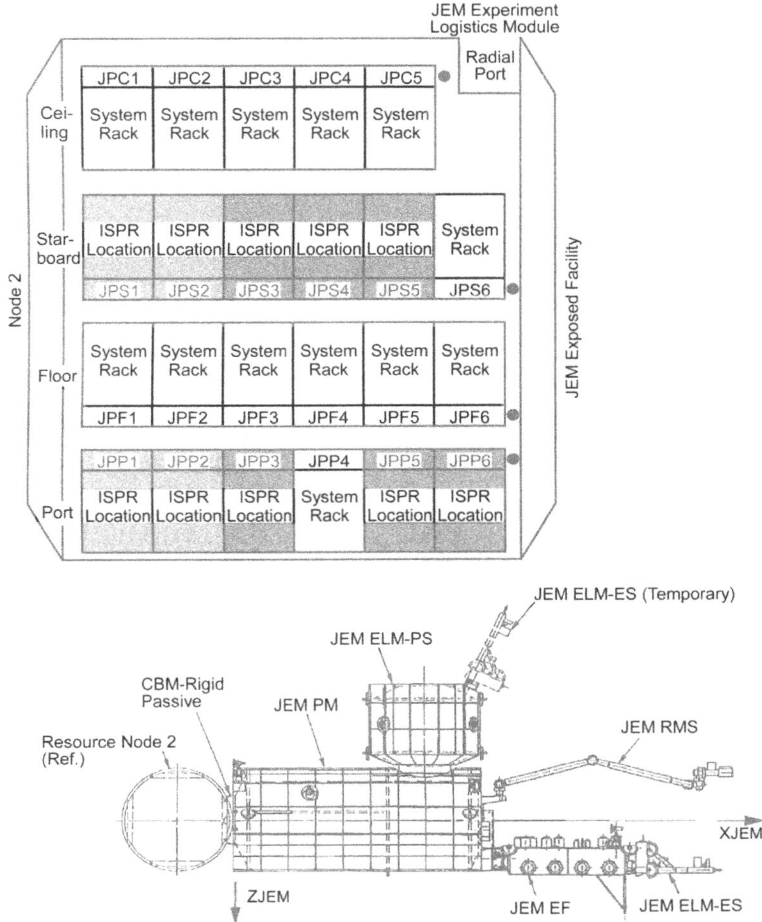

Fig. 13.5. Rack Distribution in the JEM Laboratory Module, the JEM-ELM Logistics Module, the Robotic Arm JEM-RMS, and Attachment Sites of External Payloads at the External Platform JEM-EF [ESA-Guide 96]

13.2.4 The Russian Laboratory Modules

Russian Universal Module 1 (RM1). Russia plans to contribute two research modules to ISS, namely RM1 and RM2. Module RM1 is also called the "Russian Universal Module" since it can carry pressurized internal payloads as well as external payloads (Fig. 13.6). In a way, it very much resembles the Spektr module of the space station Mir. Just like the modules described before, the Russian modules are cylindrical, but they have a smaller diameter, and for that reason they cannot carry ISPRs. Internal payloads are thus accommodated by so-called "Drawer Support Frames".

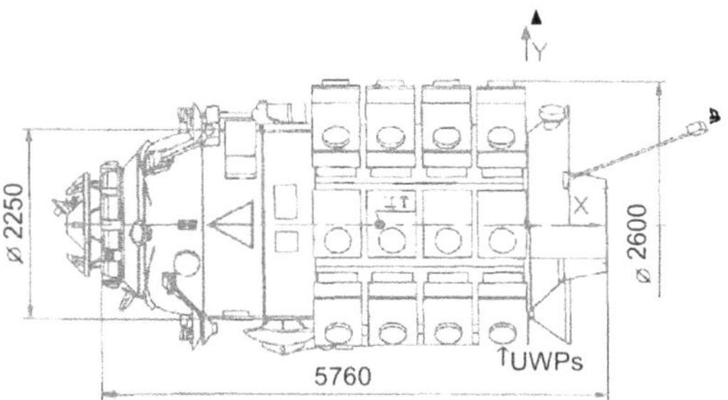

Fig. 13.6. The Russian Laboratory Module RM1, also Called Universal Research Module due to Attachment Possibilities for Both Pressurized and External Payloads

The Russian laboratory modules will be transported consecutively according to the schedule presented in Table 2.8.

Russian Universal Module 2 (RM2). The RM2 module carries mainly pressurized payloads and will be used for biological experiments. It therefore aims for a high standard of air purity. A further module, RM3, originally planned primarily for Earth observation experiments, was canceled in 1998 for budgetary reasons.

13.3 Accommodation Sites for External Payloads

Outside the pressurized modules, there are several possibilities for the installation of external payloads. There are, for example, attachment sites on the trusses, mounting sites quite near the pressurized modules, or special positions which can be reached with the station's robotic systems.

13.3.1 The US Truss Structure ITA (Integrated Truss Assembly)

On the starboard truss, the ITA offers four attachment sites for external payloads (Fig. 13.7). Two similar attachment sites on the port side are mainly used for Unpressurized Logistics Carriers and also for payloads.

Payloads attached to these sites can be installed in such way that they point towards the zenith (upwards), the nadir (downwards to Earth), the ram (flight direction) or the wake (opposite to flight direction).

Payload Attach Structure (PAS). Each attachment site will have a mechanical connecting structure, called the Payload Attach Structure, where one large payload or several small payloads (previously mounted onto a special carrier) can be installed. Each PAS offers 28 m^3 of space and can carry payloads of up to 5000 kg. It is possible to provide electrical power (3 kW at 120 V DC) and a data link (connection to the MIL-STD 1553 bus, the HRDL "high rate data link", and the onboard data storage). Thermal control, on the other hand, must be provided by the payload.

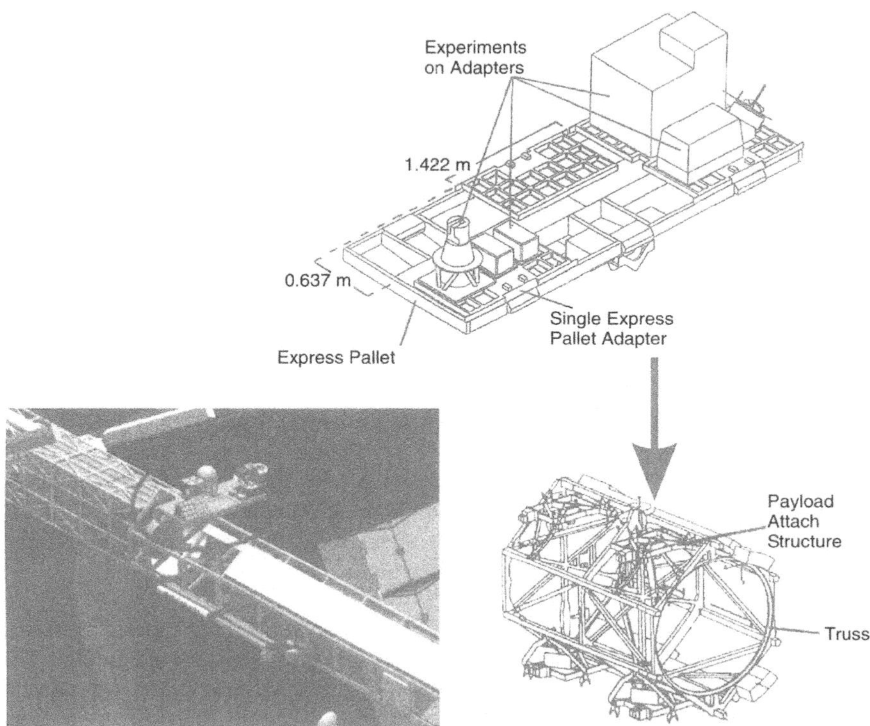

Fig. 13.7. The Externally Mounted Express Pallet with Adapters and Typical Payloads [ESA-Guide 96, ESA-Fakten 96]

Express Pallet. Such a pallet is needed to allow a set of smaller, external payloads to be accommodated onto a single PAS. Express Pallets have connecting structures for installation onto the PAS as well as resource distribution equipment for the payloads. Installation and handling of the Express Pallets and the Express Pallet Adapters (ExPA) (Fig. 13.7) are possible with the help of the Space Station Robotic Demonstrator (SSRD) and SPDM. Up to six adapters can be installed on one Express Pallet (dimensions: 1.422 m x 0.673 m); thus there are 24 ExPA on the truss. The adapters feature a payload mass capability of 225 kg per adapter; power of 2.5 kW at 120 V DC or 500 W at 28 V DC; and data links (RS422/485, MIL STD 1223B, HRDL) for analog and digital measurement data. As aforementioned, thermal control has to be provided by the payload.

The first Express Pallets will be mounted during the third assembly phase (cf. Table 2.8) and can immediately be used, carrying payloads with zenith-oriented pallets for a maximum of three years. The Express Pallet Adapters will periodically be exchanged and returned to Earth, where new payloads will be mounted.

13.3.2 The Japanese Experiment Module Exposed Facility (JEM-EF)

The Japanese external platform shown in Fig. 2.16 is 5 m long and is located at the port side end cone of the JEM. It can accommodate 10 relatively small payloads, each with a maximum mass of up to 500 kg and a volume of up to 1.5 m^3. The power distribution system delivers a total of 3 kW to each payload location. As a special feature, in contrast to the Express Pallets, this platform includes an active system for thermal control with a performance of 11 kW. Moreover, a data transmission link with a maximum of 10 Mbps and a video link are planned. With its own robotic arm and an airlock between the pressurized module and the JEM-EF, the Japanese Segment of ISS, as a whole, is a well-thought-out experiment facility, since it makes optimal use of its observational possibilities and intervention by astronauts, both internally and externally.

13.3.3 Russian External Payload Attachment Sites

Science and Power Platform (SPP). External payload attachment sites can also be found on the Russian SPP. They are standardized attachment sites with suitable electrical, mechanical and data interfaces where the European and Russian robotic systems ERA and PELICAN can install payloads.

Russian Universal Working Place (UWP). Additional UWP attachment sites (Fig. 13.6 and Fig. 13.8) will be located on the outside of the Russian modules. They will be installed before launch and connected by universal interfaces (such as vacuum-sealed cable ducts and air locks) to the interior of the modules. Later on, scientific instruments and experiments will be installed by astronauts in space; they will connect all necessary cables, as well. All experiments and equipment units mounted on these attachment sites can be operated by means of the PELICAN manipulator arm. This combination of the PELICAN manipulator arm and an airlock is planned to be installed in one of the Russian modules.

13.3.4 The Station's External Robotic Systems

The large mobile Space Station Remote Manipulator System (SSRMS) forms part of the Mobile Servicing Center (MSC); both are important parts of the Canadian Mobile Servicing System (MSS). The SSRMS can move along the truss structure up to the point where the joints of the solar arrays intersect with the truss. The SSRMS will be used mainly for assembling the station as well as for mounting and operating external payloads. Thanks to the Special Purpose Dextrous Manipulator (SPDM), it will be able to assume tasks like exchanging Orbit Replaceable Units (ORUs), containers and Express Pallets.

The JEM Remote Manipulator System (JEM-RMS) consists of a small robotic arm which can be controlled from the interior of the JEM module. It can transport payloads from the JEM air lock to the various JEM-EF attachment sites (Fig. 13.5).

The European Robotic Arm (ERA) is able to install external payloads on the Russian parts of the station, including the SPP. It is especially suitable for transporting relatively large payloads and ORUs and can also be used for inspecting the space

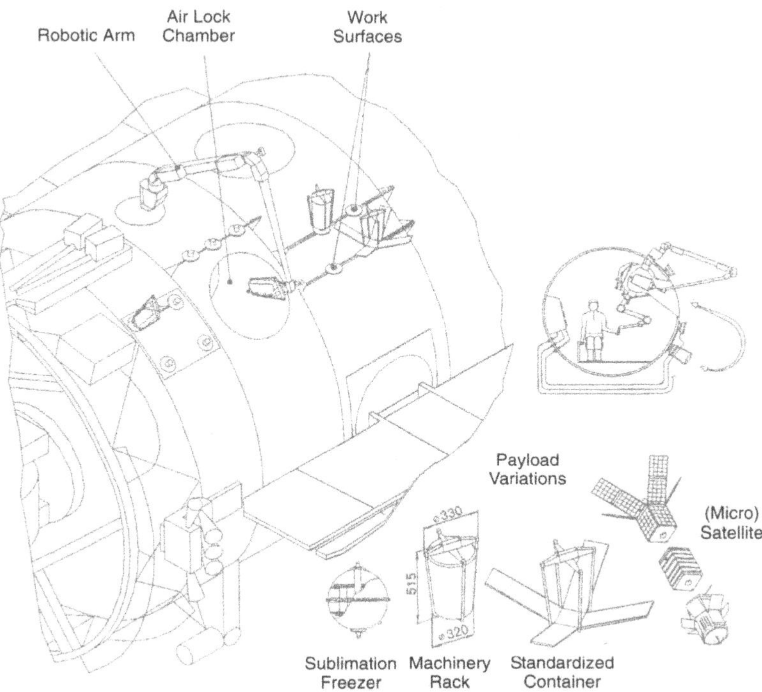

Fig. 13.8. Russian Nest For External Payloads. It can be exposed through an airlock from the inside and installed and operated with the help of the PELICAN robotic arm.

station and for supporting astronauts during EVAs. The ERA and the Russian PELICAN system, like other robotic systems, include several additional parts such as MPTE (Mission Preparation and Training Equipment), GSE (Ground Support Equipment), transportation containers, and equipment such as tools, inspection cameras, etc. Further small robotic and platform systems are planned for the PAS and Express attachment points. Figure 13.9 shows the "Hexapod" pointing platform built by ESA. In the framework of the "Early Opportunity Agreement", NASA and ESA concluded that the Hexapod is to be used for NASA research experiments investigating the upper atmosphere. It includes six electromechanical, linear actuators which are mounted between the upper and lower platforms at three fixed points. The kinematics of a hexapod are characterized by the following features: rotation and positioning with six degrees of freedom, a given pointing direction can be achieved mechanically in several ways. Compared to traditional robotic arms, this combination of linear-movement actuators and a rigid connection between the platforms enables faster tracking while at the same time allowing for a reduced mass and a more even distribution of forces and tensions.

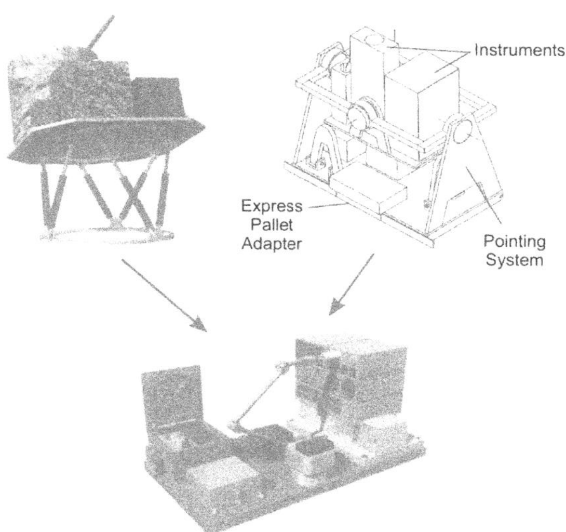

Fig. 13.9. Typical Positioning Instruments for the Express Pallets. A universal hexapod trackable in six degrees of freedom (on the left), a simpler platform trackable in two directions (on the right), and the European TEF module (bottom).

13.3.5 External Environmental Monitoring

The external Environment Monitoring Package (EMP) will measure the contamination environment outside the space station as a service to the members of the attached-payload community who require characterization of operational conditions. Specifically, the EMP will measure, either directly or indirectly, the molecular deposition, line-of-sight molecular column densities, charged plasma distributions, and the electrical potential on and around the attached payload locations on both the US truss and Japanese Exposed Facility. The EMP will include a mass spectrometer, quartz crystal microbalances, pressure gauges, plasma instrumentation, and other instruments under study. The EMP is configured such that it could be mounted on the Special-Purpose Dexterous Manipulator of the Mobile Servicing Center (MSC) in order to move around ISS and obtain measurements at multiple locations. The EMPs and their output data are recognized as facilities for potential scientific use, and when possible, are made available to researchers for scientific experiments.

13.4 Transportation Systems and Logistics Containers

The following will be the transportation systems for payloads:

- US Space Shuttle
- Ariane 5 launcher from ESA
- H-II launcher from NASDA
- Soyuz and Proton launchers from RKA

Table 13.2. Transportation and Transfer Vehicles for the International Space Station

Transportation System and Transfer Stage	Mass; Flight Rate/Year; Orbit Altitude/51.6°; Transfer Mass	Payload/Orbit/Crew; Dimensions/Duration; Propellants (kg/km); Return Mass	Cargo Carrier	Container Mass + Payload Mass [kg]
US-STS[a] Space-Shuttles Atlantis, Columbia, Discovery and Endeavour	1981 tons 5–7 200 km 128 tons	15876 kg/407 km/5 4.57 m x 18 m/9 Days 24.5 kg/km 17250 kg	Assembly Carr., MPLM+Payload 2 ULC+Payload MPLM+ULC+PL SL-Pallet Single SL-Pallet Double SL-Pallet Triple Sidewall Carr.+PL	 4780 + 8842 5355 + 9476 7457 + 8418 635 + 4128 1270 + 5874 1905 + 6350 227 + 454
Soyuz Transportation System including Soyuz TM Capsule	290 tons 7–12 80 x 220 km 7–7.8 tons	7070 kg/220 km/3 10.3 m^3/180 Days 1.37 kg/km Reboost 1073–1870 kg Propellants	Soyuz TM	283 kg/ 3 Astronauts +20–100 kg and 20–60 kg Returned
Soyuz Transportation System and Progress Capsule, Modified Soyuz Transportation System and Progress	290 tons 7–12 80 x 220 km 7–7.8 tons	2350 kg/ 1600 kg Returned 3200 kg/ 1600 kg Returned	Progress-M Progress-Rus6	880+1750 Propellants 880+2620 Propellants
RS-Heavy Lift: FGB-, UDM-, SM-, LTV-Module RS-Light Lift (about 7 tons) for Small Modules	*Proton:* 689 tons 8–18 200 km 20.9 tons; *Soyuz:* 290 tons 7–12 80 x 220 km 7.0–7.8 tons	max. 21000 kg max. 7800 kg	FGB, UDM, SM, LTV, DC, Soyuz, Progress-M, Rus, Rus6, RM1–2, LSM1–2	
Launcher Ariane-5 ATV (cf. Table 12.3)	*Ariane-5:* 710 tons 0.8 70 x 300 km 18 tons	7500 kg of 18000 kg 35 m^3 3.5 kg/km	ATV Pressurized Cargo ULC/Mixed Cargo	Payload + Propellants 7500+950 Prop. Reboost: 4250 Prop.
Japanese Transportation System H-II-Transfer System	*H-II:* 329 tons 1 150 km 13 tons	6000 kg of 10000 kg 24 m^3 3.4 kg/km	HTV	Payload = 6000 kg
Crew Transfer Vehicle (CTV)	15–21 tons	4 Astronauts + 300 kg		

[a] For Acronyms See Glossary

Since these transportation systems have been described in a very detailed way in [Isakowitz, Griffin] and in further literature, we will concentrate only on boundary conditions specific to ISS and on the logistic elements.

Transportation needs and scenarios typical for the International Space Station have already been explained in Sect. 12.1. Table 13.2 offers an overview of all transportation systems and performances.

Since its first flight in 1981, the fleet of Space Shuttles carried large pressurized payloads such as the Spacelab, small payloads like Hitchhikers, GASs, MDLs, and other unpressurized payloads. In the case of ISS logistics, there will be further containers and logistical elements for the payloads, namely MPLMs, ULCs, JEM-

ELMs and ATVs. Payloads for last-minute access can only be transported in small MDLs which require power supply in MPLMs. It is important for the launch and ascent phases that the payloads can withstand the accelerations, vibrations, high noise levels, and other mechanical loads.

Mini Pressurized Logistic Module (MPLM). The primary purpose of the MPLMs is to transport payload racks, system and subsystem elements, equipment and supplies to the station, and return exchanged parts. Easy access to the MPLM is provided through ISS-compatible hatches at the forward end (size 1.27 m x 1.27 m) and at the aft end (diameter 2.438 m). One MPLM can carry up to 16 racks. Five of the rack positions, for example, have an interface for electrical power for the operation of refrigerators, freezers or experiment units. The remaining 11 positions are passive and can, apart from ISPRS (like Express Racks, Payload Facility Racks, etc.), accommodate stowage racks, which in turn can carry SIR drawers and Mid-Deck Lockers (the latter being limited to maximum mass of 27 kg and a volume of 0.57 m^3). The internal design of an MPLM and examples for the 45 cm wide stowage drawers are presented in Fig. 13.10.

Unpressurized Logistics Carrier (ULC). The ULC structure can carry payloads (containerized or not) for transportation by the Space Shuttle or by the Ariane 5/ ATV. Apart from the ULC, Spacelab pallets (cf. Sect. 2.4 and Fig. 2.21 and Fig. 2.22) and the "Dry Cargo Carrier" (DCC) can also be used for transportation to the station and for intermediate storage of payloads outside the station during the Early-Opportunity-Utilization Phase.

JEM Experiment Logistics Module (JEM-ELM). The JEM-ELM consists of a Pressurized Section (PS) for ISPRs and an Exposed Section (ES). The ELM-ES will be located at the port side of the JEM-EF where the ELM-PS can also be kept for intermediate storage for logistics operations (Fig. 13.5).

Automated Transfer Vehicle (ATV). The European Automated Transfer Vehicle for the Ariane 5 launch vehicle has been described in detail in Sect. 12.1.3, including its mission and cargo-carrying capacity. It is able to dock automatically at ISS and, apart from tasks like transporting payloads, can also be used for reboost of the station, waste disposal by reentry and burn-up in the atmosphere, etc. Its cargo-carrying capacity is indicated as 7.5 tons for a circular orbit at 400 km altitude at 51.6° inclination. This figure has the potential to be increased to 10 tons of payload by a further enhanced version of Ariane 5. At present, ATV missions are planned for transportation of either pressurized MPLM or unpressurized ULC payloads including a 950 kg propellant resupply for a station reboost of 29 m/s, or a combination of both types of cargo (Mixed Cargo).

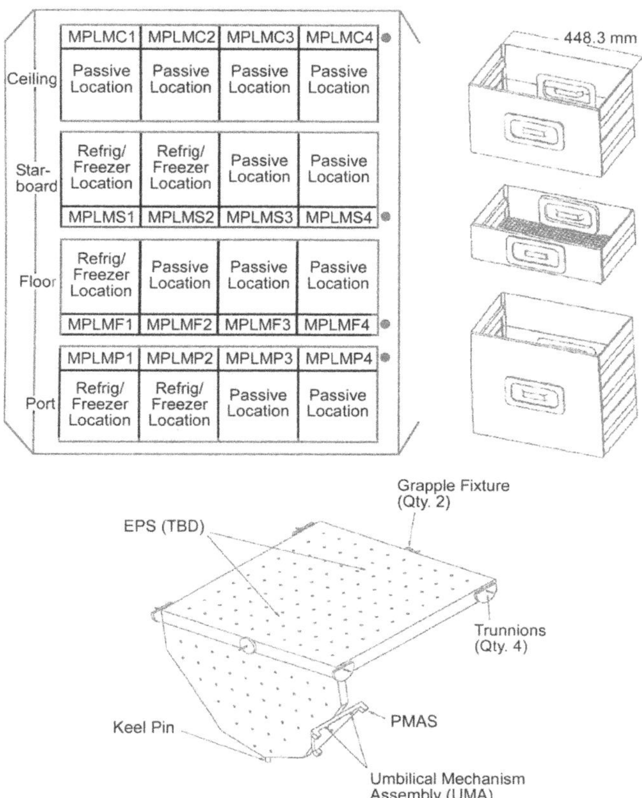

Fig. 13.10. The Mini Pressurized Logistics Module (MPLM) for Pressurized Payloads and the Unpressurized Logistics Carrier (ULC) [ESA-Guide 96]

13.5 Payloads and Payload Selection

13.5.1 Typical Payloads: Experiment Facilities and Experiments

The user-related hardware aboard the International Space Station can be divided into two classes: experiment facilities and experiments [ESA 95, ESA-Guide 96, NASA-Guide 98].

An **experiment facility**, for example, could be any of the following:

- a fluid physics laboratory
- a furnace assembly
- an Express Rack
- a material exposure facility
- a large remote sensing facility

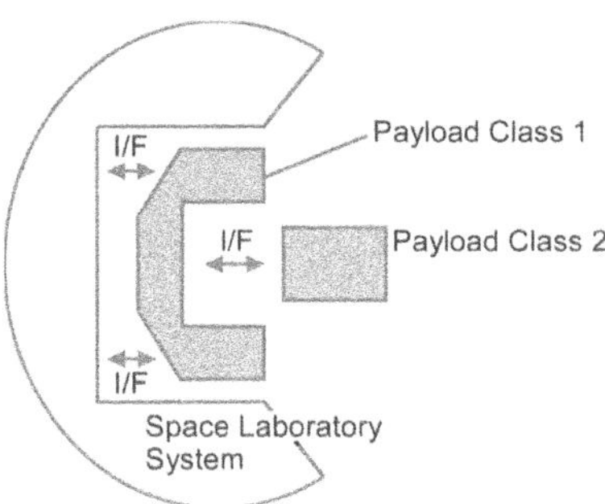

Fig. 13.11. Payload Classification

In principle, experiment facilities will be provided by NASA, ESA or other national space agencies for various scientific users. For that reason, they are also called "Multi-User Facilities". Usually, they are so large that they have direct interfaces to the station's systems at rack level and external payload attachment points. In the jargon of the International Space Station, such facilities are also called "Class 1 Payloads".

Experiments comprise many possible instruments, among them are:

- Observation cells to be included in a fluid physics laboratory
- Diagnostic instruments for a human physiology facility
- Samples containing material to be processed in a furnace facility
- Small instruments accommodated in the drawer of an Express Rack
- Collections of material samples to be placed in a material exposure facility
- Solar radiation monitors or spectrometers to be accommodated on an external instrument pointing system
- Focal plane instruments for large observation or remote sensing facilities

Experiments are either directly delivered by the users, or they are developed in close cooperation with them. Users of such experiments are e.g. scientists from universities or research institutes, researchers, and development engineers from industry. Experiments normally have no direct interfaces with the station's systems. They are located inside rack drawers or mounted on external payload adapters. The ISS jargon designates them as "Class 2 Payloads".

Figure 13.11 shows the difference between Class 1 and 2 Payloads as to their interfaces (I/F). The selection, financing and preparation processes are different for both kinds of payloads and will be described in the coming sections.

13.5.2 Selection of Class 1 Payloads and Users

Class 1 Payloads

In order to avoid duplication of hardware, Class 1 Payloads (also called "Experiment Facilities" or "Multi-User Facilities") are either selected at NASA, ESA program level, or by national space agencies in close cooperation with the other

Partners of ISS. Potential individual users may influence the selection and development of the experiment facilities for ISS either by joining the user advisory committees of NASA, ESA and the other Partners or by cooperating with them.

Once the decision is finalized for the development of an experiment facility, the corresponding hardware will be developed by industry (under contract), financial coverage being effected by the responsible Program Directorate (of NASA, ESA, etc.). Alternatively, if it is nationally financed, financial coverage will be effected by the corresponding national space agency. Definitive technical decisions and development will take place in close cooperation with the "Facility Science Group" which is composed of leading scientists in the corresponding research field. Based on the experience acquired during the Spacelab program, it should take about four to five years for the design, development and production of such facilities.

Selection of Users

With regard to the hardware necessary for scientific, technological or industrial experiments aboard ISS, the potential users are classifiable into three groups:

- *Users of Available Onboard Experiment Facilities*. This group uses the Multi-User Facilities and test beds developed by space agencies or other research organizations for scientific or technological tasks. The users provide test items or samples/specimens to be processed or exposed to space conditions, or they deliver supplementary equipment for the experiment facilities to expand their range of investigative research.
- *Users of their Own Instruments*. These users either develop their own instruments (independently or within a cooperative group) for scientific research aboard the station, or they test or qualify technical equipment in space which had previously been developed on ground.
- *Data Users*. This group exclusively uses data obtained by already existing instruments or facilities aboard ISS. This primarily applies to the disciplines of Earth observation and space sciences.

In order to select experiments to be conducted aboard ISS as well as to give access to the onboard experiment facilities (financed by space or research agencies), ESA has established a suitably adapted procedure which has already been applied in the case of various Spacelab missions and the selection of external ISS payloads in 1997 (Fig. 13.12, see also Table 13.3 for calls of proposals).

The selection process is introduced by an "Announcement of Opportunity" (AO). Such an announcement describes the mission objectives for each discipline, the system-related characteristics and constraints, and the existing onboard Class 1 Payloads available during a given period of time. Moreover, it solicits proposals from the user community concerning research projects or experiments to be carried out. AOs are regularly issued to all potential users belonging to a certain discipline. The interested user replies to such an announcement with either an "Experiment Proposal" or a "Letter of Intent". The latter will be used if the information delivered is not detailed enough to submit a formal Experiment Proposal. On the basis of a Letter of

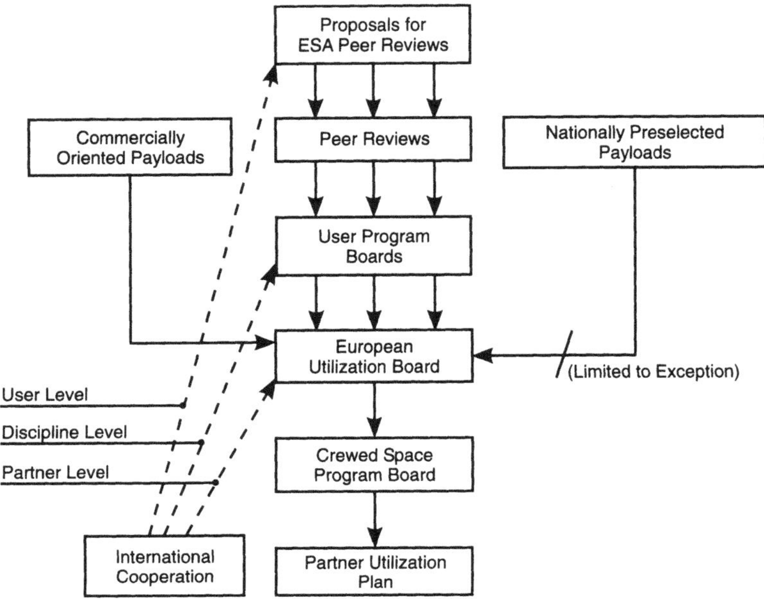

Fig. 13.12. Selection Chart for European Experiments aboard the International Space Station [ESA-Guide 96]

Intent, however, supplementary information will be provided, enabling the user to establish a formal Experiment Proposal, if requested.

All proposals submitted are examined, separately according to discipline, by Peer Review Groups. These groups are composed of leading scientists or technology experts appointed by the space agencies. They evaluate the scientific or technological quality of the experiment proposals and judge whether or not the performance aboard ISS of the activities proposed is of sufficient general interest. Simultaneously, a group of mission specific engineers (sometimes supported by engineers from industry) examines the experiments' technical feasibility as well as their compatibility with the other investigations aboard the station.

The reports containing the results of the peer reviews and technical assessments are transferred to the Programme Boards in charge of the related user disciplines. For each discipline involved, these Boards will establish priority lists for the experiments. Assessment reports and priority lists for the individual disciplines, e.g. in Europe, form the basis for the judgement of the "European Utilization Board" (EUB). This Board is composed of members from the ESA Programme Boards and agencies involved in the utilization of the International Space Station. In case of conflicts among the disciplines, the EUB or Programme Board will arbitrate, and, moreover, it will establish a list with selected experiments that will finally constitute the European payload segment. This list is handed over to the ESA Manned Space Programme Board which is responsible for funding the European utilization of the International Space Station. This Board will give approval for the set of proposed scientific and technological experiments and investigations.

According to standard operating procedures in Europe, ESA usually has neither the mandate nor the financial means to pay more than those experiment costs that arise due to the experiments' transportation into space, their integration into the station and their onboard operation. For that reason, the users are required to pay for their experiments themselves, or to envisage agreements with their respective national space agencies or a research institute for the purpose of sponsoring those experiments. In order to reach such an agreement, further examination of the proposals by the national agencies or other potential sponsors may be necessary.

Once a proposal has the approval of the Manned Space Programme Board, the selected experiments are placed "on the waiting list" for operation aboard the International Space Station. However, they do not have "seat assignments" yet. After the formal approval, they have to be integrated into the Partner Utilization Plan, and afterwards (after coordinating the plans of all space station Partners), into the overall operations and utilization plan of the International Space Station. It is not until the end of this planning process (as described in detail in Sect. 13.5.4) that they will receive a definite "time frame"; i.e. enter into the three-month planning period of station operations. This sounds more complicated than it may be in practise. There are examples where the time between idea and space experiments (returning data from Mir) was not more than 18 months.

These three-month planning periods for the utilization of ISS are called "increments" and, because of their continuous succession, a corresponding continuity in the sequence of announcements can be expected. According to how time-consuming development and preparation of experiment facilities, instruments, experiments and test items are, the periods for integrating the proposals into this cycle could vary from several months to several years. In order to fully profit from the continuous availability of the station, the plan is to create something like "last minute" access, provided that this is compatible with the available technical and operational resources as well as being able to fit into the preparation time necessary for the corresponding experiment. The Express Rack and Express Pallet concepts with their standardized interfaces with the experiment facilities of the station constitute a good basis for such "last minute" access.

Rapid access will be of particular importance if users have to react immediately to results obtained during a previous planning interval of the station's operation. Such a reaction may not only require the planning of an additional sequence of experiments, but also the transportation of new test items, sensors or actuators to the station or the exchange of an Express Rack drawer for one with a slightly modified experiment. A certain reserve of transport capacity, crew time and other station resources will be provided for this kind of interactive research aboard ISS. Additionally, the experiment selection process will, of course, be shortened accordingly.

13.5.3 Access for Commercial Users

Apart from scientific users, the International Space Station is also regarded as a new research and test facility for industrial users. Therefore, NASA, ESA and other interested space agencies will strongly encourage and facilitate access to the station for commercial organizations. However, the private sector will not only be encouraged to make use of the available resources of ISS, but also to participate in investments for experiment facilities. Of course, it is unrealistic to believe that commercial companies would invest in an unknown project like the International Space Station. Yet, national agencies and international partners have the intention of gradually increasing the role commercial companies play in utilization – and financing – during the availability of ISS. Already today we can differentiate some of the possible scenarios, for example:

- Leasing of space aboard the station for instruments and other equipment which will be provided by commercial users at their own expense. In such a case, support such as preparing and operating instruments, receiving and processing data and transporting hardware or samples to the station and returning them to Earth could be provided (and as a first step, also be promoted) by the space agencies.
- Dedicated utilization during a determined period of time made in exchange for payment of instruments and/or facilities owned by space agencies
- Commercial use of data received from space

Commercial customers will be ensured sufficient guarantees with regard to confidentiality (including intellectual property rights). It is assumed that special Announcements of Opportunity will be issued for commercial users and that the selection process will take its course according to specific regulations in order to ensure the necessary confidentiality and quick access (see Fig. 13.12).

In principle, industrial or commercial users have the opportunity to receive access either via NASA, ESA (and its established selection processes) or other national space agencies (their selection processes being subject to national jurisdiction and responsibility). The latter option will be especially interesting as aspects like industrial politics, encouragement of competition or industrial location come into play.

13.5.4 Utilization Planning

The fundamental difference between the International Space Station and earlier space flight missions lies in the continuity of scientific research operation and the routine, regular delivery of supplies, spare parts, experiment equipment, and test items as well as routine crew changes. It is exactly this difference that makes the International Space Station so attractive and that requires an operational concept which maximizes the entire utilization potential from the user's point of view.

The utilization planning for ISS comprises the following three planning levels: strategic, tactical and executional.

Strategic and Tactical Planning. The operation and utilization planning for the entire station will begin five years in advance with the yearly preparation of the "Strategic Plan", covering all activities during these five-year-periods. On the basis of this strategic plan, the long-term, high-level direction of station operation and utilization is determined and controlled. The latter applies also to the forecast of important parameters such as planning interval dates, offering and allocating resource capabilities, approved experiment facilities and experiments, transportation demands, and flight schedules of transportation vehicles.

In line with the strategic planning for the entire station, all international partners will establish their own strategic plans and their own yearly utilization plans. In the ISS jargon, those are called "Partner Utilization Plans" or "PUPs". Consequently, the European plan is designated the "European Partner Utilization Plan" or "Euro-PUP". The first plan for a given utilization period is issued five years in advance and during the subsequent years (approaching its execution), it is updated and worked out in detail. Similar plans are produced by other partner agencies (NASA etc.).

After a process of international coordination, the utilization plans of all partners are combined to form the common "Composite Utilization Plan" (CUP). This CUP and its equivalent in the field of operation (the COP = Composite Operations Plan) are then integrated into the "Consolidated Operations and Utilization Plan" (COUP) for the whole station. The COUP has to be approved by all international partners. PUPs and COUP specify the major research facilities and available resources. Detailed plans for the various three-month planning periods are worked out in the course of the tactical planning process.

The tactical plans include experiments and investigations which have to be conducted by using the available research facilities and resources. The strategic planning process for a given three-month "increment" turns into tactical planning approximately 30 months before the start of the corresponding planning increment.

As an example, the tactical planning inputs for the European Columbus Orbital Facility (COF) are developed by the COF Mission Operations Integration Team of ESA, supported by the European industry. These inputs are finally transferred to a Multilateral Tactical Planning Team led by NASA.

Executive Planning. Twelve months before the start of an "increment", the tactical planning results in the development of plans on an executive level. The first of those plans is the "Increment Operations Plan" (IOP). The IOP comprises the following points:

- Mission rules and regulations
- Plans for flight hardware and software reconfiguration, ground facility configuration, on-orbit maintenance, inventory, resupply/return flights, communications and data management, ground processing
- Flight mechanics data
- Schedule of the major system and experiment activities to be carried out throughout the increment

The IOP is worked out at the Space Station Control Center in Houston, Texas, USA, where the planning inputs of all international partners are integrated and consolidated.

It is only a few months prior to the actual flight of the experiments that the exact schedules for the onboard operations will be made known. Thanks to built-in buffers (and sometimes also to "en bloc" allocations of resources), planning at the executive level is flexible enough to allow iterative research and experiments.

On the basis of the IOP, weekly plans are established. During this decentralized process, the Space Station Control Center and the Payload Operations Integration Center in Huntsville, Alabama, USA merge the requirements and plans of all international Partners into a single integrated plan for the entire station. This integrated plan is then delivered to the control centers of the international Partners, who then use it to develop the different planning tools necessary to conduct the real-time operations of their elements. The "COF Master Timeline", for example, is an highly important planning tool for the COF at the executional level, since it automatically schedules system and experiment operations. Weekly plans are updated every 24 hours.

13.5.5 International Coordination of Space Station Utilization

As far as research in space is concerned, international cooperation is by no means an unknown issue. For over thirty years, cooperation within an increasingly international scientific community in a wide range of disciplines has contributed to the success of all major space flight programs. This cooperation includes national space agencies and other national institutions. The Spacelab missions which have been performed until present are an important, concrete example of cooperation in a utilization context comparable to the current International Space Station.

Since the concept of ISS emerged for the first time, researchers from interested disciplines have discussed many different experiments and projects that should be performed aboard the station.

Most recently, significant progress was made in the international coordination of utilization planning on the scientific discipline level; this applies especially for those disciplines using microgravity to conduct experiments.

In the fields of life sciences and physics, international strategic planning work groups have already begun to coordinate the development plans for large Class 1 Payloads for the station and to find ways and means to assure international access to all station facilities.

In the case of other scientific and technology-oriented disciplines, the coordination of multinational utilization was started by a number of International Working Group (IWG) meetings on the discipline level. Of course, the most important participants are the scientists as potential users of the International Space Station. These meetings aim for the following objectives:

- Exchanging information on the subject of experiment hardware to be developed
- Discussing strategies to cooperatively use the planned science and technology facilities
- Coordinating future cooperation in fields of common and mutual interest
- Exchanging ideas on new scientific and technological initiatives which could profit from environment and capacities of the International Space Station

The long-term boundary conditions for the utilization of ISS are determined by the allocation of accommodation and utilization resources (today's relative shares of the Partners are given in Sect. 13.1) on one hand, and by available financial means for Station use, to provide and support experiment facilities and experiments on the other. Based on these boundary conditions and the resources allocated to the disciplines according to a global long-term utilization policy, each individual discipline will, following its own proceedings, work out strategic plans for the utilization of the station.

Intensive international coordination and cooperation will have already taken place when the discipline-oriented strategic plans are being elaborated on.

The motivation to cooperate internationally in the field of fundamental research disciplines is expected to be very high, whereas in the field of application-oriented disciplines the motivation may not be as pronounced. Some industrial and commercial applications will even require confidentiality and full protection of intellectual property rights.

When elaborating on both the Euro-PUP and the CUP, advisory groups (composed of representatives from the potential utilization communities) will provide valuable multidisciplinary advice for the planning managers of the agencies involved. At the European level, the "Space Station User Panel" (SSUP) established by ESA will assume this task. On the level of the five international Partners, the International Forum for Scientific Uses of the Space Station (IFSUSS) is the spokesgroup for the international user community. The IFSUSS is composed of representatives from the user advisory bodies of all Partners involved.

13.5.6 From Conceptual Design to Qualification

Large Experiment Facilities. The large Multi-User Facilities for the International Space Station, in ISS jargon also called "Class 1 Payloads", will be provided and financed by the NASA and ESA user programs (for example, the microgravity program), or by national agencies. The selection of these facilities will be effected in coordination with the space agencies of the other international partners in order to prevent redundancy of hardware.

Large facilities include, for example, fluid physics laboratories, furnaces, material exposure facilities or remote sensing facilities. They all have direct interfaces to the station's systems – at rack level (Fig. 13.11) in the case of internal facilities or at payload accommodation level for those facilities directly exposed to space.

The decision to develop a Class 1 Payload as well as the design and development process necessary are effected in close cooperation with the Facility Science Group.

Experiments. The selection process for the experiments (ISS jargon: Class 2 Payloads) to be conducted aboard the International Space Station does not begin until the decision for the "host" of the facility has been made and its development has been introduced.

In principle, the development activities are the same as those for Class 1 Payloads, but they are less time consuming. The length of time a development cycle for a Class 2 Payload will take depends mainly on the experiment and the complexity of the equipment necessary. It is, however, estimated to range from several months to several years.

Contrary to Class 1 Payloads, experiments are usually not developed by industry, but directly by the user's institute, except for engineering and technological payloads: in this case, the interested user is by definition an industrial company or market-oriented institution. Thus, generally, the sole role of industry in the development of Class 2 Payloads is hardware production according to the specifications given by scientific users. The involvement of industry is therefore decided – and financed – by the user who developed the payload or the agency who funded it.

As it is the case with most of the equipment for crewed space flight, usually not only the flight model, but an engineering, or at least a training model for astronauts has to be delivered, as well.

Engineering Models (EM) are needed to demonstrate the functioning of payloads concerning performance and interfaces on ground, before the actual Flight Model (FM) is manufactured. After the Engineering Models have served their development purposes, they are usually reused for general operations training, development of procedures or calibration of test items.

Very early in the development process, Training Models (TM) are used to optimize the man-machine interfaces for payloads. After being used for development purposes in industry, they are transferred to astronaut training facilities, where they are intensively used to develop operating procedures for the astronauts and to allow preliminary astronaut training.

Flight models are developed and qualified by industry or directly by the user's institute, before they are transferred to NASA, ESA and other space agencies for acceptance tests and integration.

13.5.7 Development Support

The process of developing hardware for experiment facilities and experiments is accompanied by activities that support the user in preparation and later utilization. The following three different payload support activities can be distinguished:

- Support for payload accommodation and integration
- Support for payload utilization
- Support for preparation and performance of payload operations

The aim of the category *Support for payload accommodation and integration* is to ensure compatibility between payload and ground/flight infrastructures. During the development process, this task is performed by engineering analysis. Once the development process is complete, the support is continued from acceptance tests and interface verification up to testing system compatibility. Further support is provided for launch preparations and launch, itself.

The task of the category *Support for payload utilization* is to familiarize the users with all the functions and capabilities of the experiment facilities in the International Space Station. Moreover, support will be provided in development of utilization procedures, training of the crew who will conduct the experiments, calibration of test items and samples/specimens, and collection of basic data.

The main objective of the category *Support for preparation and performance of payload operations* is to ensure that all necessary procedures including software and ground support facilities are available. The corresponding activities will start simultaneously with the payload development process. During the orbital flight of a payload, this category will direct all support activities available on ground, such as collecting and interpreting all engineering and status data, distributing all scientific data to the users, taking necessary telecommand measures, and performing voice interaction with the crew.

Figure 13.13 illustrates the parallel performance of payload development and support activities. The top line represents all support activities concerning accom-

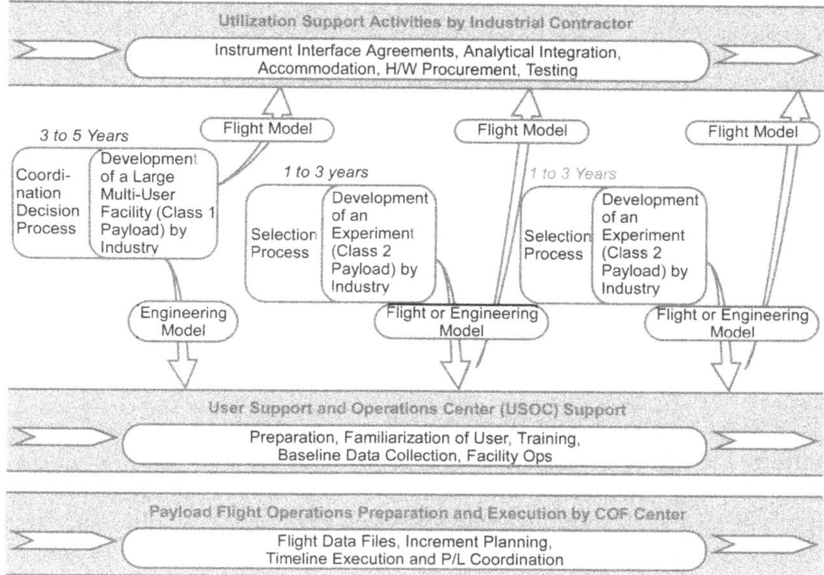

Fig. 13.13. Embedding of All Activities and Experiments into the Continuous Course of Operation of the International Space Station [ESA-Guide 96]

modation, integration and logistics of payloads. The center line stands for the support of development as well as payload operations before and after launch. The bottom line describes activities concerning the preparation and execution of flight operations. It is important to note that all three lines will not be interrupted for a period of 10–15 years because they start with the development of the very first payload for the station and terminate at the definitive end of the station's operation.

The support activities presented on the top line are performed by an industrial contractor, and those on the bottom line are performed, e.g. for an European user, by the COF Control Center which acts in a double role as COF system operations control center and as European Payload Control Center. Support in the field of payload operations (center line) is provided by the European User Support and Operations Centers (USOCs).

13.5.8 The User Support and Operations Centers

Planning expertise will be geographically distributed at control centers and user sites around the world. ISS payload operations will be multidisciplinary and will include a broad range of crew-attended, unattended, automated, and telescience operations. NASA will provide Payload Operation Integration (POI) for International Space Station Payloads. POI is the assessment and management of interactions between multiple payloads, payload facilities, and space station systems, including interactions between hardware, software and operations. The NASA POI process involves developing integrated payload operations requirements, plans, and processes and ensuring their compatibility with the flight system. This effort includes the following:

- Station-wide payload operations management and integration
- Integrated payload planning and operations
- Payload crew and ground personnel training and safety

Safety, operational effectiveness, and system-to-payload performance are the primary goals. The NASA POI process will integrate all US users, regardless of the payload's planned location on ISS.

The User Support and Operations Centers (USOCs) will play a very important role for the users of the International Space Station. They will be the main contact addresses for the users within the structure for utilization of ISS, and they will advise and support the users during the preparation and carrying out of their experiments.

USOCs will be provided engineering models of facility hardware which they use to acquaint the user with all functions and capabilities of the Class 1 Payloads, to develop operating procedures, and to support the crew training. Moreover, these centers will take care of the interfaces between experiment facilities and experiments and calibrate test items and samples/specimens which are to be processed in them.

During on-orbit operation of the scientific and technological payloads, the USOCs receive data from experiments as well as from experiment facilities. In addition to that, they support the Partners Control Center in operating the payloads for which they are responsible. USOCs also keep in touch with the user's institutes. They transfer experiment data to the users and accept requests with regard to timing and, if requested and technically feasible, also with regard to the direct control of the experiments by the users' institutes, themselves. USOCs are national facilities; they are created and financed directly by national space agencies or similar institutions since they will perform a vital task in the realization of decentralizing the user operations. Nearly all of the nations involved intend to create such a center. Incidentally, some of the USOCs have existed since Spacelab and other programs of crewed space flight and will now be reorganized and modernized according to the current requirements of the International Space Station program.

The acronym USOC, however, is only a generic term for such facilities. Most of the national USOCs have their own names and acronyms. Here is a list of those which already exist in Europe:

- CADMOS (France)
- DAMEC (Denmark)
- DUC (Netherlands)
- INTA's USOC (Spain)
- MARS (Italy)
- MUSC (Germany)
- SROC (Belgium)
- SSC's USOC (Sweden)

13.5.9 From Ground Verification to Launch

All responsible organizations firmly intend to keep the access time for the International Space Station as short as possible. As compared to earlier programs (where the access time was mainly determined by the launch date of the launch vehicle), ISS will be permanently available and thus make regular and quick access possible.

Quick access is also vital, if users wish to respond immediately to results obtained aboard ISS during an experiment. Such a response may be, e.g., to launch new test items, sensors or actuators, or it also may be to exchange a drawer containing experimental hardware. In the following section, the European payload integration and launch – typical for non-US partners – will be described.

At present, the decision has not yet been made as to whether or not **payload integration** will be effected in a centralized mode (e.g. by one single industrial facility) or in a decentralized mode (e.g. spread all over Europe). A realistic scenario would be to have one center for pressurized payloads and one for unpressurized ones.

According to the type of payload, experiments and experiment facilities are tested in a rack, drawer or external payload facility. Since the first generation of European pressurized scientific payloads are to be launched together with the European laboratory module COF, they have to be tested directly in the COF Integrations Center and integrated into the COF. The European pressurized payloads which are to be launched later on (when the COF is already in orbit), or which will be integrated into the laboratories of other partners (for example, the US Lab), are normally integrated into Express Rack drawers or ISPR racks. Payloads which are to be mounted externally on the station are either installed on an autonomous large payload adapter (which in turn will be mounted on one of the external payload attachment sites of ISS), or on an Express Pallet adapter.

Main space transportation systems for **launching European payloads** to ISS will be the Ariane 5 family in connection with the Automated Transfer Vehicle (ATV) as well as the US Space Shuttles. Thus, possible launch sites are the "Centre Spatial Guyanais" (CSG) in Kourou, French-Guyana on South America's northern coast, or the Kennedy Space Center near Cape Canaveral, Florida, USA. At the launch sites, the payloads will be handed over to the launch authorities who will install them in the corresponding launch vehicle. If necessary, the ESA Utilization Office will provide support as requested by the users.

The distribution of launches among Ariane 5/ATV, the Space Shuttle, and other space transportation systems will be determined according to the availability of the ATV and the types of payload adapters developed for it. From a technical point of view, the ATV could be produced in three different design types:

- Unpressurized Logistic Carriers (ULCs) only
- One Mini Pressurized Logistics Module (MPLM)
- A mixed version which allows the transportation of propellants and gases on an open carrier structure and the transportation of pressurized payloads in a closed payload module; this version is called "Mixed Cargo".

The distribution of launches between the European and other space transportation systems not only depends on the ATV design type the Europeans develop, but also on the question of how often the ATV will be sent to ISS. At present, ATV flights are planned for every 12 or 18 months. Transportation requirements of European experiments occurring between two ATV flights could be covered, for example, by the US Space Shuttle, presumably traded against the transportation of US payloads aboard European ATVs.

13.5.10 Performance of Experiments: From On-Orbit Installation to Return of Results to Earth

After the arrival of the logistics transportation vehicle at the International Space Station, the responsibility for further payload handling is transferred to the "Payload Operations Center" which will supervise the on-orbit installation of the payload by means of remote monitoring.

Payload Installation. Pressurized payloads will be installed by the ISS crew at rack or drawer level. Installation includes mating of all the different mechanical and functional interfaces (e.g. screws, latches, hinges, electrical connectors, fluid or vacuum lines). Normally, the associated software is installed on module level only. For that reason, the testing and verification process is limited to the verification of the interfaces.

External payloads are installed with the help of the station's remote-controlled manipulator system SSRMS, which is controlled by the ISS crew.

After successful on-orbit installation and verification, the payload routine operation can start.

Payload Operation. The basic philosophy of ISS is its decentralized mode of utilization. This means that the payload operations control centers of the international partners coordinate and control the utilization of the corresponding payloads under the overall coordination and supervision of two US institutions:

- The SSCC (Space Station Control Center) located at the Johnson Space Center in Houston, which is responsible for overall system operations
- The POIC (Payload Operation Integration Center) located at the Marshall Space Flight Center in Huntsville, which is responsible for payload and user operations

A limited ESA and NASA representation for support purposes will be provided at both centers. The POIC controls all internal and external payloads on the Station at payload system level. It distributes payload data to the corresponding payload control centers of the international partners and receives command requests from them.

The Role of the COF Control Center for Payload Operation. Payload operations control for Europe will be performed by the COF Control Center. It is proposed for installation at Oberpfaffenhofen close to Munich in Southern Germany or at Toulouse in France. The COF Control Center thus serves a double purpose: it is responsible for all systems operations of the European laboratory module COF as well as controlling the payload operations aboard the COF. Its role may be compared to that of the SSCC (Houston) and the POIC (Huntsville) on the NASA side. As to organization, the COF Control Center will have a direct interface with both centers in the USA and thus will also include the necessary communications, data and command links.

The COF Control Center is not only responsible for the operation and the payload coordination of the European payloads, but for all payloads in the COF, including those from NASA and, where appropriate, other international partners. As the European Payload Control Center it is also responsible for those European payloads which are mounted externally on the station. On the whole, it is the central point in Europe for further data distribution and payload remote controlling.

Performance of Experiments. Experiments in the COF can be carried out in four different modes of operation:

- Under the control of the onboard crew
- By internal payload automation
- In a telescience mode by the user on ground
- A combination of all three modes

The crew can control the payloads either by means of a laptop connected to the Local Area Network of the COF system or by controls, displays and computers that form part of the payload. Automated procedures can run under the control of payload-internal processors with additional support from the COF payload control unit, if desired.

A further option is telescience, i.e. the direct real-time control of the experiment by the user on the ground. Users will be able to monitor all scientific and functional data of their experiments and to influence their performance by means of a set of previously validated commands. The COF Control Center and the POIC in Huntsville will oversee these activities, but they will not intervene unless there is a potential impact on other payloads, the COF (or other ISS systems), or the onboard crew. Both centers also make sure that the payload operation does not exceed the pre-defined operational windows. Payload commands will leave the USOC of the corresponding user and will reach the COF via the COF Control Center, the POIC, the SSCC and the TDRSS of NASA.

European internal and external payloads in the US laboratory module (US Lab) will be controlled with the help of the US Lab's Data Management System (DMS), which has already been described (Sect. 12.2). The transfer of the scientific and functional payload data as well as that of the user's telescience commands is organized in a similar fashion to that for payloads in the COF.

During the performance of the experiments, the corresponding users are normally in their national USOCs, where they can monitor all interesting scientific and functional experiment data and, if necessary, influence the experiments. This does not necessarily have to be the case. If institutes have the necessary hardware and links to the USOC in charge at their disposal, the users may monitor and control the experiments directly from their institutes, called UHBs (User Home Bases) in ISS jargon. For most experimenters, this is the preferred mode of operations.

Return of Hardware to Earth. Neither the European Automated Transfer Vehicle ATV nor the Russian uncrewed freight vehicle Progress offer a chance to return payloads to Earth since both are designed to disintegrate upon reentering the Earth's atmosphere. The Progress freighter, however, may carry a small automated return capsule called "Raduga" which has a capacity of about 150 kg. The Russian crewed capsule Soyuz which is used to return crews back to Earth has only a limited carrying capacity of less than 50 kg for items which are not directly related to the crew. As a consequence, the transportation of experimental hardware from the International Space Station back to Earth will usually be carried out by the US Space Shuttle.

The transportation processes of experimental hardware, data and commands between the users and ISS are presented in Fig. 13.2. This schematic may be applied to large-scale experiments and experiment facilities. In the case of small experiments, the arrows representing the hardware flow through the payload-developing industry could be replaced by a direct connection between the user and the payload integration facility.

13.6 Preparation for ISS Utilization

13.6.1 Present NASA and ESA Early Utilization Plans

The NASA Johnson Space Center (JSC) is the lead Center for the NASA Space Station Program, while NASA Headquarters maintains a small staff responsible for relations with the US Congress and external organizations. The JSC-based Space Station Program Office contains an ISS Payloads Office which executes the Program's responsibilities regarding payload utilization planning and implementation. The Payloads Office is also responsible for overseeing the cost, schedule, and technical aspects for the design, development, and operation of major facility-class payloads. As a general rule, the NASA "Announcement of Opportunity" (AO) for experiments onboard the Space Shuttle, Mir or ISS is issued about three years before the flight date. Thus, in the past, US experimenters had ample opportunities to get ready for the ISS era. The selection of research to be performed on board the ISS is managed by the NASA research program offices, all of which are located at NASA Headquarters in Washington, DC. Research offices involved in the ISS program are:

- The Office of Life and Microgravity Sciences and Applications (OLMSA) manages microgravity research involving the life and physical sciences with a variety of programs involving both academic and commercial research.
- The Office of Space Science (OSS) manages NASA research in the astronomical, planetary, and space sciences.
- The Office of Earth Science (OES) manages NASA's program in Earth Observation and remote sensing.

Figure 13.14 shows NASA's impressive plan of the various discipline-specific facilities (Class 1) and experiment (Class 2) payload launches during the build-up of ISS [Bartoe, NASA-Guide 98].

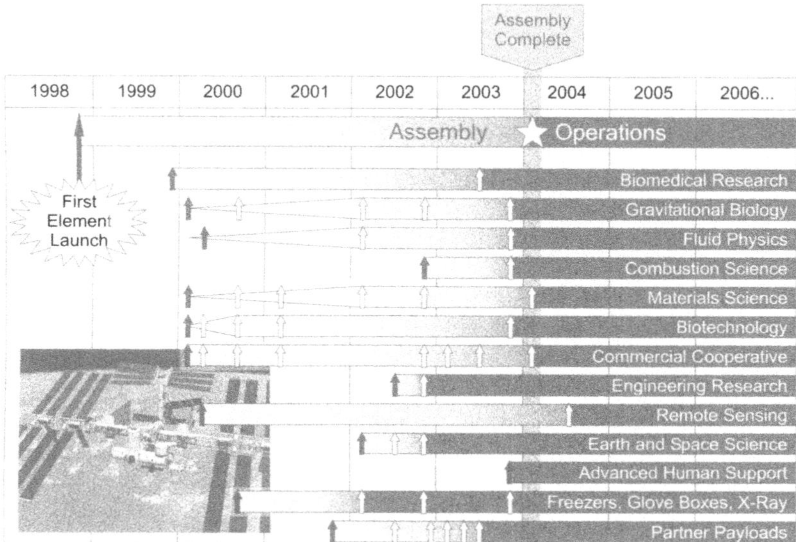

Fig. 13.14. NASA's Launches of the Discipline-Specific Facilities (Class 1, black arrows) and Experiment (Class 2, white arrows) Payloads during ISS Assembly [Bartoe]

Responding to the growing interest of European scientific disciplines in the utilization of the International Space Station, the formal AO for external payloads started in December 1996. In this context, it is important to mention the fact that both the permanent delays of the European program for crewed space flight and also the relatively remote launch date of the European laboratory module COF in 2003 had a counterproductive effect on interesting the European scientific community in research aboard ISS. Additionally, many scientific users were disappointed and had lost their confidence in such announcements since in the past, many of them had invested considerable effort and financial means into the submission of proposals for precursor missions of ISS which, in the end, never came to realization due to a lack of financial support by the ESA member states.

Based on the general interest the scientists of the different disciplines showed and also on the information available on planned or possible experiment facilities, the ESA Directorate of Manned Spaceflight and Microgravity developed generic plans for Europe's utilization of ISS during the so-called "Early Utilization Phase" and the "Initial Utilization Phase". On one hand, these plans included the information gathered from the various potential users and on the other, they contained information on the pressurized volume, external surfaces, and resources available for European utilization during the first two phases.

However, it must be kept in mind that these generic plans, as far as experiments and individual instruments are concerned, are only of a preliminary nature and must still be confirmed by the formal AO selection processes mentioned above. Table 13.3 shows the planning for ISS-related AOs at the time of the writing of this book.

Table 13.3. Planning for ISS Related Announcement of Opportunities [ESA-Guide 98]

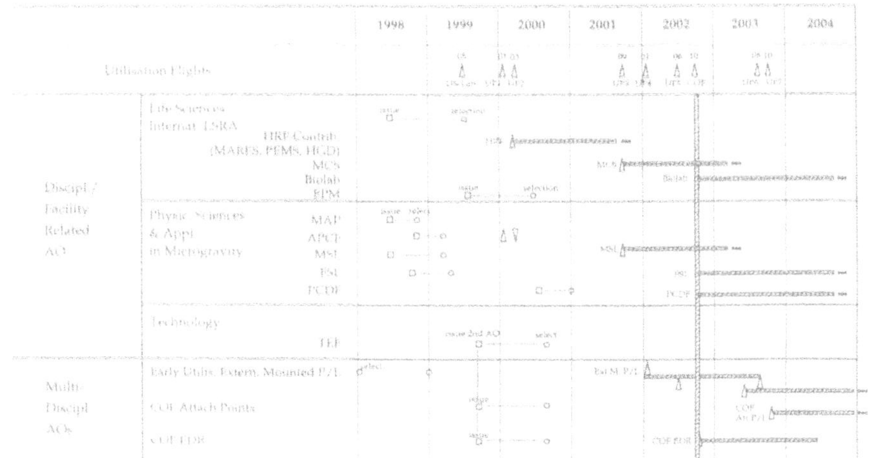

LSRA = Life Sciences Research Announcement
HRF = Human Research Facility
MARES = Muscle Atrophy Research & Exercise System
PEMS = Percutaneous Electrical Muscle Stimulator
HGD = Hand Grip Dynamometer
APCF = Advanced Protein Crystallisation Facility
MCS = Modular Cultivation System
MSL = Materials Science Lab
PCDF = Protein Crystallisation Diagnostics Facility
TEF = Technology Exposure Facility
EDR = European Drawer Rack
MAP = Microgravity Applications Promotion Projects
EPM = European Physiology Modules

13.6.2 European Facilities for the Early Utilization Phase

These plans cover the period from 2000 (beginning of ISS's utilization) to the arrival of the European laboratory module in orbit. For this period, ESA agreed with NASA on the utilization of laboratory capacities in the US Lab and also of external payload attachment sites. In return, ESA will deliver to NASA scientific support equipment.

The rack volume inside the US Lab reserved for Europe at the time of the laboratory's launch in 1999 will be used for the "Modular Cultivation System" (MCS) for biological research and by the "Protein Crystal Diagnostics Facility" (PCDF) for the growth of protein crystals under microgravity conditions located in the EDR.

Among the external payloads planned to be launched by Europe, there is a first group including environmental monitoring instruments (for dust, neutral gas, plasma, vacuum, and radiation). These will be accommodated in the "Technology Exposure Facility" provided for ESA users. This multi-purpose facility is intended for all those experiments which do not need any special orientation but which must be exposed to the space environment surrounding the space station. Also hardware of those aerospace companies who wish to test and to qualify for space flight components and systems from the fields of power generation, propulsion, sensors and actuators, life support systems, robotics, etc., can be included into this category of instruments.

The second group of external payloads dubbed Solar Monitoring Observatory (SMO) and Sky Polarization Observatory (SPORT) planned for the period from

1998–2001 are instruments for observation of the Sun and sky. By means of corresponding experiments, the following activities are planned:

- Measuring changes in solar radiation
- Monitoring the diameter of the Sun
- Performing a solar spectroscopy

These instruments will be oriented towards the Sun in regular intervals of 10 to 15 minutes. Due to the permanent attitude change of the Sun, this can only be achieved with the help of an instrument pointing facility. Current studies investigate whether a hexapod of average complexity is sufficient or whether more accurate pointing facilities will be necessary.

Radiation biology experiments are also to be found among the payloads for the Early Utilization Phase of ISS. Some respond to solar radiation and thus belong to the category of Sun experiments, whereas others which have a high sensitivity for the stronger isotropic energetic particle radiation are intended to be mounted onto Russian Facilities.

The European Earth observation instrument FOCUS and the Atomic Clock Ensemble in Space (ACES) are is also planned to be operated in the Early Utilization Phase.

Table 13.4 shows contributions of common ESA and NASA multi-user facilities, Payload Support Equipment (PSE), Laboratory Support Equipment (LSE) and Standard Payload Outfitting Equipment (SPOE) to support the early utilization. Further facilities and more detailed descriptions can be found in [ESA-Guide 98] and [NASA-Guide 98]. This applies in particular for external facilities such as the Truss Payload Attach System, JEM Exposed Facility, Express Pallet and internal (i.e. pressurized) facilities such as the Express Rack, -80°C Laboratory Freezer, Gravitational Biology Facility, Life Science Glovebox, Centrifuge, Human Research Facility, Advanced Human Support Technology Facility, Microgravity Sciences Glovebox, Materials Science Research Facility, Fluids and Combustion Research Facility, Biotechnology Facility, X-Ray Crystallography Facility, Window Observational Research Facility, Low-Temperature Microgravity Physics Facility, and many more.

13.6.3 ESA's Preparation for the Early Utilization Phase

In order to maintain the momentum of the established community and to foster concrete applications, ESA's Member States decided to participate in ISS and approved a utilization preparation program element. About 50% of its resources are earmarked for utilization promotion [Seibert 98].

Since on one hand NASA intended to terminate all research-dedicated Spacelab flights as of 1998, and RKA planned the progressive run-down of Mir-based activities in 1999, and since on the other routine operations in the COF will not start before 2003, ESA and NASA signed an Early Delivery/Early Utilization Agreement (EUA). In this agreement, NASA offers ESA:

- Access to accommodate experiment modules equivalent to a standard rack within the US Laboratory for a period of two years (2000/2001)

Table 13.4. Facilities and Payloads for Common ESA and NASA Early Utilization

		Facility	PSE/LSE	SPOE
Pressurized Payloads Accommodated in	Columbus Laboratory	• Biolab[a] • European Drawer Rack (EDR) • European Physiology Modules (EPM)[a] • Fluid Science Laboratory (FSL)[a] • Material Science Laboratory (MSL)[a] • Protein Crystal Diagnostics Facility (PCDF) (located in EDR)	*ESA Developed* • European Stowage Rack (ESR)	*ESA Developed* • Avionics Air Assembly (AAA) • Remote Power Distribution Assembly (RPDA) • Standard Payload Computer (SPLC) • Turbo Molecular Pump (TMP)
			US Developed • Mid-Deck Lockers (MDL) • International Standard Payload Rack (ISPR)	*US Developed* • Area Smoke Detector Assembly (ASDA) • Rack Main Switch Assembly (RMSA) • Interface Connectors
			Japanese Developed • International Standard Payload Rack (ISPR)	
	US Laboratory	• Advanced Protein Crystallization Facility (APCF)[b] • Handgrip Dynamometer[b] • Material Science Lab (MSL)[a,b] • Modular Cultivation System (MCS)[b] • Muscular Atrophy Research and Exercise System (MARES)[b] • Percutaneous Electrical Muscle Stimulator (PEMS)[b]	*ESA Developed* • Cryo Freezer[b] • Microgravity Science Glovebox (MSG)[b] • Minus Eighty Degree Laboratory Freezer for the ISS (MELFI)[b]	
			US Developed • Express Racks (ER) • Active Rack Isolation System (ARIS) • Mid-Deck Lockers (MDL)	
External Payloads		• Atomic Clock Ensemble in Space (ACES)[c] • Expose[c] • Fire Detection Infrared Sensor System (FOCUS)[c] • Global Transmission System (GTS)[d] • Matroshka[d] • Solar Monitoring Observatory (SMO)[c] • Sky Polarization Observatory (SPORT)[c] • Technology Exposure Facility (TEF)[c]	*ESA Developed* • Coarse Pointing Device (CPD) • Hexapod Pointing System[e]	*ESA Developed* • External Payload Computer (XPLC) • Power Distribution Unit (PDU)
			US Developed • Environmental Monitoring Package (EMP)	*US Developed* • Express Pallet Adapters (ExPAs)

[a] The facilities FSL, Biobox, MSL and EPM are also collectively known as the Microgravity Facilities for Columbus (MFC).

[b] Europe supplies this equipment to NASA under a barter agreement. The US Lab responsibility rests with NASA and ESA has utilization rights.

[c] These European facilities are externally mounted on ExPAs as part of the US Express Pallet under NASA cooperative agreements.

[d] Facility is located on a Russian element.

[e] The Hexapod is being developed by ESA in support to the Stratospheric Aerosol and Gas Experiment (SAGE III) instrument from NASA/Langley.

- Accommodation of ESA's non-pressurized payloads on three pallet adapters (about 1 m^3 each) of a NASA Express Pallet for a period of three years (2002–2005)
- A flight opportunity for two ESA astronauts

In return, ESA is delivering several -80°C Freezers, a Microgravity Glovebox and a pointing platform for external US payloads in the framework of the EUA.

In addition, ESA agreed with NASA to develop several life sciences multi-user facilities for the US Lab in the framework of its EMIR-2 Program, and it obtained the right to use all life sciences facilities of the US Lab from the start of ISS utilization (experiment selection via joint AOs in early 2000).

After the Toulouse Ministerial Conference, ESA strengthened its Space Stations User Panel by including representatives who are providing advice on technology and industrial applications.

ESA's Utilisation Preparation Programme, the fourth element of the approved Manned Spaceflight Programme, had been divided into parts of almost equal budget allocation:

- Accommodation hardware
- Development and payload integration/operations
- Utilization promotion

The design and development activities for accommodation hardware such as the procurement in industry of payload racks and their payload outfitting equipment (standard computers, power conditioning equipment, temperature control facilities), payload support equipment for external platforms, a coarse pointing device, a European Standard Drawer Rack, and associated test equipment are well underway.

The utilization promotion activities carried out so far included:

- The identification and study of typical space station experimental projects from various user disciplines which can make good use of the unique capabilities of ISS
- Creation of a number of application-oriented research projects for ISS in which research institutes and also non-space industry participate with their own funding
- An AO for externally mounted payloads was issued at the end of 1996. This AO resulted in an overwhelming demonstration of interest – almost 100 proposals were received – from all user disciplines, i.e. technology (41), space science (23), life sciences (14), physical sciences (11) and Earth observation (6). In the field of technology, eight proposals were received from industry (i.e. with industrial funding). The technical feasibility and accommodation study for experiments from this AO, after peer group selection, resulted in a demand for ten pallet adapters. Table 13.5 summarizes the results of the AO and the proposed ideas for the routine phase of the ISS utilization.
- In parallel to the AO for externally mounted payloads, ESA participated in an international AO in Life Sciences (LS) for the first two years of US Lab utilization (2000–2001). It resulted in several hundred LS experiment proposals worldwide, 65 of which were selected by international peer groups (led by the US National Institute of Health) from a scientific point of view. 36 of these 65 came from ESA Member States – an excellent result for ESA. The results of the Life Sciences Research Announcement (LSRA), as far as scientific selection is concerned, are shown in Table 13.6. The final selection, which took into account the very limited technical and operational resources available for experiments on ISS during the period 2000/2001, resulted in the selection of 27 LS experiments, 15 of which are from Europe. Regarding the final selection of externally-mounted experiments, ESA and the responsible Programme Board for Manned Spaceflight de-

Table 13.5. European ISS External Payloads

Discipline	Early Utilization Payloads (Based on AO Results)	Routine Utilization Phase	Comments
Technology	• Technology Exposure Facility (TEF) • Thruster Technology Testbed • Telecommunication Technology/Attitude Sensors	• Robotics • Solar Dynamics Power Generator • Verification and Qualification of Satellite Subsystems	• Large Number (48) of Early Utilization Proposals, of Which 8 Got Industrial Funding • TEF Could Be Used by 15 Experiments
Space Science	• 3 Solar Monitoring Packages (1 Ad.) • SPORT (High Energy Astrophysics) • UV Starlite (UV-Sky Survey) • Microwave Images (Cosmic Background Radiation) • ISSTOP (Neo-Telescope)	• Large X-ray Facility • Space Radio Telescope	• *Early Phase:* Solar Monitoring, Astrophysics Experiment, NEO Telescope, Environmental Monitoring • *Routine Phase:* Module Build-up of Large X-ray, Use of Sortie-Satellites (ATV), High Pointing Accuracy and Stability Platform (IPS-type)
Earth Observation	• FOCUS	• Wind Lidar Facility • Tropical Meteorology/ Climatology	• *Early Phase:* Imaging Spectrometer, Forest Fire Detection
Exobiology/ Radiation Biology	SEBA: 2 Facilities • Exposure to Solar UV (SEBA-Expose) • Radiation Depth Dosimetry in Simulated Human Tissue (SEBA-Matroshka)	Precursor Experiments for • Unmanned Mars Exploration • Manned Mars Mission	SEBA consists of 2 Facilities: • Exposure for Exobiology: Origin of Life, Transfer of Life • Matroshka: Human Phantom for Active Dosimetry (Radiation: Potential Limitation of Human Exploration)
Microgravity Applications	• Atomic Clock Ensemble in Space (ACES)	• Industrial R&D in Biotechnology, Crystals for Electronic Components, Combustion, etc.	•

cided to select five payloads of pallet adapter size for the three available pallet adapters (see Table 13.5):

- Technology Exposure Facility (TEF)
- Solar Packages (three experiments on one coarse pointing device)
- SPORT and SEBA-Expose (each reduced to the size of half an adapter)
- Atomic Clock Ensemble in Space (ACES)
- FOCUS

It was further decided to exchange two of the first three adapters after one and a half years in orbit and to leave the TEF either in orbit for three years with an exchange of its experiments after one and a half years (i.e. if experiment exchange at sub-adapter level is possible), or to exchange the total TEF adapter after one and a half years.

The interest shown by the various user disciplines in the early utilization promotion activities listed above can be summarized as follows:

Technology. ISS seems to be ideally suited as a technology test bed and an opportunity for verification and space qualification. In the past, technology development suffered due to the lack of flight opportunities. Initial experiments will consist of ex-

Table 13.6. Life Sciences Research Announcement (LSRA) – Proposal Summary

Research Disciplines	Canada	Europe	Japan	USA Flight	USA Ground	Other	Total
Biological Regeneration	0	0	0	3	28	1	32
Physical/Chemical Regeneration	0	0	0	1	25	0	26
Bone/Muscle Biology	6	2	0	1	23	0	32
Cardiopulmonary Physiology	1	9	1	2	28	0	41
Cellar and Molecular Physiology	1	10	0	11	32	0	54
Developmental Biology	2	6	1	7	18	0	34
Environmental Health	0	2	0	6	22	0	30
Human Factors	0	3	0	3	46	0	52
Musculoskeletal Countermeasures	1	16	0	7	25	0	49
Neurosciences	1	9	1	6	29	0	46
Plant Biology	0	15	0	1	25	0	41
Radiation Health	10	9	0	0	15	0	34
Integrated Physiology	1	19	1	8	23	0	52
Totals	**23**	**100**	**4**	**56**	**339**	**1**	**523**

posing materials, components, mechanisms, new sensors, spacecraft subsystems, etc. to the environment of space.

Later on, larger subsystems such as large antennas, new power generators, closed life support systems, etc. will be tested and qualified for routine applications. A thruster test bed was also endorsed by the Technology Peer Group.

Table 13.7 lists some details of the four technology experiments proposed and partially funded by industry.

Table 13.7. Peer Group Pre-selected Technology Experiments from Industry (Proposed in Reply to AO for Externally-Mounted Payloads)

Exp. No.	Title of Experiment	Industrial Company	Comments on Proposal
18	High Temperature Superconductors for Satellite Communication	Bosch Telecom GmbH (D)	Cryogenically cooled transponder for high power levels. Expected are lower mass and volume, better noise and performance. Aim: Lifetime testing in space
77	On-Board Testing of Large Power Space Plasma Thrusters	SEP (F)	Important for orbit transfer of large telecom satellites and future planetary missions. On Earth testing disturbed by back flow, back sputtering due to chamber walls
82	Space Astrometry Technology in-Orbit Experiment	Matra (F)	Improvement of pointing accuracy of future space science and EO satellites. Effects of aging, jitter, thermal cycles, vibration isolation, active damping
95	Ground Clock Synchronization and Time Dissemination from ISS (Global Time System)	TZ-RS (D)	Terrestrial time signal stations transmit within the scope of radio signals i.e. about 2000 km. ISS would overcome this limitation by transmitting UTC signal. Large PR effect on general public

Space Sciences. In addition to free-flyer missions with dedicated orbits, and spacecraft of high pointing and attitude stability funded by the Space Science Programme, the AO showed that this discipline is interested in using ISS for complementary missions such as solar radiation monitoring, astrophysics, sky scanning, and environment characterization. A medium-size wide-angle telescope has also been proposed to monitor asteroids and comets with the objective being to identify those that are in orbits posing a potential threat to Earth. Space science payloads for the routine phase of ISS will probably be concerned with long-term monitoring, especially in the high-energy regime such as modular build-up of a large mirror for X-ray instruments, potentially in the operations node of a free flyer (ATV) using ISS as a basis for refurbishment. A proposal to study long baseline interferometry using a radio-telescope onboard ISS has also been made.

Earth Observation. The Earth observation disciplines need predominantly dedicated missions with specific orbits and spacecraft attitude capabilities. In addition, they will use ISS as a test bed and carrier initially for one of its Explorer Missions. A peer group-selected experiment for the early ISS utilization phase is the FOCUS instrument which is an intelligent IR-sensor for the detection of forest fires. For the routine phase, a study has been started on the definition and accommodation of a large wind lidar (Aladin), an atmospheric IR-laser Doppler instrument. The instrument's measurements would help to correct a major deficiency in current meteorological operations observation mainly over oceans and tropical zones.

Life Sciences on the external platform of ISS. There are two fields of space life sciences which can use the external platforms of ISS:

- Exobiology
- Radiation biology

Exobiology studies the origin, evolution and distribution of life and chemicals (amino acids, nucleotides, sugars) necessary for life throughout the universe and on Earth. Studying the damage caused by exposure to UV radiation to macromolecules like DNA and the survival of micro-organisms like spores in space, form part of exobiology, and so does the search for traces of life on Mars. Space radiation biology investigates the effects of protons and heavy ions on humans (and other living organisms) in space. It is therefore important for EVA activities during which astronauts have less radiation protection, to measure actively (i.e. real-time transmission) and passively (dosimeters) the radiation absorption and damage in a facility which closely simulates the human body. This active radiation measurement facility should also be flown on uncrewed missions to Mars prior to long-term human spaceflights to Mars.

For the early phase of ISS utilization, the multi-user facility Space Exposure Biological Assembly (SEBA) is currently under design and development funded by the EMIR-2 program. Nine experiments have been selected for SEBA during the peer group assessment.

Microgravity Research. Preparing the utilization of COF, and European utilization of the US lab prior to the launch of the COF, is very well advanced in the frame of

ESA's MFC and EMIR programs. Multi-user facilities for all four major microgravity research disciplines materials science, fluid physics, biology, and human physiology are being designed and developed in industry, and the relevant user communities are actively involved in all reviews of these facilities (Facility Science Teams).

Microgravity Applications Promotion. The Microgravity Applications Promotion program element aims at fostering and developing a first generation of application-oriented projects with active involvement from (non-space) industry. These applications projects are expected to demonstrate that microgravity is indeed a useful tool for industrial Research & Development (R&D). The microgravity applications projects initiated so far, and their status, are listed in Table 13.8 and Table 13.9.

Table 13.8. Microgravity Application Projects (Ongoing Projects or Projects Just About to Start)

Project	Objective	Contractor	Applications
Osteoporosis	• Evaluation of accelerated bone loss in astronauts (up to 20%/6 month) • Study prevention measures by exercise, diet, drugs • Develop 3D computer tomograph for bone micro-architecture analysis • Develop bone artefact	• Medes (F) + 4 scientific partners in CH, F, I, NL • Industrial Partners: - Scano Medical (CH) - Millenium (Cdn)	• In US expenditure for osteoporosis fractures is 10 billion dollar/year • 10% of total population, mainly elderly women
Semiconductor Crystal Growth: CdTe and related compounds	• Improve perfection and homogeneity of CdTe crystal in microgravity • Systematic test of crystals as detectors by industrial partners	• Kristallogr. Inst. Freiburg (D) + 3 scient. partners in F • Industrial Partners: - AEG, Infrarot (D) - Thomson CSF-Optronique (F)	• High-resolution X-ray detectors for dental imaging, tomography, mammography • Dosimeters for industrial environment
Soret Coefficient for Crude Oil Recovery	• Prediction of extent of oil reservoirs based on precise diffusion coefficients on crude oil sample measurements on Shuttle (GAS exp. 1998) • Measurement of Soret Diffusion Coefficient in Crude Oil (hydrocarbons) on Shuttle (GAS exp. 1999)	• Microgr. Centre Brussels • TU Denmark • Industrial Partners: - Elf Acquitaine (F) - Centre of Cold Ocean Eng. (Cdn)	• High industrial relevance for oil companies, which need these diffusion coefficient measurements for their mathematical prediction models

Table 13.9. Planned Future Microgravity Application Projects

Project	Objective	Comments/Applications
Atomic Clock Ensemble in Space (ACES)	Development of ultra-stable Cs atomic clock using microgravity to improve frequency stability by a factor 100. Laser cooling of Cs atoms to temperatures below 1 μK	• ACES is highest priority proposal as external payload on the ISS in physical sciences. In-orbit comparison with other clocks (e.g. H-Maser, microwave sapphire reson.) Application in EO, Geodesy, navigation (factor 10-100 better than GPS), frequency transfers, fundamental physics, VLBI, etc.
Combustion	Study of combustion processes under microgravity conditions under which turbulent flow induced by gravity is absent	• Start after Combustion Research Workshops in Orleans, Sept. 1997 • Improvement of efficiency of engines, industrial burners, etc.
Other fields in which potentially application projects could be started: Aerosols, vapour phase crystal growth, applications of magnetic field in crystal growth and solidification, metastable phases, immiscible alloys, containerless processing, cell and tissue cultures/bioreactors		

One set of physical sciences experiments proposed in reply to the AO for externally-mounted experiments is the Atomic Clock Ensemble in Space (ACES) which obtained not only the highest scores in the peer group evaluation, but also has the highest microgravity application potential (see Fig. 7.7).

The successful performance of this experiment will push the frontier of time transfer and time dissemination which are the basis of positioning (GPS and GNSS), navigation, and telecommunication systems. It is a domain with potentially large industrial benefits.

The interest in ISS utilization during the Early Utilization Phase/Promotion Phase has been convincingly demonstrated by the overwhelmingly high number of experiment proposals from all disciplines. This had led to a situation wherein probably less than one third of the peer group-selected experiments can be performed, taking into account the limited technical and operational resources available for Europe during the Early Utilization Phase.

It is the major political objective of the Utilization Promotion Program element to pave the way for industrial applications with industry's own financial contributions to the mission costs (i.e. in addition to the payload development costs) in the Routine Phase of ISS operation [Seibert 98].

13.6.4 European Utilization Plans for the Initial Utilization Phase

Here, the relevant period ranges from 2003 to 2005, i.e. the first two or three years of European utilization of ISS after arrival of Europe's laboratory COF in 2003. Plans for this period mainly relate to the COF with its five International Standard Payload Racks (ISPRs) available for European users.

Two ISPR spaces will be occupied by those racks that contain facilities from the ESA program of microgravity research facilities for Columbus MFC; most likely there will be one gravitational biology laboratory and one fluid science research facility.

One ISPR space is reserved for elements of a research facility for human physiology which is financed by national programs and MFC. Another ISPR space will be equipped with an Express Rack which will accommodate small experiments (below rack volume). Finally, the fifth ISPR space is reserved for on-orbit storage of items that belong to the experiments (specimens, samples, test items, cartridges, etc.) but also for storage of maintenance equipment, consumables and spare parts for experiments.

Moreover, on the basis of a cooperative agreement, the inclusion of a gradient furnace developed by Europe as part of the furnace facility in the US Lab is planned.

The external payloads which were installed during the Early Utilization Phase are expected to remain aboard ISS during the Initial Utilization Phase. Robots will exchange exposed samples and sensors. Exchange of complete instruments will only take place if they fail.

13.6.5 Outlook on the Routine Phase and User Acquisition

It is expected that the utilization scheme for ISS will continue to develop after 2005 both in range and complexity; i.e. after the Early and Initial Utilization Phases, it will reach its Routine Phase. Research facilities inside the pressurized modules will be upgraded, changed, and newly designed with regard to more user-oriented tasks.

At present, there is already a clear indication (resulting from research during the Spacelab missions) of material processing and biotechnology under microgravity having concrete application potential. Moreover, it can be predicted that the spaceflight industry will benefit from the opportunities which the International Space Station offers by establishing test beds for developments as they will be needed in future spaceflight projects. The following scenario of industrial ISS utilization can be envisaged:

a) Use of ISS by space industry as a test bed for components and subsystems of application satellites (telecommunications, EO and other satellites)
b) Use of ISS as an orbital basis for periodic refurbishment (exchange of subsystems such as transponders, batteries, AOCS-sensors, etc.), periodic refuelling (cryogenics), and modular build-up of very large structures (antennas, mirros, etc.) for free-flying long-duration satellite missions
c) Use of ISS for industrial R&D activities of non-space industry as currently being prepared by the Microgravity Applications Promotion Programme of ESA and the various programs of the Office of Life and Microgravity Sciences and Applications (OLMSA) of NASA

The most important prerequisites for industrial use of ISS are short turn-around times (one year), regular access, confidentiality, and cost efficiency of the utilization operations. After the Early Utilisation Phase (2002–2005), ISS operations activities in the Routine Phase including space transportation (Shuttle, Ariane 5, Russian large carriers) are expected to be performed by industry at a much lower cost.

During the Routine Phase, the portion of cost-reimbursable utilization of ISS by industry in the areas a) to c) (see above) and by third parties is expected to be on the order of 30% to 50%.

The residual share should be used by classical space disciplines such as space science (e.g. a Horizon 2000 Plus cornerstone mission such as a large X-ray facility or radio telescope), Earth observation (e.g. Aladin Wind Lidar as Earth Explorer Mission), and basic research in microgravity research disciplines, materials processing and life sciences. In the Earth observation and microgravity activities, at least, ESA/EU cooperation is expected to positively influence the use of ISS by these disciplines.

In addition, there will probably also be "third-party" utilization of ISS e.g. for military R&D on the European level. It is already known that from the start of the ISS utilization, a considerable portion of the external pallet of ISS accommodation will be used for US military R&D.

In many respects, the International Space Station Program is a new way to continue the scientific and technological utilization of space. ISS includes many partners and demands an extensive orbital infrastructure as well as a widely distributed ground segment. As to the costs: system development and operation require substantial financial means from all parties involved in the program. For that reason it is of vital importance to guarantee a widespread and efficient utilization of ISS.

The various international and national programs suggested for participation in the International Space Station include a number of activities as to user acquisition. Yet it cannot be expected that the programs will finance most of the experiments and multi-user facilities planned for ISS. They will have to limit themselves to providing kickoff funding for promising new utilization projects in order to define them and bring them to a level of maturity where other sources of funding can be obtained more easily. Moreover, there will be activities for the support of projects that have an application potential and that, in later phases, will possibly lead to commercial use of the International Space Station. The main orientation of those activities intended for user acquisition will be presented in the following sections.

A current analysis of the utilization intended by the interested parties shows that physical and life science disciplines (in the field of which, at present, sufficient experience with human research under microgravity has already been obtained) will make ample use of the planned laboratories of the International Space Station. In the case of other disciplines that are less familiar with crewed space flight, however, it has not been proved yet how far ISS will meet their specific objectives. For example, in the case of space sciences and Earth observation there is the question of whether the station will only be used as a developmental testbed for new sensors or whether it can also serve as an observation platform for space science purposes or as an operative element within a network of uncrewed and crewed Earth observation platforms.

These and other questions are to be answered with the help of the activities for user acquisition and, as a consequence, new users will be attracted. Answers will not only be taken from theoretical studies, but they will also be sought in practice by providing flight possibilities and by supporting the pre-development of promising elements for research facilities.

Great effort will be put forth to prove the usefulness of ISS for the fields of engineering research and technology demonstration as well as for those fields that seem to have application potential for fields of industry that have no direct link to the aerospace sector.

In the past, various fields of industry showed strong interest in the performance of research and development projects in space. This interest is based on the following two objectives: better understanding of certain processes on one hand and testing of recently developed products as to their qualification for space on the other.

In both cases, industry seems to be ready to pay for the development and manufacturing of the test facilities that are to be installed in orbit, but – at least in the recent past – industry has not been ready to pay the costs for the launch of these facilities. In order to explore and support those utilization possibilities that have a commercial potential, NASA began to pay the costs for the launch but bound industry by contract to pay back a share of the profit obtained by using the results of successful experiments or qualifications in space. The European activities for user acquisition will contribute to the creation of flight possibilities on the same basis for the European industry, as well.

Coordination of international utilization of ISS proceeds along different channels for the various disciplines. While the observational sciences rely on the international structures developed within the community, life and microgravity sciences have established efficient international strategic planning groups for space station utilization. No international forum exists for the field of Engineering Sciences and Technology Development. Therefore, a group of experts, at the International Forum on the Scientific Use of the Space Station (IFSUSS), at a November 1998 meeting at ESTEC recommended the establishment of a Multilateral Consultative Working Group for Engineering Research and Technology Development with the scope of:

- Developing a coordinated international strategy for Engineering Research and Technology Development
- Coordinating opportunities for technology development and testing
- Developing technologies for enhancing space station capabilities for international utilization
- Advancing human support technologies

Such a group should follow the successful procedures established by the existing working groups.

The IFSUSS experts also recognized the critical importance of optimizing internal and external payload accommodations with respect to topology, collocation and resource allocation to maximize research return. It was felt that a multinational approach to the topology of research accommodations and resource sharing/exchange/bartering must be developed, and NASA should continue to work with other international partners to develop an approach that facilitates this amongst the partners with an eye towards facilitating the best science from the best location.

13.6.6 Preparation of Future Payloads

The first generation of multi-user facilities for the physical and life science disciplines is currently in its study and development phase. In the near future, studies for the next generation will have to be started; this applies mainly to those disciplines which intend to use external payload attachment sites. In the field of space sciences, a Phase-A study is being carried out for the planned large astrophysical facility. Scientists involved in Earth observation which, during the Early Utilization Phase, rely on small and medium-sized instruments from national institutions, want to examine and define larger ones necessary after the year 2000. Currently, the suitability of the International Space Station for a wind-lidar and a rain-radar facility is being investigated.

The user acquisition program provides the financial means for such studies. The studies themselves would have to be selected and funded by the disciplines and boards in question.

The need for constantly refined facilities in the field of human physiology can be clearly seen, and thus, the pre-development of relevant instruments for diagnosis must also be started in time.

In the field of material processing, a next generation of furnaces with higher temperatures and greater accuracy for isothermal and also for gradient applications will surely be necessary soon. The objective is to have material processing exceed those limits set by furnaces which are currently under development.

During the forthcoming definition studies, there is no doubt that the planned astrophysical and Earth observation facilities will lead to the finding that further development is not only necessary for detectors and analyzers, but also for advanced pointing and support structures.

Pre-developments for Earth observation are necessary in the field of high-energy laser technology in order to obtain a laser that is qualified for use in space, has sufficient power, and is reliable enough for the wind-lidar facility.

Pre-developments for a suitable telescope and transmission technology for high sampling rates are most likely to form part of the necessary steps to be taken, as well.

The user acquisition programs intend to support some of these pre-developments and the tests necessary until the disciplines involved have managed to exclude the largest of the developmental risks for their facilities.

After the „First European Symposium on the Utilization of the International Space Station“ (October 30 – November 2, 1996 at ESOC in Darmstadt, Germany), a number of new proposals were submitted [ESA-Symp. 96] for the utilization of instruments already planned and the manufacture of new multi-user facilities and experiments. It was surprising that most of the proposals were aimed at external accommodation sites, and that many new users from universities and large research laboratories submitted proposals. The most promising of the proposals were selected for flight (see Table 13.5 to Table 13.9). Following the "Second European Symposium on the Utilization of the International Space Station" (October 16–18, 1998 at ESTEC in Noordwijk, The Netherlands), which was attended by more than 500 participants, it is expected that the new research and application proposals will enlarge the user base and make the ISS utilization more competitive and challenging [ESA-Symp. 98].

References

Chapter 1: Introduction

[Tsiolkovsky] Tsiolkovsky, K.E.: *Raketa v mezhplanetnoe prostranstvo* (The Rocket into Interplanetary Space), Kaluga, 1923.

[von Braun] von Braun, W.; et al.: *Across the Space Frontier*, Viking Press, New York, 1952.

[Oberth] Oberth, H.: *Die Rakete zu den Planetenräumen* (The Rocket into Interplanetary Space), R. Oldenbourg, Munich, 1923.

Chapter 2: History and Current Development

[Bekey 85] Bekey, I.; Herman, D.: *Space Stations and Space Platforms - Concepts, Design and Usage.* AIAA Progress in Aeronautics and Astronautics, Vol. 99, New York, 1985, ISBN 0-930403-01-0.

[Bizony 96] Bizony, P.: *Island in the Sky – Building the International Space Station.* Aurum Press Ltd. London, ISBN 1-85410-436-5, 1996.

[Boeing 82] Boeing-Study: *Space Operations Center System Analysis - Study Extension.* NASA-CR-167555, 1982.

[ESA/PB 94] *Programme Proposal on the European Participation in the ISSA.* ESA/PB-MS(94)60, rev. 2, Paris, 15 March 1995.

[Furniss 86] Furniss, T.: *Manned Spaceflight Log.* Jane's Publishing Company Limited, London, ISBN 0-7106-0402-5, 1986.

[Gugerell 98] Gugerell, A.: *Von Gagarin zur Raumstation Mir.* Traisen, ISBN 3-9500500-0-0, 1997.

[Harland 97] Harland, D.M.: *The Mir Space Station – A Precursor to Space Colonization.* John Wiley & Sons, In Association with Praxis Publishing Ltd., ISBN 0-471-97587-7, 1997.

[Hoffmann 84] *Eugen Sänger Memorial Lecture.* AAS 84-300, DGLR/AIAA Symposium, Hamburg 1984.

[Hooper 90] Hooper, G.R.: *The Soviet Cosmonaut Team, Volume 1: Background Sections, Volume 2: Cosmonaut Biographies.* GRH Publications, Suffolk, ISBN 0-9511312-3-0, 1990.

[Laßwitz] Laßwitz, Kurd: *Auf zwei Planeten*, Felber, Weimar, 1897.

[Logsdon 85] Logsdon, J.M.; Butler, G.: *Space Station and Space Platform Concepts: A Historical Review.* In [Bekey 85]. Also: Logsdon, J.M.: *Why Has the Station Survived?* Space News, Jan. 5-11, 1998.

[Mark 87] Mark, H.: *The Space Station, A Personal Journey.* Duke University Press, Durham 1987, ISBN 0-8223-0727-8.

[MDD 95] *International Space Station*, Information Sheet GP 940021 2-95, McDonnel Douglas Space Station Marketing, Huntington Beach, CA, 1995.

[NRC 95] Committee on the Space Station: *The Capabilities of Space Stations.* Aeronautics and Space Engineering Board, National Research Council, National Academy Press, Washington, DC 1995.

[Ordway 92] Ordway III, F.I.; Liebermann, R.: *Blueprint for Space - Science Fiction to Science Fact.* Publication of Smithsonian Institution 1992, ISBN 1-56098-072-9.

[Pioneering 86] *Pioneering The Space Frontier.* The Report of the National Commission on Space, A Bantam Book, May 1986, ISBN 0-553-34314-9.

[Puttkamer 71] von Puttkamer, J.: *Raumstationen – Laboratorien im All.* Verlag Chemie, Weinheim, 1971.

[Puttkamer 85] von Puttkamer, J.: *Der zweite Tag der neuen Welt.* Umschau Verlag Frankfurt am Main, ISBN 3-524-69054-8, 1985.

[Ruppe 80] Ruppe, H. O.: *Die grenzenlose Dimension - Raumfahrt.* Econ-Verlag Düsseldorf, Vienna. Volume 1 (1. Edition 1980) ISBN 3 430 17884 7, Volume 2 (1. Edition 1982) ISBN 3 430 17849 5.

[Shapland 84] Shapland, D.; Rycroft, M.: *Spacelab - Research in Orbit.* Cambridge University Press 1984, ISBN 0-521-26077-9.

[Skylab 73] Belew, L.F.; Stuhlinger, E.: *Skylab - A Guidebook.* NASA EP-107 (1973).

[Skylab 77] Belew, L.F.: *Skylab, Our First Space Station.* NASA SP-400, NASA Science and Technical Information Office, 1977.

[Spacelab 83] *Spacelab Data Book.* ESA BR-14, September 1983.

[Spacelab 93] *Spacelab 1983-1993, Ten Years Experience in Cooperative Manned Space Activities.* Proceedings CEAS European Forum, Oct. 1993, Florence, Italy. ISBN 1-56347-073-X.

[Space News 93] *Options for the International Space Station*, Front Page, Space News, May 24-30, 1993.

[Space Times 98] *The International Space Station and the Future of Space Exploration.* Feature Article by The Honorable F. James Sensenbrenner, Jr., SPACE TIMES, Sept.-Oct. 1998.

[SSF 88] *Space Station Freedom Technical Overview.* NASA-Reston, 1988.

[Verne] Verne, J.: *Les Voyages Extraordinaires.* With illustrations from the time of the author, Volume 7, Jean de Bonnot Publisher, 1977.

[von Braun] von Braun, W.: *Across the Space Frontier*, A Series of Articles, with Artwork by Chesley Bonestell, published by Collier's Magazine, New York, 1952.

[Walter 92] Walter, W.J.: *Space Age.* QED Communications, ISBN 0-679-40495-0, 1992.

[Woodcock 86] Woodcock, G.R.: *Space Stations and Platforms.* Orbit Book Company, Malabar, 1986.

[WWW-OSF] NASA, Office of Space Flight: WWW: http://www.hq.nasa.gov/osf/mir/

Chapter 3: Orbital Environment

[Aerospace A. 88] *Standstill on Orbital Debris Policy.* Aerospace America, June 1988.

[AWST Oct. 5, 98] Aviation Week and Space Technology, Oct. 5, 1998: *Space Debris Concerns Grow as Meteoroid Storm Nears.*

[Beatty 83] Beatty, J.K.; O'Leary, B.; Chaikin, A. (Ed.): *Die Sonne und ihre Planeten.* Physik-Verlag, 1983.

[Daly 98] Daly, E.J.: *The ECSS Space Environment Standard.* ESA Space Environments and Effects Analysis Section, ESTEC, Noordwijk, The Netherlands, 1998.

[Debris 88] *Space Debris, The Report of the ESA Space Debris Working Group.* ESA SP-1109, 1988.

[Debris 95] *Interagency Report on Orbital Debris 1995.* November 1995, The National Science and Technology Council, Committee on Transportation Research and Development, Library of Congress 95-72164.

[Depth 96] Messerschmid, E.; Bertrand, R.; Hinüber, A.; Laux, M.: *Consultancy on Space Station Utilization "Definition of Environment Parameters for Technology Experiments on Space Stations".* ESA Purchase Order No. 152546, Steinbeis-Transferzentrum Raumfahrtsysteme, Reutlingen, Germany, 1996.

[Dueber 93] Dueber, R.E.; McKnight, D.S.: *Chemical Principles Applied for Spacecraft Operations.* Krieger, 1993.

[Environment 94] *Space Station Program Natural Environment Definition for Design.* NASA SSP 30425, Revision B, February 8, 1994.

[ESA Guide96] *International Space Station - A Guide for European Users - Early Issues.* ESA-SP-1202, ESA Publications Divisions, edited by Bruce Battrick, ISBN 92-9092-407-1.

[Gilmore 94] Gilmore, D.G. et al: *Satellite Thermal Control Handbook,* ISBN 1-884989-00-4, The Aerospace Corporation Press, El Segundo, CA, 1994.

[Haese 91] Haese, M.: *Kollisionsrisiken mit Debris und Mikrometeoriden für Seile im erdnahen Orbit.* Study Thesis IRS 91-S32, Space Systems Institute, University of Stuttgart, 1991.

[Haffner 67] Haffner, J.W.: *Radiation and Shielding in Space.* Academic Press, New York, 1967.

[Hallmann 88] Hallmann, W.; Ley, W. (Ed.): *Handbuch der Raumfahrttechnik.* Carl Hanser Verlag, Munich, 1988.

[Ham 87] Hamacher, H. et al.: *Fluid Sciences and Material Science in Space.* Walter, H.U. (Ed.), Springer Verlag, Berlin, 1987.

[Human-Factors 94] *Human Factors.* ESA PSS-03-70, July 1994.

[IAA Debris 95] International Academy of Astronautics: *Position Paper on Orbital Debris,* ISBN 2-9508151-1-6.

[Jursa 85] Jursa, A.S. (Ed.): *Handbook of Geophysics and the Space Environment.* Air Force Geophysics Laboratory, United States Air Force, 1985.

[Skrivanek 94] *Contemporary Models of the Orbital Environment.* AIAA Special Project Report SP-069-1994.

[Smith 82] Smith, R.E.; West, G.S. (Compilers): *Space and Planetary Environment Criteria Guidelines for Use in Space Vehicle Development.* NASA TM 82478, 1982 Revision (Volume 1).

[Tascione 88] Tascione, T.F.: *Introduction to the Space Environment.* Orbit Book Company, Malabar, 1988.

[Tribble 95] Tribble, A.C.: *The Space Environment – Implications for Spacecraft Design.* Princeton, NJ, Princeton University Press, 1995.

[Turner 82] Tuner, R.E.; Hill, C.K. (Compilers): *Terrestrial Environment Criteria Guidelines for Use in Aerospace Vehicle Development.* 1982 Revision, NASA TM 82473.

[Wertz 78] Wertz, J.E. (Ed.): *Spacecraft Attitude Determination and Control.* Kluwer Academic Publishers, Dordrecht, 1978.

Chapter 4: Environmental Control and Life Support System

[Boeing 96] *Subsystem Analyses and Analytical Models.* ISS Integrated Traffic Model Report, DAC4 Final, The Boeing Company, NASA Johnson Space Center, 1996.

[Eckart 96] Eckart, P.: *Spaceflight Life Support and Biospherics.* Microcosm Press, Torrance, CA; Kluwer Academic Publishers, Dordrecht, The Netherlands, 1996.

[Eckart 97] Eckart, P.: *The Lunar Base Handbook.* NASA Space Technology Series [to be published].

[ESA PSS-03-401] *Atmosphere Quality Standards in Manned Space Vehicles.* ESA PSS-03-401, Issue 1, ESA-ESTEC, June 1992.

[ESA PSS-03-70] *Human Factors.* Life Support and Thermal Control Division, ESTEC, ESA, 1994.

[Gustavino 94] Gustavino, S.R.; McFadden, C.D.; Davenport, R.J.: *Concepts for Advanced Waste Water Processing Systems.* 24th International Conference on Environmental Systems, Friedrichshafen, Germany, June 20-23, 1994.

[Gutzenberg 93] Gutzenberg, A.S.: *Air Regeneration in Spacecraft Cabins.* In Nicogossian, A. E. (Ed.) et al.: Space Biology and Medicine, Vol. II Life Support and Habitability, AIAA, Washington, DC, 1993.

[Hallmann 88] Hallmann, W.; Ley, W. (Eds.): *Handbuch der Raumfahrttechnik.* Carl Hanser Verlag, Munich, Vienna, ISBN 3-446-15130-3, 1988.

[Kimble 94] Kimble, M. C. et al.: *Molecular Sieve CO_2 Removal Systems for Future Missions: Test Results and Alternative Designs.* 24th International Conference on Environmental Systems, Friedrichshafen, Germany, June 20-23, 1994.

[Kumagai 94] Nacheff-Benedict, M.S.; Kumagai, G.H. et al.: *An Integrated Approach to Bioreactor Technology Development for a Regenerative Life Support Primary Water Processor.* 24th International Conference on Environmental Systems, Friedrichshafen, Germany, June 20-23, 1994.

[MacElroy 90] MacElroy, R.: *The Controlled Ecological Life Support Systems Research Program.* AIAA Space Programs and Technologies Conference, AIAA-90-3730, 1990.

[Malkin 93] Malkin, V.B.: *Barometric Pressure and Gas Composition of Spacecraft Cabin Air.* In Nicogossian, A. E. (Ed.) et al.: Space Biology and Medicine, Vol. II: Life Support and Habitability, AIAA, Washington, DC, 1993.

[Meleshko 93] Meleshko, G. I. et al.: *Biological Life Support Systems.* In Nicogossian, A. E. (Ed.) et al.: Space Biology and Medicine, Volume II Life Support and Habitability, AIAA, Washington, DC, 1993.

[NASA-STD-3000] *Man-Systems Integration Standards.* NASA-STD-3000, Rev. A, October 1989.

[Rethke 94] Rethke, D. et al.: *Testing of Russian ECLSS - Sabatier and Potable Water Processor.* 24th International Conference on Environmental Systems, Friedrichshafen, Germany, June 20-23, 1994.

[Rygh 93] Rygh, K.; Flindh, J.: *Fire Safety Research in Micro-Gravity - How to Detect Smoke and Flames You Cannot See.* Preparing for the Future, Vol. 3, No. 4, Dec. ESA Publication Division, ESA-ESTEC, The Netherlands, 1993.

[Silbernagl 88] Silbernagl, S.; Despopoulos, A.: *Taschenatlas der Physiologie.* Thieme Verlag, Stuttgart, Germany, 1988.

[SSP 41000D] *System Specification for the International Space Station.* NASA Space Station Program Office, Johnson Space Center, Houston, TX, Nov. 1995.

[Tan 93] Tan, G. B. T.: *Air Revitalisation.* Preparing for the Future, Vol. 3, No. 2, June 1993, ESA-ESTEC, Noordwijk, The Netherlands, 1993.

[Woodcock 86] Woodcock, G.: *Space Stations and Platforms.* Orbit Book Company, Malabar, FL, 1986.

[Wydeven 88] Wydeven, T.: *A Survey of Some Regenerative Physico-Chemical Life Support Technology.* NASA TM 101004, NASA-Ames Research Center, 1988.

Chapter 5: Power and Thermal Control System

[Ames 92] Ames, B.: *Evaluation of Space Station ATCS Evolution Growth Paths.* SAE 932153, SAE International Conference on Environmental Systems, 23rd, Colorado Springs, CO, July 12-15, 1993.

[Audy 99] Audy, Ch.; Fischer, M.; Messerschmid, E.: Nonsteady Behaviour of Solar Dynamic Power Systems with Stirling Cycle for Space Stations. *Aerospace Science and Technology*, No. 1, 1999.

[Bekey 85] Bekey, I.; Herman, D. et al.: *Space Stations and Space Platforms - Concepts, Design and Usage.* Chap. IV, AIAA Progress in Aeronautics and Astronautics, Vol. 99, New York, NY, 1985.

[Belew 73] Belew, L.F.; Stuhlinger E.: *Skylab - A Guidebook.* NASA EP - 107, 1973.

[Cannon 95] Cannon T. M., Winters B.A.: *International Space Station Alpha External Thermal Control System Overview.* SAE 951649, SAE International Conference on Environmental Systems, 25th San Diego, California, July 10-13, 1995.

[Chambliss 94] Chambliss, J. et al.: *An Overview of the Redesigned Space Station Thermal Control System.* SAE Technical Paper 941495, SAE International, Commonwealth Drive, Warrendale, PA, 1994.

[CNES 89] Carré, F.; Proust, E.; Keirle, P.: *CNES- CEA Comparative Evaluation Study of Various Candidate 20 kWe Space Power Systems.* Proceedings of the European Space Power Conference, Madrid, Spain, 2 - 6 October 1989, ESA SP - 294.

[DASA-TB 85] *Thermal Control Handbook.* RT523, DASA-RIT, Daimler-Benz Aerospace, Bremen, 01.06.1985.

[ECSS-E-30] *Mechanical Engineering of Space Programs - Thermal Control Standard.* ECSS-E-30-00, ECSS Secretariat, ESA-ESTEC, Requirements & Standards Division, Noordwijk, The Netherlands, 10.12.1996.

[ESA PSS-03-108] ESA, Thermal Control & Life Support Divison, *Spacecraft thermal control design data*, ESA PSS-03-108, Noordwijk, The Netherlands, 1989.

[ESA PSS-03-406] ESA, Thermal Control & Life Support Divison, *Life Support and habitability manual*, ESA PSS-03-406, Noordwijk, The Netherlands, 1991.

[ESA PSS-03-401] ESA, Thermal Control & Life Support Division, *Atmosphere quality standards in manned space vehicles*, ESA PSS-03-401, Noordwijk, The Netherlands, 1992.

[Fachinetti 89] Fachinetti, F.; Levins, D.: *120 V 10 A SSPC for the Columbus Programme.* in Proceedings of the European Space Power Conference, Madrid, Spain, 2 - 6 October 1989, ESA SP - 294, 10/89.

[Gilmore 94] Gilmore, D.G.. et al: *Satellite Thermal Control Handbook*, ISBN 1-884989-00-4, The Aerospace Corporation Press, El Segundo, CA, 1994.

[Glines] Glines, A.: *Space Station Challenge: On-Orbit Assembly of a 75-kW Power System.* American Chemical Society, Paper 8412-0986-3860869-430, 1986.

[Griffin 91] Griffin, M.D.; French, J.R.: *Space Vehicle Design.* AIAA Education Series, 1991, ISBN 0-930403-90-8.

[Griegoriev 97] Griegoriev, Y.I.; Grigorov, E.I.: *Russian-segment Two-phase Thermal Control System.* Space Technology & Applications International Forum - Conference of Applications of Thermophysics in Microgravity, 1st Conference on Commercial Development of Space, Jan. 26-30, 1997.

[Haas 89] Haas, R.J.; Koehler, C.W.: *Space Station Freedom Energy Storage System Design and Development.* In: Proceedings of the European Space Power Conference, Madrid, Spain, 2 - 6 October 1989, ESA SP - 294.

[Hajela 92] Hajelas, G.: *Thermal Control Subsystem for the Photovoltaic Module on the International Space Station Alpha.* AIAA Paper 94-4061-CP, AIAA Proceedings of the Intersociety Energy Conversion Engineering Conference Volume 2, NJ, 1994.

[Hanford 96] Hanford, A.J.; Ewert, M.K.: *An Assessment of Advanced Thermal Control System Technologies for Future Human Space Flight.* SAE 961480, SAE International Conference on Environmental Systems, 26th, Monterey, CA, July 8-12, 1996.

[Kawamura 89] Kawamura, Y.: *Japanese Experiment Module Electrical Power System.* in Proceedings of the European Space Power Conference, Madrid, Spain, 2 - 6 October 1989, ESA SP - 294.

[Krüger 97] Krüger, J.: *Thermal Control: Fundamentals.* Lecture Series at IRS, University of Stuttgart, Germany, January, 1997.

[Kurtz 89] Kurtz, H.: *Energieversorgungssysteme für die Raumfahrt.* Lecture Handout, Space Systems Institute, University of Stuttgart, 1989.

[Laurini 96] Laurini, D.; Hovland, S. et al.: *Development of Columbus Orbital Facility Thermal Mathematical Models for Integrated International Space Station Thermal Analysis.* SAE 961540, SAE International Conference on Environmental Systems, 26th, Monterey, CA, July 8-12, 1996.

[Leisten 88] Leisten, V.: *120 VDC-Energieversorgung für COLUMBUS.* DGLR Jahrbuch 1988-I, p. 219.

[Longhurst 89] Longhurst, F.: *The ESA Module: COLUMBUS*, AIAA/NASA International Space Station Freedom Technical Symposium - Speaker Presentation Materials, Vienna, VA, AIAA 1989.

[Martinez 88] Martinez-Sanchez, M.; Hastings, D. E.; Estes, R. D.: *A Systems Study of a 100 kW Electrodynamic Tether.* Final Report - Executive Summary, MIT, 12/88, NASA Contract NAS 3 - 24669.

[NASA 8105] *Spacecraft Thermal Control.* NASA-SP-8105. Space Vehicle Design Criteria. May 1973.

[Oberle 93] Oberle, B.: *Numerische Simulation des instationären Betriebsverhaltens eines solardynamischen Energieversorgungssystems für Raumfahrtmissionen.* Zeitschrift für Flugwissenschaft und Weltraumforschung 17, 1993, see also DLR-FB 92-09 (PhD Thesis).

[Ollendorf 83] Ollendorf, S.: *Recent and Planed Developments at the Goddard Space Flight Center in Thermal Control Technology*, Proceedings of the International Symposium on Environmental and Thermal Systems for Space Vehicles (ESAP SP 200), Toulouse, France, 4-7 October 1983.

[Peter 92] Peter, J. F.: *A Technical Overview of the Passive Thermal Control System for the Space Station Freedom.* SAE Paper 921242. SAE International Conference on Environmental Systems, 22nd, Seattle, WA, July 13-16, 1992.

[Pohlemann 94] Pohlemann, F.: *Zur Dynamik eines elektrodynamischen Tethers.* Zeitschrift für Flugwissenschaft und Weltraumforschung, 18, 1994.

[Reysa 97] Reysa, R. P., Thurman R.L.: *International Space Station Environment Control and Life Support and Thermal Control System Overview*, Proceedings of the 6th European Symposium on Space Environmental Control Systems, Noordwijk, The Netherlands, 20-22 May 1997.

[SDR 87] *Energieversorgungssysteme für die Raumfahrt – Solardynamische Energieversorgungssysteme.* Executive Summary, DLR Stuttgart, Germany, August 1987.

[SEMDA 96] Ellis, G. L.; White, M. K.: *Lockheed Martin Systems Engineering Modelling and Design Analysis Laboratory.* http://l43i4-3.jsc.nasa.gov/docs/mod_dat.html, May 21, 1996.

[Sorensen] Sorensen, A.A.: *Space Station Electrical Power Systems.* AAS Paper 84-125.

[Sprengel 97] Sprengel, U., DLR-Stuttgart, private communication, 1997.
[SSF 88] *Space Station Freedom Technical Overview.* NASA - Reston, 1988.
[SSODB 98] Nicholls, M.: *Space Station Operations Data Book (preliminary): Thermal Control System*, Flights 2A, 3A, 4A, 5A, 6A, 7A, TC AIT, Boeing, USA, March 1998.
[Woodcock 86] Woodcock, G.R.: *Space Stations and Platforms.* Orbit Book Company, Malabar, 1986.

Chapter 6: Attitude and Orbit Control System

[Anselmo 96] Anselmo, J. C.: *Shelved for Years, Arc Jets Reappear.* Aviation Week and Space Technology, May 20, 1996.
[Bacarat 86] Bacarat, W.A.; Butner, C.L.: *Tethers in Space Handbook*, NASA contract NASW-3921, 1986.
[Belew 73] Belew, L.F.; Stuhlinger, E.: *Skylab - A Guidebook.* NASA EP-107, 1973.
[Bertrand 96] Bertrand, R.; Glocker, B.; Messerschmid, E.: *Synergetic Orbit Control of the ISS using Waste Pyrolysis.* Proceedings of the First Symposium on the Utilisation of the International Space Station, ESA SP-385, Darmstadt, September 30 - October 2, 1996.
[Burkhardt 96] Burkhardt, J.; Zimmermann, F.; Schöttle, U.: *Consultancy on Space Station Utilization: Analysis of a Reentry Capsule for Space Station Sample Retrieval.* ESTEC Purchase Order No. 152667, IRS 96-P-2, Space Systems Institute, University of Stuttgart, 1996.
[Caesar 98] Caesar, T., Nentwig, G., Peters, K., Glocker, B., Wilhelmi, H., Messerschmid, E.: *Experimental and Theoretical Investigation on the Mixing of Reacting Liquids in a Thermal Arc-Heated Steam Plasma.* 5th International Conference on Thermal Plasma Processes, St. Petersburg, Russia, July 13-18,1998.
[DARA 95] *Statusbericht Internationale Raumstation, Programm und Technik.* Deutsche Agentur für Raumfahrtangelegenheiten, Bonn, Germany, March 08, 1995.
[ESA 95] *Utilization of the International Space Station.* ESA Directorate of Manned Spaceflight and Microgravity, Doc. No. MSM-4785, 22 June 1995.
[Foley 96] Foley, T.M.: *Engineering the Space Station.* Aerospace America, ISSN 0740-722X, October 1996.
[Glocker 98] Glocker, B., Nentwig, G., Hug, S., Messerschmid, E.: *A Test Plant for Treating Halogenated Hydrocarbons Using Steam Plasma Technology.* 5th International Conference on Thermal Plasma Processes, St. Petersburg, Russia, July 13-18,1998.
[Hans 95] Hans, M.: *Verwendung von Magnetspulen zur Lageregelung von Raumstationen.* Thesis, IRS-95-S1, Space Systems Institute, University of Stuttgart, 1991.
[Hitachi 92] Hitachi Ltd.: *Hitachi Magnetic Torquers.* Product Information, Tokyo, Japan.
[ISS 95] *Technical Data Book.* http://issa-www.jsc.nasa.gov/ss/techdata.html, based on SSP 50037-19A, July 19, 1995.
[Jones 87] Jones, R.E.; Meng, P.R. et al.: *Space Station Propulsion System Technology*, Acta Astronautica 9/87, 1987.
[Laible 95] Laible, T.: *Synergetische Bahnregelungsstrategien für die 'International Space Station'*, IRS 95-S-16, Space Systems Institute, University of Stuttgart, 1995.
[Marcos 94] Marcos, F.A. et al.: *Neutral Density Models for Aerospace Applications.* in: AIAA Special Report, Contemporary Models of the Orbital Environment, AIAA SP-069-1994, America Institute of Aeronautics and Astronautics, Washington, DC, 1994.

[Messerschmid 96] Messerschmid, E., Zube, D., Meinzer, K., Kurtz, H.L.: *Arcjet Development for Amateur Radio Satellite*, Journal of Spacecraft and Rockets, Vol. 33, No. 1, Jan.-Feb. 1996.

[Pohlemann 92] Pohlemann, F.: *Systemanalyse eines elektrodynamischen Tetherantriebs für Anwendungen im niederen Erdorbit.* Dissertation, Space Systems Institute, University of Stuttgart, 1992.

[SSP 30425] *Space Station Program Natural Environment Definition for Design.* SSP 30425, NASA Space Station Program Office, Johnson Space Center, Houston, TX, 1994.

[SSP 41000D] *System Specification for the International Space Station.* SSP 41000D, NASA Space Station Program Office, Johnson Space Center, Houston, TX, Nov. 1995.

[Wertz 78] Wertz, J.E. (Ed.): *Spacecraft Attitude Determination and Control.* Kluwer Academic Publishers, Dordrecht, 1978.

[Woodcock 86] Woodcock, G.R.: *Space Stations and Platforms*, Orbit Book Company, Malabar, 1986.

Chapter 7: Utilization

[ACCESS 90] Aachener Centrum für Erstarrung unter Schwerelosigkeit, Statusbericht 89/90, 1990.

[Beardsley 96] Beardsley, T.: *Science in the Sky.* Scientific American, June 1996.

[D1 85] Merbold, U.; Furrer, R.; Messerschmid, E.; Ockels, W.; Hahn, H.-M.; Siefarth, G.: *D1 – Unser Weg ins All.* Georg Westermann Verlag GmbH, Braunschweig 1985, ISBN 3-07-508886-2.

[D2 93] Sahm, P.R.; Keller, M.H.; Schiewe, B. (Ed.): *Research in Space – The German Spacelab Missions.* Wissenschaftliche Projektführung D-2 c/o DLR Cologne, Germany, 1993, ISBN 3-89100-021-9.

[D2 95] Sahm, P.R.; Keller, M.H.; Schiewe, B. (Ed.): *Scientific Results of the German Spacelab Mission D-2.* Norderney Proceedings, Wissenschaftliche Projektführung D-2 c/o DLR Cologne, Germany 1995, ISBN 3-89100-025-1.

[DARA 95] *Nutzungskonzept der Raumstation.* Edition 4, May 8, 1995.

[ESA 95] *Utilization of the International Space Station.* ESA Directorate of Manned Spaceflight and Microgravity, Doc. No. MSM-4785, 22 June 1995.

[ESA DMSM 98] *Microgravity: A Tool for Industrial Research.* Applied Research on the International Space Station. ESA Directorate of Manned Spaceflight and Microgravity. May 1998.

[Feuerbacher 86] Feuerbacher, B.; Hamacher, H.; Naumann, R.J.: *Materials Sciences in Space.* Springer-Verlag Berlin, Heidelberg, ISBN 3-540-16358-1.

[Feuerbacher 96] Feuerbacher, B.: Private Communication.

[Foley 96] Foley, T.M.: *Engineering the Space Station.* Aerospace America, ISSN 0740-722X, October 1996.

[Haas 96] Haas, B.L., et al.: *Simulating Rarefied Aerothermodynamics.* Aerospace America, June 1996.

[Littke 86/92] Littke, W.; John, Ch.: *Protein Single Crystal Growth under Microgravity.* J. of Crystal Growth 76 (1986) 663-672, [An Historical Collection in Celebration of 25 Years of the Journal of Crystal Growth (special issue, edited by D.T.J. Hurle)], 1992.

[Messerschmid 95] Messerschmid, E.; Huber, F.; Roesgen, Th.; Littke, W.: *Technologieexperimente zur magnetischen Levitation in transparenten Ferrofluiden MagLev2 für die Raumflugmission MIR 96/97.* TZ-Raumfahrtsysteme, TZ-10036.

[Moore 96] Moore, D.; Bie, P.; Oser, H.: *Biological and Medical Research in Space.* Springer Verlag Berlin, Heidelberg New York, ISBN 3-540-60636-X, 1996.

[MSFC 94] Torr, M.R.; Chief Scientist, Payload Projects Office: *Scientific Product of Spacelab Missions: Status.* MSFC-JA01-Report, June 30, 1994.

[NASA 88] *Science in Orbit: The Shuttle and Spacelab Experience 1981-1986.* NP-119 US Government Printing Office, Washington DC 20402.

[NASA 95] *A Science and Technology Institute in Space.* NASA Fact Book. 1995.

[NASA NRP 98] *The NASA Research Plan: Improving Life on Earth and in Space*, NP-1998-02-232-HQ.

[NRC 95] Committee on the Space Station: *The Capabilities of Space Stations.* Aeronautics and Space Engineering Board, National Research Council, National Academy Press, Washington, DC 1995.

[Preu 96] Preu, P.; Binnenbruck, H.; Kuhl, R.: *Forschung im Weltraum für die Erde und geplante nationale Aktivitäten im Vorfeld der Raumstation.* DGLR Kongreß 1996, Dresden, Germany.

[Walter 87] Walter, H.U. (Editor): *Fluid Sciences and Materials Science in Space.* Springer-Verlag Berlin, Heidelberg, ISBN 3-540-17862-7, 1987.

Chapter 8: Microgravity

[Alexander 90] Alexander, J.I.D.: *Low Gravity Experiment Sensitivity to Residual Acceleration: A Review.* Microgravity Science and Technology, Volume 3, 1990.

[ANSI 92] ANSI/AIAA R-004-1992: *Recommended Practice for Atmospheric and Space Flight Vehicle Coordinate Systems.* American Institute of Aeronautics and Astronautics, American National Standards Institute, 1992.

[ESA 95] *Utilisation of the International Space Station.* European Space Agency, Directorate of Manned Spaceflight and Microgravity, Doc. No. MSM-4785, 22 June 1995.

[ESA SP-1116] *Facilities for Microgravity Investigations in Physical Sciences supported by ESA.* ESA SP-1116 Rev. 2, ESA Publications Division, ESTEC, Noordwijk, The Netherlands, 1995.

[Feuerbacher 88] Feuerbacher, B.; Hamacher, H.; Jilg, R.: *Compatibility of Microgravity Experiments with Spacecraft Disturbances.* 333-88/1, DLR, Institut für Raumsimulation, Köln, 1988.

[Feuerbacher 86] Feuerbacher, B.; Hamacher, H.; Naumann, R.J.: *Materials Sciences in Space.* ISBN 3-540-16358-1, Springer-Verlag, Heidelberg, 1986.

[Greger 87] Greger, G.: *Microgravity Research - A Challenge for Fundamental Research and Applications within the German Space Programme.* Applied Microgravity Technology, Vol. 1, Issue 1, 1987.

[Hamacher 87] Hamacher, H.: *Mikrogravitationsbedingungen eines Raumfahrzeugs.* Lecture Notes, 1. Sommerschule Mikrogravitation, DLR IB 33-87/2, 1987.

[Hamacher 88] Hamacher, H.: *Simulation of Weightlessness.* In [Feuerbacher 88].

[Hamacher 96] Hamacher, H.; Öry, H.: *Microgravity Characterisation and Microgravity Improvement for the Columbus Orbiting Facility.* Proceedings of the First Symposium on the Utilisation of the International Space Station, ESA SP-385, Darmstadt, September 30 - October 2, 1996.

[Huber 96] Huber, F. et al.: *Technology Experiment for Magnetic Levitation in Transparent Ferrofluids.* Proceedings of the First Symposium on the Utilisation of the International Space Station, ESA SP-385, Darmstadt, 30.09. - 02.10.1996.

[ISS 94] *International Space Station Reference Guide.* Boeing Missiles & Space Division, Defense & Space Group, http://issa-www.jsc.nasa.gov/ss/techdata/ISSAR/ISSAReferenceGuide.html, 1994.

[Laible 95] Laible, T.: *Synergetische Bahnregelungsstrategien für die 'International Space Station'*, IRS 95-S-16, 1995.

[MBB] *Flight Opportunities and Facilities for Microgravity Research and Applications.* MBB-ERNO, Space Systems Group, Orbital Systems and Launcher Division, Bremen.

[Monti 87] Monti, R.; Langbein, D.; Favier, J.J.: *Influence of Residual Accelerations on Fluid Physics and Materials Science Experiments.* 1987.

[Mori 93] Mori, T.; Goto, K.; Oshashi, R.; Sawaoka, A.B.: *Capabilities and Recent Activities of Japan Microgravity Center (JAMIC).* Microgravity Science and Technology, Volume 4, February 1993.

[Ockels 87] Ockels, W.J.: *Adsorbitve Tethers - A First Test in Space.* ESA Journal, Vol. 11, No. 3, 1987.

[Otto 87] Otto, G. (Hrsg.): *Vortragskurzfassungen der 1. Sommerschule Mikrogravitation.* Burghotel Schellenberg, July 13 - 17, 1987, IB 333-87/2, Institut für Raumsimulation, DLR, Cologne, Germany, 1987.

[Walter 87] Walter, H.U.: *Fluid Sciences and Materials Science in Space.* Springer-Verlag, Berlin, Heidelberg, New York, Tokyo, 1987.

[ZARM 96] *Zentrum für angewandte Raumfahrttetchnologie und Mikrogravitation (ZARM).* http://www.zarm9.zarm.uni-bremen.de.

Chapter 9: System Engineering

[Bertrand 98] Bertrand, R.: *Conceptual Design and Flight Simulation of Space Systems.* Space Systems Institute, Dissertation, University of Stuttgart, 1998.

[Casani 97] Casani, E. K.; Thomas, N. W: *The Flight System Testbed.* WWW-document http://fst.jpl.nasa.gov/papers/SPIE-94/SPIE-94.html, Jet Propulsion Laboratory, California Institute of Technology, 1997.

[De Kruyf 91] De Kruyf, J.: *Analysis Modelling and Space-Specific Analyses Using "ESA-BASE".* ESA Journal, Vol. 15, No. 3/4, 1991.

[ECSS-E-10A] *Space Engineering - System Engineering.* ECSS-E-10A, ECSS Secretariat, ESA-ESTEC, Requirements and Standards Division, Noordwijk, The Netherlands, 19 April 1996.

[ECSS-M-30A] *Space Project Management - Project Phasing and Planning.* ECSS-M-30A, ECSS Secretariat, ESA-ESTEC, Requirements & Standards Division, Noordwijk, The Netherlands, April 19, 1996.

[Eichler 90] Eichler, P.; Rex, D.: *The Risk of Collision Between Manned Space Vehicles and Orbital Debris - Analysis and Basic Conclusions.* Zeitschrift für Flugwissenschaften und Weltraumforschung, 14, 1990, pp. 145-154, Springer Verlag.

[ESA FFP/PS/753] Sachs, P.: *Generic Review Plan.* ESA FFP/PS/753, Ref. 91115EST0095, Rev. 2, ESA-ESTEC, Noordwijk, The Netherlands, 1991.

[Foley 96] Foley, T. M.: *Engineering the Space Station.* Aerospace America, October 1996, AIAA, Reston, VA, ISSN 0740-722X, 1996.

[Hunger 95] Hunger, J. W.: *Engineering the System Solution: A Practical Guide to Developing Systems.* Prentice Hall PTR, Englewood Cliffs, NJ, ISBN 0-13-594524-0, 1995.

[IEEE 1220] *IEEE Trial-Use Standard for Application and Management of the Systems Engineering Process.* IEEE Std 1220-1994, Institute of Electrical and Electronics Engineers, New York, NY, 1995.

[Larson 92] Larson, W. J.; Wertz, J. R. (Ed.): *Space Mission Analysis and Design.* Kluwer Academic Publishers, Dordrecht, The Netherlands, 1992.

[McCaffrey 88] McCaffrey, R.W.: *Space Station/Platform Configurations.* Advances in the Astronautical Sciences, AAS 84-114.

[NASA SP-6105] *NASA Systems Engineering Handbook.* NASA SP-6105, National Aeronautics and Space Administration, USA, June 1995.

[NASA TM 87383] *Conceptual Design and Evaluation of Selected Space Station Concepts.* NASA Technical Memorandum 87383, Johnson Space Center, Houston, TX, 1983.

[NRC 95] Committee on the Space Station: *The Capabilities of Space Stations.* Aeronautics and Space Engineering Board, National Research Council, National Academy Press, Washington, DC 1995.

[Powell 84] Powell, L. E.: *Space Station Concept Development Group Studies.* IAF Paper 84-27, 35th International Astronautical Congress, Lausanne, Switzerland, Oct. 7-13, 1984.

[Shaw 92] Shaw, T. E.; Lacy, J. A.: *An Overview of Systems Engineering Management Methods and Processes.* AIAA paper no. 92-1539, Space Programs and Technologies Conference, Huntsville, AL, Mar. 24-27, 1992.

[Smith 96] Smith, D. B.: *Reengineering Space Projects.* Computer Tools, Systems Engineering and Competitiveness Symposium, March 5-7, Paris, France, 1997, WWW document http://fst.jpl.nasa.gov/DNP/Paris.html, November 1996.

[Rittel 73] Rittel, H. W.; Webber, M. M.: *Dilemmas in a General Theory of Planning.* Policy Sciences 4, 1973, pp. 155-169.

[Weiser 96] Weiser, Thomas: *Raumstation.* Thesis, Institut für Darstellen und Gestalten, Space Systems Institute, University of Stuttgart, 1996.

[Wöhlke 88] Wöhlke, W.: *Modelling the Microgravity Environment of the Man-Tended Free Flyer (MTFF).* 25th Space Congress, Cocoa Beach, FL, 1988.

[Woodcock 86] Woodcock, G.R.: *Space Stations and Platforms*, Orbit Book Company, Malabar, 1986.

Chapter 10: Synergisms

[Bertrand 96] Bertrand, R.; Glocker, B.; Messerschmid, E.: *Synergetic Orbit Control of the ISS Using Waste Pyrolysis.* Proceedings of the First Symposium on the Utilisation of the International Space Station, ESA SP-385, Darmstadt, September 30 - October 2, 1996.

[ERDT 96] *Engineering Research and Technology Development on the Space Station.* National Academy Press, Washington, DC, 1996.

[Haken 83] Haken, H.: *Synergetics. An Introduction.* Springer Series in Synergetics, Vol. 1. Springer, 3rd edition, Berlin, 1983.

[Heckert 87] Heckert, B.J.: *Space Station Resistojet System Requirements and Interface Definition Study.* NASA Contractor Report 179581, RI/RD 87-109, 1987.

[RSW 92] *Regenerative Stoffwirtschaft in der Raumfahrt.* Final Report 50 RS 8902, Dornier GmbH, Friedrichshafen, Germany; DLR-Institut für Technische Thermodynamik, Stuttgart, Germany, 1992.

[Webster 83] Webster's New World Dictionary. Warner Books, Inc., New York, 1983.

Chapter 11: Human Factors

[Britannica 90] *Encyclopaedia Britannica (The New)* – Macropaedia Vol. 21, Encyclopaedia Britannica Inc., Chicago etc., 1990.

[DLR 95] Weyer, T.H. (Red.): *DLR Jahresbericht 1994/95.* ISSN 0938-2194, DLR, Public Relations Department, Cologne, Germany, October 1995.

[Duden 82] Duden Fr*emdwörterbuch.* Bibliographisches Institut, Mannheim, Germany, 1982.

[ESA 90] *Announcement of Opportunity to ESA Member States for Selection of European Candidate Astronauts* (ESA/PB-COLUMBUS(90)16). European Space Agency Columbus Programme Board, Paris, 1990.

[ESA 91] Mortensen, U. K.; Bagiana, F.: *Interactive Graphical Simulation of Humans in Space.* ESA bulletin 67, August 1991.

[ESA 93] Novara, M.: *The Habitability Mini-Laboratory: Testing the Tools of Space Habitat Architecture.* ESA bulletin 76, November 1993.

[ESA PSS-03-70] *Human Factors.* ESA PSS-03-70 Issue 1, July 1994.

[Friess 91] Friess, P.: *Les Facteurs Humains dans la Preparation d'un Vol Habité.* Thesis, TU München, ENSAE Toulouse, CNES Toulouse, 1991.

[Kohl 86] Kohl, R. L.; Calkins, D. S.; Mandell, A. J.: *Arousal and Stability: The Effects of Five New Sympathomimetic Drugs Suggest a New Principle for the Prevention of Space Motion Sickness.* Aviation, Space and Environmental Medicine 57, February 1986.

[NASA STD-3000] *Man-Systems Integration Standards.* NASA-STD-3000, Rev. A, October 1989.

[Nixon 87] Nixon, D. et al.: *Low-Cost Prototypes for Human Factors Evaluation of Space Station Crew Equipment.* 38th IAF Congress, IAF/IAA-87-553, Brighton, UK, 1987.

[Sanders 87] Sanders, M. F.; McCormick, E. J.: *Human Factors in Design and Engineering.* McGraw Hill Book company, New York, NY, 1987.

[SSMTF 96] *Space Station Mockup and Trainer Facility (SSMTF).* NASA-Johnson Space Center, http://www-sa-jsc.nasa.gov, 1996.

[NASA CP 2426] Cohen, M. et al. (Hrsg.): *Space Station Human Factors Research Review.* NASA Conference Publication 2426, NASA Ames Research Center, Moffett Field, CA, 1985.

Chapter 12: Logistics, Communications and Operation

[Amadieu 98] Amadieu, P.; Heloret, J. T.: *The Automated Transfer Vehicle.* ESA Bulletin, November 1998.

[AW&ST 98] Aviation Week and Space Technology, August 10, 1998.

[Boeing 96] *Subsystem Analysis and Analytical Models, ISS Integrated Traffic Model Report (DAC 4 Final).* September 13, 1996, Boeing-Report for NASA-JSC Contract No. NAS15-10000.

[D2 95] Feuerbacher, B.: *Telescience*; Padeken, D. et al.: *Telemedicine - Spin-off from Anthrorack on D-2*; Hirzinger, G.: *ROTEX - The First Remotely Controlled Robot in Space.* In: Scientific Results of the German Spacelab Mission D-2. Norderney Proceedings, Wissenschaftliche Projektführung, 1995, ISBN 3-89100-025-1.

[DPG 90] Deutsche Physikalische Gesellschaft (Ed.): *Entschließung der DPG zur bemannten Raumfahrt.* Bad Honnef, December 1990.

[ESTEC 92] *Definition Study of a Payload Servicing Breadboard.* Final Report, ESTEC Contract No. 9421/61/NL/JSC, 1992.

[ESTEC 93] *Equipment Manipulation and Transportation System EMATS.* System Description Handbook, ESTEC Contract No. 7412/87/NL/MAC(SC), 1993.

[Feuerbacher 90] Feuerbacher, B.; Mirra, C.; Napolitano, L.G.; Wittmann, K.: *Experience and Concepts in Microgravity User Support.* Space Technology Volume 10, No. 1/2, pp. 109-115, 1990.

[Feustel 96] Feustel-Büechl, J.: *Die Internationale Raumstation: Europas Sprungbrett in die Zukunft.* Presentation at the DGLR Space Congress, Dresden, Germany, September 25, 1996.

[Griffin 91] Griffin, Michael D., French, James, R.: *Space Vehicle Design.* AIAA Education Series, 1991, ISBN 0-930403-90-8.

[Hartl 88] Hartl, Ph.: *Fernwirktechnik der Raumfahrt.* Berichtigter Nachdruck der 2. Auflage, Springer-Verlag, ISBN 0-387-18851-7, 1988.

[Hirzinger 91] Hirzinger, G.; Dietrich, J.; Brunner, B.: *Sensor Controlled Space Robots.* SPACE COURSE, Aachen, Germany, 1991.

[LuR 91] *Pro und Contra: Bemannte Raumfahrt.* DGLR-Zeitschrift „Luft- und Raumfahrt", Heft 1/2-91.

[LuR 93] *Bemannte Raumfahrt als Kulturaufgabe.* DGLR-Zeitschrift „Luft- und Raumfahrt", Heft 5-93.

[Maral 86] Maral, G.; Bousquet, M.: *Satellite Communications Systems.* John Wiley & Sons, 1988, ISBN 0 471 90220 9.

[Messerschmid 92] Messerschmid, E.: *The role of man in space.* Z. Flugwiss. Weltraumforsch. 16 (1992) pp. 1-7, Springer-Verlag 1992.

[NRC 95] National Research Council, Committee on the Space Station, *The Capabilities of Space Station.* National Academy Press, Washington, DC, 1995.

[Ockels 95] Ockels, W.J.; Van der Heide, E.J.; Kruijff, M.: *Space Mail and Tethers Sample Return Capability for Space Station Alpha.* 4th IAF Conf., IAF-95-T4.10, Oct. 2-6, 1995.

[Renner 88] Renner, U; Nauck, J.; Balteas: *Satellitentechnik.* Springer-Verlag 1988, ISBN 0-387-18030-3.

[Schmidt 87] Schmidt, H.P.; Feuerbacher, B.; Messerschmid, E.: *Telescience: a Concept for Scientific Experiment Operations in Space.* Z. Flugwiss. Weltraumforsch. 11, 1987, pp. 71-77, Springer-Verlag.

[SSP 41000D] *System Specification for the International Space Station.* NASA Space Station Program Office, Johnson Space Center, Houston, TX, Nov. 1995.

[Wertz 91] Wertz, James R. and Larson, Wiley J.: *Space Mission Analysis and Design.* Space Technology Library, Kluwer Academic Publishers Group, 1991, ISBN 0-7923-0971-5.

[Woodcock 86] Woodcock, Gordon R.: *Space Station and Platforms.* Orbit Book Company, Malabar, Florida, 1986.

[Zimmermann 96] Zimmermann, F.; van der Heide, E.; Messerschmid, E.; Schöttle, U.: *Application of Tethers to Space Station Utilization.* Proceedings of the First Symposium on the Utilisation of the International Space Station, ESA SP-385, Darmstadt, September 30 - October 2, 1996.

Chapter 13: The International Space Station

[Bartoe] Bartoe, J.-D.: *International Space Station – Program Progress and Research Prospects.* 2nd European Symposium for the Utilization of the ISS, ESTEC, Noordwijk, Nov. 16, 1998.

[DARA-ISS 96] *Internationale Raumstation.* With a preface by the federal minister for science, research and technoloy, Dr. Jürgen Rüttgers. Edited by Kesberg & Büfering, Bonn, Germany, published by DARA GmbH, Edition 1996.

[DARA-Nutzer 96] *Die Internationale Raumstation. Ein Wegweiser für den Nutzer: Experimentakkommodation, Nutzerunterstützung, Experimentvorschlag.* Redaktion INTOSPACE GmbH (Thomas Hauschild), Herausgeber DARA GmbH., Ausgabe 1996.

[ESA-Fakten 96] *Die europäische Beteiligung an der Internationalen Raumstation.* Overall concept and text: Dieter Isakeit, Images: D. Ducros. ESA Directorate for Manned Spaceflight and Microgravity, Doc. MSM-PI/8041, February 17, 1995.

[ESA-Guide 96] *International Space Station - A Guide for European Users - Early Issues.* ESA-SP-1202, ESA Publications Divisions, edited by Bruce Battrick, ISBN 92-9092-407-1.

[ESA-Guide 98] *The International Space Station – European Users Guide.* ESA Space Station Utilization Division, Directorate of Manned Spaceflight and Microgravity, prepared by G.T. Biddis and G. Peters, Nov. 11, 1998.

[ESA-Symp. 96] *Proceedings of the First European Symposium on the Utilisation of the International Space Station.* 30 Nov. – 2 Oct.1996, ESOC, Darmstadt, ESA SP-385, 1996.

[ESA-Symp. 98] *Proceedings of the Second European Symposium on the Utilisation of the International Space Station.* 16–18 Nov., 1998, ESTEC, Noordwijk, ESA SP-433, 1998.

[ESA 95] *Utilisation of the International Space Station.* European Space Agency, Directorate of Manned Spaceflight and Microgravity, Doc. No. MSM-4785, 22 June 1995.

[Griffin] Griffin, M. D., French, J. R.: *Space Vehicle Design.* AIAA Education Series, ISBN 0-930403-90-8.

[Isakowitz] Isakowitz, S. J.: *International Reference Guide to Space Launch Systems.* AIAA 2nd Edition, ISBN 1-56347-002-0, 1994.

[NASA-Guide 98] NASA Research & Technology Utilization Guide (Draft); integrated by Futron Corporation, Version June 21, 1998.

[Ripken 95] Ripken, H.W. et al.: *Nutzungskonzept der Raumstation.* DARA-Doc. HWR: NK-V4COV.DOC May 9, 1995, and personal communications of December 12, 1996.

[Seibert 98] Seibert, G.: *ESA's International Space Station (ISS) Utilization Preparation.* ESA Microgravity News, Vol. 11, No. 2, August 1998.

[Traffic 96] *Subsystem Analysis and Analytical Models, ISS Integrated Traffic Model Report (DAC 4 Final.* Boeing-Report NASA-JSC-Contract NAS15-10000, September 13, 1996.

Fundamental Constants

Physical Constants

Atomic Mass Unit	u	=	$1.66057 \cdot 10^{-27}$ kg
Boltzmann Constant, $k = R / N_A$	k	=	$1.380662 \cdot 10^{-23}$ J/K
Electron Mass	m_0	=	$9.10953 \cdot 10^{-31}$ kg
Elementary Charge	e_0	=	$1.60219 \cdot 10^{-19}$ C
Planck Constant	h	=	$6.626176 \cdot 10^{-34}$ J s
Speed of Light in Vacuum	c	=	$2.99792458 \cdot 10^{8}$ m/s
Stefan-Boltzmann Constant	σ	=	$5.67051 \cdot 10^{-8}$ W/m^2 K^4

Astronomical Constants

Astronomical Unit	AU	=	$1.49597870 \cdot 10^{11}$ m
Earth Albedo	a	=	0.35
Earth Equatorial Radius	R_0	=	$6.378140 \cdot 10^{6}$ m
Earth Gravitation Constant, $\mu = \gamma \cdot M$	μ	=	$3.986005 \cdot 10^{14}$ m^3/s^2
Earth Mass	M	=	$5.9742 \cdot 10^{24}$ kg
Earth Standard Free Fall	g_0	=	9.81 m/s^2
Gravitational Constant	γ	=	$6.6720 \cdot 10^{-11}$ m^3/kg s^2
Inclination of Earth Equator to Ecliptic	ε	=	23.439°
Sidereal Day Earth, Epoch 1980	d_s	=	23 h 56 min 4.1 s
Sidereal Year Earth, Epoch 1980	a_s	=	365.256 days
Solar Radiation at 1 AU	S	=	1371 (±10) W/m^2

Glossary

2BMS	Two-Bed Molecular Sieve
3D	Three-dimensional
4BMS	Four-Bed Molecular Sieve

A

A/V	Audio/Video
AAA	Avionics Air Assembly
AC	Alternating Current
ACES	Atomic Clock Ensemble in Space
ACRV	Assured Crew Return Vehicle
ACS	Attitude Control System
ACS/UCS	Audio Control Switch
ADA	Area Smoke Detector Assembly
AFD	Aft Flight Deck
AIDS	Acquired Immune Deficiency Syndrome
AM	Amplitude Modulation
AMS	Alpha Magnetic Spectrometer
AO	Announcement of Opportunity
AO	Atomic Oxygen
AOCS	Attitude and Orbit Control System
AP	Astrophysics
APCF	Advanced Protein Crystallization Facility
APDS	Androgynous Peripheral Docking System
APFR	Articulating Portable Foot Restraints
APL	Attached Pressurized Laboratory
APM	(Columbus) Attached Pressurized Module
AR	Acceptance Review
AR	Atmospheric Research
ARIS	Active Rack Isolation System
ASAP	Ariane Structure for Auxiliary Payloads
ASI	Agenzia Spaziale Italiana (Italian Space Agency)
ASL	Aligned with Sun Line
ASTP	Apollo-Soyuz Test Project
ATCS	Active Thermal Control System
ATF	Autonomous Thruster Facility
ATLAS	Atmospheric Laboratory for Applications and Science
ATM	Apollo Telescope Mount
ATV	Automated Transfer Vehicle
AU	Astronomical Unit

B

b.p.	Boiling Point
BAT	Battery
BCDU	Battery Charge/Discharge Unit
BCR	Battery Charge Relay
BDR	Battery Discharge Relay
BMR	Body-Mounted Radiators
BOL	Begin of Life
BPSK	Binary Phase Shift Keying
BSCCM	Battery Signal Conditioning and Control Module

C

C&T	Communications and Tracking
CAD	Computer-Aided Design
CADMOS	Centre d'Aide en Développement de la Micropesanteur et aux Opérations Spatiales; (= Center of Assistance for the Development of Microgravity and for Operations in Space)
CAL	Columbus Attached Laboratory
CAM	Centrifuge Accommodation Module
CC	Control Center
CCHP	Constant Conductance Heat Pipe
CCSDS	Consultive/Consultative Committee for Space Data Systems
CCTV	Close Circuit Television
CDC	Charge/Discharge Controller
CDG	Concept Development Group
CDMA	Code Division Multiple Access
CDMS	Command and Data Management System
CDR	Critical Design Review
CELSS	Controlled Ecological Life Support System
CETA	Crew and Equipment Translation Aid
CEV	Crew Escape Vehicle
CF	Centrifuge Facility
CFFL	Columbus Free Flying Laboratory
CFR	Carbon Formation Reactor
CFU	Carbon Formation Unit
CHeCS	Crew Health Care System
CHX	Condensing Heat Exchanger
CIRA	COSPAR International Reference Atmosphere
CIS	Commonwealth of Independent States
CK	Carrier Keying
CM	Center of Mass
CMD	Command
CMG	Control Moment Gyro
CNV	Converter
COF	Columbus Orbital Facility
COF-CC	COF Control Center
COP	Composite Operations Plan
COSPAR	Committee of Space Research

COUP	Consolidated Operations and Utilization Plan
CP	Center of Pressure
CP	Critical Point
CPD	Coarse Pointing Device
CRISTA	Cryogenic Infrared Spectrometer Telescope for Atmosphere
CRRES	Combined Release and Radiation Effects Satellite
CRT	Cathode Ray Tube
CRV	Crew Return Vehicle
CSG	Centre Spatial Guyanais, Kourou
CTV	Crew Transfer Vehicle
CUIS	Columbus User Information System
CUP	Composite Utilization Plan

D

D&C	Display and Control Systems
DAM	Double-Aluminized Mylar®
DAMEC	Danish Aerospace Medical Center of Research A/S
DASA	DaimlerChrysler Aerospace AG
DC	Direct Current
DC	Docking Compartment
DCC	Direct Cargo Carrier
DCSU	Direct Current Switching Unit
DDCU	DC-to-DC Converter Unit
DL	Design Life
DLR	Deutsches Zentrum für Luft- und Raumfahrt e.V. (German Aerospace Center)
DM	Docking Module
DM	Double Module
DMS	Data Management System
DMS	Docking and Stowage Module (Russian)
DNA	Desoxyribonucleic Acid
DoD	Department of Defense
DOD	Depth of Discharge
DPSK	Differential Phase Shift Keying
DR	Decommissioning Review
DRM	Design Reference Mission
DRS	Data Relay Satellite
DRSS	Data Relay Satellite System
DSMC	Direct Simulation Monte-Carlo
DSMP	Defense Meteorological Satellite Program
DUC	Dutch Utilization Center

E

EASE	Experimental Assembly of Structures in Extravehicular Activity
EATCS	External Active Thermal Control System
ECLS	Environmental Control and Life Support
ECLSS	Environmental Control and Life Support System
ECS	Environmental Control Subsystem (Spacelab)
EDC	Electrochemical Depolarized CO_2 Concentration

EDR	European Drawer Rack
EEATCS	Early External Active Thermal Control System
EHF	Extra High Frequency
EIRP	Equivalent Isotropic Radiated Power
EL	Electrolyzer
ELM	Experiment Logistics Module
ELM-ES	Experiment Logistics Module – Exposed Section
ELM-PS	Experiment Logistics Module – Pressurized Section
EM	Engineering Model
EMATS	Equipment Manipulation and Transportation System
EMI	Electromagnetic Interferences
EMP	Environment Monitoring Package
EMR	Electromagnetic Radiation
EMU	Extravehicular Mobility Unit
EO	Earth Observation
EOL	End of Life
EPC	Equipment Power Converter
EPDB	Experiment Power Distribution Box
EPM	European Physiology Module
EPS	Electrical Power Supply
EPS	Electrical Power System
ER	Express Rack
ERA	European Robotic Arm
ERM	Extendible/Retractable Mast
ERS-1	European Radar Satellite 1
ES	Exposed Section
ESA	European Space Agency
ESOC	European Space Operations Centre
ESR	European Stowage Rack
ESRO	European Space Research Organization
ESSC	Evolutionary Space Station Concept
ESTEC	European Space Research and Technology Center, Noordwijk, The Netherlands
ETCS	External Thermal Control System
EUA	Early Utilization Agreement
EUB	European Utilization Board
EURECA	European Retrievable Carrier
EUV	Extreme Ultraviolet
EVA	Extravehicular Activity
EVAS	Extravehicular Attach Structure
EVM	Mir Computer System
EVOLVE	NASA Atmosphere Prediction Model
ExPA	Express Pallet Adapter

F

FC	Fuel Cell
FDMA	Frequency Division Multiple Access
FFDB	Functional Flow Block Diagram
FGB	Functional Cargo Block, named "Zarya" (Acronym Derived from Russian Term)

FM	Flight Model
FM	Frequency Modulation
FOCUS	European High-Temperature Environment Disaster Recognition System
FOM	Figures of Merit
FRR	Flight Readiness Review
FSK	Frequency Shift Keying
FSL	Fluid Science Laboratory
FSS	Fire Suppression System
FWD	Forward

G

GAS	Get-Away Special
GAS-CAN	Get-Away Special Canister
GAUSS	Spherical Mirror Camera, SL-D2
GCR	Galactic Cosmic Radiation
GDR	German Democratic Republic
GEO	Geosynchronous/Geostationary Earth Orbit
GIS	Geomagnetically Induced Storm
GISSAD	Graphical Interface for Space Station Analysis and Design
GLONASS	Global Navigation Satellite System
GN&C	Guidance, Navigation and Control
GNC	Guidance, Navigation and Control
GPC	General Purpose Computer
GPS	Global Positioning System
GRAM	Global Reference Atmosphere Model
GSC	Ground Station Control
GSE	Ground Support Equipment
GTO	Geostationary Transfer Orbit
GTS	Global Time System

H

H/W	Hardware
HDRR	High Data Rate Recorder
HEO	High Eccentricity Orbit
HERP	High Efficiency Radiator Panel
HF	High Frequency
HFE	Human Factors Engineering
HIV	Human Immunodeficiency Virus
HP	Heat Pipe
HP	High Pressure
HRDL	High-Rate Data Link
HRDM	High Rate Demultiplexer
HRM	High Rate Multiplexer
HRTS	Heat Rejection Transport System
HTV	H-II Transfer Vehicle
HZE	Highly charged (i.e. high Z) Elementary particles

I

I/F	Interface
IATCS	Internal Active Thermal Control System
ICE	Interconnecting Elements
ICO	Intermediate Circular Orbit (Communication System)
IDR	Incremental/Initial Design Review
IFSUSS	International Forum for Scientific Uses of the Space Station
IGA	Intergovernmental Agreement
IGS	Instrument Ground Support Equipment
IML	International Microgravity Laboratory
IMPLM	International Mini Pressurized Logistics Module
IMV	Intermodule Ventilation
INMARSAT	International Maritime Satellite
INTA	Instituto Nacional De Técnica Aerospacial
IOC	Initial Operational Capability or Initial Operational Configuration
IOP	In Orbit Plane
IOP	Increment Operations Plan
IPS	Instrument Pointing System
IPV	Individual Pressure Vessel
IR	Infrared
IRI	International Reference Ionosphere
IRS	Institut für Raumfahrtsysteme, University of Stuttgart, Germany
ISO	International Organization for Standardization
ISPR	International Standard Payload Rack
ISS	International Space Station
ISSA	International Space Station Alpha
ITA	Integrated Truss Assembly
ITCS	Internal Thermal Control System
IVA	Intravehicular Activity
IWG	International Working Group

J

JAMIC	Japan Microgravity Center
JEM	Japanese Experiment Module
JEM-EF	JEM-Exposed Facility
JEM-ELM	JEM Experiment Logistics Module
JEM-PM	JEM Pressurized Module
JEM-RMS	JEM Remote Manipulator System
JPC, JPF, JPP, JPS	Japanese Ceiling, Floor, Port, Starboard Side Location
JPL	Jet Propulsion Laboratory
JSC	NASA Johnson Space Center

K

KSC	NASA Kennedy Space Center
KuSP	Ku Band Signal Processors

L

LAC, LAS, LAP	Lateral Areas at Ceiling, Starboard, Port Side
LAF, LAX	Lateral Areas at Floor Side, X-Direction (US Lab)
LAN	Local Area Network
LDEF	Long Duration Exposure Facility
LEO	Low Earth Orbit
LET	Linear Energy Transfer
LF	Low Frequency
LH	Local Horizontal
LMS	Life and Microgravity Spacelab
LOX	Liquid Oxygen
LS	Life Sciences
LSE	Laboratory Support Equipment
LSM	Life Support Module (Russia)
LSRA	Life Sciences Research Announcement
LT HX	Low Temperature Heat Exchanger
LTL	Low Temperature Loop
LTV	Logistics Transfer Vehicle
LV	Local Vertical
LVLH	Local Vertical Local Horizontal

M

MAGLEV	Magnetic Levitation
MARES	Muscular Atrophy Research and Exercise System
MARS	Microgravity Advanced Research and Support
MAXUS	European ballistic rocket (derived from TEXUS)
MBB	Messerschmitt-Bölkow-Blohm (now part of DaimlerChrysler Aerospace)
MBS	Mobile Base System
MBSU	Main Bus Switching Unit
MCC	Mission Control Center
McD	McDonnel-Douglas
MCD	Molecular Column Density
MCR	Mission Concept Review
MCS	Modular Cultivation Facility
MCV	Microbial Check Valve
MD	Molecular Deposition
MDE	Mission-Dependent Equipment
MDL	Mid-Deck Locker
MDM	Multiplexer/Demultiplexer
MDR	Mission Definition Review
MELFI	Minus Eighty Degree Laboratory Freezer for the ISS
MELISSA	Modular Environment for Life Support Systems Simulation and Analysis
MERP	Medium Efficiency Radiator Panel
MET	Marshall Engineering Thermosphere Model
MF	Multifiltration
MFC	Microgravity Facilities for Columbus
MFR	Manipulator/Mobile Foot Restraint
MIL-STD	Military Standard
MLI	Multi-Layer Insulation

MMH	Mono-Methyl Hydrazine
MMI	Man/Machine-Interface
MMU	Manned Maneuvering Unit
MMU	Mass Memory Unit
MOL	Manned Orbital Laboratory
MOMS	Mobile Optical Measurement System
MORL	Manned Orbiting Research Laboratory
MOS	Metal-Oxide Semiconductor
MOSC	Manned Orbital Systems Concept
MPBM	Multiple Power Bus Management
MPD	Magneto-Plasma-Dynamic
MPE	Mission Peculiar Equipment
MPLM	Mini-Pressurized Logistics Module
MPP	Maximum Power Point
MPTE	Mission Preparation and Training Equipment
MS	Materials Science
MSC	Mission Support Center
MSC	Mobile Servicing Center
MSFC	NASA Marshall Space Flight Center
MSG	Microgravity Science Glovebox
MSIS	Mass Spectrometer and Incoherent Scatter
MSK	Minimum Shift Keying
MSL	Materials Science Laboratory
MSS	Mobile Servicing System
MT	Mobile Transporter
MT HX	Moderate Temperature Heat Exchanger
MTS	Moderate Temperature Loop
MUA	Most Unfavorable Attitude
MUSC	Microgravity User Support Center

N

N/A	Not Applicable or Not Available
NAA	North American Aviation Company
NASA	National Aeronautics and Space Administration
NASA-MSC	NASA Mission Support Center
NASDA	National Space Development Agency (Japan)
NCAR	Global Ionosphere-thermosphere Model
Neurolab	Neuro Science Laboratory
NIH	National Institute of Health
NIZEMI	Slow Rotating Centrifuge with Microscope, IML-2
NRZ-PCM	No Return to Zero - Pulse Code Modulation
NTA	Nitrogen Tank Assembly
NTO/UDMH	Nitrogen Tetroxide/Unsymmetrical Dimethylhydrazine
NTSC	National Television Standards Committee

O

OAST	NASA Office of Aeronautics and Space Technology
OCS	Orbit Control System

OFT	Orbital Flight Test
OLMSA	NASA Office of Life and Microgravity Sciences and Applications
OMA	Onboard Management
OMV	Orbit Maneuvering Vehicle
OPS	Operations
ORR	Operational Readiness Review
ORU	Orbital Replaceable Unit
OSI	Open Systems Interconnect
OSI	Operator/System Interface
OSM	Orbital Support Module
OSS	NASA Office of Space Science
OSTA	NASA Office of Space and Terrestrial Applications
OTD	ORU (Orbit Replacable Unit) Transfer Device
OWS	Orbital Workshop

P

P/L	Payload
PAS	Payload Attach Structure
PCA	Polar-Cap Absorption
PCB	Printed Circuit Board
PCDF	Protein Crystal Diagnostics Facility
PCMMU	Pulse Code Modulation Master Units
PCOU	Power Connecting Unit
PCWQM	Process Control Water Quality Monitor
PDR	Preliminary Design Review
PDU	Power Distribution Unit
PEGASUS	Satellite
PEMS	Percutaneous Electrical Muscle Stimulator
PEP	Power Extension Package
PFD	Power Flux Density
PFR	Portable Foot Restraint
PICLIB	Visualization Software using PICASSO Libraries
PIU	Power Incoming Unit
PL	Payload
PLCU	Payload Control Unit
PM	Phase Modulation
PM	Pressurized Module
PMA	Pressurized Mating Adapter
PMU	Power Management Unit
POCC	Payload Operations Control Center
POI	Payload Operation Integration
POIC	Payload Operation Integration Center, Huntsville
POIC	Payload Operations Interface Control
POP	Perpendicular to Orbit Plane
PP	Plasma Physics
PPSL	Perpendicular to the Projected Sun Line
PR	Public Relations
PRR	Preliminary Requirements Review
PSE	Payload Support Equipment
PSK	Phase Shift Keying

PSL	Perpendicular to Sun Line
PSU	Power Switching Unit
PTCS	Passive Thermal Control System
PUP	Partner Utilization Plan
PV	Photovoltaic
PV SPM	Photovoltaic Solar Power Module
PVA	Photovoltaic Array
PVCU	Photovoltaic Controller Unit
PVM	Photovoltaic Module
PVTCS	Photovoltaic Thermal Control System

Q

QPSK	Quadrature Phase Shift Keying
QR	Qualification Review

R

R&D	Rendezvous and Docking
R&D	Research & Development
RASU	Russian American Switching Unit
RAU	Remote Acquisition Unit
RBE	Relative Biological Effectiveness
RBVM	Radiator Beam Valve Module
RCS	Reaction Control System
RF	Radiofrequency
RFC	Regenerative Fuel Cell
RKA	Russian Space Agency
RKK	Rocket Space Corporation
RM	Research Module (Russian)
RMS	Remote Manipulator System
RMS	Root Mean Square
RMSA	Rack Main Switch Assembly
RO	Reverse Osmosis
ROM	Rough Order of Magnitude
ROS	Russian Orbit Section
ROTEX	Robotic Technology Experiment
ROTEX	Space Robot Technology Experiment
RPCM	Remote Power Control Mechanism
RPDA	Remote Power Distribution Assembly
RS	Russian
RS	Russian Segment
RS-LSS	Russian Segment Life Support System
RSA	Russian Space Agency
RSP	Resupply Stowage Platform
RSR	Resupply Stowage Rack
RSTCS	Russian Segment Thermal Control System
RUS	Russian
RvD	Rendezvous and Docking

S

S/S	Subsystem
SA	Solar Array
SAA	South Atlantic Anomaly
SAB	Sabatier Reactor
SAGE	Stratospheric Aerosol and Gas Experiment
SAR	System Acceptance Review
SARA	Solar Array Reflection Assembly
SARJ	Solar Alpha Rotary Joint
SAS	Space Adaptation Syndrome
SAWD	Solid Amine Water Desorption
SCR	Solar Cosmic Radiation
SCSI	Small Computer Systems Interface
SD	Solar Dynamic
SDR	Standard Drawer Rack
SDR	System Definition Review
SE	Systems Engineering
SEBA	Biology Facility for Space Exposure
SFE	Static Feed Electrolyzer
SFOG	Solid-Fuel Oxygen Generator
SFWE	Static Feed Water Electrolyzer
SHF	Super-High Frequency
SID	Sudden Ionospheric Disturbances
SIR	Spacelab Interface Rack
SIR	Standard Interface Rack
SL	Spacelab
SLDP	Spacelab Double Pallet
SLP	Spacelab Payload / Pallet
SLS	Spacelab Life Sciences
SM	Service Module (Russian)
SM	Single Module
SMAC	Spacecraft Maximum Allowable Concentration
SMO	Solar Monitoring Observatory
SOC	Space Operations Center
SOLAR MAX	Satellite
SPAS	Shuttle Pallet Satellite
SPCN	Satellite Personal Communication Network
SPDM	Special Purpose Dextrous Manipulator
SPLC	Standard Payload Computer
SPM	Solar Power Module
SPOE	Standard Payload Outfitting Equipment
SPORT	Sky Polarization Observatory
SPP	Science Power Platform (Russia)
SPS	Solar Power Satellites
SR	Sabatier Reactor
SR	Solar Research
SR	Storage Rack
SRL	Space Radar Laboratory
SROC	Space Remote Operations Center

SRR	System Requirements Review
SS	Space Segment
SS	Spread Spectrum
SSAP	Science and Space Application Platform
SSSAS	Space Station System Analysis Study
SSC	Swedish Space Corporation
SSCC	Space Station Control Center, Houston
SSDW	Space Station Design Workshop
SSF	Space Station Freedom
SSM	Second Surface Mirror
SSMTF	Space Station Mockup and Trainer Facility
SSO	Sun Synchronous Orbit
SSPC	Solid State Power Controller
SSRD	Space Station Remote Demonstrator
SSRMS	Space Station Remote Manipulator System
SSU	Sequential Shunt Unit
SSUP	Space Station User Panel
STDN	Space Tracking and Data Network
STS	Space Transportation System (Space Shuttle)
SW	Software

T

TACS	Thruster Attitude Control System
TAS	True Air Speed
tbd.	to be determined
TC	Telecommand
TCC	Trace Contaminent Control
TCM	Trace Contaminent Monitoring
TCS	Thermal Control System
TDMA	Time Division Multiple Access
TDRSS	Tracking and Data Relay Satellite System
TEA	Torque Equilibrium Attitude
TEF	Technology Exposure Facility
TEMPUS	Materials Processing Facility for High Temperatures Under Microgravity
TEXUS	Technologische Experimente unter Schwerelosigkeit
TIMES	Thermoelectric Integrated Membrane Evaporation System
TLA	Time Line Analysis
TLM	Telemetry
TM	Telemetry
TM	Training Model
TM	Transport Module
TMP	Turbo Molecular Pump
TOC	Total Organic Content
TPS	Thermal System (in this context, not Thermal Protection System)
TRIAD	Project name
TRRJ	Truss Rotary Joint
TSS	Tethered Satellite System
TTC	Telemetry, Tracking and Communications

TZ-RS	Steinbeis-Transferzentrum Raumfahrtsysteme, Reutlingen, Germany

U

UDM	Universal Docking Module (Russian)
UDMH	Unsymmetrical Dimethylhydrazine
UHB	User Home Base
UHF	Ultra High Frequency
ULC	Unpressurized Logistics Carrier
ULYSSES	Name of ESA spacecraft
UMA	Umbilical Mechanism Assembly
UOP	User Operations Panel
UOP	Utility Outlet Panel
US-STS	US Space Transportation System (Space Shuttle)
USML	United States Microgravity Laboratory
USMP	United States Microgravity Payload
USOC	User Support and Operations Center
USOS	United States On-Orbit Segment
USS	Unique Support Structure
USSR	Union of Socialist Soviet Republics
USU	Time-dependent model of the global atmosphere
UV	Ultraviolet
UWP	Universal Working Place

V

VCD	Vapor Compression Distillation
VCHP	Variable Conductance Heat Pipes
VDC	Volts of Direct Current
VDU	Visual Display Unit
VLBI	Very Large Baseline Interferometry
VP	View Point
VPCAR	Vapor Phase Catalytic Ammonia Removal

W

WAN	Wide Area Network

X

XPLC	External Payload Computer

Y

Z

ZARM	Zentrum für angewandte Raumfahrttechnologie und Mikrogravitation

Index

— D —

— F —

— N —

— O —

— P —

— Q —

— R —

— S —

— T —

— X —

— Y —

— Z —

GPSR Compliance
The European Union's (EU) General Product Safety Regulation (GPSR) is a set of rules that requires consumer products to be safe and our obligations to ensure this.

If you have any concerns about our products, you can contact us on

ProductSafety@springernature.com

In case Publisher is established outside the EU, the EU authorized representative is:

Springer Nature Customer Service Center GmbH
Europaplatz 3
69115 Heidelberg, Germany

www.ingramcontent.com/pod-product-compliance
Ingram Content Group UK Ltd.
Pitfield, Milton Keynes, MK11 3LW, UK
UKHW021901190726
13853UKWH00003B/1376